Rocket Propulsion Primer

Subramaniam Krishnan · Jeenu Raghavan

Rocket Propulsion Primer

Foreword by Prof. Luigi Tonino DeLuca

Springer

Subramaniam Krishnan
Department of Aerospace Engineering
Indian Institute of Technology Madras
Chennai, Tamil Nadu, India

Jeenu Raghavan
Solid Propulsion Combustion Research Group
Vikram Sarabhai Space Centre
Indian Space Research Organisation
Thiruvananthapuram, Kerala, India

ISBN 978-981-97-5643-8 ISBN 978-981-97-5644-5 (eBook)
https://doi.org/10.1007/978-981-97-5644-5

This Springer imprint is published by the registered company Springer Nature Singapore Pte Ltd.
The registered company address is: 152 Beach Road, #21-01/04 Gateway East, Singapore 189721, Singapore

Foreword

With great pleasure I accepted to write this foreword to the primer book on rocket propulsion by two esteemed colleagues from the Indian Institute of Technology in Madras (Subramaniam Krishnan) and the Indian Space Research Organization in Thiruvananthapuram (Jeenu Raghavan). I believe this book will be greatly appreciated by undergraduate and graduate students interested in rocket propulsion in general throughout the world.

Chapter 1 is a detailed history of the development of rocket propulsion from scratch to the latest achievements. As such, it should appeal to readers of all ages and cultural interests. Chapter 2 is a broad classification of aerospace propulsion systems, including electric and nuclear propulsion. As such, it should appeal to everyone who is interested in flight techniques on our planet or interstellar missions. Chapter 3 focuses on the fundamental parameters useful for evaluating the performance of the rocket. Overall, these three chapters make a good introduction for undergraduate rocket propulsion students and readers with little or no background.

The second part of the book is technically more demanding and requires adequate background. Chapter 4 is based on a previous textbook by the authors (Chemical Rockets: Performance Prediction and Internal Ballistics Design, Springer, 2019) and provides an advanced discussion of the composition of combustion products, under conditions of frozen and shifting equilibrium, for solid and liquid propellants. Chapter 5 offers a detailed discussion on the optimization of multistage rocket vehicles based on structural coefficients. This part is aimed at more advanced students or professionals working in the field of rocket propulsion.

A commendable feature of this book is the inclusion of an abstract and synopsis at the beginning of each chapter, as well as conceptual questions and numerical problems with answers at the end of each chapter. Moreover, in each chapter, there are several worked examples to help understand the material and grasp the basic concepts. Finally, each chapter contains a list of Further Reading in addition to the standard bibliography. It is worth mentioning that the volume is enriched by numerous clear illustrations, sometimes even in color. All these features make the book very suitable for its intended audience and different from most competing educational publications.

The treatment of frozen and equilibrium flow, two-phase flow, equilibrium composition of the mixture, and stage optimization goes far beyond that of similar books. This is a welcome addition, useful not only for graduate students but also for practicing engineers. At the end of my quick read, I feel impressed by the excellent work carried out by the two authors. I would like to congratulate them wishing the highest international success for their outstanding primer book on rocket propulsion.

April 2024

Luigi Tonino DeLuca
Professor of Aerospace Propulsion
(Retired), Department of Aerospace
Science and Technology
Politecnico di Milano
Milan, Italy

Preface

This textbook is principally based on the lectures delivered by the first author to undergraduate- and graduate-level students, and to engineers and scientists in industries. The book is titled as a primer in the sense that the topics selected are basic and a few. But, these are important to be understood by the students and professionals interested in rocket engineering. The book is arranged in five chapters: "History of Rockets," "Propulsion Systems Classification," "Qualitative-Performance and Technology," "Quantitative Internal-Ballistics Prediction and Design," and "Stage Optimization and Trajectories." Depending on the analytical rigor, the chapters are arranged. Chapters 1–3 are applicable to a first course in rocket propulsion to undergraduate students. Chapters 4 and 5 are of interest to senior undergraduate and graduate students and to practicing engineers and scientists in the rocket industry.

Rocket propulsion history covered in Chap. 1, being a popular one, could also be of interest to the general public having curiosity in astronautics and rocketry. To the first-year engineering students at the undergraduate level and also to the senior-level students of high schools, the material presented formed a motivational one to tell them about the pioneers of rocket engineering, the trinity: Tsiolkovsky of Russia, Goddard of the USA, and Oberth of Germany. With the undaunted determination and dedicated perseverance, these eminent men worked tirelessly against many odds such as delayed recognition or total disapproval, acute shortage of funding, rebuke, criticism, and personal ill health. It is amazing that in spite of their different origins, all three had one common vision that rocket technology will be more useful to mankind as the tool for space exploration.

Chapter 2 on the classification of aerospace propulsion systems has focused on rocket propulsion with added emphasis on the electric and nuclear propulsion. Flight-proven electric thrusters with their characteristics are tabled. Electric propulsion with nuclear power has become mandatory for deep-space missions. While the chemical rockets are energy-limited devices, in contrary the electric thrusters are power-limited ones. But, for the interplanetary travels and beyond by humans, relatively unrestricted power with unlimited energy is needed, and this is possible only with nuclear propulsion. The solution to the human-rated nuclear safety appears to be the main issue.

Typical rocket performance characteristics (characteristic velocity c^*, thrust coefficient C_F, and specific impulse I_{sp}) and their efficiency factors are detailed in Chap. 3. Nozzle-flow divergence and separation, thrust oscillations, side loads, and frozen and equilibrium flows are all discussed. A two-phase flow model is introduced. Nozzle contour design (real flow effects and turn-back angle) is discussed along with Rao's charts for optimum contours. Plug- and expansion-deflection nozzles and their altitude characteristics are included. Nozzle-construction details of solid and liquid rockets are considered. Polymeric materials, composites, and structural as well as high-temperature alloys, and their applications are presented. Thus, in this chapter, added stress has been on two-phase flow, advanced nozzle concepts, nozzle contour design, nozzle construction, and materials.

Chapter 4 is a revised and condensed version of the Part I of our previous textbook *Chemical Rockets: Performance Prediction and Internal Ballistics Design* (2019, Springer). The pedagogical presentation of the energy minimization procedure to calculate the equilibrium composition of the combustion products and subsequently the quantitative evaluation of the internal ballistics are the principal objectives of this chapter. For a specified chamber pressure, adopting Lagrange multipliers, the governing equilibrium equations are solved through Gaussian elimination with partial pivoting to obtain the composition of the combustion products and thus c^*. This then leads to the calculation of the characteristics C_F and I_{sp} for the given pressure- or area ratio of the nozzle under equilibrium- or frozen-flow assumption. Using these steps, the evaluation of internal ballistics and engine envelope of a proven engine is presented. In addition, classification, composition selection, and ingredient details are discussed for propellants. Typical solid-propellant compositions and data on proven LOX LH2 and LOX Kerosene engines are presented. Thermodynamic data of many propellant ingredients and additives are tabled.

Chapter 5 addresses the optimization of multistage rocket vehicles. In preparation to this, the data on current satellite launch vehicles are tabled. And, the graphs presented for the variations of structural coefficients ε's with the flight-proven stages of different masses and propellants indicate a high and a near constant ε's with dispersion for solid propellant stages. The time-independent gravity- and drag-free (loss-free) optimization procedure is a preliminary step to the design of multistage rocket vehicles. Two variants of this procedure, marked-up terminal velocity and scaled-up stage mass, are discussed. While these variants give a reasonable estimate of the optimum vehicle mass, the stage mass distribution is better with the latter. However, for the chosen payload mass, the final design of a satellite launch vehicle is reached by finding the optimum trajectory to reach the terminal conditions with a minimum vehicle mass. Hence to enable trajectory analyses, the characteristics of Earth's atmosphere to estimate the density- and the gravity variation are detailed. In the inertial frame of reference, the two-dimensional equations of motion in polar coordinates for the launch vehicle of zero lift are drawn. And, to facilitate numerical calculation, their incremental forms are detailed. An optimum trajectory comprises a judicious selection of different maneuvers: vertical climb, constant pitch-rate maneuvers, gravity turn, bi-linear and linear tangent steering, and also coasting. The related equations are drawn. As a simple analysis of the trajectory, neglecting drag and assuming average gravitational acceleration, the vertical ascent of sounding rockets, powered as well as coasting, is considered. Equations are derived for the burnout velocity, burnout altitude, and coasting height.

We are grateful to Group Director C. Ravikumar (Aerospace Fight Dynamics) of Vikram Sarabhai Space Centre of Indian Space Research Organisation for his thorough checking of Chap. 5: "Stage Optimization and Trajectories." Also, we are thankful to Prof. S. Varunkumar of Indian Institute of Technology Madras for his careful reading of the book draft and valuable comments.

Sufficient number of worked examples in every chapter and a number of concept questions and numerical problems with answers at the end of each chapter are included. As indicated previously, at the university level, the first three chapters are applicable to a first course in rocket propulsion at the undergraduate level and Chaps. 4 and 5 can be covered for the graduate and the senior undergraduate students. It is hoped that this textbook will fit the needs of the faculty for their instructions and be useful to the students of aerospace and mechanical engineering and to the practicing engineers and scientists in rocket industries.

Chennai (Madras), India
Thiruvananthapuram, India
April 2024

Subramaniam Krishnan
Jeenu Raghavan

Contents

About the Authors

Subramaniam Krishnan received degrees in mechanical engineering from University of Madras (B.E., 1964), Indian Institute of Science Bangalore (M.E., 1968), and in aerospace engineering from Indian Institute of Technology Madras (Ph.D., 1976). He was a research assistant at Gas Turbine Research Establishment, Bangalore, from 1964 to 1966. He joined Indian Institute of Technology Madras as a lecturer in 1968 and rose to the position of professor in 1987. He retired from IIT Madras in 2002. He taught a wide variety of courses in aerospace propulsion from 1968 to 2013. His research interest lies in solid propulsion: solid rockets, solid fuel ramjet, and hybrid rockets. He has several publications to his credit in these areas. He is an honorary fellow of the High Energy Materials Society of India, a fellow of the Aeronautical Society of India, and an associate fellow of the American Institute of Aeronautics and Astronautics.

Dr. Jeenu Raghavan received degrees in chemical engineering from Calicut University (B.Tech., 1984) and in aerospace engineering from IIT Madras (M.Tech., 1986, and Ph.D., 1992). He served as Senior Project Officer and Research Associate (CSIR Fellowship) at IIT Madras from 1987 to 1992. Later, he did his post-doctoral studies at Georgia Institute of Technology, Atlanta (1997–1999). After working for a short period of time in Bharath Dynamics Limited, Hyderabad, as Assistant Manager, Propulsion, he joined Vikram Sarabhai Space Centre, Thiruvananthapuram (VSSC) of Indian Space Research Organisation (ISRO) in year 1993 and retired from there in 2022 as the Director of Solid Propulsion and Combustion Research Group. He has more than 35 years of experience in solid rocket propulsion-related fields and has several publications to his credit in this area. He got the Team Excellence Award from the Department of Space, ISRO in 2009, the Raghavaiah Memorial Team Award from the High Energy Materials Society of India in 2014, and the GSLV Lifetime Achievement Award in 2021.

Chapter 1
History of Rockets

Synopsis

The experimental demonstrations of action–reaction principle in the first century B.C. by the Greek philosopher Archytas of Tantrum through his flying toy pigeon and by the Greek mathematician Hero of Alexandria through his self-rotating sphere aeolipile are presented. The Chinese were the first to develop rockets as early as the first century B.C., and they called them as arrows of flying fire. The first recorded usage of rockets as weapons was in 1232 at the battle of Kai-fung-fu, where the Chinese Sung Dynasty repelled the invading Mongols. The rocket-technology route evolved through battles: from Chinese to Mongols, from Mongols to Arabs, from Arabs to European Countries, and as a closure from Europe to America at the battle of Bladensburg (Maryland) in 1814. Thus, from the battle of Kai-fung-fu in 1232, it took about 600 years for the rockets to travel around the world.

Modern astronautics and rocketry owes its development to the pioneering works of three great scientists: Konstantin Eduardovitch Tsiolkovsky of Russia, Robert Hutchings Goddard of the United States of America, and Hermann Oberth of Germany. With undaunted determination and dedicated perseverance, these eminent men worked tirelessly against many odds such as delayed recognition or total disapproval, acute shortage of research funding, rebuke, criticism, and personal ill health. Biographical sketches of these great men are drawn.

The use of Tsiolkovsky equation to determine the velocity increments of space vehicles is described along with problems with answers at the chapter end.

S. Krishnan and J. Raghavan, *Rocket Propulsion Primer*,
https://doi.org/10.1007/978-981-97-5644-5_1

1.1 Introduction

Take any technology and you always find that its practical demonstration had been realized much before the theory was established. However, you may note that the fast and effective refinement of a technology begins only after its theory, explaining the underlying basic principles, has been established. The action–reaction principle that is fundamental to jet propulsion, which includes air-breathing- as well as rocket propulsion, was theoretically explained only in 1687 by the English scientist Sir Isaac Newton by his famous publication "Philosophiae Naturalis Principia Mathematica" ("Mathematical Principles of Natural Philosophy"). But, approximately 2100 years before this, Archytas, a Greek philosopher, mathematician, astronomer, statesman, and strategist, had demonstrated the action–reaction principle by his toy pigeon in the city of Tarentum, Fig. 1.1. Most information given in this chapter was gathered from [1–14]. Any needed specific reference is mentioned in the text of this chapter.

Fig. 1.1 Archytas demonstrating the action–reaction principle

1.1.1 Action–Reaction Principle

Archytas suspended on a wire his wooden pigeon that contained hot steam at an elevated pressure in its belly cavity. The other end of the wire was hooked onto the top of a tall pole. On releasing a plug, a jet of steam escaped through a hole from the rear of the pigeon to produce a thrust that made the toy pigeon fly in circles around the pole. Thus Archytas mystified and amused the citizens of Tarentum by his flying toy pigeon and demonstrated the fundamental principle of propulsion: "every force has an equal and opposite reaction."

The second recorded demonstration of the action–reaction principle was in the first century B.C. Hero of Alexandria, a Greek mathematician and scientist, constructed a device known as aeolipile. The device consisted of a kettle of water and a spherical chamber that had been horizontally pivot-mounted at the ends of the two conduits rising from the kettle, Fig. 1.2. The spherical chamber had two L-shaped tubes fixed 180 degrees apart with their outlets pointing in the opposite directions in a plane perpendicular to the rotational axis of the chamber. On heating the water to boil, the steam entered the spherical chamber through the conduits and escaped as jets in opposite directions from the two outlets. The two equal but opposite components of thrusts that the exiting jets produced gave a torque to rotate the spherical chamber. The aeolipile did not produce any useful power. The pressure in the not-so-strong kettle would have been a little above atmospheric and quite a lot of steam could have leaked through the two crude pivot joints, but the aeolipile effectively demonstrated the action–reaction principle. The garden water sprinklers of today are nothing but the water aeolipiles that demonstrate the working principles of air-breathing- as well as rocket-propulsion systems!

Fig. 1.2 Hero engine

1.1.2 Arrows of Flying Fire

As noted previously, jet propulsion includes air-breathing propulsion and rocket propulsion. As its name implies, air-breathing propulsion "breathes" air and uses the oxygen contained in the air as the oxidant for combustion with the fuel stored within. Examples of air-breathing propulsion systems are the ramjets and the aircraft gas turbines in various forms such as turboshaft, turbofan, and turbojet with reheat. As these propulsion systems need air they cannot operate outside the atmospheric air layer. On the other hand, rocket propulsion systems, or non-air-breathing propulsion systems, store the required oxidant and fuel within. These systems, therefore, can operate independently of the environment—within the air layer, underwater, or in space.

In general literature, the term rocket is used for the rocket propulsion system as well as the rocket vehicle that comprises a rocket propulsion system, a payload, and a guidance and control system. In this book, we use the term rocket engine or rocket motor to represent rocket propulsion system. The rocket engine-propelled vehicle is written simply as a rocket or where more clarity is required as a rocket vehicle. Having noted the earliest demonstrations of action–reaction principle, let us now briefly look at the history of how rocket technology was born and how it was spread all over the world.

Chinese were the first to develop rockets and they called them as arrows of flying fire. As early as 1300 B.C. fireworks were used in China during religious and other festive celebrations. For firecrackers, the Chinese used to fill crude explosives in bamboo tubes and seal them with fuses. They tossed the firecrackers into fires during the celebrations. The explosive powder, on ignition, burned and instantaneously produced hot combustion gases. The resulting steep rise in the internal pressure exploded the sealed bamboo tubes with big noise. The Chinese believed that the explosions would drive away evil spirits.

It is conjectured by many that the invention of rockets could be by accident. Possibly some of the firecrackers were not properly sealed. In such a typical case, the weakest point of the sealed tube could be the fuse hole that was sealed after the fuse had been inserted. At the consumption of the fuse on ignition and the subsequent immediate-enlargement of the weak fuse hole, the hot high-pressure combustion gases escaped as a jet without causing explosion. Such unsuccessful firecrackers got propelled and skittered out of the fire due to the escaping jets and these gave the clue for the arrows of flying fire.

In order to transform the firecrackers into the arrows of flying fire, burnt clay closures with holes were provided at the rear ends of the bamboo tubes, which were otherwise sealed all over. The holes were used as the access passages for the loosely mounted fuses. Long bamboo sticks attached to the explosive-filled tubes provided stability in flight. No sooner had they been lit, the arrows of flying fire were launched from bows. Before long it was found that the arrows of flying fire need not be launched from bows but could launch themselves by the power of the escaping jets. Thus, the first avatar of a rocket was born.

Fig. 1.3 A Chinese soldier igniting an "arrow of flying fire" (Image sketched from the one found in [15])

A typical arrow of flying fire consisted of a bamboo-tube rocket motor as its propulsion system, Fig. 1.3. It could be with a payload of a pack of explosive, which, at the end of flight, would give with or without an explosion a shower of glowing cinders. But for the long bamboo stick to provide stability in flight, there was no guidance and control system.

Although the Chinese were using some crude explosives in their fireworks for many centuries before Christ, the first recorded development of explosives happened during the rule of the Emperor Wu Di (156–87 B.C.) of the Han dynasty. Under the emperor's sponsorship, the Chinese alchemists experimented with sulfur and saltpeter (a mineral containing potassium nitrate, KNO_3) to heat up and transform substances. In the eighth century, during the Chinese Tang dynasty, sulfur and saltpeter were combined with charcoal powder to create an explosive called the "huoyao" or gunpowder. The modern gunpowder—also known as black powder—is a mixture of about 75% potassium nitrate, 15% charcoal, and 10% sulfur.

1.1.3 The Rocket Route

Undoubtedly, the arrows of flying fire were used during religious and other festive celebrations for many centuries before Christ. Considering this fact, we may say that the arrow of flying fire is also one of the earliest systems to demonstrate the action–reaction principle. It is not known when the Chinese started using the arrows of flying fire in battles. The first recorded-usage of them as weapons was in 1232 at the battle of Kai-fung-fu where the Chinese Sung Dynasty repelled the invading Mongols.

The Mongols, repelled by the Chinese at the battle of Kai-fung-fu in 1232, realized the power of the arrows of flying fire and started developing their own rockets. In 1241, at the battle of Sejo, Mongols used their rockets against Magyar (Hungarian) forces to capture Budapest. And in 1258, the Mongols used their rockets against Arabs to capture the city of Baghdad. Arabs in turn quickly learned to construct their own rockets and used them during the Seventh Crusade in 1268 against the French Army of King Louis IX. In the late 1260 s Bacon Roger, an English philosopher and scientist, worked on improved forms of gunpowder. Thus, within four decades after the battle of Kai-fung-fu the rocket found its route to the entire Europe.

In 1370 s, Jean Froissart, a French chronicler, noted that more accurate flights could be achieved by launching rockets through tubes or canisters and this paved the way for the modern bazooka. In 1420, Joanes de Fontana of Italy designed a surface-running rocket-powered torpedo for setting enemy ships on fire. Almost in the last phase of the Hundred Years' War between the French and the English (1337–1453), the French army led by Saint Joan of Arc broke the siege of Orléans in 1429, and, in this, the French were reported to have used rockets.

In 1591, a German firework-maker Von Schmidlap invented the "step rocket," a large rocket having on its top a small rocket motor. The large rocket on ignition climbed to a high altitude, and just before its burnout and separation, ignited the small rocket motor to climb still higher altitude. At its burnout, the small rocket showered the sky with glowing cinders. Schmidlap's idea of step rocket is the forerunner to the present day multistage rockets used for satellite-launch vehicles and intermediate range ballistic missiles (IRBMs) and intercontinental ballistic missiles (ICBMs). In 1650, Kazimierz Siemienowicz, a Polish artillery expert, released drawings for a staged rocket.

After the battle at the siege of Orléans, there appears to be no record of rockets being extensively used in battles until the war in India between the Muslim ruler Tippu Sultan and the British forces at the two battles of Srirangapatnam, 1792 and 1799. The British forces won the battles but they had initially experienced serious reverses due to the usage of a large number of rockets by Tippu Sultan, Fig. 1.4. Although the British learned the use of rockets in battles as early as the thirteenth

Fig. 1.4 Rockets of Tippu Sultan [16]

century, the superiority of Tippu's rockets made the British realize once again the importance of rocket weaponry. How the technology of rocket weaponry entered India is not clear. The Mongol emperor Genghis Khan invaded India in the early thirteenth century and the Mongols were in India for the next three centuries. From Kabul, the emperor Baber, a descendant of Tamerlane and Mongol rulers, invaded India and established the Mughal dynasty (1526–1858). Baber was known to have used superior artillery in his battles. Either the Mongols or the Mughals or both could have brought rockets into India.

The initial success of Indian rocket barrages against British forces at the two battles of Srirangapatnam resulted in the revival of rocket development in England. In 1800s, Sir William Congreve, an English artillery-expert, began development of rockets weighing up to about 140 kg with iron cases and almost 5 m long stability sticks. In 1806, Claude Ruggiere, an Italian living in Paris, launched small animals in rockets equipped with parachutes. Congreve rockets were used in the Napoleonic Wars as the British attacked Copenhagen and Denmark in 1807. On August 24, 1814, British forces used their rockets against American forces at Bladensburg, Maryland to capture Washington DC and the White House. Again, on September 13th the same year the British forces unsuccessfully bombarded with their rockets the Fort McHenry in Baltimore harbor. The rocket attacks by the British against Americans in the War of Independence inspired Francis Scott Key to write the words "rockets' red glare" in his famous poem "Star Spangled Banner" which became the National Anthem of the United States of America.

In 1821, sailors found another use for rockets by using rocket-propelled harpoons to hunt whales. In 1826, Sir William Congreve, continuing his rocket developments of three decades, experimented with staged rockets as originally set out by Von Schmidlap in 1591.

In those days, the devastating nature of war depended mainly on the number of rockets used and not on their precision. More than the damages, it was the fear and confusion caused by the fire and explosions of the descending rockets that demoralized the enemies in the battles. Despite the six centuries of rocket-weaponry usage since the battle of Kai-fung-fu, the long sticks were the only means to provide stability to the rockets and they were messy and did not provide the desired accuracy. In 1844, a British inventor William Hale invented "stickless rocket." In stickless rockets, the rocket jet was made to impinge on small vanes fixed at the jet exit. The reactive force due to the impingement made the rocket to spin and provided better stability and precision in rocket flight. Many variations of this spin-stabilization concept are used in present day rockets.

After the War of Independence, Americans developed their own rockets for use in their war against Mexicans in 1847 at the Siege of Veracruz and the capture of Chapultepec Castle. Also, Americans used rockets to fight among themselves during their Civil War from 1861 to 1865.

Thus, from the battle of Kai-fung-fu in 1232, it took about 600 years for the rockets to travel around the world.

1.2 The Trio of Rocket Technology

The modern astronautics and rocketry owes to its development to the pioneering works of three great scientists: Konstantin Eduardovitch Tsiolkovsky (1857–1935) of Russia, Robert Hutchings Goddard (1882–1945) of the United States of America, and Hermann Oberth (1894–1989) of Germany. With undaunted determination and dedicated perseverance, these eminent men worked tirelessly against many odds such as delayed recognition or total disapproval, acute shortage of research funding, rebuke, criticism, and personal ill health. Tsiolkovsky, Goddard, and Oberth stand out as pioneers because of their greatness in evolving the underlying scientific principles, projecting technically the application of rocket propulsion for space exploration, and visualizing the associated challenges. Brief biographical sketches of these great men are given here. The fourth pioneer in this line could be Wernher von Braun (1912–1977), the designer of German V-2 bombs and the space age architect of the United States of America. The drawing of his adventurous biography that is a World War II sensation is left as an exercise for the reader.

1.2.1 Konstantin E. Tsiolkovsky (1857–1935)

Konstantin Eduardovitch Tsiolkovsky was born to a Polish immigrant on September 17, 1857 in the village of Ijevskoe, Russia. His portrait is shown in Fig. 1.5. Tsiolkovsky's parents were not rich and had a large family of 18 children. At the age of 10, Tsiolkovsky lost his hearing almost completely as the result of scarlet fever. Throughout his life, this disability made him work very hard to prove to himself and

Fig. 1.5 Konstantin E. Tsiolkovsky (Courtesy: The Memorial House Museum of Konstantin Tsiolkovsky [17]): "The Earth is the cradle of the mind, but we cannot live forever in a cradle"

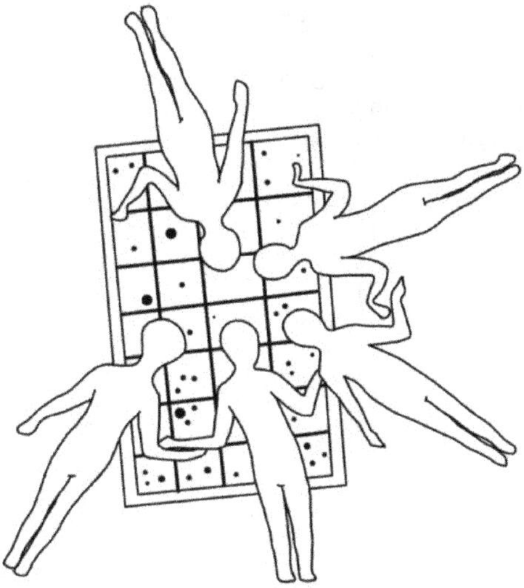

Fig. 1.6 A sketch of the original drawing of Tsiolkovsky showing astronauts in weightlessness looking through their window at the stars (Courtesy: The Memorial House Museum of Konstantin Tsiolkovsky [17])

others that he was cleverer than others. Tsiolkovsky never had any formal education but was a ferocious reader. He was greatly inspired by the novels of the French author Jules Verne (1828–1905), who is considered as the father of science fiction.

Young Tsiolkovsky was dreaming of human space travel not only for humanity to go into outer space but to establish space civilization there. At the age of 22, he wrote his first manuscript "Astronomical Drawings" to explain the Solar system indicating the planets, and the distances between the planets and their satellites. The self-taught Tsiolkovsky obtained his Teacher's Certificate and went to Borovsk, Kaluga Province in 1880 to work there as a mathematics teacher for the next 12 years. During his spare time, he continued to work on his favorite subject of space travel. In 1883, he wrote his second manuscript "Free Space" in which he described the life and ways of motion in space, zero gravity, a spacecraft with cosmonauts floating within due to weightlessness, gyroscopes for attitude control, and an airlock for exit from the spacecraft into free space, Figs. 1.6 and 1.7.

On a promotion in his teaching job, Tsiolkovsky moved to Kaluga in 1892 and lived there until his death. During the early 1890s, he wrote articles on a metal dirigible (1892) and an airplane (1894). It was in Kaluga that Tsiolkovsky became a well-known scientist by publishing his studies on space flight and inter-planetary travels.

Inspired by the Eiffel Tower in Paris, Tsiolkovsky proposed in 1895 a "celestial castle" at the geosynchronous (or geostationary) altitude, which is equal to 35786 km. A tower with an elevator would connect the castle and the ground station right "below." The castle at the geosynchronous altitude would be static with respect to its ground station as the castle's orbital period would be exactly equal to one day. The celestial castle would serve as a geosynchronous space station. Cosmonauts and objects could be sent to the castle through an elevator. The objects could be assembled into spacecraft and launched into outer space from the castle. Following

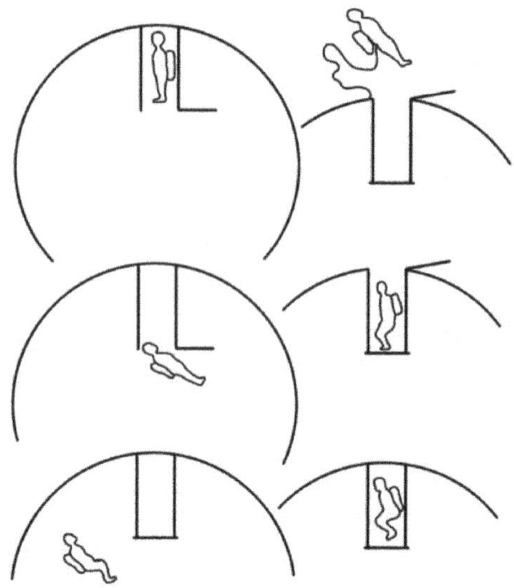

Fig. 1.7 A sketch of the original drawing of Tsiolkovsky showing the cosmonaut with a space suit using an airlock to exit a space vehicle in weightlessness (Courtesy: The Memorial House Museum of Konstantin Tsiolkovsky [17])

this proposal many interesting studies have appeared on the possibility of building the Tsiolkovsky's space tower or its variants. However, up till now, such a concept remains only fictional for want of super strength materials and cheaper ways of executing such a colossal venture.

In 1897, Tsiolkovsky wrote the manuscript entitled, "Exploration of the Universe with Reaction Machines," in which he accurately explained the state of weightlessness, the need to employ rockets for space exploration, and the importance of using liquid propellants instead of the conventional solid propellants. Furthermore, in this manuscript, he derived the famous equation governing rocket motion in gravity-free vacuum:

$$\Delta v = u_e \ln\left(\frac{m_0}{m_0 - m_p}\right) = u_e \ln\left(\frac{m_0}{m_s + m_l}\right) \tag{1.1}$$

where Δv is the velocity increment of rocket vehicle, u_e is the velocity of exiting gases from rocket engine, which also represents the most often talked about energy parameter, namely, specific impulse, m_0 is the initial mass of rocket vehicle, m_p is the propellant mass consumed or expelled into the vacuum by way of exiting gases, m_l is the payload, and m_s is the structural mass, which includes engine and other accessories. The **Tsiolkovsky equation**, Eq. 1.1, clearly brings out the two fundamental principles of vehicle motion in zero-gravity vacuum space: (1) the required velocity increment to the vehicle can be achieved by spending less propellant if the velocity of the exiting gases or the specific impulse is more, and (2) the velocity increment to the vehicle can be greater than the velocity of the exiting gases. Tsiolkovsky submitted this manuscript to the journal, "Science Review" in St. Petersburg in 1897 but the editors took almost 6 years to publish it. The use of Tsiolkovsky equation is demonstrated in Example 1.1.

Example 1.1
A cryogenic upper stage has an initial velocity of 7 km/s in a space-field of negligible gravity. It is powered by a liquid oxygen and liquid hydrogen engine delivering a specific impulse of 4200 Ns/kg. The stage has an initial mass of 215000 kg, structural mass of 25700 kg, and a payload mass of 81100 kg. Determine (1) the propellant mass and (2) the velocity with which the payload can be hurled at the burnout of this stage.

Solution
The upper stage carrying the payload has been injected into a gravity-free vacuum with a velocity of 7 km/s. The value of specific impulse, which is nothing but the exit velocity of combustion products from the rocket-engine nozzle, is given as 4200 Ns/kg = 4200 m/s. The initial mass of the stage, the structural mass, and the payload mass are all given. From these, propellant mass can be calculated. On calculating the possible velocity increment Δv, the final velocity can be calculated.

Assumptions
The stage is assumed to be flying in a gravity- and drag-free field. At burnout, it is assumed that all the propellant is consumed instantaneously.

Analysis

$$u_e = 4200 \text{ m/s}; \quad m_0 = 215000 \text{ kg}$$

$$m_s = 25700 \text{ kg} \quad m_l = 81100 \text{ kg}$$

$$m_0 - m_p = m_l + m_s \qquad \Longrightarrow m_p = m_0 - (m_l + m_s)$$

$$= 215000 - (25700 + 81100) = 108200 \text{ kg}$$

mass of propellant $m_p = 108200$ kg

$$\Delta v = u_e \ln\left(\frac{m_0}{m_0 - m_p}\right) = u_e \ln\left(\frac{m_0}{m_l + m_s}\right) \tag{1.1}$$

$$\Delta v = 4200 \times \ln\left(\frac{215000}{25700 + 81100}\right) = 2938.66 \text{ m/s}$$

$$v_{final} = v_i + \Delta v = 7000 + 2938.66 = 9938.66 \text{ m/s}$$

Velocity of the hurled payload = 9938.66 m/s.

Discussion
We have used Tsiolkovsky equation in this Example. This equation is the most powerful one and finds application in multistage optimization, trajectory analysis, and spacecraft dynamics. We will learn more on the application of this equation in Chap. 5.

Tsiolkovsky was made a member of the Soviet Academy of Science in 1919. Although Tsiolkovsky had been writing technical papers and science fictions on space travel from 1879, his works did not attract serious attention in Russia until 1923, in which year Hermann Oberth in Germany published his book "Die Rakete zu den Planetenraumen" (The Rocket into Interplanetary Space). In this work, Oberth explained in great detail the application of rocket propulsion to spaceflight. Following the publication by Oberth, Germans started to take great interest in the development of rockets. As this news spread to Russia, Tsiolkovsky's earlier works were sought out and avidly studied by Russians.

In 1926, Tsiolkovsky published the book, "Plan of Space Exploration" in which he gave sixteen steps for the humanity to leave the finite Earth's "cradle" to the far-off less-finite worlds of "other suns" in the Milky Way Galaxy and perhaps beyond. The interesting sixteen steps are the following.

1. Creation of rocket airplanes with wings.
2. Progressively increasing the speed and altitude of these airplanes.
3. Production of real rockets without wings.
4. Ability to land on the surface of the sea.
5. Reaching the orbital velocity about 8 km/second, and the first flight into the Earth orbit.
6. Lengthening rocket flight times in space.
7. Experimental use of plants to make an artificial atmosphere in spaceships.
8. Using pressurized space suits for activity outside of spaceships.
9. Making orbiting greenhouses for plants.
10. Constructing large orbital habitats around the Earth.
11. Using solar radiation to grow food, to heat space quarters, and for transport throughout the Solar System.
12. Colonization of the asteroid belt.
13. Colonization of the entire Solar System and beyond.
14. Achievement of individual and social perfection.

15. Overcrowding of the Solar System and the colonization of the Milky Way Galaxy.
16. The Sun begins to die and the people remaining in the Solar System's population go to other suns.

The vision charter by Tsiolkovsky for space exploration is amazing in that except for the last four steps, space scientists have seriously addressed the other steps and achieved success in most of them.

In 1929, Tsiolkovsky wrote "The Space Rocket Trains" in which he proved that multi-stage rockets are required to reach velocity to orbit around the Earth. The drawings in the "Album of Space Travels" written in 1932, Tsiolkovsky revealed his brilliant ideas about life in space, including zero gravity, air pressure locking, space habitats, and rocket guidance.

In addition to the subject of space travel, Tsiolkovsky was interested in philosophy and aptly it was the philosophy of space. His main work on this subject was "Ethics or the Natural Foundations of Morality" (1902–1918). In 1932 Tsiolkovsky wrote "The Cosmic Philosophy" to achieve "Universal Happiness," the happiness not only forhumanity but also for all the living beings in the Cosmos, for the entire Universe.

Tsiolkovsky wrote over 500 scientific papers. Although he never constructed any rocket himself, he influenced many young Russian engineers and designers. Konstantin E. Tsiolkovsky died in Kaluga at the age of 78 on September 19, 1935. He was buried in the old Kaluga Cemetery, not far from the Museum that honors his life and work. A large crater on the far side of the moon of about 240 km in diameter, first identified by the Russian unmanned spacecraft Luna III (1959), is named after Tsiolkovsky.

1.2.2 Robert Hutchings Goddard (1882–1945)

Robert Hutchings Goddard (Fig. 1.8) was born at Maple Hill, Worcester, Massachusetts, USA on October 5, 1882 in a family of British ancestry. As a teenage boy, he was greatly inspired by the science fiction of H. G. Wells (1866–1946), an English author and political philosopher. On October 19, 1899, climbing on a tall cherry tree to trim its dead limbs, young Goddard daydreamt of making some device which had even the possibility of ascending to Mars. When he descended the tree he was a different boy and started to find how to get off the planet Earth. The day of cherry-tree dream became Goddard's "Anniversary Day" and he remembered it throughout his lifetime.

During 1904–1908 Goddard studied at the Worcester Polytechnic Institute for his bachelor's degree. Here he conducted experiments on solid rocket motors with steel casings and convergent nozzles. After his bachelor's degree and while working as an instructor in physics at the Institute, Goddard determined that liquid hydrogen and liquid oxygen would be a very efficient propellant combination. Goddard went to Clark University in 1909 to do his PhD. The PhD thesis that he submitted in 1911 was "On the Conduction of Electricity at Contacts of Dissimilar Solids." For the next

Fig. 1.8 Robert Hutchings Goddard (Courtesy NASA, Public domain, via Wikimedia Commons [18]): "It is difficult to say what is impossible, for the dream of yesterday is the hope of today and the reality of tomorrow"

one year, he was at Clark University as an honorary fellow in physics and for the academic year 1912–1913 he was at Princeton University as a Research Instructor in Physics. In 1912, Goddard explored mathematically the possibility of using rocket propulsion to reach high altitudes and attain escape velocity.

In 1913, Goddard took ill with tuberculosis and many including his doctor thought that the illness would kill him. However, Goddard became better after battling it for many months. After recovering from his illness, Goddard returned to Clark University in 1914 as an Instructor in Physics and rose to the position of full Professor in 1920 and then in 1923 became Professor and Director of the Physical Laboratories there to serve until 1943. In 1914, Goddard was awarded two patents, the first one was on "A Multistage Step Rocket" using solid propellants and the second was on "Liquid Fuel Gun Rocket." Although the second patent was mainly concerned with his concept of cartridge rockets, a brief section fully outlined the concept of liquid rocket propulsion. During 1915 and 1916, using a portion of his own pay and the money provided by Clark University, he tested rockets using convergent–divergent or DeLaval nozzles with different types of gunpowder and measured their exhaust velocities. Thus, it appears that although Tsiolkovsky had sketched his rocket engines with divergent nozzles, Goddard was the first to have adopted convergent–divergent nozzles in working rocket engines.

In 1916, Goddard applied to Smithsonian Institution for a grant to support his rocket research and in January next year, the Institution gave him a grant of 5000$. During 1917–1918, he was working on the construction of various types of rocket weaponry; important among them was a small rocket weapon that could be shot from a handheld launcher. By this time, the use of double base propellant (a homogeneous propellant containing essentially two chemicals, nitroglycerin and nitrocellulose) was found to be better than that of gunpowder. In November 1918, Goddard demonstrated his rockets before the US armed services at the Aberdeen Proving Ground. In this demonstration, to the great disappointment of Goddard, only the shooting of the rocket weapon from a handheld launcher was of interest to them and all the other concepts of rocketry were ruled out to be of any relevance to military applications. However, this handheld rocket launcher, after further development, became quite successful in World War II as the modern bazooka.

In January 1920, Smithsonian Institution published Goddard's closure report of the project it funded. "A Method of Reaching Extreme Altitudes" was the title of the report, where Goddard explained mathematically the theory of rocket propulsion and the methods of using rockets to measure atmospheric conditions at altitudes higher than that possible at that time using balloons. This was the concept of the present day meteorological sounding rockets. The report contained the results of other research carried out under the project. Also at the end of the report, Goddard indicated the possibility of a rocket reaching the moon and exploding a load of flash powder there to mark its arrival.

"The possibility of a rocket reaching the moon" caught the attention of many, and several newspapers published articles about the report. This resulted in an unfortunate yet historic journalistic controversy. On January 13, 1920, the New York Times printed an editorial that laughed at Goddard and his assertion that a rocket could produce its thrust in a vacuum beyond the Earth's atmosphere. The most pejorative statement in the editorial was: "Professor Goddard, with his 'chair' in Clark College and the countenancing of the Smithsonian Institution, does not know the relation of action to reaction; and of the need to have something better than a vacuum against which to react. Of course, he only seems to lack the knowledge ladled out daily in high schools." In his inimitable style Goddard responded to this editorial and his other critics saying, "Every vision is a joke until the first man accomplishes it; once realized, it becomes commonplace." On July 17, 1969, more than 49 years after its editorial that the rocket fraternity frequently cited and laughed for the impetuous criticism and three days before man's first landing on the moon, the New York Times published "A Correction" stating, "It is now definitely established that a rocket can function in a vacuum as well as in an atmosphere. The Times regrets the error."

It is believed by many that several scores of the copies of the Smithsonian report of Goddard and the related newspaper articles reached Europe in the early 1920s and this goaded quite a few European countries, particularly Germany, to enhance their activities of rocket developments.

In June 1924, Goddard married Esther Christine Kisk. Esther was the secretary of Clark University's president. After their marriage, Esther devoted her time in documenting Goddard's rocket research by filming and photographing his experiments.

During 1920–1925, Goddard was developing a liquid propellant rocket engine using liquid oxygen and gasoline. In December 1925, he successfully tested on a test bed a small liquid-propellant rocket engine of dry mass of about 6 kg and pressure of around 6 atmospheres. During the test, the engine could lift itself by about 25 mm

Fig. 1.9 Goddard and a liquid oxygen-gasoline rocket in the frame from which it was fired on March 16, 1926, at Auburn, Massachusetts (Courtesy Esther C. Goddard, Public domain, via Wikimedia Commons [19])

by its own thrust and Goddard noted in his diary, "This test, being the first in which a liquid propelled rocket had operated satisfactorily and lifted its own weight, is of much significance, for it shows that a larger rocket constructed on the same plan could raise itself to considerable altitudes."

A few months later, on March 16, 1926, Goddard had the momentous first flight of a liquid-propellant rocket at the farm owned by his distant relative "Aunt" Effie Ward in Auburn, Massachusetts, Fig. 1.9. The vehicle rose about 13 meters and traveled a distance of around 56 meters in 2.5 s after the lower half of the nozzle had burned off. Although his first rocket had the engine above the propellant tanks to keep the "the nozzle above the center of gravity of the rocket" the flight was not stable and Goddard reverted to having the nozzle at the bottom of his later liquid-rockets.

In July 1929, at Aunt Effie's farm in Auburn, Goddard launched the first liquid rocket with a scientific payload, a camera and a barometer. This attracted a wide public attention. Sensing the danger involved in rocket launches in crowded places the local fire marshal banned further rocket launches in Auburn. However, as a positive result, Charles A. Lindberg, a famous American-aviator, engineer, and writer, became an avid supporter of Goddard's rocket experiments. Until 1929, Goddard funded his research through his own pay, the support coming from Clark University, and the grants from the Smithsonian Institution and the Carnegie Institution. Evidently, the funds coming from these sources were fairly small. Now, with the recommendation

Fig. 1.10 Goddard examining one of his P-series rocket (turbopump fed liquid oxygen and gasoline engine) in Roswell, New Mexico, in 1940 (Courtesy NASA [20])

of Lindberg, Goddard could get relatively liberal funding from the Daniel and Florence Guggenheim Foundation. In the summer of 1930, Goddard shifted his rocket experiments from his Aunt Effie's farm in Auburn to Mescalero Ranch, a 10-acre farm at the edge of Roswell, New Mexico, where "we would not bother anyone and no one would bother us," said Mrs. Goddard. A photograph of Goddard's workshop in Roswell is shown in Fig. 1.10. Note the most inadequate facilities that existed in Goddard's workshop where many path-breaking achievements were made. Also note the use of a convergent–divergent nozzle, perhaps with a low half-cone angle.

For the next 12 years, Goddard, with the support of Guggenheim Foundation and the Smithsonian Institution, achieved many firsts by his rocket experiments in Roswell and elsewhere, and important among them are: film cooling of rocket combustion chamber (1930, Goddard called it "curtain cooling"); gyro stabilization and deflector vanes at the rocket exhaust for control and guidance (1932); liquid propellant rocket attaining supersonic speeds (1935); and turbo pump fed liquid rockets (1938–'41).

In 1936, Smithsonian Institution published Goddard's classic report on "Liquid Propellant Rocket Development," reviewing his liquid-propellant rocket research and flight-testing since 1919. The same year, US Military sent Lt. John Sessums to assess the military value of Goddard's rocket developments. Again to the disappointment of Goddard, Lt. Sessums reported that there was little military value in the rocketry, but combustion gases produced by burning rocket propellants would appear useful to drive turbines. Undeterred by this negative report, Goddard offered in 1940 all his research data, patents, and facilities for use by the armed services at a meeting with

US military representatives arranged by his main sponsor, Harry Guggenheim. The result of the meeting was uninspiring except an expression of possible use of liquid rocket engines in jet-assisted take-offs (JATOs) of aircraft.

By the end of 1941, Goddard and his team completed in all 103 static tests of liquid rocket engines and attempted 48 liquid rocket flights of which 31 resulted in flights. With the determination and optimism of an undeterred scientist, Goddard used to refer to the failed rocket tests as "valuable negative information." From 1942 to 1945, Goddard was the Director of Research at the Naval Engineering Experiment Station at Annapolis, Maryland, where in 1943, he and his team successfully developed and flight-tested a liquid-propellant JATO rocket engine. However, the US Navy abandoned this project in 1944. While at Annapolis, Goddard also developed liquid rocket engines with variable thrust capability, which were later successfully used on the Bell X-2 rocket planes during 1955–'56. Apparently, due to the ill health of Goddard, the work of Goddard's rocket team came to an end by July 1945. After fighting throat cancer for several months, Robert Goddard died on August 10, 1945, 4 days after the first atomic bomb was dropped on Japan.

Robert H. Goddard had a rare talent in rocketry in both theoretical and practical engineering realization. Rocket scientists of his time had known many concepts in rocket propulsion, such as canister launching (Jean Froissart 1370s), multistaging (Von Schmidlap 1591), and spin stabilization (William Hale, 1844). However, the eminence of Goddard lies in the engineering realization of rocket hardware involving these old as well as his many new concepts. The dedicated perseverance of this self-effacing scientist went largely unrecognized in the United States during his lifetime. It is reported that when Wernher von Braun and his team of 500 German rocket experts, who had surrendered to an American private, were brought to America after the World War II and questioned about rocketry, many stared in amazement and asked why American officials did not inquire of Goddard, from whom they had learned virtually all they knew.

Goddard is credited with 214 patents, of which his widow Esther Goddard filed 131 after his death. In 1951, Mrs. Goddard, who worked tirelessly at keeping alive the memory of her husband's pioneering research, and the Guggenheim Foundation, which largely funded Dr. Goddard's research, filed a joint claim against the US government for infringing upon Goddard's patents. In June 1960, Mrs. Goddard and the Guggenheim Foundation were given a $1,000,000 settlement, at that time the largest patent settlement that the government had ever given.

In memory of Professor Robert H. Goddard, a major space science laboratory, NASA's Goddard Space Flight Center, Greenbelt, Maryland, was established on May 1, 1959. On September 16, 1959, the 86th Congress authorized the issuance of a gold medal in the honor of the brilliant scientist.

1.2.3 Hermann Julius Oberth (1894–1989)

Hermann Julius Oberth (Fig. 1.11) was born on June 25, 1894, in the city of Hermannstadt in Hungary. Hermannstadt is the old German name, and the city's present name is Sibiu and it is in central Romania. When Hermann was 2 years old, his family moved to Schaessburg where his father, a physician, had been appointed Director of the City Hospital. Like Tsiolkovsky, also Oberth was greatly inspired by the science fiction of Jules Verne. Hermann's mother, who had keen interest in scientific and technical matters, was mainly responsible for kindling the scientific interest in him. She bought the novels of Jules Verne and gave them to the 12 years old Hermann to read. In his science fiction of 1865, "From the Earth to the Moon," Jules Verne had written that spacecraft containing astronauts were shot to the escape velocity using large gun barrels. Hermann at the age of 13 calculated from fundamentals that the escape velocity mentioned by Verne found it to be correct to his great exhilaration and self-assurance. He went on further to compute the acceleration that the Verne's astronauts would experience—a point not considered by Verne—and found it to be 47000 times the Earth's gravitational acceleration. Hermann then concluded that the space travelers would be flattened into pancakes with such acceleration, and hence Verne's space gun was not the solution but rockets should be used to launch the spacecraft.

In his spare time during his school days in Schaessburg, Hermann studied rockets and space flight and came out with the details of a manned spacecraft to be launched by rockets. He noted that solid propellant rocket motors were not suitable for launching

Fig. 1.11 Hermann Julius Oberth [21]: "To make available for life every place where life is possible. To make inhabitable all worlds as yet uninhabitable, and all life purposeful"

space ships but rocket engines using liquid hydrogen and liquid oxygen were to be adopted. Furthermore, he concluded that spacecraft were to be launched by multistage rocket vehicles.

After completing his school education in 1912, Hermann went to the University of Munich to study medicine. During the First World War (1914–1918), Hermann Oberth served in the medical corps, mostly in army hospitals in Hungary. Here again, he spent most of his free hours on the study of rockets. At the end of the First World War, a part of Hungary with Schaessburg was given to Romania and hence Oberth became Romanian citizen.

On returning from the World War I assignment, Oberth, in line with his insatiable interest in rockets and space travel, left medicine to study physics and mathematics in Klausenburg, Munich, and Gottingen. Professors in these places considered Oberth's ideas on rockets and space travel to be mostly fantasies. However, Oberth consolidated his research findings in a concise manuscript entitled "The Rocket into Interplanetary Space" (Die Rakete zu den Planetenraumen) and submitted it as a doctoral thesis to the University of Heidelberg. The University rejected the thesis outright, and the examiners criticized and even ridiculed Oberth's work. Professors of physics told him that rockets did not work in empty space, liquid propellants would explode on contact with liquid oxygen, and flight could not be controlled at supersonic speeds. One bright exception in this deluge of negative comments was Professor Ludwig Prandtl, the giant and genius of aerodynamics in Gottingen. He gave Oberth encouragement and praise, and he urged him to continue his work, regardless of the critics.

Against many adversaries, Oberth continued his studies on rockets and space travel. In 1922, with great respect and recognition for Professor Goddard's work and accomplishments in the United States, Oberth wrote to him for a copy of his 1920 book, "A Method of Reaching Extreme Altitudes." Goddard, basically a shy and secretive person, did send him a courtesy copy although not without some apprehension.

When Oberth wanted to publish his doctoral thesis, no publisher was willing to accept it. However, Oberth's wife Mathilde with her indomitable faith in her husband took the task of publishing her husband's work using her personal fund that she had saved over the years from the monthly provisions that she received for household expenses. The book, "The Rocket into Interplanetary Space" was published in 1923. In this book of 87 pages, Oberth discussed all the aspects of rocket propulsion (combustion, pump- and pressure-fed systems, and design of propellant tanks), flight mechanics (air and jet vanes, gyro control, inertial guidance, and aerodynamics), and space travel (life support systems, zero-gravity effects, re-entry, meteorite danger, and radiations in space). Among the uses of rockets and satellites, he mentioned mail service, meteorology, the Earth observations, astronomy, radio relays, astronautical research, solar and planetary studies, and general space sciences. The book contained for the first time a serious scientific non-fictional treatment of space travel.

Against the backdrop of Goddard's 1920 Smithsonian report and the related newspaper articles that had possibly reached Europe, the book by Oberth got the attention of many in Europe and caused much discussion about rocket and space technology. In fact, the popularity of the book goaded Russians to look at the reports of Tsiolkovsky that were a few decades old and start serious studies on rockets.

In the year 1923, Hermann Oberth became a professor of physics and mathematics at the high school in Schaessburg. Around this time, in their firm belief in open discussions and exchange of information, Oberth and Tsiolkovsky exchanged their papers and had a series of very friendly letters. In June 1927, the Society for Space Travel (Verein fuer Raumschiffahrt or VfR) was formed in Breslau, Germany and Oberth was one of its early members.

In 1928, UFA, a Berlin film studio, requested Oberth to be its technical adviser for its science fiction "The Woman in the Moon" ("Frau im Mond"). The film studio asked Oberth to construct one 2-meter oxygen-gasoline rocket to be launched in 1929 at the movie premiere. Oberth lost sight in his left eye while working on an experiment related to this project. Although the rocket launch was a failure at the premiere, the movie was a box-office hit and created quite a lot of awareness among people on rockets and space travel.

Oberth published in 1929 his second book entitled, "Path to Space Travel" (Wege zur Raumschiffahrt). Even though this book was considered to be an elaboration of his first book, it contained quite a few innovative ideas such as electric propulsion and methods of fabrication in orbits including that of large reflective mirrors. The book fetched the coveted French REP-Hirsch Prize and gave international recognition to Oberth's scientific contribution in the field of space technology.

In July 1930, Oberth under VfR successfully tested a liquid oxygen gasoline rocket engine of 70 N thrust for a thrusting time of 90 s. The engine used only a convergent nozzle. The development of the 70 N engine was a tremendous step forward in the German rocket development. After this success, Oberth worked for a brief period in Berlin under VfR with a group of enthusiastic assistants. One of these assistants was an 18-year-old student by the name of Wernher von Braun who went on to successfully develop many rocket weaponries for Germany and later became the space age architect of the United States. Because Oberth had to support a large family, he could not continue his work under VfR and had to return to his teaching job in Romania.

By 1930, German Army was looking for a flying weapon that did not contravene the 1919 Treaty of Versailles, a peace settlement signed after World War I. As per the treaty, Germany was not allowed to have an air force. Therefore, the country was interested in developing a rocket force in place of the prohibited air force. In 1932, VfR successfully fired Mirak II rockets fitted with engines of 600 N thrust. Following this, the German Army Ordinance Office formalized its rocket development by placing Captain Walter Dornberger as the officer-in-charge. He in turn enlisted the help of Wernher von Braun. The association of Dornberger and von Braun was historic and led to the successful development of the Vengeance Weapon Number 2 or popularly known as the V-2 bomb.

In Romania, Oberth was teaching and continuing his rocket development work in his free time. Then he moved to Vienna Institute of Technology to initiate a research program on rockets. When World War II broke out, Oberth moved to Dresden, Germany. In 1941, after acquiring German citizenship, Oberth joined the group of

Fig. 1.12 Hermann Oberth (forefront) at the Army Ballistic Missile Agency at Huntsville, Alabama in 1956. Clockwise starting from Oberth: Dr. Ernst Stuhlinger, an electric-propulsion pioneer (seated); Major General H.N. Toftoy, who took rocket scientists and engineers out of Germany after World War II to design rockets for American military use; Dr. Eberhard Rees, and Dr. Wernher von Braun. (Courtesy NASA [20])

his once assistant Wernher von Braun in Peenemuende and contributed to the success of V-2 bomb development. At the end of World War II, Oberth moved to a small town near Nuremberg, Feucht, where his family had been living during the war years. For brief periods, Oberth worked in Switzerland (1948) and also in Italy (1950–'53), where he developed a smokeless ammonium nitrate rocket.

In 1955, Wernher von Braun, now the Director of Rocket Development Operations, invited his mentor to join his team in Huntsville, Alabama, USA. Oberth participated in the team's program on satellite launch development, Fig. 1.12. In February 1958, this culminated in the successful launch of Explorer I, the first Earth satellite of the USA. Soon afterwards, Oberth returned to Germany to lead a pensioner life at Feucht, Nuremberg. However, on several occasions, he visited the USA after 1958. In 1961 for 9 months, he served as a consultant to Convair in Santiago, USA. Also, he came on invitation to Cape Canaveral in Florida in July 1969 to see the launching of Saturn Apollo 11 rocket with astronauts Neil A. Armstrong, Michael Collins, and Edward E. Aldrin that enabled the first landing of man on moon.

When he advanced in age, Oberth, like Tsiolkovsky, became interested in philosophy and wrote many treatises on this subject. Most notable among these is the book entitled "Primer for Those who Would Govern." Hermann Oberth died in Nuremberg on December 28, 1989 at the age of 95.

In appreciation of the contributions of Hermann Oberth, the Hermann Oberth Society founded in 1971 the Hermann Oberth Space Museum in Feucht, Nuremberg. A crater on the moon is named after Oberth.

1.3 Summary

The practical demonstrations of a technology mostly precede its theory. However, the fast and effective technology refinement begins only after the theory, explaining the underlying basic principles, is established. The technology of jet propulsion, which includes air-breathing- and rocket technology, is not an exception to this.

The Chinese were the first to develop rockets many centuries before Christ. However, the first recorded use of their rockets as weapons was in 1232 at the battle of Kai-fung-fu against the invading Mongols. The rocket technology, mostly as its application to weapon technology, spread to the entire world in about six centuries.

Modern astronautics and rocketry owe its development to the pioneering works of three great scientists: Konstantin Eduardovitch Tsiolkovsky of Russia, Robert Hutchings Goddard of the United States of America, and Hermann Oberth of Germany. With undaunted determination and dedicated perseverance, these eminent men worked tirelessly against many odds such as delayed recognition or total disapproval, acute shortage of research funding, rebuke, criticism, and personal ill health. It is amazing that in spite of their different origins, all the three had one view in common, namely, the rocket technology will be more useful to mankind as the tool for space exploration to "colonize the entire solar system and beyond."

Problems

1.1 For about 1000–2000 words, write a brief biography of Dr. Werner von Braun.

1.2 Realizing that the engine exit velocity u_e represents effectively the specific impulse I_{sp} (Sect. 3.4), from Eq. 1.1 derive the following relations. Note that the final mass m_f includes the structural mass m_s comprising engine mass and other accessories, and mass of payload m_l.

$$\frac{m_0}{m_f} = \frac{m_0}{m_s + m_l} = exp\left(\frac{\Delta v}{I_{sp}}\right) \tag{1.2}$$

$$m_p = m_0\left[1 - exp\left(\frac{-\Delta v}{I_{sp}}\right)\right] \tag{1.3}$$

$$m_p = m_f\left[exp\left(\frac{\Delta v}{I_{sp}}\right) - 1\right] \tag{1.4}$$

$$m_l = m_0\, exp\left(\frac{-\Delta v}{I_{sp}}\right) - m_s \tag{1.5}$$

1.3 A spacecraft of burnout mass $m_f = 1025$ kg in a geo-transfer orbit is required to give itself a "kick"—velocity increment—of 0.425 km/s to place itself into geosynchronous orbit. Two propulsion systems are being considered: (1) a mono-propellant hydrazine thruster of $I_{sp} = 2260$ Ns/kg and (2) a bi-propellant N_2O_4/MMH thruster of $I_{sp} = 3040$ Ns/kg. Calculate the mass of propellants required by these two thrusters for the maneuver.

Answers:
$\left(m_p\right)_{hydrazine} = 212.07\,\text{kg}$; $\left(m_p\right)_{N_2O_4/MMH} = 153.798\,\text{kg}$

1.4 The total mass of a three-stage satellite launch vehicle is $m_0 = 701$ ton. Its first stage is presently of LOX/RP1 engine. Including the other two stages and the actual payload, the "payload" of the first stage $m_l = 252$ ton. The structural coefficient ϵ of the present LOX/RP1 first stage is 0.093. Let us suppose that the first stage is to be replaced by a LOX/LH2 engine. The specific impulse of the LOX/LH2 engine is expected to be 1.44 times that of the LOX/RP1 engine. But, the mass of the LOX/LH2 engine is expected to be 35% heavier than that of the LOX/RP1 engine. Estimate for both the cases (1) the mass values of m_s (dry stage: engine with its tank and other accessories) and m_p, (2) the ratio of burnout velocities between LOX/LH2 and LOX/RP1 stages. Assume for simplicity the vehicle is moving in a gravity-free vacuum.

The structural coefficient is defined as $\epsilon \equiv m_s/\left(m_s + m_p\right)$

Answers:
(1) For LOX/RP1 stage: $m_s = 41757\,\text{kg}$ & $m_p = 407243\,\text{kg}$. For LOX/LH2 stage: $m_s = 56372\,\text{kg}$ & $m_p = 392628\,\text{kg}$. (2) $\Delta v_{LOX/LH2}/\Delta v_{LOX/RP1} = 1.3596$.

References

Cited

1. https://en.wikipedia.org/wiki/Main_Page. Cited 10 June 2021
2. http://genesismission.jpl.nasa.gov/people/biographies/. Cited 10 June 2021
3. https://www.flickr.com/photos/nasacommons. Cited 10 June 2021
4. http://history.nasa.gov/. Cited 10 June 2021
5. http://inventors.about.com/od/rstartinventions/a/Rockets.htm. Cited 10 June 2021
6. http://scienceworld.wolfram.com/physics/Rocket.html. Cited 10 June 2021
7. http://www.astronautix.com/. Cited 10 June 2021
8. http://www.clarku.edu/offices/library/archives/ExhibitionRoom.htm. Cited 10 June 2021
9. http://www.kiosek.com/oberth/. Cited 10 June 2021
10. http://www.meaus.com/spacePioneer.html. Cited 10 June 2021
11. http://www.nso.lt/history/hermann.htm. Cited 10 June 2021
12. http://www.oberth-museum.org. Cited 10 June 2021
13. http://www.roswellmuseum.org/. Cited 10 June 2021
14. http://www.solarviews.com/eng/rocket.htm. Cited 10 June 2021
15. https://www.grc.nasa.gov/www/k-12/TRC/Rockets/history_of_rockets.html. Cited 10 June 2021

16. Charles H. Hubbell (1898–1971), Public domain, via Wikimedia Commons. https://upload.wikimedia.org/wikipedia/commons/c/c6/Rocket_warfare.jpg. Cited 10 June 2021
17. The Memorial House Museum of Konstantin E Tsiolkovsky (2021) http://www.informatics.org/museum/tsiol.html. Cited 10 June 2021
18. https://upload.wikimedia.org/wikipedia/commons/3/3f/Dr._Robert_H._Goddard_-_GPN-2002-000131.jpg. Cited 10 June 2021
19. https://upload.wikimedia.org/wikipedia/commons/7/7c/Goddard_and_Rocket.jpg. Cited 10 June 2021
20. https://www.flickr.com/photos/nasacommons. Cited 10 June 2021
21. https://upload.wikimedia.org/wikipedia/commons/2/2b/Hermann_Oberth_1950s.jpg. Cited 10 June 2021

Further Reading

22. https://www.grc.nasa.gov/www/k-12/TRC/Rockets. Cited 15 Jan 2014
23. https://en.wikipedia.org/wiki/History_of_rockets. Cited 15 Jan 2024
24. https://web.archive.org/web/20010306020430 http://napoleonic-literature.com/Articles/Rockets/History_of_Rockets.html. Cited 15 Jan 2024
25. Major D (1794) A narrative of the campaign in India which terminated the war with Tippoo Sultan in 1792, Bulmer and Co, London. https://books.google.co.in/books?id=WOdAAQAAMAAJ&printsec=frontcover&source=gbs_ge_summary_r&cad=0#v=onepage&q&f=false Cited 15 Jan 2024
26. Goddard R (1919)A method of reaching extreme altitudes, Smithsonian Institution. https://edan.si.edu/transcription/pdf_files/8542.pdf. Cited 15 Jan 2024
27. Topics of the Times, New York Times, 13 Jan 1920. https://web.archive.org/web/20080209230323/ http://it.is.rice.edu/~rickr/goddard.editorial.html. Cited 15 Jan 2024
28. The A-4 (V-2) ballistic missile https://www.russianspaceweb.com/a4.html. Cited 15 Jan 2024
29. https://www.sciencelearn.org.nz/resources/1868-brief-history-of-rockets-timeline. Cited 15 Jan 2024

Chapter 2
Propulsion System Classification

Synopsis

In a propulsion system, the fluid momentum rate at exit ($\dot{m}_e u_e$) being greater than that at entry ($\dot{m}_a u$) is the basis for the thrust production. The thrust can be produced by the acceleration of either the open stream as in propellers for low speeds or the enclosed streams as in nozzles for high speeds.

The broad classification of jet propulsion into air-breathing propulsion and rocket propulsion is brought out. Under air-breathing systems, the working principles of ramjets and gas turbines and their variants are explained. Ramjets can be liquid fuel ramjets and solid fuel ramjets. Also ramjets are classified as subsonic combustion ramjets and supersonic combustion ones (scramjets). Under gas turbine propulsion, its variants are turboshaft, turboprop, turbofan, and turbojet with reheat. How the suitable air-breathing propulsion systems for different flight ranges—subsonic, transonic, supersonic, and hypersonic—change from turboshaft at one end to scramjet at the other is addressed.

Rocket propulsion is broadly classified into chemical rockets, electric rockets, and nuclear rockets. Chemical rocket propulsion has its limitations that the chemical has to serve as the source for energy as well as working fluid. Chemical rocket propulsion is further classified into solid rocket motor, liquid rocket engine, and hybrid rocket engine. Merits and demerits of these three systems are discussed. Because of their simplicity and ever readiness for firing, solid rocket motors find wide applications in booster stages of satellite launch vehicles and warfare missiles. On the other hand, although not as simple as solid ones, the liquid rocket engines find extensive applications—large boosters, medium and large upper stage and space engines, and mini control thrusters. This is because of (1) the cooling provision for extended thrusting, (2) the thrust control and start-stop capability, (3) the separate tanks facilitating the broader choice on oxidizer and fuel and the consequent wider and higher specific energy (specific impulse), and (4) the auto-ignition capability. The

S. Krishnan and J. Raghavan, *Rocket Propulsion Primer*,
https://doi.org/10.1007/978-981-97-5644-5_2

hybridization of solid and liquid engines, with their respective merits and demerits, results in hybrid rocket engines. However, all chemical propulsion systems are energy limited devices although not limited by power.

As the energy source is separated from the propellant, a wide variety of electric propulsion systems is feasible with a specific impulse more than 10 times to as low as a fraction of the value possible with chemical systems. Despite being power limited, the advantage of the unlimited energy source that is possible with electric propulsion is explained. With suitable sketches, the working principles of resistojets, arcjets, ion engines, Hall thrusters, pulsed plasma thrusters, magneto plasmadynamic thrusters, and pulsed inductive thrusters are explained. For interplanetary travel and beyond by humans, a relatively large power with unlimited energy in space propulsion is the requirement. This is possible through nuclear propulsion and working principles of a nuclear rocket engine are explained.

Number of concept questions and numerical problems with answers are included at the chapter end.

2.1 Introduction

Propulsion systems are those that are required to propel the vehicles to which they are attached. By this definition, propulsion systems are many and varied, starting from our muscles using which we propel ourselves in walking, running, swimming, and so on. In engineering, let us first consider the classical reciprocating engines, which can be the steam-based external-combustion piston engines or the petroleum fuel-based internal-combustion piston engines. The steam engines propel locomotives and ships while the internal-combustion piston engines propel automobiles, ships, locomotives, and small low-speed aircraft. The propulsion principle by which the reciprocating engines propel locomotives and automobiles is different from the one in ships and low-speed aircraft. In the former, the wheels, powered by the engine, attempt to push the road or the rails through friction. By the reactive force (Newton's Third Law), the vehicles to which the wheels are attached are pushed in the opposite direction or propelled. In the case of ships and low-speed aircraft, the propellers draw in and push (accelerate) the fluid stream (water in the case of ships and air in the case of aircraft) to a velocity greater than the vehicle velocity; that is, the momentum rate of the fluid at exit ($\dot{m}_e u_e$) is greater than the momentum rate at entry ($\dot{m}_a u$) to the propellers. Here again, by the reactive force, the vehicles are pushed or propelled.

This acceleration of fluid stream is also the means for high-speed propulsion that we find in high-speed aircraft, rockets, and spacecraft. But the high-speed propulsion systems employ nozzles to accelerate the fluid streams and these systems are known as jet-propulsion systems. The nozzle accelerates and ejects the fluid stream due to the pressure difference across the nozzle's entry and exit. While propellers accelerate "open" streams, nozzles accelerate the "enclosed" streams and eject them as jets. Be it propellers or nozzles, the thrust is produced essentially due to the difference in the momentum rates at the entry and the exit of the propulsion systems. In the case of a propeller, we have to consider the momentum rates ($\dot{m}u$) at the propeller's entry and exit. But in the case of jet propulsion systems, we have to consider the momentum rates at the entry and the exit of the jet propulsion engine. We consider in detail the derivation of the thrust equation in Chap. 3. Here, neglecting the pressure forces, we can write the thrust as equal to the difference in the momentum rates at the exit and the entry.

$$F = \dot{m}_e u_e - \dot{m}_a u \tag{2.1}$$

where $\dot{m}_a$ is the mass flow rate of atmospheric air entering the engine with the flight velocity u and $\dot{m}_e$ is the mass flow rate of combustion gases exiting the nozzle with the velocity u_e.

The jet propulsion systems can be broadly classified into two types: (1) **air-breathing jet propulsion** systems and (2) **non air-breathing jet propulsion** systems or **rocket propulsion** systems. *The subject matter of this book is rocket propulsion systems.* However, for the sake of completeness and continuity, let us briefly consider the classification of air-breathing propulsion systems.

2.2 Air-Breathing System Classification

As its name implies, the air-breathing propulsion system breathes (inducts) air from the Earth's atmosphere. For the ease of comparison, let us consider simultaneously two principal air-breathing propulsion systems. The first one is the **ramjet** and the other is the basic propulsion system using a gas turbine, Figs. 2.1 and 2.2. Ramjets propel supersonic missiles. **Gas turbine propulsion systems** propel subsonic and supersonic aircraft. Images of Thor ramjet that powered Bloodhound Missiles of the United Kingdom are shown in Figs. 2.3 and 2.4. Image of a basic gas turbine propulsion system (turbojet) is shown in Fig. 2.5.

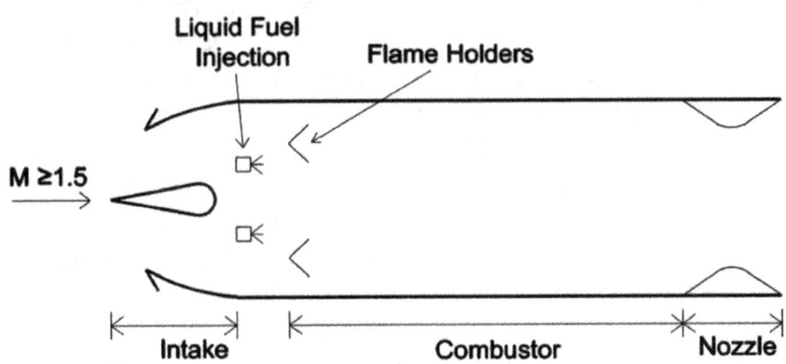

Fig. 2.1 Schematic sketch of ramjet

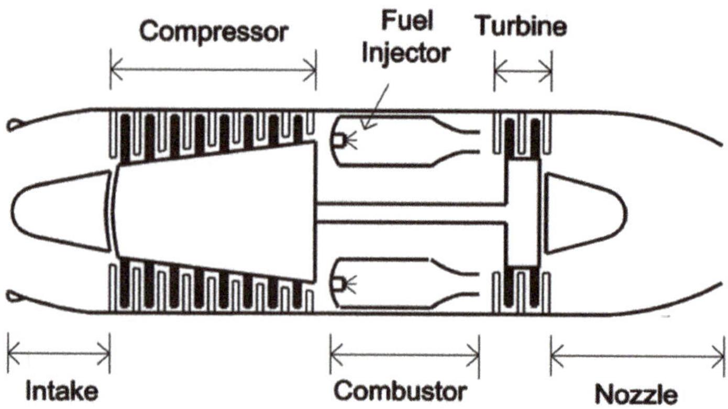

Fig. 2.2 Basic gas turbine propulsion system, turbojet

Fig. 2.3 Thor ramjet of Bristol Aero Engines, United Kingdom (Courtesy Wikipedia [1])

In the ramjet as well as the gas-turbine propulsion system, air is inducted at a momentum rate corresponding to the flight velocity. The oxidant required for combustion is the oxygen in air. The inducted air is decelerated (gas-dynamically compressed) in the intake (or diffuser). This results in the increase in the static pressure and static temperature of the air stream with certain losses in total pressure due to surface friction, Figs. 2.1 and 2.2. The process is essentially adiabatic. Ramjets mostly power missiles flying at supersonic speeds. In Fig. 2.1, therefore, the ramjet intake lip and center body tip are sketched sharp to negotiate the resulting shock front. The sketch of the turbojet shown in Fig. 2.2 is intended to fly at subsonic speeds, and hence the intake lip and the center body front are shown rounded. This sketching procedure is also followed for other types of engines.

In the gas turbine propulsion system, the decelerated air is further mechanically compressed by a compressor that results in the increase in the total pressure and total temperature, Fig. 2.2. The compression process is essentially adiabatic.

Fig. 2.4 Bloodhound missile fitted with Thor ramjets, RAF Museum, London (Courtesy Wikipedia [1])

Fig. 2.5 J85-CE-17A General Electric turbojet engine (Courtesy Wikipedia [1])

In the ramjet as well as gas turbine propulsion system, the liquid fuel, generally a type of kerosene, is injected into the compressed air and burned in the combustion chamber to result in the increase in total temperature and a small loss in the total pressure due to mass addition, friction, and heat addition.

As the gas turbine propulsion system employs a compressor, Fig. 2.2, the high-pressure and high-temperature combustion products are mechanically expanded through a turbine that results in the reduction in total pressure and total temperature. The expansion process is essentially adiabatic. The turbine power is used to run the compressor. There is no mechanical compression in ramjet and hence there is no need for the turbine.

In the ramjet as well as gas turbine propulsion system, the high pressure and high temperature combustion products are gas-dynamically expanded in a nozzle to accelerate the stream of the combustion products to a momentum rate that is substantially greater than the inducted value, Figs. 2.1 and 2.2. The nozzle expansion results in the decrease in the static pressure and static temperature with certain loss in total pressure due to friction. The nozzle expansion is essentially adiabatic.

As the average Mach number of the fluid through the compressor, combustion chamber, and turbine is generally low around 0.3–0.4, the differences between the total and static values of pressure and temperature are low, around or less than 3% between the total and static temperatures and 12% between the total and static pressures. Note, however, that the differences between the total and static values are substantial in the intake and the nozzle.

We have previously discussed simultaneously two principal air-breathing propulsion systems, ramjet and gas-turbine propulsion systems. The ramjet has only the gas-dynamic compression and expansion, Fig. 2.1. The gas turbine propulsion system, on the other hand, has gas-dynamic and mechanical compression and expansion, Fig. 2.2. Although we have discussed the differences between the two principal types of air-breathing propulsion systems, it is interesting to note that the ramjet is a particular case of the gas-turbine propulsion system. Without the mechanical compression and expansion—without the compressor and the turbine—the gas turbine propulsion system simplifies into the ramjet. Furthermore, we note that essentially due to the positive difference between the exiting and inducted momentum rates in the air-breathing propulsion systems, the thrust is produced to propel the vehicle, Eq. 2.1.

In view of the point that ramjet is only a particular case of gas turbine propulsion system, let us consider the gas turbine propulsion system as a general case of the air-breathing propulsion. The fluid that passes through the propulsion system changes in composition and mass flow rate as per the following.

1. The low-temperature air enters the intake, gets decelerated in the intake, and further compressed in the compressor. The composition of air in this can change due to the increase in pressure and temperature.
2. Fuel is injected in the combustion chamber and the fuel–air mixture is of composition and mass flow rate different from that of the air through the intake and compressor.
3. High-temperature products of combustion are formed in the combustion chamber and these are of composition different from that of the fuel–air mixture.
4. High-temperature combustion products are expanded through the turbine. Because the pressure and temperature drop across the turbine, the composition of the gas can change from the one at the turbine entry to that at its exit.

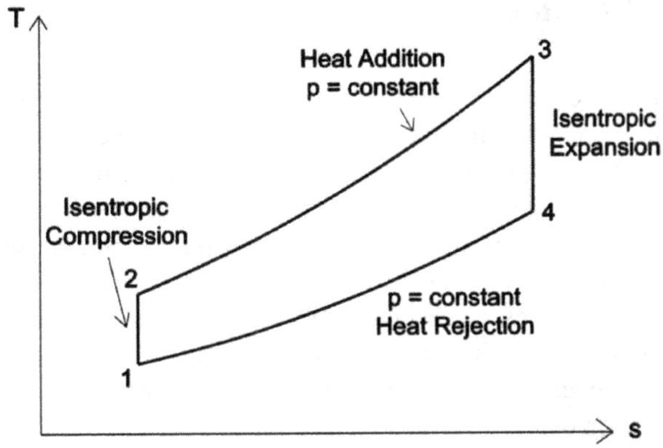

Fig. 2.6 Brayton or Joule cycle

5. Because of the reduction in static temperature and pressure from the nozzle inlet to the exit, the composition of the combustion products at the exit can be different from that at the nozzle inlet.

In spite of the previously discussed realities, for the purpose of idealization, we can assume (1) the composition and the mass flow rate of the fluid stream to be invariant, (2) the processes of compressions and expansions to be isentropic, (3) the fuel addition and the combustion to be at constant pressure, (4) compared to the mass flow rate of air, the fuel mass flow rate to be negligible, (5) the differences between the total and static pressures and between the total and static temperatures to be negligible from the outlet of the intake to the inlet of the nozzle, and (6) the nozzle-exit static pressure to be equal to the atmospheric pressure. Under these idealized conditions, the processes can be represented by those in the Brayton or Joule cycle, Fig. 2.6.

With respect to the Brayton cycle (Fig. 2.6), the process 1–2 is the isentropic compression in the intake of a ramjet propulsion system, or that in the intake and the compressor of a gas turbine propulsion system. The process 2–3 is the fuel addition and combustion at constant pressure. The process 3–4 is the isentropic expansion in the nozzle of a ramjet propulsion system or that in the turbine and the nozzle of a gas turbine propulsion system. The process 4–1 is the cooling of the ejected combustion products in the outside atmosphere at constant pressure. However, realize that in the case of the ideal ramjet propulsion system while the total pressure remains constant throughout, only static pressures vary, 1–2 and 3–4. But in the case of the ideal gas turbine propulsion system, while the total pressures in the intake and the nozzle remain constant with static pressure variations, the total pressures vary in the compressor and turbine. Therefore, the isentropic processes compression 1–2 and expansion 3–4 can be effectively taken to represent the total pressure variations in the compressor and turbine, respectively.

2.2.1 Ramjet

Because there is no moving component within the system, the ramjet is the simplest of all air-breathing propulsion systems. The ramjet can be of different types. Based on the fuel type, it is classified as **liquid-fuel ramjet** (Fig. 2.1) and **solid-fuel ramjet**, Fig. 2.7. And, based on the type of combustion, it can be the **subsonic-combustion ramjet** or the **supersonic-combustion ramjet** or **scramjet**. The previously discussed ramjets (Figs. 2.1 and 2.7) are subsonic combustion ramjets since combustion in these takes place under subsonic conditions. Considering now the scramjet, the inducted air at hypersonic speeds in its intake (Mach number greater than around 6) gets decelerated to a supersonic value (Mach number around 2 or less), Fig. 2.8. Combustion takes place under supersonic conditions. Thereafter, the combustion products expand in the nozzle to hypersonic velocities. The scramjet is the simplest system that is most suitable for powered hypersonic flights within the Earth's atmosphere and hence it is actively being developed.

Although the ramjet is the simplest of all the air-breathing propulsion systems, it suffers from a major disadvantage of its inability to produce thrust under static condition and hence it cannot propel its vehicle from zero speed. The ramjet works on the principle of utilizing the kinetic energy contained in the in-flight inducted air. Through gas-dynamic compression, the kinetic energy of the inducted air is reduced to increase the static pressure of the air. This compressed air after combustion is expanded through the nozzle to produce thrust. When the ramjet is not in flight it cannot induct air on its own and hence it cannot produce a thrust.

Fig. 2.7 Schematic sketch of solid fuel ramjet

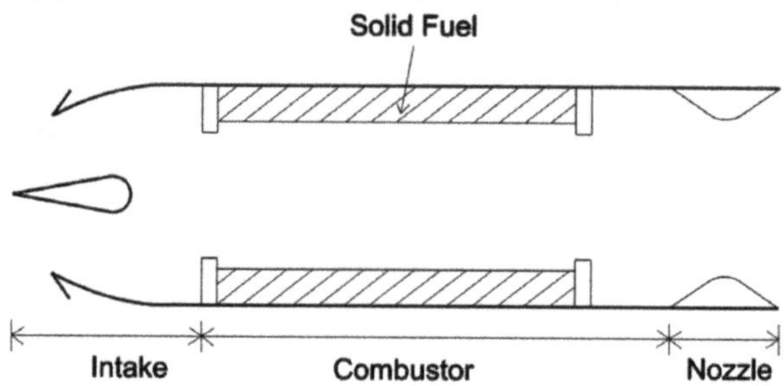

Fig. 2.8 Schematic sketch of scramjet engine

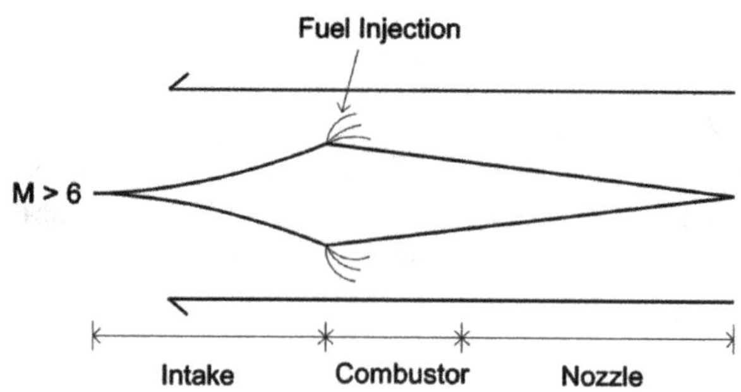

The quality of ramjet performance depends on the quantity of the specific kinetic energy in the inducted air. To have an acceptable performance parameter, say an acceptable thrust per unit mass flow rate of inducted air—this represents the size of an air-breathing propulsion system for a specified thrust—or an acceptable fuel consumption per unit thrust representing the fuel economy, there are lower and upper limits on flight velocities.

Let us consider the subsonic combustion ramjet and first look at the lower velocity limit for flight. There is a minimum flight velocity and correspondingly a minimum kinetic energy that will overcome the internal drag due to various total-pressure losses in the intake, combustion chamber, and nozzle and still have sufficient total- and static pressure for expansion in the nozzle to produce a positive thrust. This condition of producing a positive thrust starts around a flight Mach number of 0.9. For the application of powered flight, this positive internal thrust should be sufficiently higher than the external drag due to the flow over the external surfaces of the vehicle including that of the propulsion system. Therefore, from the point of view of acceptable performance parameters, the minimum flight Mach number for the ramjet is around 1.5.

Let us now look at the upper limit on flight velocity for the subsonic combustion ramjet. The total temperature of the inducted air increases as one increases the flight velocity. We have seen earlier that there is not much of a difference between total and static values from the intake outlet to the nozzle inlet. As we do not have materials to withstand temperatures beyond a certain limit, there is an upper limit on the maximum allowable temperature in the combustion chamber. With the increase in the flight velocity, the gap between the total temperature of the inducted air and the maximum allowable temperature reduces. This gap truly represents the quantity of fuel that can be injected into the combustion chamber to produce a positive thrust after negotiating the external drag. The reduction in fuel addition means the reduction in thrust. This, beyond a certain flight Mach number, leads to a condition of uneconomical fuel consumption and then unacceptably low thrust per unit mass flow rate of air. Furthermore, close to the maximum allowable temperature, the dissociation in combustion products increases. Thus, for the fuel injected the dissociation absorbs a significant combustion energy without any effective rise in chamber temperature. Therefore, for subsonic combustion ramjets, this upper limit appears around a flight Mach number of 5.

2.2.2 *Gas Turbine Propulsion System*

The gas turbine propulsion system, unlike the ramjet propulsion system, is a complex one. Its major component, the rotating compressor-turbine assembly, is a dynamic one with respect to the engine structure. The design and the operation of the compressor are complicated because the compressor is not only a dynamic system but also an adverse pressure gradient one—total- and static-pressure increase along the stream. Furthermore, the turbine, although a supporting pressure gradient one, has to withstand temperatures of around 1500 K or more of the combustion gases. Its blades have to be designed with complex cooling arrangements. These blades are mostly of high temperature low-creep nickel chromium super alloys and have to be fabricated adopting advanced manufacturing techniques.

Although the gas turbine propulsion system is a complex one it has the advantage of its ability to produce thrust when its flight vehicle is stationary—the compressor can still induct air from the quiescent atmosphere. Furthermore, the gas turbine propulsion system offers a wide range of variants that are suitable for different regimes of flight Mach numbers. Maximum flight Mach number that is possible by the gas turbine propulsion system is around 3. Beyond this number, the kinetic energy contained in the inducted air is so high that the mechanical compression and expansion become redundant. And, as indicated earlier, without the compressor and turbine, the gas turbine propulsion system simplifies into the ramjet.

The basic gas-turbine propulsion system we considered earlier, Figs. 2.2 and 2.5, is known as turbojet. This configuration is rarely adopted in flights. The major variants of gas turbine propulsion system adopted in flights are (1) **turboshaft**, (2) **turboprop**, (3) **turbofan**, and (4) **turbojet with afterburner or reheat**.

2.2.2.1 Turboshaft Engine

In the turboshaft engine used in helicopter propulsion, the turbine produces power in excess of what is required for the compressor and this excess power through a reduction gearbox runs a power shaft. This shaft power can be used to rotate the helicopter blades, the vertically arranged propellers, Fig. 2.9. The static pressure at the turbine exit is essentially atmospheric, and there is very little jet power that can be tapped by way of expanding the turbine exhaust through a nozzle. The maximum forward speed that a typical helicopter can reach is around 300 km/hr and this roughly corresponds to a flight Mach number of 0.25.

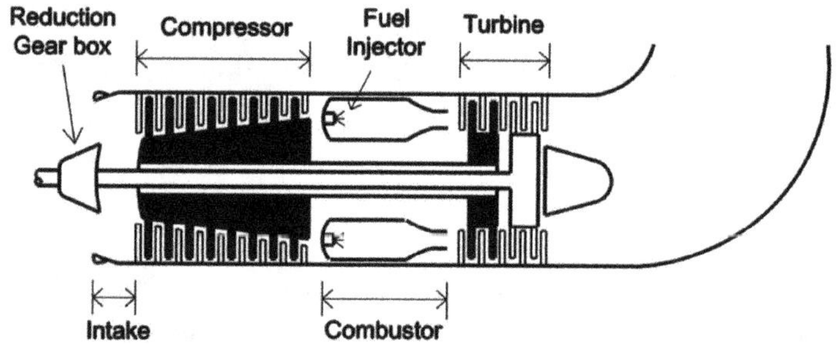

Fig. 2.9 Schematic sketch of turboshaft engine

Apart from their application in helicopter propulsion, turboshaft gas turbines are widely used in electric power generation. Here, the power shaft runs an electric generator. The application also exists in the propulsion of ships where the power shaft runs the ship propellers. Also, we find a few applications of this class of propulsion systems in locomotives and racing cars.

2.2.2.2 Turboprop Engine

In the turboprop engine, schematically shown in Fig. 2.10, the propulsion is partly by the propeller power and partly by the jet power. The turbine power, produced in excess of compressor power, is used to run the propellers through a reduction gear-box. The pressure at turbine exhaust is substantially greater than the atmospheric pressure to enable the expansion through a nozzle to produce a jet thrust. Thus, the propulsion of the aircraft is by two thrust components, one the "cold thrust" from the propeller accelerating the "open" stream and the other the "hot thrust" from the nozzle accelerating the "enclosed" stream. The maximum flight Mach number that is possible for a turboprop-powered aircraft is around 0.6. However, recent developments in the concept of turboprop propulsion have enabled the use of turboprop propulsion to higher flight Mach numbers and altitudes. This concept of turboprop engine with an improved propeller configuration is known as propfan engine and these engines power most of the recent short haul aircraft.

2.2.2.3 Turbofan Engine

In the turbofan engine, schematically shown in Fig. 2.11, the propulsive force is obtained through two jets, "cold" jet and "hot" jet. The turbine power, produced in excess of the high-pressure compressor-power, is used to run a low-pressure compressor and an enclosed fan; hence the engine is named as the turbofan engine.

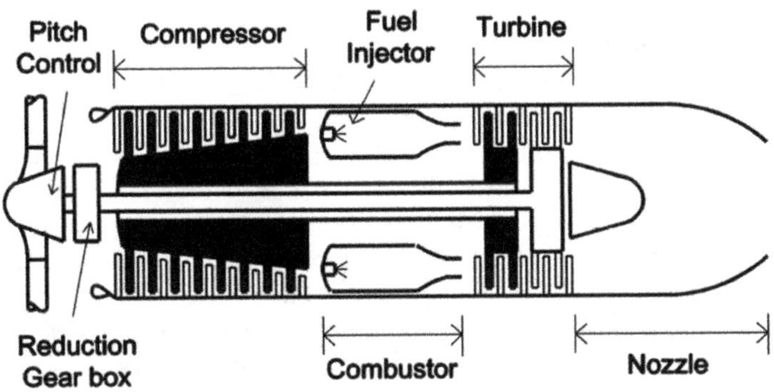

Fig. 2.10 Schematic sketch of turboprop engine

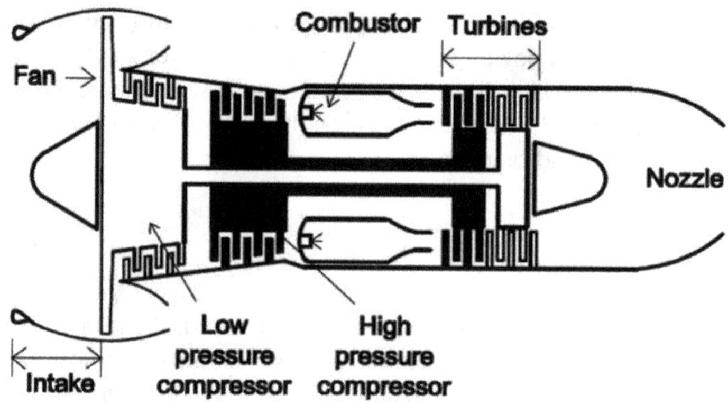

Fig. 2.11 Schematic sketch of turbofan engine used for transonic flight Mach number, M $\approx$ 0.9

Air is drawn-in by the fan at a momentum rate corresponding to the flight velocity. After its deceleration in the intake, the fan compresses the inducted air. A part of the fan-compressed air is bypassed into the first nozzle to be expanded to a velocity greater than the flight velocity. Because of this bypass, the turbofan engine is also known as the bypass turbojet engine. The balance of the fan-compressed air is further compressed by the low-pressure and high-pressure compressors and fed into the combustion chamber. The combustion products expand through the turbine to supply power to the compressors and the fan. The relatively hot turbine exhaust is expanded through the second nozzle. As the turbine exhaust is hotter than the fan-compressed air, the first nozzle is known as cold nozzle and the second one as hot nozzle. Correspondingly, the two jets are characterized as cold and hot jets. Thus, the propulsion of the aircraft is by two thrust components, one from the cold nozzle and the other from the hot nozzle. Turbofan engines can fly aircraft with maximum fuel economy in the Mach number ranges around 0.9. Such engines have the bypass flow rates about four to five times the flow rates through the basic compressor. Or in other words, the major part of the inducted air gets expanded through the cold nozzle. Almost all the present-day transport planes fly in the Mach number ranges around 0.9 and turbofan engines power these.

For flight in the supersonic regime, the turbofan engine has to have an additional system known as **afterburner** or **reheat**. Due to material limitations, there is an upper limit on the maximum allowable temperature at the turbine entry. Because of this, not all the oxygen contained in the inducted air is used up in the combustion chamber. This means that the adopted fuel–air ratio in the combustion chamber is far less than the stoichiometric value. After the turbine expansion, as the temperature has substantially come down, the unused oxygen in the inducted air can be used by injecting and burning additional fuel in the turbine exhaust. This added combustion system is known as afterburner or reheat system. The additional energy added to the turbine exhaust gives increased thrust in the nozzle expansion without much variation in the engine envelope and mass. Such additional thrust is needed for crossing the sonic barrier and flying the aircraft at supersonic speeds. Turbofan engines with afterburners can fly aircraft up to a Mach number of 1.5. The schematic sketch of a turbofan engine with an afterburner is shown in Fig. 2.12. In this variant, note that the bypass air is ducted all along the engine to get mixed with the turbine exhaust. This engine is known as **ducted turbofan-engine** or **ducted bypass-engine**, and in this the bypass flow rate is only a fraction of the total inducted air. Such an arrangement is needed for turbofan engines that fly in supersonic regimes.

2.2.2.4 Turbojet Engine

As the flight Mach number in the supersonic regime increases, for the suitable turbofan engine the ratio of the bypass flow rate to the flow rate through the basic compressor keeps decreasing. And beyond a certain value of flight Mach number it becomes zero. A turbofan engine with zero bypass flow rate is nothing but a turbojet engine. The schematic sketch of a turbojet engine with an afterburner is shown in Fig. 2.13. The processes that take place in a turbojet engine are previously explained while considering gas turbine propulsion systems, Fig. 2.2. In Fig. 2.13, additionally we have an afterburner, which is required for supersonic propulsion as explained previously under the turbofan engine, Fig. 2.12. The nozzle configurations required

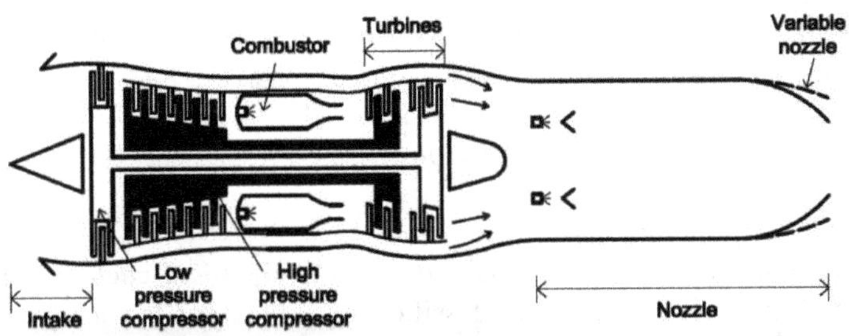

Fig. 2.12 Schematic sketch of turbofan engine with afterburner used for supersonic flight Mach number, M $\approx$ 1.5

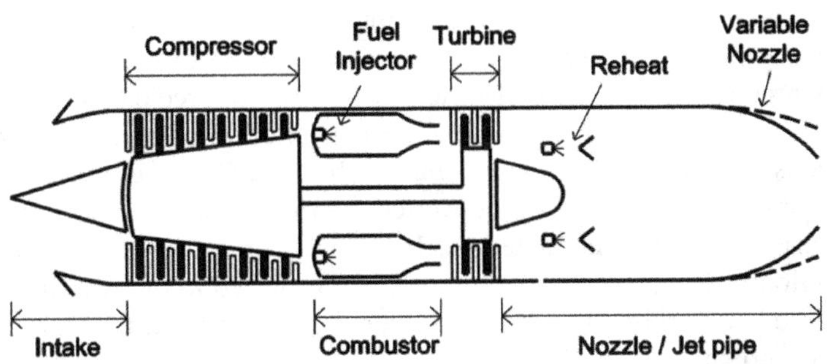

Fig. 2.13 Schematic sketch of turbojet engine with afterburner for supersonic propulsion

for optimum performance with and without reheat are different, turbofan and turbojet engines with afterburners will have nozzles of variable configurations as shown in Figs. 2.12 and 2.13.

With this treatment, we close our discussion on air-breathing propulsion systems. For a more detailed discussion on this topic, see [2, 3].

2.3 Rocket Propulsion System Classification

The rocket propulsion systems do not draw any air from the atmosphere and the working fluid is generated or stored within. The working fluid is generated within the chemical rockets, whereas it is stored in electric- and nuclear rockets. By this unique quality, two characteristics of rocket propulsion are evident. First, since the working fluid is generated or stored within, the production of thrust by a rocket propulsion system is not dependent on its surrounding medium. Second, since no part of the working fluid is drawn for free from the surrounding medium, the production of thrust by a rocket propulsion in the Earth's atmosphere is more expensive than that by an air-breathing propulsion. The first characteristic is the most striking one, because be it in water, air, space (vacuum), or the atmosphere of another planet where neither fuel nor oxidant is available, a rocket propulsion system can be designed to produce the required thrust.

As the rocket propulsion system does not draw any fluid from its surroundings, the simple thrust equation, Eq. 2.1, takes a simpler form, Eq. 2.2.

$$F = \dot{m}_e u_e \tag{2.2}$$

where $\dot{m}_e$ is the exiting mass flow rate.

Rocket propulsion is broadly classified into (1) **chemical rocket**, (2) **electric rocket**, and (3) **nuclear rocket**. This classification is based on the energy source for the propulsion. In the chemical rocket propulsion, chemicals on combustion release the working fluid as well as thermal energy. The hot working fluid, namely the combustion products, acquires a high kinetic energy by being expanded through a nozzle—a transformation of thermal energy into kinetic energy. In the electric rocket propulsion, it is the electrical energy that is responsible for the kinetic energy addition to the working fluid. The required electrical energy is produced either through solar- or nuclear energy. The solar energy can be converted into electricity using photovoltaic cells. Regarding the nuclear source, the heat from the radioactive decay or nuclear fission can be directly converted into electricity by the use of either thermoelectric- or thermionic-materials. In the nuclear rocket propulsion, nuclear fission reactions release thermal energy that is transferred to the working fluid. By expanding the hot working fluid through a conventional nozzle, the thermal energy is converted into kinetic energy.

2.3.1 *Chemical Rocket Propulsion*

Chemical rocket propulsion is further classified into (1) **solid rocket motor**, (2) **liquid rocket engine**, and (3) **hybrid rocket engine**. In the chemical rocket propulsion, a chemical that releases heat on exothermic reaction or combustion is known as propellant. Apart from releasing heat, as previously said, the propellant by way of combustion also produces the working fluid. Depending on the phase of the chemical, it is solid propellant or liquid propellant. It is generally an accepted practice to term the rocket propulsion systems using solid propellants as motors and the ones using liquid propellants as engines. Correspondingly, the respective systems are known as solid propellant motors and liquid propellant engines. In an attempt to take advantage of the merits of two different related-systems, it is usual in science and technology to resort to hybridization. Rocket technology is not an exception to this. The hybrid rocket propulsion uses one of its two propellants, generally the oxidant, in liquid phase and the other in solid phase.

2.3.1.1 Solid Rocket Motor

The schematic sketch of a solid rocket motor is shown in Fig. 2.14. There are four major components: (1) **igniter**, (2) **grain**, (3) **casing**, and (4) **nozzle**. The solid block of propellant, housed in the casing, is known as the grain. The grain contains principally the ingredients of oxidant, termed as oxidizer, and fuel. Usually, the central cavity in the grain, termed as **port**, serves as the combustion chamber. On initiation, the igniter ignites the propellant grain. The grain burns to produce hot combustion products, which are principally in gaseous phase by volume. These expand through the nozzle to exit at a high momentum rate. The nozzle is generally a convergent–divergent type. There is no incoming momentum rate since the working fluid is generated in-house. The thrust is produced because of the high momentum rate of the exiting combustion products against the zero incoming momentum rate.

In order to protect the casing walls from hot combustion products a suitable **insulator** and **liner** are provided between the propellant grain and the casing. The liner is usually a thin film between the grain and the insulator to prevent the migration

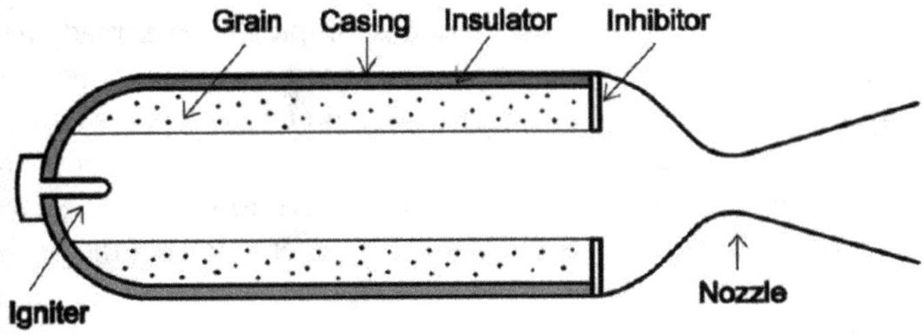

Fig. 2.14 Schematic sketch of a solid propellant rocket motor

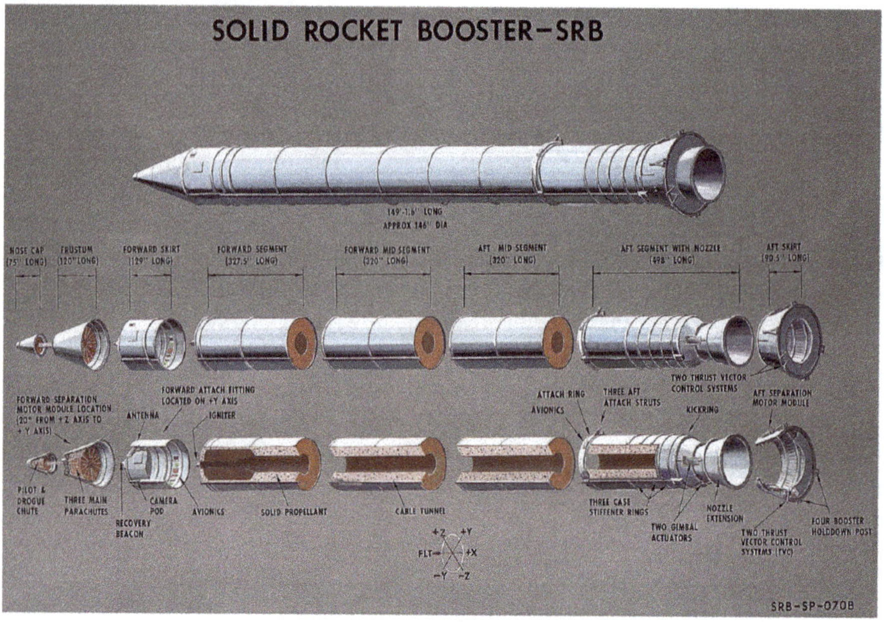

Fig. 2.15 Views of one of the two solid rocket boosters of the Space Shuttle (Courtesy Cutaway Drawings [4])

of chemicals between them. Furthermore, certain areas of the grain may have to be protected from burning for the desired thrust variation with time. This is done by covering such areas with a suitable **inhibitor**. In the sketch shown in Fig. 2.14, you see the insulator between the casing wall and the propellant grain and the inhibitor applied at the aft ends of the grain. As the convergent–divergent nozzle is handling the hot combustion gases, its walls should also be protected by proper insulation and adoption of high-temperature resistant materials.

Solid rocket motor is the simplest of the three chemical propulsion systems because both the fuel and oxidizer are contained in the grain. The major disadvantage of a solid rocket motor is that, once ignited, the thrust is delivered as per the grain configuration and cannot be varied at will. Furthermore the motor cannot be easily stopped and restarted.

Because the solid rocket propulsion system is simple and ever ready to be fired it finds extensive applications in warfare missiles. The energy content per unit mass of propellant in solid motors is less than that in liquid engines. Nevertheless, because of the system simplicity, invariably huge solid rocket motors are adopted as satellite launch vehicle boosters to overcome the colossal gravitational force experienced at launch. The classic example is the Solid Rocket Booster (SRB) of the Space Shuttle, Figs. 2.15 and 2.16. Each motor, the primary component of the Space Shuttle's twin solid rocket boosters, generates an average thrust of 11520 kN and is just over 38.5 m long and 3.7 m in diameter, and weighs about 590 tons.

Fig. 2.16 Test firing of the Space Shuttle's solid rocket booster in February 2010 in Promontory, Utah. (Courtesy NASA via Wikimedia Commons [1])

2.3.1.2 Liquid Rocket Engine

The schematic sketch of a liquid rocket propulsion system is shown in Fig. 2.17. The system has two major subsystems, the propellant feed system and the engine. For the purpose of cooling, one of the propellants, generally fuel, is circulated around

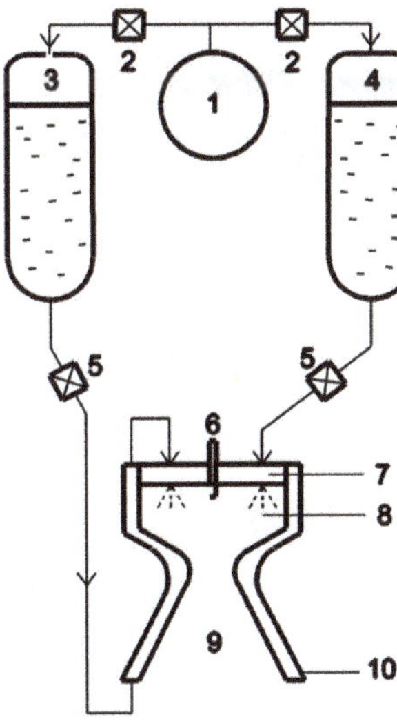

Fig. 2.17 Schematic sketch of a gas pressurized liquid rocket engine (1: high-pressure inert gas, 2: pressurization control valves, 3: fuel tank, 4: oxidizer tank, 5: propellant-flow control valves, 6: igniter, 7: injector plate, 8: combustion chamber, 9: convergent–divergent nozzle, and 10: regenerative cooling jacket)

the nozzle and the combustion chamber. The engine comprises of igniter, injector, combustion chamber, and nozzle. The others form the feed system. Based on the type of feed system adopted for the propellant flow, liquid rockets can be classified into **pressure-fed liquid rocket** and **turbo-pump-fed liquid rocket**. Note that the one shown in Fig. 2.17 is a pressure-fed liquid rocket. The inert gas pressurizes the propellants and injects them into the combustion chamber through flow control valves. The atomized sprays of the injected propellants get mixed and evaporated in the combustion chamber. On ignition, the vapor mixture of propellants burns in the combustion chamber. The combustion chamber length is designed to allow sufficient residence time to the burning propellants such that the combustion is essentially complete by the time the products of combustion arrive at the nozzle entry.

For large total-impulse (thrust × thrusting time; $\int Fdt$), as the pressure-fed system will become very heavy due its high overall feed-system pressure, a turbo-pump-fed system is employed. The schematic sketch of a turbo-pump-fed liquid rocket propulsion system is shown in Fig. 2.18. The image of a turbopump fed liquid rocket engine is shown in Fig. 2.19. Unlike in pressure-fed systems, in pump-fed systems it is sufficient that the propellant-tank pressures be kept low just to maintain the pressures at the pump-inlets adequately above the vapor pressures of the propellants. Therefore, the propellant tanks are lighter with thinner walls.

The turbo-pumps draw up the propellants from the tanks and pressurize them for injection into the combustion chamber. The turbo-pumps are powered by a turbine that is run by expanding the fuel-rich combustion gases from the **gas generator**. The

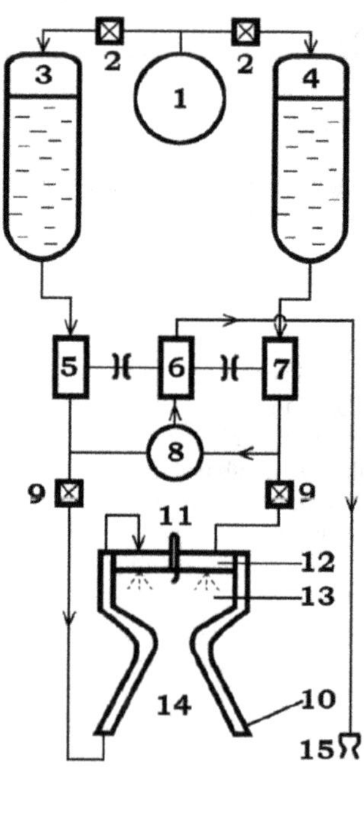

Fig. 2.18 Schematic sketch of a turbo-pump fed liquid rocket engine (1: low pressure inert gas tank, 2: pressurization control valves, 3: fuel tank, 4: oxidizer tank, 5: fuel pump, 6: turbine, 7: oxidizer pump, 8: gas generator, 9: propellant-flow control valves, 10: regenerative cooling jacket, 11: igniter, 12: injector plate, 13: combustion chamber, 14: convergent–divergent nozzle, and 15: control thruster)

Fig. 2.19 RL10A-4 liquid oxygen and liquid hydrogen engine used in Centaur upper stage. The engine delivers 93 kN with the specific impulse of 4400 Ns/kg. (Courtesy Purdue University [5])

fuel-rich propellant combination is fed into the gas generator from the main feed lines of propellants. The exhaust from the turbine is frequently expanded through a small nozzle, the thrust of which can be used for control purposes.

Mostly, liquid rocket engines are required to operate for a long duration of several minutes. Therefore, a cooling system has to be employed. Usually, the fuel is found to have better cooling characteristics. Furthermore, the oxidizer as a coolant may oxidize the metallic surfaces of the coolant passages, resulting in premature engine failure. In the cooling system, shown in Figs. 2.17 and 2.18, the fuel circulates around the nozzle and combustion chamber, cools them, and thus picks up the heat that would have otherwise been lost to the surroundings, if cooled by other means. The picked-up thermal energy in the heated coolant-propellant is fed back into the combustion chamber. Hence, the cooling system adopted is known as the **regenerative cooling** system. Small liquid rocket engines that operate in pulses or for short durations are known as liquid thrusters and they are used for control purposes. Generally **radiative cooling** is adopted for such thrusters and also for nozzle extensions adopted in large upper stage engines (see Sect. 3.5.8).

Liquid rocket engines, with the additional systems needed for propellant supply, propellant injection, and cooling, are not as simple as solid rocket motors. But liquid engines have many advantages over solid motors. The most striking one is that the thrust of liquid propellant engines can be easily controlled by controlling the propellant flow rates. Furthermore, the engines can be stopped and restarted. As the propellants are in liquid phase, one of the propellants, as explained previously, can be effectively used for cooling and hence the liquid engines can be operated for a longer duration than that is possible with solid motors. Unlike in solid motors, the propellants in liquid engines are stored away from the combustion chamber and hence the assembly of the thrust chamber and nozzle is not that heavy and can be easily gimbaled for thrust vectoring.

In the case of solid motors, the propellant ingredients have to be compatible with each other, as they are required to remain mixed and be stable in the grain for a long time. But in the case of liquid engines the oxidizer and the fuel are stored in separate tanks and only when needed get mixed in the combustion chamber. This "de-linking" gives a wider choice in selecting separately the most suitable oxidizer and fuel. "Wider the choice for its components better is the system." This is a universal rule and hence the performance of liquid rocket engines ought to be better than that of solid rocket motors. An important figure of merit in this respect for all rocket propulsion systems is the **specific impulse** and it is the thrust delivered per unit mass flow rate of propellant. From Eq. 2.2,

$$I_{sp} = \frac{F}{\dot{m}} = u_e \tag{2.3}$$

The specific impulse values of liquid engines are always higher than that of solid motors. Under SI units that we follow in this book, the dimension of specific impulse is Ns/kg or simply m/s. In other words, the specific impulse is nothing but the *effective* nozzle exit velocity. While Ns/kg or m/s is the dimension for the specific impulse under SI units, under the MKS system of units it is kgf-s/kgm and under the FPS system of units it is lbf-s/lbm. The numerical value of specific impulse under the MKS or FPS system of units will be (1/9.8065) times the value that we get under SI units. Commonly the unit mentioned for specific impulse in literature that adopts MKS or FPS system of units is simply seconds, with the perception that the specific impulse is the thrust force per unit *weight* flow rate of propellant. In Chap. 3, we will discuss in detail about the specific impulse.

2.3.1.3 Hybrid Rocket Engine

The schematic sketch of a hybrid rocket engine is shown in Fig. 2.20. One of the propellants, usually the oxidizer, is in liquid phase and it has its feed system as in liquid rocket engine. The other propellant, fuel, is in solid phase and is housed in a casing as in a solid rocket motor. The injected liquid oxidizer gets atomized and vaporized, and flows through the fuel-grain port. On ignition the combustion takes place between the oxidizer and the fuel vapor blowing from the wall of the grain port. The combustion products expand through the nozzle to exit with a high momentum rate to produce thrust.

Hybrid rocket engines combine the advantages of both solid rocket motors and liquid rocket engines. They are simpler than liquid rocket engines, though not as simple as solid motors. Hybrid rockets can be stopped and restarted and also have

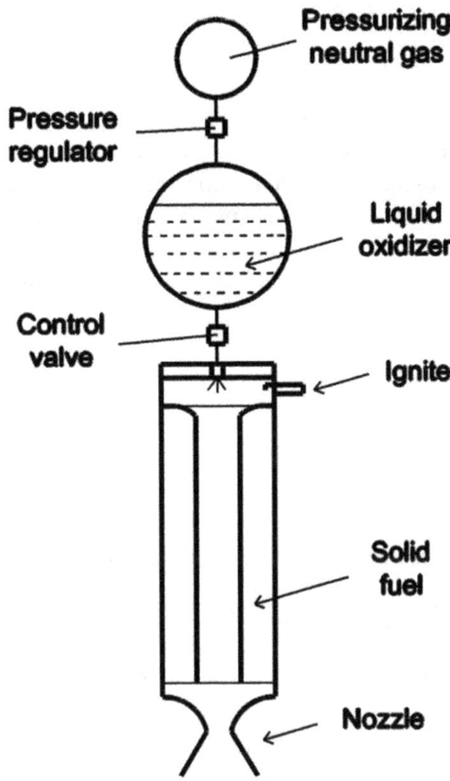

Fig. 2.20 Schematic sketch of a hybrid rocket propulsion system

the provision for thrust control by the regulation of the oxidizer flow rate. Presently developed hybrid rocket engines are of short duration operation essentially due to the limited web thickness possible for the fuel grain. Therefore, these engines do not require an active cooling system. Nevertheless, provision always exists to use the liquid propellant in a hybrid engine for active cooling. Developmental studies indicate that the hybrid rocket engines can attain, with lower overall system mass, the specific impulse values comparable to those of liquid engines using the Earth storable or semi-cryogenic propellant combinations. Compared to the number of flight-proven solid rocket motors and liquid rocket engines, the numbers of hybrid rocket engines that are flight-proven are meager. However, quite a few research and developmental activities on hybrid rocket engines have been reported in recent years. The motivation for this has been the growing interest towards (1) lower developmental and operational costs without much loss in specific impulse and density specific impulse, (2) safer operational characteristics, and (3) better environmental friendly exhaust. However, low regression rate of fuel, resulting in very high engine L/D ratios for the required thrust, has been the major impediment in large scale use of hybrid rocket engines. Nevertheless, active research studies are widely in progress to solve this problem.

2.3.2 Electric Rocket Propulsion

In realizing this Section, frequent references are drawn to [6]. Apart from the specific impulse, the **thrust efficiency** η_F is the second important figure of merit for electric propulsion systems, Eq. 2.4. It is defined as the ratio of the kinetic power of the exiting jet to the input electric power to the propulsion system. The working fluid in electric propulsion is also termed as propellant, although it, in most cases, does not possess an energy source of its own. The required electric energy is produced independently through solar- or nuclear-source, and applied suitably to the propellant.

$$\eta_F \equiv \left(\frac{\dot{m}u_e^2}{2}\right) \Big/ P_e \tag{2.4}$$

By using Eq. 2.3, the thrust efficiency can also be written as,

$$\eta_F = \frac{Fu_e}{2P_e} = \frac{FI_{sp}}{2P_e} \tag{2.5}$$

It is interesting to look at the basic reason behind electric propulsion systems of different types having vastly different values of specific impulse against the chemical ones. In solid motors, the propellant ingredients have to be compatible and stable with each other, and remain mixed in a single solid grain that may have to be stored for many years. On ignition, the burning grain produces the working fluid as well as the heat energy. In liquid engines, the "constraints" on the propellant are somewhat relaxed in the sense that the oxidizer and the fuel are separated and stored in different tanks. Only when required, the oxidizer and the fuel are mixed, and the working fluid along with heat energy is released on combustion. In electric propulsion, however, the propellant serving only as a working fluid and the energy source being independent, the constraints are further relaxed. As the energy-source requirement is separated from the propellant, a wide variety of electric propulsion systems is feasible with the specific impulse more than ten times to as low as a fraction of the value possible with chemical systems. Also, the thrusting times can be as short as a few milliseconds to as long as many months. Some electric propulsion systems have been demonstrated to have a continuous thrusting time of nearly two years, and there are attempts to extend the thrusting time even further.

The thrust–mass ratios of electric propulsion systems are very low at four to five orders of magnitude less than the chemical-system values. For example Hall thruster BPT-2000 has a thrust–mass ratio at 2.3×10^{-2} N/kg while the liquid-oxygen kerosene engine RD-58M has a thrust–mass ratio at 362 N/kg. As a result of this, for large thrust requirements and high accelerations as required in satellite-launch operations (or in other words for large power requirements), the electric propulsion systems are not at all suitable and the chemical systems are the only alternatives. However, for small thrust levels and low accelerations that are needed or sufficient for in-space operations—spacecraft maneuvers and interplanetary missions—the propellant mass required by a high specific-impulse electric system is approximately one-tenth of that required by a chemical system. Hence, even with its high hardware mass (due to very low thrust–mass ratios), an electric system has its overall mass (mass of hardware and propellant) less than that of a chemical system. Furthermore, chemical propulsion systems cannot be used for requirements of very long thrusting time.

In consideration of the previous discussion, with its relatively unlimited specific energy (specific impulse) but with low thrust capabilities (power) the electric propulsion system is considered to be a **power limited device**. On the other hand, with its relatively limited specific energy (specific impulse) but with unlimited thrust capabilities (power), the chemical propulsion system is considered to be an **energy limited device**.

In a typical spacecraft, the overall mass of a conventional chemical propulsion system is substantial. It varies from about 25% in small spacecraft of low Earth orbit to about 75% in interplanetary missions. Replacing the chemical system with a high specific-impulse electric system will result in considerable mass savings that may lead to the following advantages. For the given spacecraft mass, more propellant can be stored to enhance the spacecraft lifetime. In addition, for this given condition, with the saving on overall propulsion-system mass, the spacecraft can be equipped with improved- and additional-facilities to give better mission capability and flexibility. Alternatively, for the specified lifetime, less propellant needs to be stored, and this can lead to a lighter spacecraft and hence a smaller launch vehicle with a reduced launch cost. Furthermore, on a different application front such as long duration interplanetary travel and beyond, the electric propulsion system is the only alternative available at present.

In certain textbooks in the past, electric rocket propulsion is classified into three, namely, electrothermal, electrostatic, and electromagnetic systems. There is no confusion in listing thrusters under electrothermal propulsion. However, it is invariably seen that propulsion systems listed under electrostatic propulsion do involve or substantially depend on the effects of electromagnetic field, and vice versa. Therefore, without following this past classification, we first list thrusters that can be grouped under electrothermal propulsion and then list other important electric thrusters, which operate on the effects of electric and magnetic fields.

Figure 2.21 shows in schematic form the principal elements of an electric propulsion system and its interfaces with other systems in the spacecraft. Typically the electric power system (EPS) receives its power from a nuclear- or solar-source and it supplies regulated DC power to a power processor unit (PPU), as well as to other auxiliary elements, such as valves and heaters. The PPU processes this raw power into a specific form required by the electric thrusters. A regulated pressure fed system or a blow-down one can be used for the propellant supply. The propellant flows to be handled are extremely small (a few $\mu g/s$) and precise and are to be regulated without leaks for very long periods of months or sometimes for a few years. Commands to the

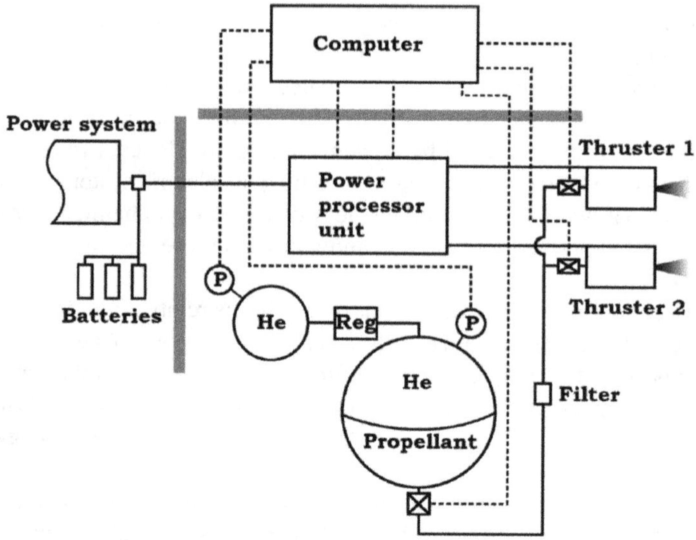

Fig. 2.21 Schematic diagram of a typical electric propulsion system (adopted from [6])

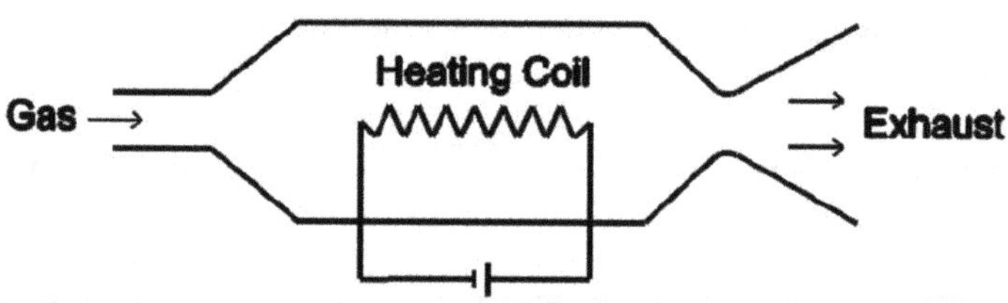

Fig. 2.22 Schematic sketch of a resistojet

various power switches and valves are supplied by the spacecraft computer. Apart from the mass of the electric power source, typically the mass of a thruster forms a minor portion of the total electric propulsion system. For example, a one kW Hall thruster may be of mass 2.5 kg in the total electric propulsion system of 14 kg and a one kW ion engine may be mass 5 kg in the total electric propulsion system 18 kg.

2.3.2.1 Electrothermal Propulsion

In electrothermal propulsion, the propellant is electrically heated and expanded through a conventional nozzle. The electrothermal propulsion systems can further be classified as **resistojets** and **arcjets**.

A schematic sketch of a resistojet is shown in Fig. 2.22. The resistojet is the simplest device among all the electric propulsion systems that use fluids as propellants.

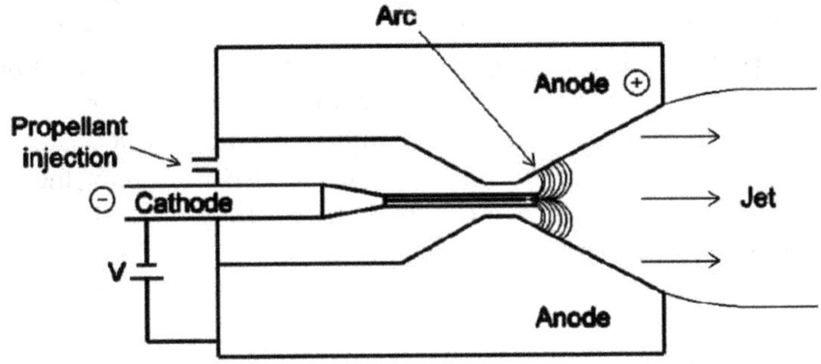

Fig. 2.23 Schematic sketch of an arcjet

Fig. 2.24 Esex Arcjet engine (Aerojet, USA) [7]

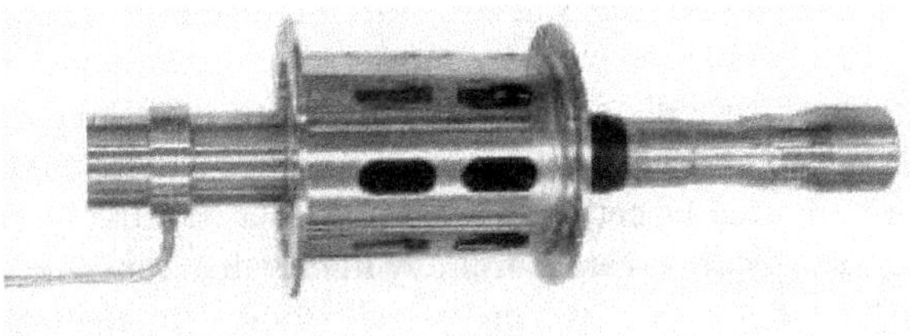

In a resistojet, an electric current is passed through a high-resistance material to generate the heat that convects into the propellant. Usually, catalytically-decomposed hydrazine (N_2H_4), ammonia, water, or nitrous oxide is adopted as a propellant. The heated propellant gas expands through the nozzle to produce a thrust.

A schematic sketch of an arcjet is given in Fig. 2.23. In an arcjet, an arc is struck between a cathode and an anode, and the arc heats up the cross-flowing propellant to a very high temperature. The adopted propellant is usually decomposed hydrazine, ammonia, or hydrogen. The high-temperature propellant gas expands through the nozzle to produce thrust. The image of Esex arcjet engine of Aerojet, USA is shown in Fig. 2.24. It uses ammonia as its propellant and delivers a thrust of 2 N for 15 minutes with maximum of 10 restarts at a specific impulse of 7850 Ns/kg.

Resistojets are simpler and of higher thrust efficiency than arcjets, and they can use any inert liquid or gas as a propellant. For example, even the wastewater or unwanted gas in a space station can be suitably used. However, on the point of view of specific impulse, because the propellants in them can be heated to higher temperatures, arcjets deliver higher specific impulse than resistojets (5000–10000 Ns/kg as against around 3500 Ns/kg). This is essentially due to the difference between the heating patterns. In a resistojet, the wall of high resistance material convects the heat to the propellant. Due to the limitation on the maximum allowable material temperature, around 3000 K, the hottest part in a resistojet is the wall that heats up the relatively cooler propellant. But in an arcjet, the hottest zone is the arc region of temperatures in the range of 10000–20000 K and it is away from the cathode- and anode surfaces. These surfaces, though relatively cooler, can be of temperatures equal to the maximum allowable values around 3000 K. Therefore, the propellants in an arcjet can be heated to higher temperatures to deliver higher specific impulse (you will learn later that specific impulse is proportional to the square root of chamber temperature divided by molar mass of the chamber gas). Furthermore, as the heating in a resistojet is along the walls, not localized as in the case of an arcjet, the propulsion system length of a resistojet tends to be very high resulting in a larger system mass and an enhanced pressure loss in the heating chamber.

2.3.2.2 Lorentz Force and Hall Effect

Before studying different types of other electric propulsion systems, we have to first understand the **Lorentz force** and the resulting **Hall effect**. Negatively and positively charged particles move under the influence of electric and magnetic fields in electric propulsion systems. Negatively charged particles are mostly electrons and the positively charged particles are electron deficient propellant atoms, namely **ions** (more correctly cations); an **anion** is a negatively charged electron-excess atom. Because electrons are extremely lighter than ions, the effect of a force due to electric and magnetic fields is a great deal stronger for electrons than for ions. Lorentz force is the force $\vec{F}$, on a point charge q, moving with a velocity $\vec{v}$, due to electric field $\vec{E}$ and magnetic field $\vec{B}$. It is given by the following equation.

$$\vec{F} = q\left(\vec{E} + \vec{v} \times \vec{B}\right) \tag{2.6}$$

Let us first consider the influence of magnetic fields. Figure 2.25 depicts the effect of magnetic force on positively and negatively charged particles projected with a velocity $\vec{v}$. The magnetic field of strength $\vec{B}$ is uniform throughout, perpendicular to the plane, and pointing towards you. A neutral particle, projected with a velocity $\vec{v}$ perpendicular to the magnetic field, is not affected by it—shown as an arrow with a solid line in Fig. 2.25. But, if a positively charged particle of charge q is projected with a velocity $\vec{v}$ perpendicular to the uniform magnetic field of strength $\vec{B}$, then, the particle will experience a magnetic force $q\vec{v} \times \vec{B}$ in the direction normal to both $\vec{v}$ and $\vec{B}$—shown as an arrow with dotted lines curving to the right, Fig. 2.25. If you recall the principles of the vector product operations, the direction of the magnetic force can be easily determined. Alternatively, the direction of the force can be found by the right-hand rule that can be stated as follows. *For a positively charged particle, with the thumb of right hand pointing along $\vec{v}$ and the index finger along $\vec{B}$, the middle finger kept perpendicular to both $\vec{v}$ and $\vec{B}$ indicates the direction of the magnetic force $q\vec{v} \times \vec{B}$. For a negatively charged particle of charge $-q$, the magnetic force will be in the opposite direction. This is represented in vectorial form as $-q\vec{v} \times \vec{B}$.* Therefore, the negatively charged particle projected with a velocity will have to curve leftward as shown in Fig. 2.25. The deflected particle in the uniform magnetic field will be moving in a circle, the positive one clockwise and negative one anti-clockwise.

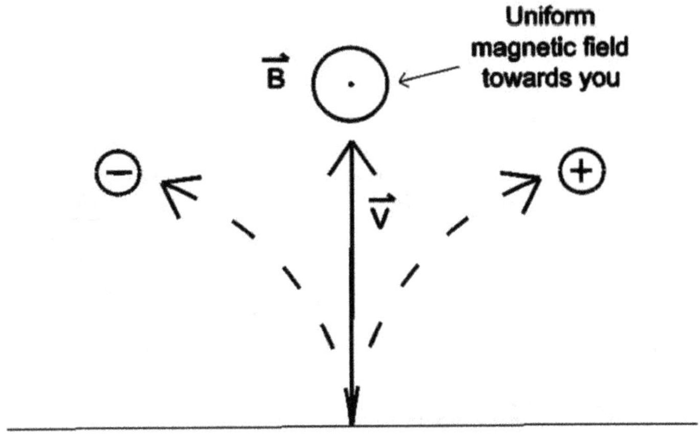

Fig. 2.25 Lorentz force on the moving charged particle in a uniform magnetic field

Now considering both the electric and magnetic fields, the total Lorentz force on the positively charged particle of charge q with velocity $\vec{v}$ will be $q\left(\vec{E}+\vec{v}\times\vec{B}\right)$. On the other hand, the total force will be in the opposite direction for the negatively charged particle. This is represented in vectorial form as $-q\left(\vec{E}+\vec{v}\times\vec{B}\right)$.

It follows therefore, with the uniform electric and magnetic fields perpendicular to the plane and pointing towards you, the projected positive particle will be traveling in a clockwise helical path towards you and the negative one will be in an anti-clockwise path away from you in the opposite direction. It is important to recall that the effect of a force due to electric and magnetic fields is a great deal stronger for lighter electrons than for heavier ions.

The basic physical principle underlying the Hall effect is the Lorentz force. *Hall effect is the generation of an electric potential (**Hall voltage**) perpendicular to both the electric current flowing along a conducting material and the external magnetic field applied at right angles to the current.* Polarity of the Hall voltage depends on the moving charge carriers. In most cases, charge carriers are electrons. Figure 2.26 depicts the Hall effect in a conductor where the charge carriers are predominantly electrons—negatively charged particles. The current I flows from right to left through a thin flat conductor, and the charge carrier electrons move in the opposite direction. The uniform magnetic field is applied perpendicularly downward. The Lorentz force on the charge carrying electrons pushes the electrons towards the edge of the conductor that faces you. This excess negative charge on one side of the conductor results in the Hall voltage and the current that flows across is the **Hall current**. If the charge carriers are positive as in some semiconductors, the charge carriers move in the same direction as the current I, and the polarity of the Hall voltage will be opposite to that shown in Fig. 2.26.

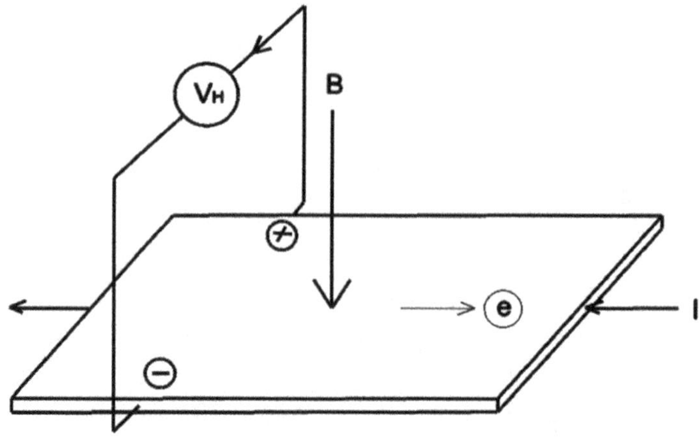

Fig. 2.26 Hall effect in a conductor, where the charge carriers are predominantly electrons

2.3.2.3 Ion Engine

In an ion engine or ion thruster, an electrostatic field is principally responsible for the acceleration of ionized propellant gas. In 1959, Herald Kaufman, an engineer at NASA, first built an engine based on this concept and therefore the engine is also known as **Kaufman thruster**. Although the vapors of cesium and mercury were initially used as propellants, due to the spacecraft-contamination concerns from the use of these liquid metals, the noble gas xenon is presently the most preferred one. Xenon is having a high atomic mass with a low ionization potential. The other noble gases that can be adopted are krypton and argon.

A schematic sketch of an ion engine is shown in Fig. 2.27. A neutral propellant gas enters the ionization chamber. Electrons are emitted by the heated hollow cathode

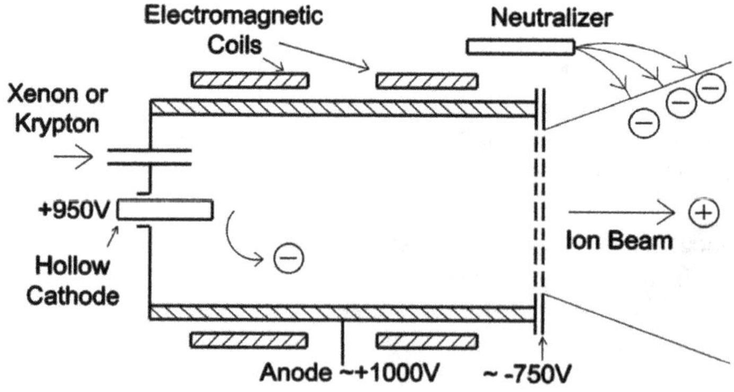

Fig. 2.27 Schematic sketch of an ion engine

at the center. Attracted by the anode rings arranged at the periphery, the emitted electrons start drifting radially towards them. Also positioned outside the engine chamber are magnetic coils that impose an axial magnetic field to the radially drifting electrons. In the presence of the magnetic field, the electrons, due to Lorentz force, take up an elongated spiral path (approximately similar to the threads on a highly-tapered screw) that improves the probability of successful collision and ionization of propellant atoms. On a successful collision or bombardment, an electron in the outer orbit of the neutral propellant atom is knocked off. Consequently, the reflected electron after collision, the knocked-off electron, and the electron-deficient propellant atom of positive charge, namely ion, are all obtained. Note that although the magnetic field can influence both the electrons and the ions that are of equal but opposite charges, it is the electrons that are mostly affected because they are much lighter than the ions. A pair of metal grids charged with positive and negative potentials is positioned near the exit and it sets up a strong electric field in the engine chamber. This field pulls and accelerates the propellant ions through the grid holes to exit the engine as a beam with velocities in the range of 25000–40000 m/s, which also corresponds to a specific impulse in the range of 25000–40000 Ns/kg. Finally, the exiting ion beam is neutralized by a stream of electrons that are injected by the neutralizer positioned at the very end of the engine exit. This neutralization is necessary to keep the spacecraft with its engine electrically neutral. Without neutralization, the spacecraft will become negatively charged and the exiting ion beam may be attracted back into the engine. Figure 2.28 shows the ion engine of DS1 mission being tested in a high vacuum chamber.

Note that plasma is a neutral fluid mixture of positive-, negative- and neutral-particles and it can be influenced by electric and magnetic fields. When an ion engine is operating at a steady state, a stable dilute plasma cloud exists in the engine chamber.

Fig. 2.28 DS1 mission's xenon ion engine (mass 8 kg, diameter 40 cm, and length 40 cm) delivers a thrust of 90 mN at a specific impulse of 30500 Ns/kg [8]

Through the cloud, streaming in at the cathode end are the emitted electrons and the neutral-propellant gas, and streaming out are the previously emitted electrons and the newly formed knocked-off electrons (both essentially moving radially in spirals towards the anode), and the propellant ions (moving axially towards the grids). At equilibrium, the numbers of electrons, ions, and neutral atoms in the plasma cloud are essentially constant.

2.3.2.4 Hall Thruster

The Hall effect thruster or simply **Hall thruster** is also known as **stationary plasma thruster** or **magnetic layer thruster**. As its name implies, the thruster uses the Hall effect to accelerate the ionized propellant gas.

A schematic sketch of a Hall thruster is shown in Fig. 2.29. As in ion engines, in Hall thrusters too, xenon is the most preferred propellant gas. Two coaxial dielectric cylinders form an annular passage. At the entry to the annular passage (the anode end) is a perforated anode through which the neutral propellant gas is injected. Outside the annular passage, at the exit end, is a cathode neutralizer. A potential difference between the anode and the cathode, usually around 300 V, sets up an axial electric field. Magnetic coils are arranged around the axis within the inner cylinder and at the periphery of the outer cylinder. These coils introduce a radial magnetic field B, strong at the exit region and weak at the anode end. Here, we see the current I flowing through the coaxial cylindrical conductor experiencing the radial magnetic field B, that is perpendicular to the current I.

The operating principle of Hall thruster is as follows. The cathode neutralizer injects electrons, and a portion of these enters the annular passage from the exit end

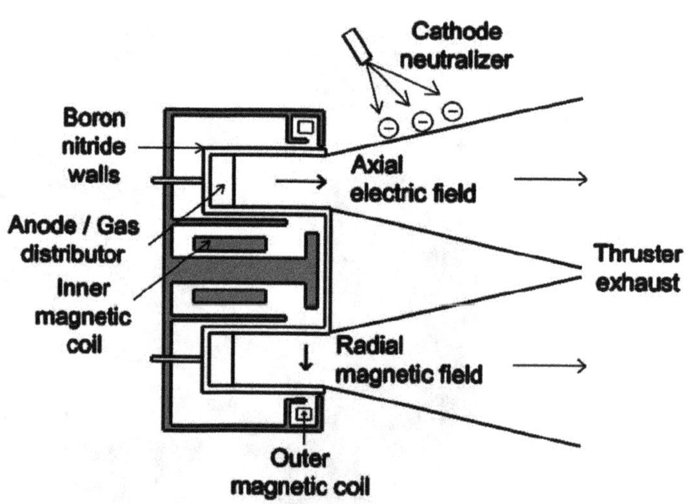

Fig. 2.29 Schematic sketch of a Hall effect thruster

and attempts to drift towards the anode. But, because of the strong radial magnetic field perpendicular to the annular passage, the electrons attempting to move from the cathode neutralizer to the anode are magnetized and trapped at the exit region to orbit around the axis to give Hall current. At equilibrium, a finite-width annular ring of azimuthally rotating electrons is formed at the exit region, and this ring acts as a virtual cathode grid. Although trapped, due to collisions and instabilities, a few of the orbiting electrons do get released from the annular ring to drift towards the anode. In this drift, they may ionize the oncoming propellant atoms, but the major ionization and subsequent acceleration of the propellant ions occur as the propellant enters the annular ring of rotating electrons. Due to the high electron number density there, the ionization of the propellant atoms is practically complete. Although the radial magnetic field can influence and trap both the electrons and the ions, it is the electrons that are significantly affected because they are much lighter than the ions. The heavy ions, practically unaffected by the magnetic field, are accelerated by the electric field to exit the thruster with velocities in the range of 16000 m/s, which also corresponds to a specific impulse range of 16000 Ns/kg. The other portion of the electrons from the cathode neutralizer joins and neutralizes the exiting ion beam.

Let us compare the plasma discharge in an arcjet with that in a Hall thruster. In an arcjet, it is a non-magnetized plasma discharge essentially perpendicular to the propellant gas flow. In such a discharge, due to the abundant and fast-moving light electrons towards the anode, the electron current forms the major part of the discharge and the balance is due to the scarce and slow moving heavy ions towards the cathode—most heavy ions are swept downstream by the propellant gas flow. The transfer of arc heat to the propellant makes it possible to accelerate the propellant gas in the nozzle to produce thrust. In a Hall thruster, it is a perpendicularly magnetized plasma discharge essentially along the propellant gas flow. In such a discharge, the portion of the electrons injected by the cathode neutralizer is trapped in the magnetized annular ring and only a few of the electrons get released at a time to drift towards anode. Consequently, the electron current forms the minor part of the discharge current. However, it does not contribute to the thrust production. On the other hand, the passage of heavy ions is unhindered by the magnetic field, and hence the ion current forms the major part. And favorably, this part is responsible for the ion acceleration to produce thrust.

In comparison with an ion engine, a Hall thruster works with a lower system voltage that makes the power processing simpler. Since ion acceleration takes place in a quasi-neutral plasma, Hall thrusters are not limited by space-charge build up as in the case of the ion engine. A major disadvantage in using a Hall thruster is that the exiting beam is quite divergent, that is, the velocity vectors are not quite parallel to the engine axis. This results in a considerable thrust loss. Furthermore, the divergent exiting beam may hit the other parts of the spacecraft and cause operational problems.

In ion engines as well as Hall thrusters, both electric and magnetic fields are responsible for the propellant acceleration. Traditionally, however, ion engines are categorized under electrostatic propulsion since the magnetic field plays only a minor role. In certain literature, both these engines are grouped under electrostatic propulsion. Example 2.1 brings out the operational characteristics of the Hall thruster PPS-1350.

Example 2.1
SMART-1 spacecraft of mass 367 kg was used for a 16 months transfer (2003–'04) from a Geosynchronous (or Geostationary) Transfer Orbit (GTO) to a lunar orbit. The solar panels of the spacecraft supplied a total power of 1850 W to the electric power system (EPS), Fig. 2.21. Of this power the power processor unit (PPU) supplied 1388 W to the Hall thruster PPS-1350 that delivered an average thrust of 70 mN at a specific impulse of 15700 m/s. The spacecraft contained 82 kg of supercritical xenon at a pressure of 15 MPa. Calculate the xenon mass flow rate for this thrust. What is the expected total operating time in hours for this thruster? And, compare it with the listed operating time of 5000 h. Estimate the thrust efficiency η_F.

Solution
This is the straightforward application of the derived equations Eqs. 2.2 and 2.4.

Analysis
Using Eq. 2.2,

$$\dot{m} = \frac{70 \times 10^{-3}}{15700} = 4.4586 \times 10^{-6} \text{ kg/s}$$

Total thrusting time.

$$\Delta t = \frac{82}{4.4586 \times 10^{-6}} = 18.39142 \times 10^{6}\ s = 5108.7 \text{ hr}$$

Using Eq. 2.4,

$$\eta_F = \frac{4.4586 \times 10^{-6} \times 15700^2}{2 \times 1388} = 0.396$$

Discussion
Note that the specific impulse of the Hall thruster is far greater than that of a chemical propulsion system. Appreciate the advanced technology involved in regulating a few $\mu g/s$ flow rate of xenon over a period of about 6 months without any leak.

2.3.2.5 Pulsed Plasma Thruster

In the pulsed plasma thrusters, the current through the propellant plasma interacts with the self-induced magnetic field to create a force to accelerate the propellant plasma. A pulsed plasma thruster is schematically shown in Fig. 2.30. Using a spring, the face of a solid inert propellant is forced against a stopper at the beginning of the ionization chamber. Usually, a chlorofluorocarbon such as Teflon® is used as a propellant. The ionization chamber is generally of rectangular cross section formed by a pair of flat-plate electrodes and a pair of insulator plates. The flat-plate electrodes are generally arranged parallel for a short length and then diverging towards the exit. The electrodes are connected to the power supply across a high-voltage (1–2 kV) high-energy capacitor. A spark plug, which acts as the igniter, is positioned close to the propellant face.

The operating principle of the pulsed plasma thruster is as follows. When the high energy capacitor is fully charged, the igniter, by its small discharge, triggers the capacitor to discharge through a high-current arc (of an instantaneous value around a few tens of kA) across the face of the propellant. By this high current, a thin slice of propellant is instantly ablated and ionized. Instantly, because of the spring, the new face of the propellant is automatically pushed forward against the stopper. The high current flowing in the loop connecting the plates of anode and cathode introduces a self-induced magnetic field. The Lorentz force $\vec{j} \times \vec{B}$ created by the interaction of the electric arc current with a self-induced magnetic field accelerates the ionized plasma. The exiting plasma gives a thrust pulse of about 10 μs duration at specific impulse values in the range 10000–15000 Ns/kg. The capacitor gets fully charged within a few milliseconds for the next cycle of operation to start.

As the pulsed plasma thruster has its propellant in solid phase, there is no need for the complicated propellant feed system and associated ground-handling scheme as required for liquid- or gaseous-propellant thrusters. We noted previously that the resistojet is the simplest device among all the electric propulsion systems that use

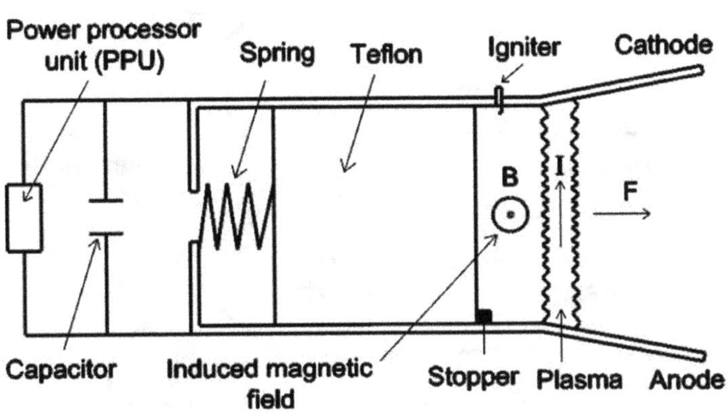

Fig. 2.30 Schematic sketch of a pulsed plasma thruster

fluids as propellants. With its utmost simplicity in handling its propellant in solid phase, we can say that the pulsed plasma thruster is the simplest of all the electric propulsion systems. Although the pulsed plasma thrusters are among the earliest electric thrusters used in spacecraft propulsion, their thrust efficiency is quite low around 10%, Eq. 2.5.

2.3.2.6 Magneto Plasmadynamic Thruster

A basic version of the magneto plasmadynamic thruster is schematically shown in Fig. 2.31. A long cylindrical cathode rod, usually of tungsten, is coaxially positioned inside a short dielectric cylinder to form an annular passage. Surrounding this cylinder is the anode sleeve with a circular shoulder ring at the right exit end. The shoulder ring projects perpendicularly a bit into the annular passage and is positioned a little ahead of the cathode tip.

The operating principle of the magneto plasmadynamic thruster is as follows. Propellant gas (ammonia, hydrogen, argon, hydrazine, nitrogen, or lithium vapor) is injected into the annular passage through the cover plate at the left end. A high current arc, struck between the anode shoulder ring and the cathode, ionizes the propellant gas and sets up a radially inward electric field. The returning high current through the long cathode rod induces an azimuthal magnetic field. The resulting Lorentz force, due to the radially inward electric field and the azimuthal magnetic field, acts on the propellant plasma. Because of the previously explained positioning of the anode shoulder ring with respect to the cathode tip, the resulting Lorentz force has two components: the radially inward component due to the electric field and the axially outward one caused by the magnetic field. The radial component constricts the flow and is known as the pumping force. And, the axial component accelerates the

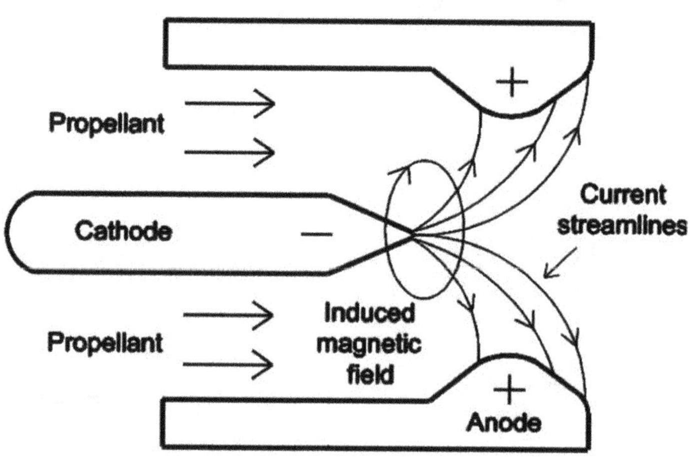

Fig. 2.31 Schematic sketch of a magneto plasmadynamic thruster

propellant plasma to produce thrust and is known as the blowing force. The specific impulse that can be achieved by the magneto plasmadynamic thruster varies from 20000 to 50000 Ns/kg.

The basic version of the thruster, schematically shown in Fig. 2.31, is known as **self-field magneto plasmadynamic thruster**. Additional magnetic field coils can be arranged around the anode to introduce radial and axial magnetic fields to the accelerating propellant plasma. The thruster with the added magnetic fields is found to deliver higher thrust efficiency and is known as the **applied-field magneto plasmadynamic thruster**.

2.3.2.7 Pulsed Inductive Thruster

The basic principle of operation of pulsed inductive thruster is that a radial magnetic field interacts with an induced azimuthal current in the propellant plasma to create a force to accelerate the propellant plasma. A schematic sketch of a pulsed inductive thruster is shown in Fig. 2.32. A thin dielectric circular plate covers a flat spiral coil. By a sharp "puff" of injection from a conical nozzle, a thin layer of propellant gas (argon or ammonia typically of mass 2 mg) spreads at supersonic speed across the plate with azimuthal symmetry. At the same time, a bank of capacitors, charged to about 15 kV, discharges a high current through the spiral coil lasting only a few

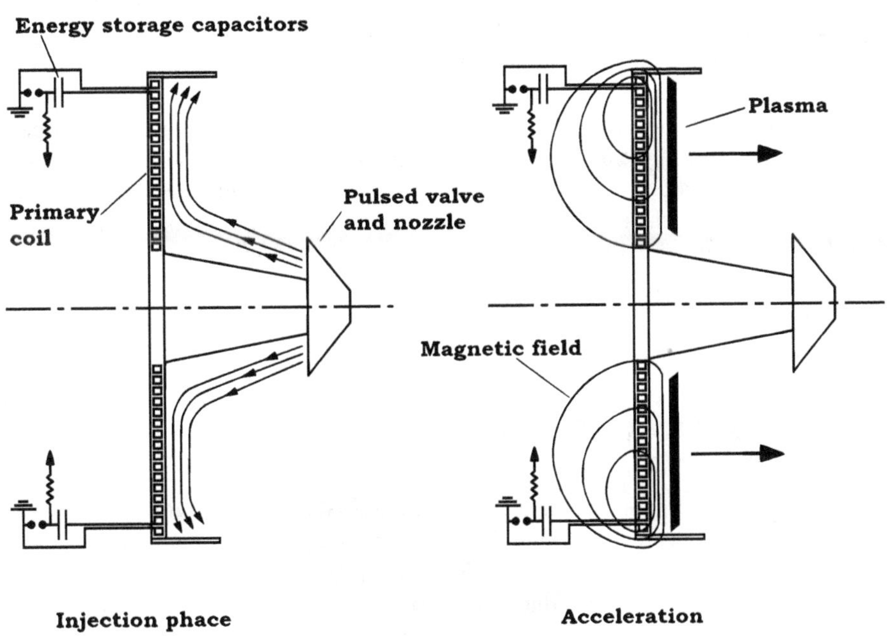

Fig. 2.32 Schematic sketch of a pulsed inductive thruster

microseconds. This discharge generates a time varying radial magnetic field, which in turn induces an azimuthal electric field. This electric field ionizes the propellant gas layer into a conductive plasma. Now, the time varying radial magnetic field that is still rising induces in the plasma an azimuthal current that flows in the direction opposite to that through the spiral coil. This azimuthal current, being perpendicular to the radial magnetic field, gives rise to a magnetic pressure between the plasma and the spiral coil accelerating the former to a high velocity in the range of 20000–80000 m/s (20000–80000 Ns/kg) to produce a thrust pulse.

The main advantage of the pulsed inductive thruster is that it is a system without grids and electrodes. The ion engine has electrodes in the form of grids through which the ionized propellant is accelerated. The grids are obstructive to the flow. Furthermore, the charged particles hitting the electrodes cause erosion and limit the life of the propulsion system. Hall thrusters, pulsed plasma thrusters, and magneto plasmadynamic thrusters have "gridless" electrodes. As a further improvement, the pulsed inductive thrusters are "electrodeless" ones thereby completely removing the problem of obstruction as well as erosion of electrodes.

As a final remark on electric rocket propulsion systems, we should make a mention about the grouping under a category and the naming of individual electric thrusters. Under electrostatic propulsion and electromagnetic propulsion, the grouping and naming are more historical and traditional than the exclusive application of electric field or electromagnetic field. Under these two propulsion concepts, one finds in almost all thrusters the interaction of electric and magnetic fields. In certain literatures, Hall thruster is grouped under electrostatic propulsion. In all electric thrusters, barring resistojets, the working fluid at a certain stage becomes plasma. Features of flight-proven electric thrusters and the satellites that adopted them are given in Table 2.1.

Most electric thrusters are well developed and are used in more than hundred operational spacecraft. Magneto plasmadynamic thrusters and pulsed inductive thrusters under electromagnetic propulsion system are under development and yet to be fully flight qualified. With the continuous improvement in the reliability of electric rocket propulsion systems, the number of electric thrusters used in spacecraft propulsion is constantly increasing with time. For detailed analyses of electric propulsion, see [2, 6, 9–11].

2.3.3 Nuclear Rocket Propulsion System

In a nuclear rocket engine, the heat from a **nuclear-fission** reaction is used to heat up, accelerate, and eject a propellant gas, say hydrogen, at high speeds. A schematic sketch of a nuclear rocket engine is shown in Fig. 2.33. The engine usually has a graphite-based fission reactor containing a mixture of uranium/zirconium carbide in a graphite matrix. The propellant is pumped to circulate around the nozzle and the reactor chamber for the purpose of cooling. A part of the heated propellant gas runs the turbine that powers the pump. Thereafter, the propellant gas enters the heat

Table 2.1 Features of flight-proven electric thrusters [6]

Thruster	Resistojet	Resistojet	Arcjet	Arcjet	Hall thruster	Ion engine	Pulsed plasma thruster
	(N_2H_4)	(NH_3)	(N_2H_4)	(NH_3)	(Xe)	(Xe)	$(Teflon)$
Power range (W)	500–1500	500	300–2000	500–30000	300–6000	200–4000	1–200
I_{sp} typical (m/s)	3000	3500	5000–6000	5000–8000	16000	28000	10000
η_F%	80	80	35	27–36	50	65	7
Plume divergence	<20°	–	<20°	<20°	30–40°	<20°	–
Peak voltage (V)	28	28	100	–	300	900	1000–2000
Thruster power ratio (kg/kW)	1–2	–	0.7	0.7	2–3	3–6	120
PPU[a] mass (kg/kW)	1	–	2.5	2–3	6–10	6–10	110
Miscellaneous[b] (kg per thruster)	1	–	5	–	10	10	Small
Lifetime (h)	500	–	>1000	1500	>7000	10000	$>10^7$ pulses: 4000 Ns (total impulse)
Application	NSSK[c], EWSK[d], orbit insertion, & deorbit	Orbit corrections	NSSK	NSSK & orbit raising	NSSK & orbit raising (medium Δv)	NSSK & Orbit transfer (large Δv)	Small orbit corrections (precision)
Flown in	INTELSAT-V, Satcom-1R, Gstar-3, & Iridium	GOMS (Electro), US Navy satellite, & Meteor 3-1	Telstar-4, GE-1, & DRTS	AMSAT P3D	Kosmos, Luch, Gals, Express, Stentor, & Meteor-18	ETS-6, COMETS, PAS-5, Galaxy-8i, Galaxy-10, ARTEMIS, New Millennium, & DS-1	Zond-2, LES-6, TIP, NOVA, & MightySat II.1

[a] Power processing unit
[b] Cable, gimbal, thermal control, structure, and propellant feed (power range 0.5–2 kW)
[c] North-south station keeping
[d] East-west station keeping

Fig. 2.33 Schematic sketch of a nuclear rocket

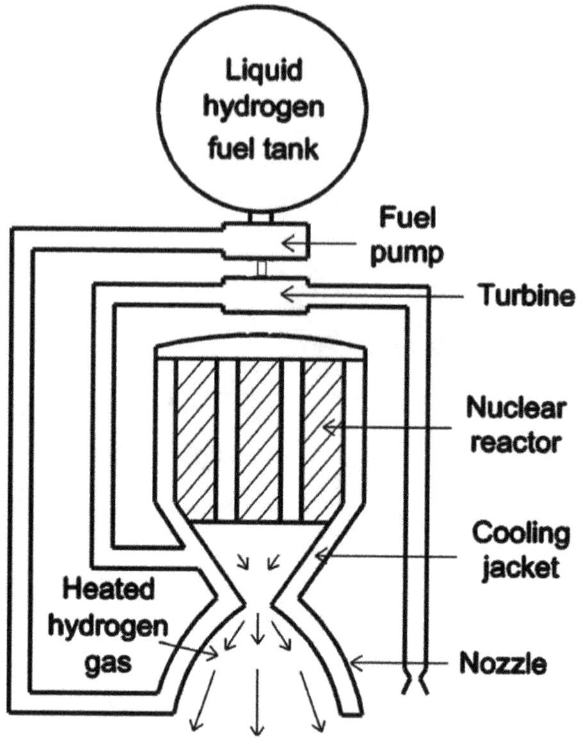

exchanger tubes that run through the reactor. The nuclear heat is transferred to the propellant gas. The hot propellant gas is accelerated by the conventional nozzle expansion.

The specific impulse of a nuclear rocket engine can be more than twice that is possible in chemical propulsion systems. A chemical rocket, with its characteristic high thrust–mass ratio, can produce practically unlimited jet power, or in other words unlimited thrust force. But, it can operate only for a very short duration—at the maximum for a few tens of minutes. This performance of the chemical rocket (practically unlimited power but limited energy) makes it ideally suitable for launching a spacecraft into an orbit against a large gravitational force. But, it does not make it suitable for powering a spacecraft on a deep-space mission. Globally, the most important next step in space exploration is to realize a deep-space mission of a large payload and not too long a mission time (say, a manned spacecraft to Mars with a mission time of a few years). The key to this step lies in finding a propulsion system of reasonably high thrust to mass ratio that can produce fairly a large thrust force continuously for a long duration (many months). Among the near-term technologies, a nuclear rocket engine, which basically has no limitation on its energy, will be the best to meet this requirement.

Problems

2.1 How is ramjet propulsion classified? What are the advantages and disadvantages of ramjet propulsion?

2.2 Briefly explain the processes in subsonic-combustion ramjet propulsion. Practically, what are its lower and upper flight Mach numbers? Discuss the reasons for the same.

2.3 With a suitable sketch, briefly discuss the operating principles of a scramjet. How is the upper flight Mach number limit by subsonic combustion ramjet solved by adopting scramjet engine?

2.4 Compare the advantages and disadvantages of gas turbine propulsion with ramjet propulsion.

2.5 Maximum flight Mach number limit for turbojets with afterburner is around 3. Why?

2.6 Discuss the classification of gas turbine propulsion system.

2.7 What are the application areas for turboshaft engines?

2.8 What are cold thrust and hot thrust in turboprops and turbofans?

2.9 How is turbofan different from turboprop? What are the flight Mach number ranges for these?

2.10 What is a ducted turbofan or ducted bypass engine? Where is the necessity for such an arrangement?

2.11 Turbojets have to be fitted with afterburners for supersonic flights. Why?

2.12 Briefly discuss the classification of rocket propulsion. What is the energy source for each of the classification?

2.13 Discuss the classification and the sub-classification of chemical rocket propulsion. Provide sketches.

2.14 Identify the major components of a solid propellant rocket motor. Distinguish the purposes of insulator, liner, and inhibitor.

2.15 Classify liquid propellant rocket engines.

2.16 What are the application areas for pressure fed liquid engines and pump fed liquid engines? Under what conditions, pump fed propellant injection is adopted?

2.17 What is vapor pressure? What is cavitation in liquids? Why should the liquid propellant tank pressure be sufficiently above the vapor pressure of propellant?

2.18 Oxidizer: fuel ratio in a gas generator of a pump fed propellant feed system is kept highly fuel rich. Why?

2.19 Basically, there are three kinds of cycles in the turbopump fed liquid propellant engines: gas generator, staged combustion, and expander. Write a brief note with suitable sketches on these cycles (Study Tables 4.2 and 4.3).

2.20 Briefly compare the advantages and disadvantages between solid propellant rocket motors and liquid propellant engines.

2.21 From the operational safety point of view, hybrid rocket propulsion is better than solid- and liquid-propulsion. Why?

2.22 Define thrust efficiency in electric propulsion, η_F. How is it linked to the specific impulse, I_{sp}?

2.23 What is the fundamental reason for electric propulsion system to have very high specific impulse values?

2.24 Chemical propulsion is energy limited whereas electric propulsion is power limited. Explain.

2.25 Identify the differences between resistojet and arcjet?

2.26 Explain through suitable diagrams of Lorentz force and Hall effect.

2.27 Explain through a suitable sketch the operating principles of ion engines? What are the propellant gases adopted in ion engines?

2.28 Explain the operating principles of a Hall thruster. What are its other names?

2.29 Explain the operating principles of a pulsed plasma thruster.

2.30 Compare the performance of magneto plasmadynamic thruster with that of Hall thruster.

2.31 Explain the operating principles of a nuclear rocket. Why is its development important as a future propulsion for manned interplanetary travels?

2.32 A large quad-engined passenger aircraft is flying at $\mathrm{M} = 0.85$ at 11 km altitude (216.8 K and 0.227 bar). The inlet diameter of the high bypass ratio engine is 2.95 m. If the effective exhaust velocity from the engine is 1.7 times the flight velocity, calculate the thrust produced by each engine. Since the overall fuel:air ratio is significantly less than 1%, neglect the fuel flow rate with respect to the overall air mass flow of the engine.
Answer: Thrust = 109.856 kN

2.33 An air launched cruise missile is powered by a subsonic combustion ramjet engine that has an inlet diameter of 500 mm. The missile is flying at M = 2.6 at an altitude of 13 km (216.65 K and 0.1417 bar). If the TSFC (fuel consumption rate per unit thrust force) is 71 g/kNs, and the fuel:air ratio adopted is 0.055, calculate the fuel (kerosene) flow rate and the thrust.
Answers: Fuel flow rate = 1.888 kg/s. Thrust = 26.588 kN.

References

Cited

1. https://en.wikipedia.org/wiki/Main_Page. Cited 17 2021
2. Hill PG, Peterson CR (1999) Mechanics and thermodynamics of propulsion, 2nd edn. Addition-Wesley, Reading
3. Oates GC (1989) Aircraft propulsion systems technology and design, vol 3. AIAA Inc, AIAA Education Series
4. https://conceptbunny.com/tag/rocket. Cited 20 2021
5. https://engineering.purdue.edu/~propulsi/propulsion/rockets. Cited 20 2021
6. Martinez-Sanchez M, Pollard JE (1998) Spacecraft electric propulsion—an overview. J Propulsion and Power 14:688–699
7. http://www.astronautix.com/. Cited 20 June 2021
8. https://www.jpl.nasa.gov. Cited 20 June 2021
9. Jahn RG (1968) Physics of electric propulsion. McGraw Hill Book Co, New York
10. Goebel DM, Katz I (2008) Fundamentals of electric propulsion: ion and hall thrusters. Wiley, Hoboken
11. Polzin KA (2011) Comprehensive review of planar pulsed inductive plasma thruster research and technology. J Propulsion and Power 27:513–531

Further Reading

12. https://en.wikipedia.org/wiki/Spacecraft_electric_propulsion. Cited 16 2024
13. https://en.wikipedia.org/wiki/Nuclear_propulsion. Cited 16 2024
14. Olsson U (2012) Aerospace propulsion from insects to spaceflight, 2nd edn. Volvo Aero Corporation, Stockholm
15. Choueiri EY (2004) A critical history of electric propulsion: the first fifty years (1906–1956), AIAA-2004-3334
16. Jahn RG, Choueiri EY (2001) Electric propulsion, In Encyclopedia of physical science and technology, 3rd edn. The Academic Press, San Diego
17. Crumpty N (1997) Jet propulsion. Cambridge University Press, Cambridge
18. Cohen H, Rogers GFC, Saravanamuttoo (1996) Gas turbine theory, 4th edn. Longman, Essex
19. Mattingly JD (1996) Elements of gas turbine propulsion. MaGraw-Hill, Singapore
20. Kerrbrock JL (1992) Aircraft engines and gas turbines. The MIT Press, Massachusetts
21. Oates GC (1985) Aerothermodynamics of aircraft engine components. AIAA Education Series, AIAA, New York
22. Herman TC (1981) Gas turbine engineering. Mac Millan Press, London
23. Bussard RW, DeLauer RD (1965) Fundamentals of nuclear flight. McGraw-Hill, New York
24. Stuhlinger E (1964) Ion propulsion for space flight. McGraw-Hill, New York
25. Bussard RW, DeLauer RD (1958) Nuclear rocket propulsion. McGraw-Hill, New York

Chapter 3
Qualitative-Performance and Technology

Synopsis

The thrust equation of a general thrust producer and its different operating conditions of under-expansion, adaption, and over-expansion are discussed. As against an air-breathing propulsion system, a rocket propulsion one, being independent of its environment, can propel its vehicle at flight speeds far greater than the nozzle-exit velocity. The thrust loss in a conical nozzle due to its exit velocities not being parallel to the nozzle axis is accounted for through the divergence factor. The importance of the specific impulse and the density of a propellant in the overall mass of a propulsion system are discussed.

By the deliberate simplification that the molecular mass of combustion products $\overline{m}$ and the specific heat c_p are constants, the qualitative performance of rocket propulsion systems is discussed. Nozzle-flow equations for mass-flow rate, velocity, and area ratio are derived. The criteria for the nozzle-flow separation and the altitude performance are treated in detail. Equations for the thrust coefficient and its variants are derived. That the nozzle-adapted condition at an altitude provides the maximum thrust is proved. Thrust controls through nozzle extensions, traversing plug, and fluid injection, and the wider altitude operation through stepped nozzle, trip ring, ejectable secondary-nozzle, and fluid injection are all discussed. However, due to flow transitions, the thrust oscillations and the side loads can result from many of these thrust control and altitude compensation measures.

The importance of the three characteristic parameters characteristic velocity, thrust coefficient, and specific impulse in comparing the realized experimental values with the theoretical ones towards rocket engines' design and development is discussed. The two limiting assumptions in performance calculations, frozen flow and shifting-equilibrium flow are introduced and noted that only $\overline{m}$ is constant in the former while both $\overline{m}$ and c_p are variants in the latter. Two-phase flow issues are discussed along

S. Krishnan and J. Raghavan, *Rocket Propulsion Primer*,
https://doi.org/10.1007/978-981-97-5644-5_3

with the derivation of simple governing equations that account for velocity lag and temperature difference between the flows of particulates and gaseous mixture.

The procedures for fixing nozzle contours in the convergent and throat region are specified. Taking into account the real flow effects, the methods to arrive at the optimum contour for the divergent portion of nozzle are explained along with the limitation on turn-back angle for two-phase flows. Rao's charts for optimum nozzle contours, drawn for two-dimensional inviscid isentropic flows, are presented to enable quick designs and classroom exercises.

Plug nozzles and expansion-deflection nozzles are of advanced concepts and these are expected to render acceptable performance practically for an unlimited range of altitudes with attractive savings on nozzle mass and length. Discussion on these and the flow characteristics in them with altitude variations are presented.

In a rocket-nozzle construction, there are typically three sub-components to be attended to: liner, insulator (or alternatively cooling system), and structure. Construction details of solid rocket nozzles (external as well as submerged) are presented. As composites are extensively adopted in nozzle construction, brief notes on (1) polymeric materials and their classification into elastomers and plastics (thermoplastics and thermoset plastics) and (2) composites (polymer matrix, ceramic matrix, metal matrix) are presented along with their applications. Apart from the composites, the details of special structural as well as high-temperature erosion-resistant metal alloys are discussed. Materials adopted for the nozzle sub-components of solid rocket nozzles as well as liquid rocket ones are listed.

Worked examples are introduced under the sections. A large number of concept questions and numerical problems with answers are added at the chapter end.

3.1 Thrust Equation

A **general thrust producer** is a system that receives some air from the atmosphere; adds to the air a certain mass of fuel for burning; and then throws the resulting combustion products into the atmosphere at a higher velocity. Note here that the working fluid is initially air and it then changes finally to combustion products. In certain types, the thrust producer may have mechanical energy transfer to and from the working fluid by way of compression (using a compressor) and expansion (using a turbine). Three different thrust producers are schematically shown in Fig. 3.1. In the case of a turbojet engine in flight [Fig. 3.1a], the air is drawn in at a **momentum rate** (mass-flow rate multiplied by velocity) corresponding to the **flight speed**. The air is compressed to a high pressure by the compressor and sent to the combustion chamber. Fuel from the storage within is injected at a certain rate into the combustion chamber to burn with the oxygen in the air. The combustion products thus produced expand through the turbine to produce mechanical power just sufficient to run the compressor. Thereafter the combustion products, as the static pressure is still higher than the atmospheric pressure, are expanded through the nozzle and ejected at a momentum rate greater than that at the entry. Because of this positive momentum–rate

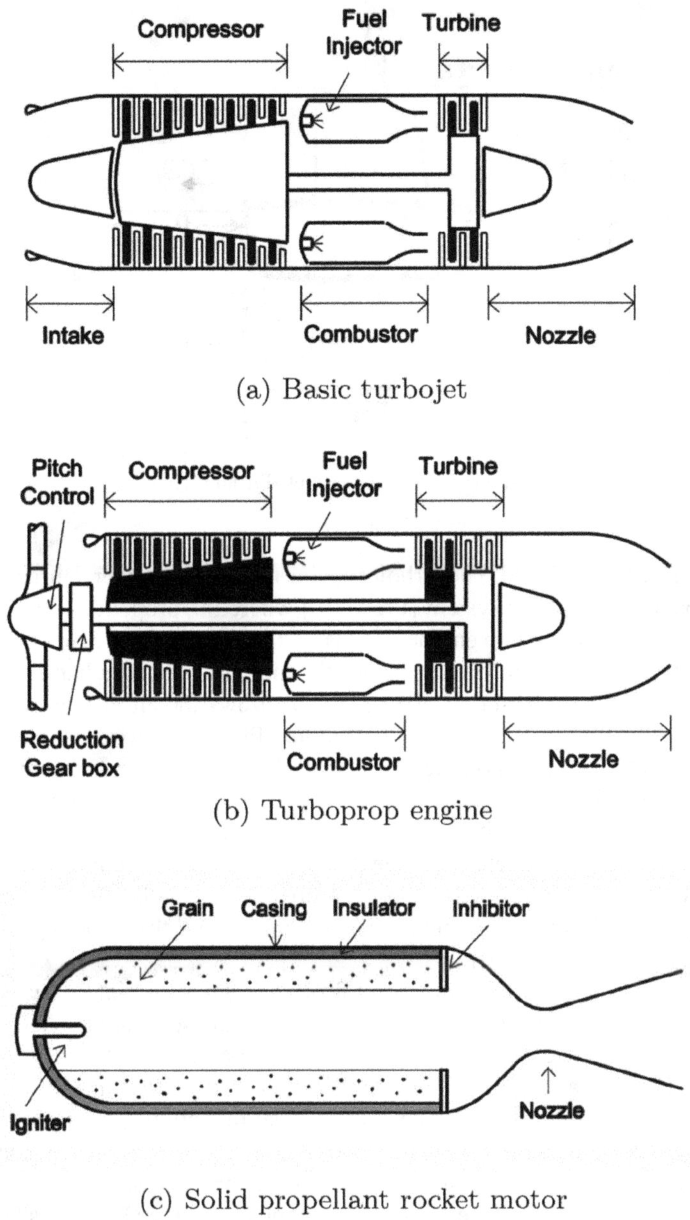

Fig. 3.1 Different thrust producers

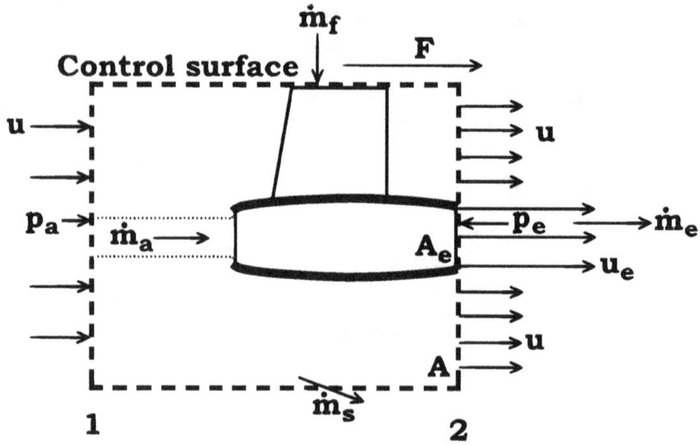

Fig. 3.2 Schematic of a general thrust producer (adopted from [1])

difference thrust is produced. The turbojet engine represents a general thrust producer. Let us consider a propeller system [Fig. 3.1b] where the propeller is powered by a piston engine or a gas turbine engine. In this system, a particular case of the general thrust producer, there is no fuel addition. By imparting the mechanical energy at a certain rate to the drawn-in air, the propeller throws the air at a higher velocity, producing a thrust. In the case of a rocket engine [Fig. 3.1c], no air is drawn into the engine. Or in other words, nothing is drawn in from the environment in which the engine operates. The propellant stored within the engine system is burned. And, the combustion products thus produced are accelerated through the nozzle and ejected at a momentum rate far greater than the "zero" incoming momentum rate. Thus we note that whether it is a turbojet engine, a propeller, or a rocket engine the thrust is produced because the exiting momentum rate is greater than the incoming momentum rate.

Based on one-dimensional and steady flow considerations, let us derive a thrust equation for the general thrust producer, moving in a quiescent atmosphere, as represented in Fig. 3.2. Various operating conditions are possible for the thrust producer. The vehicle (an aircraft, a missile, or a rocket vehicle) fitted with the thrust producer may be in any one of the following modes: take-off, climb, constant-level flight (cruising), descent, or landing. The thrust equation that we will derive is applicable to any of these modes that are assumed to be steady or quasi-steady.

In order to derive the thrust equation, we adopt a technique known as "black box technique." Using this technique, many flow systems can be easily analyzed and quickly understood. In the black box technique, we do not bother about the details of the processes (compression, combustion, expansion, etc.) that may be happening within the control volume, namely, the "box." We assume that the box surfaces are painted black and hence opaque, and the internal processes are "not visible." Only the flow properties at the surfaces of the "box" are considered.

In the schematic that we have, the thrust producer is within the control volume. Air enters the control volume at a velocity equal to the flight velocity. Equivalently, we can assume that the vehicle and its thrust producer are stationary and the air flows towards the control volume of the thrust producer. From the storage within, the fuel is added to the air drawn into the thrust producer and burned. The resulting combustion products are expelled at a momentum rate greater than that at the entry. As we are interested in calculating the thrust produced by the **internal flow** through the thrust producer, we assume that the **external flow** is reversible, that is, frictionless. The **control surface** in front of the thrust producer, station 1, is far upstream where **free stream conditions** can be assumed to prevail. Similarly, the control surfaces on the top and bottom sides are again far removed from the thrust producer, such that the mass-flow rates of air escaping through the top and bottom sides do not have appreciable velocity components perpendicular to the flight direction. At station 2 in the rear, the control surface passes through the nozzle-exit section of the thrust producer. At this station, the velocity of the burned products at the nozzle-exit section is greater than the flight velocity, but the velocity of external air flow in the remaining area is equal to flight velocity as we have assumed the external flow to be reversible.

The **static pressure** at the nozzle-exit section of the thrust producer can be equal to, less than, or greater than the atmospheric pressure. This depends on the velocity of the exiting flow. Depending on the type of thrust producer and its operating conditions, the **nozzle exit** or **exhaust velocity** can be **subsonic**, **sonic**, or **supersonic**. If u_e is subsonic, the condition that p_e is equal to p_a is the only possibility, Eq. 3.1.

$$p_e = p_a \tag{3.1}$$

$$p_e < p_a \tag{3.2}$$

$$p_e > p_a \tag{3.3}$$

If the exhaust velocity is sonic, p_e can be greater than or equal to p_a. And, for the case of the exhaust velocity being supersonic, p_e can be greater than, equal to, or less than p_a, Eqs. 3.1–3.3. The operating condition represented by Eq. 3.1, $p_e = p_a$ is known as **adapted flow**, since the nozzle-exit pressure is adapted to the pressure of the atmosphere. Similarly, the condition by Eq. 3.2, $p_e < p_a$ is identified as **over-expanded flow**, since the nozzle flow is "over-expanded" below the atmospheric pressure. And, the condition by Eq. 3.3, where $p_e > p_a$ is identified as **under-expanded flow**.

By Newton's second law, and considering the components of force and momentum rate only in the flight direction x, we have

$$\sum F_x = \int_{cs} u_x d\dot{m} \tag{3.4}$$

The left-hand side of Eq. 3.4 comprises **pressure forces** and the reaction of the thrust F. At Station 1, the pressure is atmospheric and at Station 2, it is again atmospheric except that at the nozzle-exit plane of the thrust producer. At the exit plane, as discussed previously, the pressure can be equal to, less than, or greater than the atmospheric pressure p_a. The pressure forces at the top and bottom sides are perpendicular to the flight direction and being opposite in direction, they cancel each other. The pressure forces, therefore, can be written as

$$F_p = Ap_a - (A - A_e)\, p_a - A_e p_e = A_e\,(p_a - p_e) \tag{3.5}$$

Thus the left-hand side of Eq. 3.4

$$\sum F_x = A_e\,(p_a - p_e) + F \tag{3.6}$$

The right-hand side of Eq. 3.4 can now be written as

$$\int_{cs} u_x d\dot{m} = (\rho u)\,(A - A_e)\,u + (\rho_e u_e A_e)\,u_e + \dot{m}_s u - (\rho u A)\,u \tag{3.7}$$

where $\dot{m}_s$ is the mass-flow rate of air, escaping at the sides. By the **mass conservation** for the control volume,

$$\dot{m}_s + \rho_e u_e A_e + \rho u\,(A - A_e) = \rho u A + \dot{m}_f$$

That is,

$$\dot{m}_s = -\rho_e u_e A_e + \rho u A_e + \dot{m}_f \tag{3.8}$$

We note, again by mass conservation, that the mass-flow rate of fuel injected into the thrust producer is the difference between the mass-flow rate ejected and the mass-flow rate of air captured by the thrust producer. Therefore,

$$\dot{m}_f = \rho_e u_e A_e - \dot{m}_a \tag{3.9}$$

Substituting Eq. 3.9 into Eq. 3.8 for $\dot{m}_f$, we get

$$\dot{m}_s = \rho u A_e - \dot{m}_a \tag{3.10}$$

Substituting this result of the mass conservation for $\dot{m}_s$ in Eq. 3.7, we get

$$\int_{cs} u_x d\dot{m} = (\rho_e u_e A_e)\,u_e - \dot{m}_a u = \dot{m}_e u_e - \dot{m}_a u \tag{3.11}$$

This is the right-hand side of Eq. 3.4 and we have already obtained the terms for its left-hand side in Eq. 3.6. Using these quantities into Eq. 3.4, we get

$$F = \dot{m}_e u_e - \dot{m}_a u + A_e (p_e - p_a) \tag{3.12}$$

Or,

$$F = \left[\left(\dot{m}_a + \dot{m}_f\right) u_e - \dot{m}_a u\right] + A_e (p_e - p_a) \tag{3.13}$$

$$F = \dot{m}_a [(1 + f) u_e - u] + A_e (p_e - p_a) \tag{3.14}$$

Equations 3.12–3.14 are the thrust equations of the general thrust producer in various forms; here f is the **fuel–air ratio** defined as

$$f = \frac{\dot{m}_f}{\dot{m}_a}$$

Let us consider the thrust producer of the propeller system and see how the thrust equation gets modified. There is no fuel addition to the propeller stream, which is always subsonic: $f = 0$ and $p_e = p_a$. Therefore, the thrust equation for the propeller system reduces to

$$F = \dot{m}_a (u_e - u)$$

Next, let us consider a rocket engine in flight. We note that for a rocket engine, no air is drawn from the atmosphere, $\dot{m}_a = 0$. And, the total propellant flow rate $\dot{m}$ can be taken equivalent to $\dot{m}_f$. The nozzle-exit velocity is mostly supersonic. In the case of a **liquid propellant engine** $\dot{m}$ is the mass-flow rate of combustion gases resulting from the burning of a liquid fuel and a liquid oxidizer. And, in the case of a **solid propellant motor**, it is that resulting from the burning of a solid propellant containing oxidizer and fuel ingredients. Therefore, the thrust equation for a rocket engine is

$$F = \dot{m} u_e + A_e (p_e - p_a) \tag{3.15}$$

Equations 3.12–3.14 are the thrust equations of a general thrust producer, which "breathes" air. The rocket engine is a particular case of the general thrust producer that does not "breathe" air and hence a "non-air-breathing" engine.

The thrust comprises two quantities, one the thrust due the difference in the momentum rates, and the other due to the pressure difference. The thrust due to the pressure difference is same for air-breathing as well as rocket engines, namely,

$$A_e (p_e - p_a)$$

But, the thrust due to the difference in momentum rates is

$$\left(\dot{m}_a + \dot{m}_f\right) u_e - \dot{m}_a u$$

for the **air-breathing engine** and it is simply

$$\dot{m}u_e$$

for the **rocket engine**. The thrust due to the difference in momentum rate is known as **momentum thrust** and that due to pressure difference is known as **pressure thrust**. Generally the pressure thrust is quite small compared to the momentum thrust. However, in the case of rocket engines, the magnitude of pressure thrust can be substantially more than that in air-breathing engines.

By comparing Eqs. 3.12–3.14 with Eq. 3.15, we note two important differences between the air-breathing engines and the rocket engines (the non-air-breathing engines). The first difference is that for an air-breathing engine the thrust depends on flight velocity u whereas for a rocket engine it is not. Consequently, satellite launch vehicles, satellites, and spacecraft powered by rocket engines can fly at velocities far greater than the engine-exit velocities. For example, satellite launch vehicles powered by rocket engines can reach the needed flight velocities of 8000–12000 m/s against the maximum possible engine-exit velocities of about 4500 m/s. But, aircraft and missiles powered by air-breathing engines cannot reach flight velocities greater than the engine-exit velocities. Mostly they fly at velocities around half the engine-exit velocities, as this condition provides maximum overall efficiency for the air-breathing engines.

$$u \approx \frac{1}{2}u_e$$

The second difference between the air-breathing engines and the rocket engines is that, by the very fact that a rocket engine does not draw-in any working fluid from the atmosphere, its surrounding medium, it can produce thrust in any medium, say underwater or vacuum. Note torpedoes that are launched under water are powered by rocket engines. In vacuum conditions, rocket engines navigate satellites around the Earth and propel spacecraft to outer planets.

Let us recall that engine-exit pressure can be equal to, less than, or greater than ambient pressure. The operating condition at which the **exit plane pressure** is equal to ambient pressure, namely, the adapted condition, has a special significance. Given an altitude, a nozzle entry stagnation pressure, and a nozzle geometry, the thrust producer delivers its maximum thrust at the adapted condition [see Sect. 3.5.6]. Under an adapted condition the thrust equation for a rocket engine takes a simpler form,

$$F_{p_e=p_a} = \dot{m}u_e \tag{3.16}$$

By defining an **equivalent exhaust velocity**, namely, , $\overline{u}_e$, we can modify the general thrust equation of a rocket engine, Eq. 3.15, to a simpler form similar to Eq. 3.16. That is,

$$F = \dot{m}\overline{u}_e \tag{3.17}$$

where

$$\overline{u}_e = u_e + \frac{A_e\left(p_e - p_a\right)}{\dot{m}} \tag{3.18}$$

Note that at the adapted condition,

$$\left(\overline{u}_e\right)_{p_e = p_a} = u_e \tag{3.19}$$

3.2 Propulsive Efficiency

We shall evaluate the propulsive efficiency of an air-breathing engine and that of a rocket engine, and draw the difference between the two types of engines. Let us consider that an air-breathing engine or a rocket engine is operating under **cruising condition** (altitude and flight velocity at constant values: thrust = drag). For simplicity, let us assume that for both the types the nozzles are adapted, $p_e = p_a$. As far as the air-breathing engine is concerned, again for simplicity, we can neglect the mass-flow rate of fuel when compared to the mass-flow rate of air—the ratio of fuel-flow rate to air-flow rate generally varies from about 0.02 to a maximum of around 0.07 and the latter value corresponds to the one around stoichiometric fuel–air ratio. Under these simplifying considerations, the thrust equation of the air-breathing engine, Eq. 3.12, becomes

$$F = \dot{m}_a\left(u_e - u\right) \tag{3.20}$$

Similarly, the thrust equation of the rocket engine, Eq. 3.15, becomes

$$F = \dot{m}u_e \tag{3.21}$$

Under the cruising condition, the useful power P_U is the product of thrust and flight velocity. And, the available power P_A is the sum of the useful power and the power lost in the exiting flow from the engine. The power lost in the exiting flow is to be taken considering the absolute velocity of the exiting flow with respect to a static observer. The absolute velocity of the exiting flow is equal to $(u_e - u)$. Therefore, the available power for the air-breathing engine:

$$\left(P_A\right)_{ab} = \dot{m}_a\left(u_e - u\right)u + \dot{m}_a\frac{\left(u_e - u\right)^2}{2} = \frac{\dot{m}_a}{2}\left(u_e^2 - u^2\right) \tag{3.22}$$

And, the available power for the rocket engine,

$$\left(P_A\right)_r = \dot{m}u_e u + \dot{m}\frac{\left(u_e - u\right)^2}{2} = \frac{\dot{m}}{2}\left(u_e^2 + u^2\right) \tag{3.23}$$

The propulsive efficiency is defined as the ratio of useful power to the available power,

$$\eta_p \equiv \frac{P_U}{P_A} \tag{3.24}$$

Using Eqs. 3.20 and 3.22 into Eq. 3.24 for the air-breathing engine and Eqs. 3.21 and 3.23 into Eq. 3.24 for the rocket engine, we get

$$\left(\eta_p\right)_{ab} = \frac{2\nu}{1+\nu} \tag{3.25}$$

for the air-breathing engine, and

$$\left(\eta_p\right)_r = \frac{2\nu}{1+\nu^2} \tag{3.26}$$

for the rocket engine. Here, the **velocity ratio** ν is defined as

$$\nu \equiv \frac{u}{u_e} \tag{3.27}$$

From Eqs. 3.25 to 3.26, we note that for an air-breathing engine as well as for a rocket engine the maximum propulsive efficiency is reached when the flight velocity is equal to the engine-exit velocity: $u = u_e$ or $\nu = 1$. In the case of air-breathing engines, this condition of maximum propulsive efficiency leads to an impractical situation of the thrust becoming zero, Eq. 3.20. The impractical situation of zero thrust at the maximum propulsive efficiency at $u = u_e$ is obtained using the simplified thrust equation, Eq. 3.20. Evidently, the impractical situation remains unaltered even when we use the full thrust equation, Eq. 3.12. The only difference is that the thrust does not become zero, but becomes very small due to the non-zero value of fuel–air ratio and the small value of a possible pressure thrust. Let us now look at the case of a rocket engine. Contrary to what we see in the case of an air-breathing engine, the attainment of maximum propulsive efficiency is practical as the thrust here is independent of flight velocity, Eq. 3.21. As mentioned previously, remember that rocket vehicles can fly at a much higher velocity than the engine exhaust velocity, evidently at a lower propulsive efficiency after crossing the maximum. Example 3.1 calculates the propulsive efficiency of a rocket engine, when the nozzle-exit velocity (specific impulse) is less than the vehicle velocity.

Example 3.1 A satellite launch vehicle is powered by a cryogenic rocket engine having an equivalent exhaust velocity of 4500 m/s. The mass-flow rate of the propellant = 50 kg/s. The flight velocity of the launch vehicle is 6000 m/s. Calculate the thrust and the propulsive efficiency of the engine.

Solution
This is the straightforward application of derived equations. Note that the flight velocity is more than the equivalent exhaust velocity, the important capability of rocket engines in contrary to air-breathing propulsion.

Analysis
Using Eq. 3.16, thrust,

$$F = 50 \times 4500 = 225000N$$

The velocity ratio,

$$\nu = \frac{6000}{4500} = 1.333$$

By Eq. 3.26, the propulsive efficiency,

$$\eta_p = \frac{2 \times 1.333}{1 + 1.333^2} = 0.96$$

3.3 Divergence Factor

We have derived the thrust equation, Eq. 3.15, under one-dimensional considerations.

$$F = \dot{m}u_e + A_e\,(p_e - p_a) \tag{3.15}$$

For small- and low-performance rocket engines, in order to keep the system cost down, the nozzles are chosen to be of a conical shape. For these **conical nozzles**, the exit velocity vectors are not truly along the nozzle axis. To be closer to the real conditions, we can consider these velocity vectors to be uniform and orthogonal to the spherical surface formed at the exit of the nozzle cone, Fig. 3.3. By this, we will be able to introduce a correction factor to the one-dimensional thrust equation, Eq. 3.15.

Let the **half cone angle** of the conical nozzle = α_d. The thrust delivered along the nozzle axis can be written as

$$F = \int_{A_e} u_{e_x} d\dot{m} + A_e\,(p_e - p_a) \tag{3.28}$$

Here u_{e_x} is the velocity component along the nozzle axis and $d\dot{m}$ is the mass-flow rate exiting from the elemental area of the annular strip on the spherical surface of the nozzle exit.

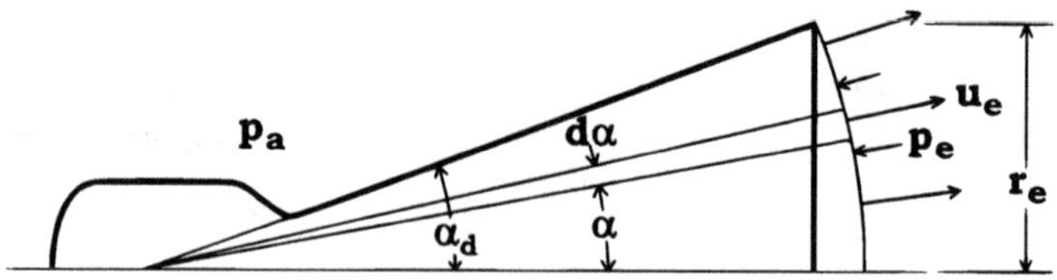

Fig. 3.3 Nozzle divergence factor for a conical nozzle

Then,

$$d\dot{m} = \rho_e u_e dA = \rho_e u_e \left[2\pi \left(\frac{r_e}{\sin \alpha_d} \right) \sin \alpha \left(\frac{r_e}{\sin \alpha_d} \right) d\alpha \right] \tag{3.29}$$

The total mass-flow rate $\dot{m}$,

$$\dot{m} = \int_0^{\alpha_d} \rho_e u_e \left[2\pi \left(\frac{r_e}{\sin \alpha_d} \right)^2 \sin \alpha d\alpha \right]$$

Thus,

$$\dot{m} = 2\pi \rho_e u_e \left(\frac{r_e}{\sin \alpha_d} \right)^2 (1 - \cos \alpha_d) \tag{3.30}$$

Similarly,

$$\int_{A_e} u_{e_x} d\dot{m} = \int_{A_e} (u_e \cos \alpha) \left[\rho_e u_e 2\pi \left(\frac{r_e}{\sin \alpha_d} \right)^2 \sin \alpha d\alpha \right]$$

Thus,

$$\int_{A_e} u_{e_x} d\dot{m} = 2\pi \rho_e u_e^2 \left(\frac{r_e}{\sin \alpha_d} \right)^2 \int_0^{\alpha_d} \sin \alpha \cos \alpha d\alpha \tag{3.31}$$

After integrating and using Eq. 3.30 for $\dot{m}$,

$$\int_{A_e} u_{e_x} d\dot{m} = \left(\frac{1 + \cos \alpha_d}{2} \right) \left[2\pi \rho_e u_e \left(\frac{r_e}{\sin \alpha_d} \right)^2 (1 - \cos \alpha_d) \right] u_e$$

Thus,

$$\int_{A_e} u_{e_x} d\dot{m} = \left(\frac{1 + \cos \alpha_d}{2} \right) \dot{m} u_e \tag{3.32}$$

Substituting Eq. 3.32 into Eq. 3.28,

$$F = \left(\frac{1 + \cos \alpha_d}{2} \right) \dot{m} u_e + A_e (p_e - p_a) \tag{3.33}$$

Let us define the correction factor that we have found by this analysis as the divergence factor λ.

$$\lambda \equiv \frac{1+\cos\alpha_d}{2} \tag{3.34}$$

Thus, after introducing the divergence factor, the thrust equation becomes

$$F = \lambda \dot{m} u_e + A_e \left(p_e - p_a\right) \tag{3.35}$$

As stated previously, conical nozzle shapes are adopted for simple- and low-performance rocket engines and the values of α_d for such nozzles are generally kept less than 15° and the corresponding values of $\lambda \leq 0.983$.

3.4 Specific Impulse

Specific impulse is the most important performance parameter in rocket propulsion. Its theoretical value depends on the energy contained in the propellant composition and its experimental value depends on the efficiencies involved in the combustion and the nozzle expansion, and the environmental pressure.

Before understanding the significance of specific impulse, we should first look at a related characteristic quantity, **total impulse**, I. Based on its mission, every rocket vehicle requires a certain type of thrust variation with respect to time. This is known as **thrust–time trace** or thrust–time history. For a fixed nozzle geometry, the chamber pressure represents the thrust to be delivered [see Sect. 3.5.5]. A typical pressure–time trace is shown in Fig. 3.4. For the thrust–time trace shown, we note that the vehicle requires from its engine a high thrust initially, and thereafter the thrust is to reduce with time. Somewhere in the middle, the thrust is to increase with time to attain again a high value before burnout. The space shuttle requires a similar thrust–time trace from its two solid rocket booster motors. Some more typical thrust–time traces are given in Fig. 3.5. The time integral of the thrust that the engine is required to deliver is known as the total impulse, I. Or in other words, it is the area under the thrust–time trace. Therefore,

$$I = \int_0^{t_b} F dt \tag{3.36}$$

Substituting for F from Eq. 3.17,

$$I = \int_0^{t_b} \dot{m}\overline{u}_e dt \tag{3.37}$$

For the chosen engine configuration and propellant composition, the equivalent exhaust velocity $\overline{u}_e$ is essentially constant—its variation with the environment is

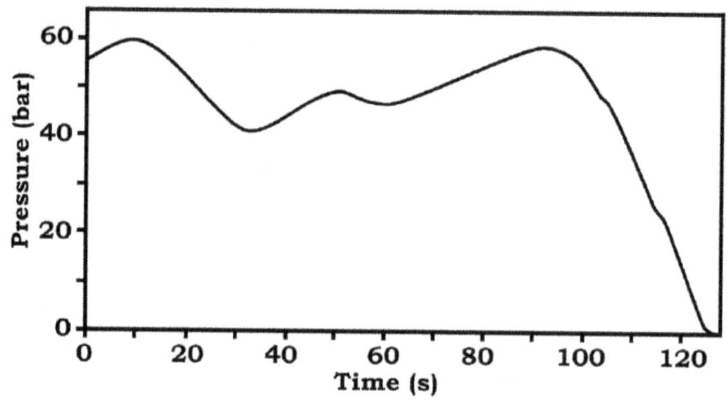

Fig. 3.4 A typical pressure–time trace of a large solid propellant rocket motor

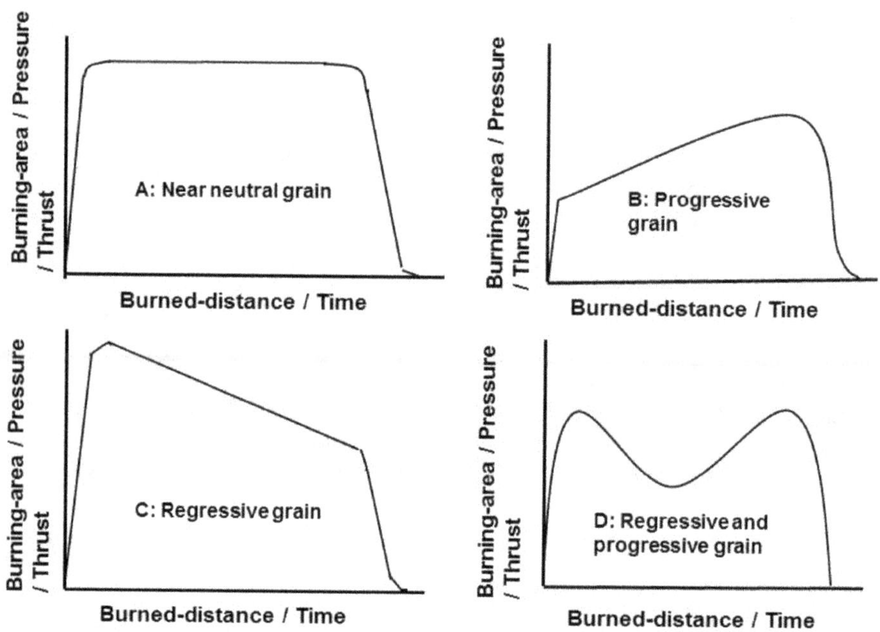

Fig. 3.5 Different thrust–time traces

very small under normal operating conditions. Therefore, the total impulse can be written as

$$I = \overline{u}_e \int_0^{t_b} \dot{m} dt = m_p \overline{u}_e \tag{3.38}$$

where m_p is the total propellant mass stored. The dimension of the total impulse is [Ns] or if we express it in fundamental units, it is [kg-m/s].

As the total impulse is the product of total-propellant mass and equivalent exhaust velocity, the specific impulse, I_{sp}, is the impulse that can be delivered per unit propellant mass

$$I_{sp} = \frac{I}{m_p} = \frac{\int_0^{t_b} F dt}{\int_0^{t_b} \dot{m} dt} = \frac{F}{\dot{m}} = \overline{u}_e \tag{3.39}$$

Also note from Eq. 3.39, I_{sp} is the thrust per unit mass-flow rate of propellant. The dimension of the specific impulse is [Ns/kg]. In fundamental units, the dimension of I_{sp} is [m/s], as you rightly see the velocity term on the right-hand side of Eq. 3.39.

In rocket industries, it has been a practice all along to refer to the dimension of I_{sp} as seconds, [s]. That is the thrust force per unit "weight" rate of propellant. Since "weight" is also a force, it has been a practice to use seconds [s] as the dimension for the specific impulse. Strictly, if we differentiate between force and mass, the dimension of specific impulse should be [kgf-s/kgm] as per the old metric system of units or [lbf-s/lbm] as per the British system of units. The values of specific impulse of a particular propellant are all the same in seconds, [kgf-s/kgm], or [lbf-s/lbm]. Therefore whenever you read seconds, [kgf-s/kgm], or [lbf-s/lbm] as the dimension of specific impulse, understand that to get the value in [Ns/kg] it should be multiplied by the gravitational conversion constant, 9.8065 $[kgm - m/(s^2 - kgf)$ or $N/kgf]$. As we follow SI units in this book, the dimension of specific impulse is given in [Ns/kg] throughout.

According to its required thrust–time trace, every rocket vehicle has a value of total impulse. From Eq. 3.39, the total impulse can be put into a form

$$I = m_p I_{sp} = m_p \overline{u}_e \tag{3.40}$$

Equation 3.40 reveals the importance of having a high value for specific impulse. By using a propellant of higher specific impulse for a given mission the mass of total propellant to be carried becomes lower. But if a propellant of higher specific impulse is of lower density, although the mass of total propellant is lower, the volume of total propellant to be carried becomes higher resulting in a larger and massive propellant tank. Therefore, a related characteristic parameter to be considered along with I_{sp} is the **density specific impulse**,

$$\rho_p I_{sp} = \frac{I}{V_p} = \frac{F}{\dot{V}_p} = \rho_p \overline{u}_e \tag{3.41}$$

The dimension of the density specific impulse is $[Ns/m^3]$. For the required total impulse, a propellant combination that has a higher density specific impulse will have a more compact propulsion system. Higher the required total impulse, more significant is the advantage of compactness. In Example 3.2, we consider the calculation of total impulse, specific impulse, and total volume of the propellant to be carried for a large solid propellant booster motor.

Example 3.2 A very large solid rocket booster, such as the one used in the Space Shuttle, stores 525,000 kg of solid propellant of density 1700 kg/m^3 and delivers an average thrust of 11.5 MN for 120 s. Calculate the total impulse, specific impulse, and density specific impulse for the booster. Also, estimate the storage volume required for the total propellant.

Solution
This is the straightforward application of derived equations. Note the terminology "average thrust." Although the thrust may be varying with time, the average thrust is the total impulse divided by the thrusting time.

Analysis
The total impulse,

$$I = \int_0^{t_b} F dt \tag{3.36}$$

Therefore

$$I = 11.5 \times 10^6 \times 120 = 1380 \times 10^6 \text{ Ns}$$

The specific impulse,

$$I_{sp} = \frac{I}{m_p} = \frac{1380 \times 10^6}{525000} = 2628 \frac{\text{Ns}}{\text{kg}}$$

The density specific impulse,

$$\rho_p I_{sp} = 1700 \times 2628 = 4.4676 \times 10^6 \frac{\text{Ns}}{\text{m}^3}$$

The propellant storage volume

$$V_p = \frac{525000}{1700} = 308.82 \text{ m}^3$$

Another characteristic parameter, rarely used though, is the **thrust specific propellant consumption**, which is the reciprocal of specific impulse.

$$TSPC \equiv \frac{1}{I_{sp}} = \frac{m_p}{I} = \frac{\dot{m}}{F} \tag{3.42}$$

A similar characteristic parameter widely used in air-breathing propulsion is the **thrust specific fuel consumption**, TSFC.

$$TSFC = \frac{\dot{m}_f}{F} \tag{3.43}$$

These two characteristics represent the economy of producing the required thrust. In rocket propulsion, TSPC varies from about 2.3×10^{-4} kg/Ns (cryogenic rocket engines) to 4×10^{-4} kg/Ns (solid propellant motors). In the case of air-breathing propulsion, TSFC varies from about 10^{-5} kg/Ns (subsonic turbofan propulsion) to 7×10^{-5} kg/Ns (supersonic ramjet propulsion). Evidently producing the thrust by air-breathing engines is less expensive because we do not pay for the oxygen contained in air!

In the case of air-breathing engines, we have a characteristic parameter equivalent to the specific impulse in rocket engines. This quantity is *conventionally* known as **specific thrust**—there is nothing wrong in calling also this as specific impulse because impulse and thrust are basically one and the same. Specific thrust is the ratio of thrust to mass-flow rate of air and it varies from 400 to 900 Ns/kg. The specific impulse for rocket engines varies from about 2500 to 4500 Ns/kg. As the flow rate of working fluid, air or propellant, will depend on the characteristic engine area and hence the engine volume, we can say that in general for the given thrust requirement, the engine volume of the air-breathing system is always larger than that of a rocket engine. Furthermore, as the engine volume represents through a certain multiplying factor the engine mass and hence the cost, for the given thrust requirement, the engine cost of an air-breathing system is mostly higher than that of a rocket engine.

3.5 Nozzle Flow

In high-performance rocket engines, because of the **losses due to combustion ("heat addition") and mass addition**, the stagnation pressure reduces from the **head end** of the combustion chamber to its **aft end** or **nozzle end** (that is, entry to the nozzle). The stagnation pressure loss is around 3–5 percent. However, in this chapter, we will assume for simplicity that there is uniform stagnation pressure p_{0c} in the combustion chamber and this is equal to the nozzle entry stagnation pressure p_{0n}. Furthermore, we assume that the entire engine is an adiabatic system of constant stagnation temperature T_{0c} equal to the adiabatic flame temperature. The uniform gaseous flow through the convergent divergent nozzle is assumed to be (1) one dimensional, (2) satisfying ideal gas equation of state, (3) isentropic, and (4) of constant specific heat c_p and molecular mass $\overline{m}$, which means that the ratio of specific heats γ is also constant. The assumption of constant c_p and $\overline{m}$ makes the rocket performance calculations simpler to get the base values quickly. But in Chap. 4 we will discuss the details of the performance calculations to a next higher level of rigor by adopting the two properties c_p and $\overline{m}$ to be dependent on local equilibrium composition of combustion products. For the present, however, we continue with our analysis with the assumptions (1)–(4). By the ideal gas equation of state,

$$p = \rho \frac{R_u}{\overline{m}} T = \rho R T \tag{3.44}$$

where R_u is the universal gas constant equal to 8314.51 J/(kgmolK) and R is the specific gas constant of dimension J/(kgK). Mach number is given by

$$M = \frac{u}{a} = \frac{u}{\sqrt{\gamma R T}} \tag{3.45}$$

where a is the local sonic velocity. For isentropic process,

$$\frac{p_x}{p} = \left(\frac{T_x}{T}\right)^{\frac{\gamma}{\gamma-1}} = \left(\frac{\rho_x}{\rho}\right)^{\gamma} \tag{3.46}$$

The subscript x represents any arbitrary station. Applying this relationship to stagnation (or total) values,

$$\frac{p_x}{p_{0n}} = \left(\frac{T_x}{T_{0n}}\right)^{\frac{\gamma}{\gamma-1}} = \left(\frac{\rho_x}{\rho_{0n}}\right)^{\gamma} \tag{3.47}$$

Note stagnation values are constant everywhere for the isentropic flow. The following relations resulting from the above assumptions will be frequently employed in the nozzle flow.

$$\frac{T_0}{T} = 1 + \frac{\gamma - 1}{2} M^2 \tag{3.48}$$

$$\frac{p_0}{p} = \left(1 + \frac{\gamma - 1}{2} M^2\right)^{\gamma/(\gamma-1)} \tag{3.49}$$

$$c_p = \frac{R\gamma}{\gamma - 1} \tag{3.50}$$

3.5.1 Mass-Flow Rate

Mass-flow rate of gas at any section of the nozzle,

$$\dot{m} = \rho_x u_x A_x = \rho_t u_t A_t \tag{3.51}$$

where the subscript t refers to the throat section. From Eq. 3.51, the mass-flow rate at any section is given by

$$\dot{m} = \frac{p_{0n} A_x}{\sqrt{R T_{0c}}} \left[\sqrt{\gamma} M_x \left(1 + \frac{\gamma - 1}{2} M_x^2\right)^{\frac{-(\gamma+1)}{2(\gamma-1)}}\right] \tag{3.52}$$

At the throat, the flow is choked: $u_t = a_t$ and $M_t = 1$. Therefore,

$$\dot{m} = p_{0n} A_t \frac{\sqrt{\gamma}\left(\frac{2}{\gamma+1}\right)^{(\gamma+1)/[2(\gamma-1)]}}{\sqrt{RT_{0c}}} \tag{3.53}$$

Denoting,

$$\Gamma = \sqrt{\gamma}\left(\frac{2}{\gamma+1}\right)^{(\gamma+1)/[2(\gamma-1)]} \tag{3.54}$$

Equation 3.53 becomes

$$\dot{m} = p_{0n} A_t \left(\frac{\Gamma}{\sqrt{\frac{R_u}{\overline{m}} T_{0c}}}\right) \tag{3.55}$$

Maximum mass flux $\dot{m}/A_x$ occurs at the throat. From Eq. 3.55, we note that the mass-flow rate through the choked nozzle is dependent on (1) nozzle entry stagnation pressure p_{0n}, (2) throat area A_t, and (3) the quantity within the brackets involving the characteristics of combustion: T_{0c}, $\overline{m}$ and γ. The significance of this combustion characteristic will be discussed in Sect. 3.6. For the present, we note that it involves only the combustion characteristics, and its value is dependent largely on the type of propellant and much less on the engine design. From Eq. 3.55, we also note that for a given class of rocket engines using a particular propellant—the quantity within the brackets being essentially constant—as the mass-flow rate is linearly proportional to p_{0n} and A_t, the thrust can be varied by changing either one or both of them, Eq. 3.15.

3.5.2 Nozzle-Flow Velocity

By conservation of energy,

$$h_0 = h + \frac{u^2}{2} \tag{3.56}$$

From this we can show that the velocity at any section,

$$u_x = \sqrt{\frac{2\gamma}{\gamma-1} RT_{0c}\left[1 - \left(\frac{p_x}{p_{0n}}\right)^{\frac{\gamma-1}{\gamma}}\right]} \tag{3.57}$$

At the choked throat, by Eq. 3.39,

$$\frac{p_t}{p_{0n}} = \left(\frac{2}{\gamma+1}\right)^{\frac{\gamma}{\gamma-1}} \tag{3.58}$$

Substituting Eq. 3.58 into Eq. 3.57, it is easy to show

$$u_t = a_t = \sqrt{\gamma R T_t} \tag{3.59}$$

It follows from Eq. 3.57 that the velocity at the exit section of the nozzle,

$$\begin{aligned} u_e &= \sqrt{\frac{2\gamma}{\gamma-1} R T_{0c} \left[1 - \left(\frac{p_e}{p_{0n}}\right)^{\frac{\gamma-1}{\gamma}}\right]} \\ &= \sqrt{\frac{2\gamma}{\gamma-1} \frac{R_u}{\overline{m}} T_{0c} \left[1 - \left(\frac{p_e}{p_{0n}}\right)^{\frac{\gamma-1}{\gamma}}\right]} \end{aligned} \tag{3.60}$$

The nozzle-exit velocity u_e forms the principal part of $\overline{u}_e$ or the specific impulse I_{sp}. Therefore there is always an interest to achieve as high a value as possible for u_e. Equation 3.60 indicates that u_e is dependent on T_{0c}, $1/\overline{m}$, p_{0n}/p_e, and γ. Let us look at the effects of these on the engine performance.

By using high-energy propellant ingredients of "less negative if not more positive" **heats of formation** (refer Sect. 4.6) one can get high T_{0c}. But there are two problems associated with high T_{0c}. Firstly, with high T_{0c} we have to (1) use or develop materials of high-temperature resistance and/or (2) use advanced cooling techniques. Both of these turn out to be expensive. Furthermore, by the adoption of advanced cooling techniques, the inert or structural mass of the engine may also increase. Secondly, on the point of view of energy audit, the use of high-energy ingredients turns out to be an inefficient proposition, because the energy contained in these special ingredients is used more for the energy absorption by **dissociated species** rather than for the raise in T_{0c}. For the present, it appears that T_{0c} of 3500 K is about the maximum that is practical for many operating engines.

In order to get high u_e, it is beneficial to have propellants that give combustion products of low molecular mass. In this respect the use of fuel ingredients containing more hydrogen is advantageous. For the same adiabatic flame temperature of around 3500 K, the conventional liquid propellants—**Earth storables**—have combustion products of molecular mass of around 25 kg/kgmol, but the **cryogenic propellant** combination of liquid oxygen and liquid hydrogen gives the molecular mass of around 10 kg/kgmol. And, this is the reason why the attainable specific impulse with the latter is around 4500 Ns/kg while that with the former is around 3500 Ns/kg.

The third factor that influences the variation of u_e is the **nozzle pressure ratio** p_{0n}/p_e. A high value of this ratio gives enhanced values of u_e. Maximum nozzle pressure ratios for the rocket engines are constrained by (1) limit on over-expansion in order to avoid nozzle-flow separation [see Sect. 3.5.4], (2) maximum acceptable nozzle-exit area imposed by the vehicle configuration (for example, the nozzle-exit diameter cannot be more than the maximum diameter of the total vehicle or cannot foul the specified spacecraft envelope), (3) maximum permissible nozzle mass (high nozzle pressure ratio will need high nozzle area ratio and consequently large nozzle length and mass), and (4) maximum permissible pressure in the combustion chamber. For rocket engines operating only in vacuum, the first constraint is not applicable. All in all, the pressure ratios of rocket engines operating in Earth's atmosphere vary from 30 to 60. For the engines operating only in vacuum, these values are much higher.

The effect of change in γ on the value of u_e is not substantial. Also, there is no workable means to change the value of γ, as the combustion products of most propellants, being mostly populated with poly-atomic gases, have its value in the narrow range around 1.25.

3.5.3 Nozzle Area Ratio

The nozzle area ratio is the ratio of nozzle-exit area to throat area, A_e/A_t. By the continuity equation,

$$\rho_t u_t A_t = \rho_e u_e A_e \tag{3.61}$$

Noting for isentropic flow assumptions, $p_{0n} = p_{0t} = p_{0e}$,

$$\frac{\rho_t}{\rho_e} = \left(\frac{p_t}{p_e}\right)^{1/\gamma}$$

and for choked flow,

$$\frac{p_{0t}}{p_t} = \left(\frac{\gamma+1}{2}\right)^{\gamma/(\gamma-1)}$$

At the throat,

$$u_t = a_t = \sqrt{\gamma R T}$$

And, using Eq. 3.60 in Eq. 3.61, we get

$$\frac{A_e}{A_t} = \frac{\Gamma}{\left(\frac{p_e}{p_{0n}}\right)^{1/\gamma}\sqrt{\frac{2\gamma}{\gamma-1}\left[1-\left(\frac{p_e}{p_{0n}}\right)^{(\gamma-1)/\gamma}\right]}} \tag{3.62}$$

Alternatively, Eq. 3.62 can be written in a different form which is convenient to calculate nozzle pressure ratio, p_{0n}/p_e, given the area ratio of nozzle, A_e/A_t.

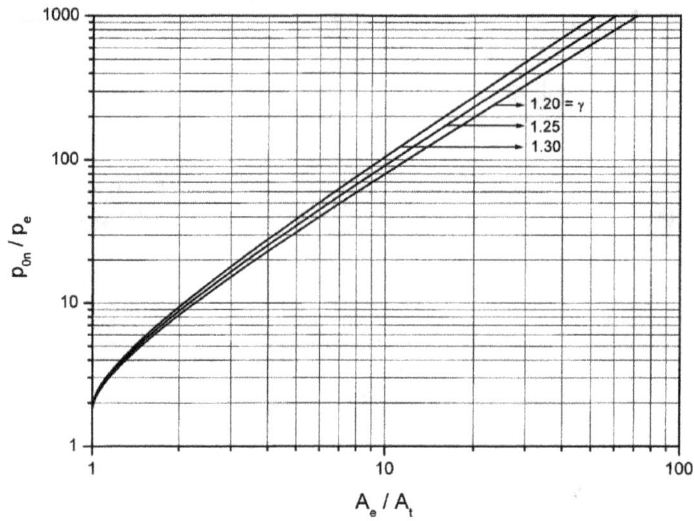

Fig. 3.6 Nozzle pressure ratio versus nozzle area ratio

$$\frac{p_{0n}}{p_e} = \left\{ \frac{A_e}{A_t} \sqrt{\frac{2\gamma}{\gamma - 1} \left[1 - \left(\frac{p_{0n}}{p_e} \right)^{-(\gamma-1)/\gamma} \right]} / \Gamma \right\}^{\gamma} \tag{3.63}$$

From Eqs. 3.62 and 3.63, we note that for the given value of γ, the nozzle area ratio is an unique function of p_{0n}/p_e. This means that so long as the nozzle of fixed configuration is flowing full (that is, the nozzle flow is without separation) the change in the nozzle entry stagnation pressure (or, in other words, the change in the combustion chamber stagnation pressure) does not change the nozzle pressure ratio p_{0n}/p_e. Furthermore, for the assumed isentropic flow in a fixed-configuration nozzle, as p_{0n}/p_e is function of Mach number and γ, for the nozzle full-flow condition, the change in chamber stagnation pressure does not have any effect on M_e [Eq. 3.65]. In Fig. 3.6, the variations of nozzle pressure ratio for three different typical values of ratio of specific heats are given against the nozzle area ratios.

Generally for solid propellants and the Earth-storable liquid propellants, the ratio of specific heats is higher (around 1.25) than that for the traditional cryogenic propellant combinations of liquid hydrogen and liquid oxygen (around 1.15). From Fig. 3.6, it is clear that for the given A_e/A_t, the nozzle pressure ratio p_{0n}/p_e is lower for cryogenic propellant combinations of liquid hydrogen and liquid oxygen. This effect is substantial at higher A_e/A_t.

Given the values of p_{0n}/p_e and γ, it is a straightforward substitution to calculate the value of A_e/A_t, Eq. 3.62. But, many times, it is required to calculate the values of p_{0n}/p_e for the given values of A_e/A_t, Eq. 3.63. This requires a simple iterative procedure [see Example 3.3]. In such situations, Fig. 3.6 becomes handy to have an initial estimate for p_{0n}/p_e.

Similar to the derivation of Eq. 3.62, we can derive equations for area ratio in terms of Mach number at an arbitrary location x and exit section. The derived equations are

$$\frac{A_x}{A_t} = \frac{1}{M_x}\left[\frac{2}{\gamma+1}\left(1+\frac{\gamma-1}{2}M_x^2\right)\right]^{\frac{\gamma+1}{2(\gamma-1)}} \tag{3.64}$$

For the exit section,

$$\frac{A_e}{A_t} = \frac{1}{M_e}\left[\frac{2}{\gamma+1}\left(1+\frac{\gamma-1}{2}M_e^2\right)\right]^{\frac{\gamma+1}{2(\gamma-1)}} \tag{3.65}$$

Note for the given A_e/A_t, Eq. 3.64 or Eq. 3.65 has two solutions of Mach numbers, the first one the subsonic value and the other the supersonic value. These values can be easily obtained through the iterative procedures explained in Example 3.3. For rocket nozzle flows, we are considering the flow in the diverging portion of the convergent–divergent nozzle. Therefore, the supersonic Mach number is the applicable solution. Also note that in Eq. 3.64, A_x/A_t becomes unity for $M_x = 1$, that is, for the requirement at the choked throat of the nozzle.

3.5.4 Nozzle-Flow Separation

Now let us look at the situation at which the nozzle flow may not flow full. To understand the phenomenon of nozzle-flow separation, let us imagine a rocket with its engine of constant chamber pressure and fixed nozzle configuration is descending from a high altitude where the ambient pressure p_a is lower than the nozzle-exit plane pressure p_e. For this condition that the exit plane pressure is greater than ambient pressure $p_e > p_a$ (under-expanded), the pressure equalization of the exiting jet with the atmosphere is only "down the hill" process (**supporting pressure gradient**). The flow has a system of oblique expansion waves at the nozzle exit. This has no effect on the flow inside the nozzle because the supersonic flow is "ignorant" of what is happening downstream. This "comfortable" situation continues during the descent until the exit plane pressure is *slightly less* than ambient pressure $p_e < p_a$ (over-expanded). The nozzle still flows full but the pressure equalization process has become the "up the hill" process (**adverse pressure gradient**). The exiting flow with pressure p_e less than the ambient pressure p_a has to climb to the ambient pressure through an oblique shock. However, the shock is of small angle, and hence weak and of small pressure discontinuity. As the rocket descent continues, the degree of over-expansion increases, p_e is further less than p_a, and the oblique shock angle grows. The resulting strong shock can enter the nozzle and interact with the nozzle-wall boundary layer. Even in supersonic flow, the boundary layer close to the wall is subsonic and the "message" of the large adverse pressure discontinuity downstream is conveyed upstream of the wall-shock interaction point. At a condition that p_a is *substantially*

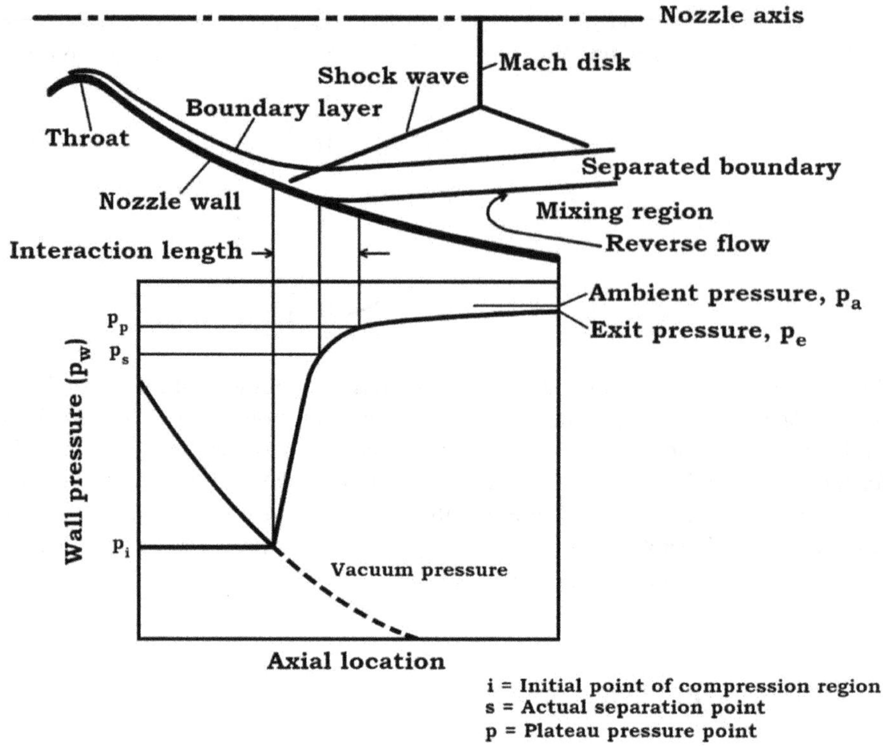

Fig. 3.7 Flow field and pressure distribution in an over-expanded nozzle with flow separation [2]

greater than p_e the adverse pressure discontinuity is no longer manageable without separation. The flow separates at a point upstream of the nozzle exit where the "new" exit plane pressure and the corresponding oblique shock pressure discontinuity is within the manageable limits. Generally, the new exit plane pressure p_e is found to be slightly less than the ambient pressure p_a. The phenomenon of the nozzle-flow separation is depicted in Fig. 3.7.

Rocket engines, from launch to final destination, are required to operate at varying altitudes. And, ambient pressure reduces with altitude. Therefore, the condition of nozzle flowing full is an important operational requirement. For simple conical nozzles whose half cone angle α_d is around 15°, it has been experimentally observed that the nozzle flow may separate for values of p_a/p_e greater than or equal to 2.5 [3, 4]. Therefore, for the full flow of the conical nozzles of α_d of around 15°, the condition is

$$\frac{p_a}{p_e} < 2.5 \tag{3.66}$$

Here p_e is the value calculated using isentropic one-dimensional flow assumptions, Eq. 3.63. Reference [5] presented a correlation of the results obtained by different experimental studies on conical nozzles having α_d around 15°. The studies included liquid as well as solid propellant engines. The correlation for the p_a/p_e, at or above which there is a possibility of separation, can be put in the form

$$\frac{p_a}{p_e} = \frac{1}{\left\{1 - \frac{p_{0n}}{p_a}\left[0.00854 + 0.08142exp\left(-0.0579\frac{p_{0n}}{p_a}\right)\right]\right\}} \tag{3.67}$$

As also confirmed by many other later studies on different types of nozzles, Eq. 3.67 indicates that higher the value of p_{0n}/p_a higher is the limit on p_a/p_e for separation.

Even as conical nozzle-flow separation can be taken safely for $p_a/p_e > 2.5$, it has been observed that **contoured nozzles** can tolerate still higher pressure difference and the flow generally separates for p_a/p_e greater than 3.5 where p_e is again obtained by one-dimensional considerations through Eq. 3.63. For the same area ratio A_e/A_t, the actual wall pressure at the nozzle exit of a contoured nozzle is observed to be more than that calculated using one-dimensional flow assumptions, Eq. 3.63. As discussed previously, the phenomenon of nozzle-flow separation is very much dependent on the wall static pressure. Although a contoured nozzle will produce an exit flow, which is considerably nonuniform both in velocity and in flow direction [6], its average velocity at the exit will be more uniform as well as parallel to the nozzle axis than that of a conical nozzle. We will discuss contoured nozzles in Sect. 3.9.

By an elaborate compilation of flow separation data for different nozzles, it is seen that the flow generally separates for p_a/p_{e_w} greater than M_e, where p_{e_w} is the wall pressure, [7]. As indicated previously, the actual wall pressure near the nozzle exit of a contoured nozzle is observed to be more than p_e calculated using one-dimensional flow assumptions. On the whole however, the nozzle-flow separation of the over-expanded nozzles being a complex phenomenon, the separation limit of p_a/p_e at 2.5 for conical nozzles and 3.5 for contoured nozzle appears to be a safe assumption unless otherwise one has a higher value through experiments conducted for the specific nozzle configuration at its typical operating conditions. Example 3.3, apart from evaluating the motor dimensions and other characteristics, explains the iteration procedure required to calculate nozzle pressure ratio p_{0n}/p_e given the nozzle area ratio A_e/A_t and checks the nozzle separation criterion.

Example 3.3 A solid propellant rocket motor, a typical of one used as a strap-on booster to a satellite launch vehicle, has the following data. Average chamber pressure = 48 bar. Sea level thrust (average) = 426 kN. Specific impulse at sea level = 2324 Ns/kg. Motor outer diameter = 1020 mm. Nozzle area ratio = 9.29. Total burning time = 56 s. This motor uses an aluminized ammonium perchlorate hydroxyl-terminated polybutadiene propellant whose combustion products are calculated to have the following properties. Ratio of specific heats $\gamma = 1.25$. Molecular mass $\overline{m}$ = 25 kg/kgmol. Adiabatic flame temperature $T_{0c} = 3430$ K.

Assume (1) the nozzle entry stagnation pressure = chamber stagnation pressure and (2) the nozzle flow is isentropic.

Estimate the following: (1) total propellant mass, (2) nozzle throat diameter, (3) nozzle pressure ratio p_{0n}/p_e, (4) nozzle-exit plane diameter and check whether it is fouling the motor outer diameter, and (5) thrust and specific impulse in vacuum. Assuming that the nozzle is of simple conical configuration, check whether there is any danger of nozzle-flow separation for the sea level launch.

Solution

For the given nozzle area ratio, we have to calculate the pressure ratio. This has to be obtained through iteration. Otherwise, the solution is obtained through straightforward application of derived equations.

Analysis

Thrust at sea level = 426 kN
Sea level specific impulse = 2324 Ns/kg
Therefore, the equivalent exhaust velocity $\overline{u}_e = 2324$ m/s

Average propellant mass-flow rate

$$F = \dot{m}\overline{u}_e \tag{3.17}$$

$$426000 = \dot{m} \times 2324 \Rightarrow \dot{m} = 183.3 \text{ kg/s}$$

Burning time = 56 s
Total propellant mass $m_p = 56 \times 183.3 = 10265$ kg.
(After accounting for possible losses near the burnout, the actual propellant mass may be a few percentages more than this value.)

$$\dot{m} = p_{0n} A_t \frac{\sqrt{\gamma}\left(\frac{2}{\gamma+1}\right)^{(\gamma+1)/[2(\gamma-1)]}}{\sqrt{RT_{0c}}} \tag{3.53}$$

Assuming nozzle entry stagnation pressure is equal to chamber stagnation pressure: $p_{0n} = p_{0c}$

$$183.3 = 4.8 \times 10^6 A_t \frac{\sqrt{1.25}\left(\frac{2}{2.25}\right)^{2.25/0.5}}{\sqrt{\frac{8314.51}{25}3430}}$$

$$A_t = 0.06198 \text{ m}^2 \Rightarrow D_t = 0.281 \text{ m}$$

$$\frac{p_{0n}}{p_e} = \left\{ \frac{A_e}{A_t} \sqrt{\frac{2\gamma}{\gamma - 1} \left[1 - \left(\frac{p_e}{p_{0n}} \right)^{(\gamma-1)/\gamma} \right]} / \Gamma \right\}^{\gamma} \tag{3.63}$$

$$\frac{p_{0n}}{p_e} = \left\{ 9.29 \sqrt{\frac{2.5}{0.25} \left[1 - \left(\frac{p_{0n}}{p_e} \right)^{-0.2} \right]} / 0.6581 \right\}^{1.25}$$

$$\frac{p_{0n}}{p_e} = 115.394 \left[1 - \left(\frac{p_{0n}}{p_e} \right)^{-0.2} \right]^{0.625}$$

This is of the form

$$x = f(x)$$

In many equations where the solution is to be obtained by iteration, we can bring the equation to the form $x = f(x)$. Solution is obtained when the value x is equal to the value of $f(x)$. Assume a trial value of x. Substitute this value into $f(x)$. If this value is not the assumed trial value, re-substitute this value into $f(x)$ until convergence is reached. Most equations are amenable to this procedure to give a fast convergence. If the solution is not converging bring the equation to the form

$$f(x) - x = 0$$

and resort to the method of regula falsi or the method of bisection. In the present problem, the solution is obtained from the form $x = f(x)$ with the initial estimate of $x = 70$. The converged solution to the above equation is

$$\left(\frac{p_{0n}}{p_e} \right) = 82.67$$

Therefore, the nozzle-exit plane pressure at sea level

$$p_e = \frac{4.8 \times 10^6}{82.67} = 58065.3 \text{ Pa}$$

For the nozzle to flow full

$$\frac{p_a}{p_e} < 2.5 \tag{3.66}$$

For the present rocket motor

$$\frac{p_a}{p_e} = \frac{101325}{58065.3} = 1.74 < 2.5$$

Therefore, there is no danger of nozzle-flow separation.

Nozzle-exit area

$$A_e = 9.29 \times 0.06198 = 0.5758 \text{ m}^2 \Rightarrow D_e = 0.856 \text{ m}$$

Therefore, even after providing some allowance for the nozzle-wall thickness, the calculated nozzle-exit diameter of 0.856 m is not fouling the motor outer diameter of 1.02 m.

$$u_e = \sqrt{\frac{2\gamma}{\gamma-1} R T_{0c} \left[1 - \left(\frac{p_e}{p_{0n}}\right)^{\frac{\gamma-1}{\gamma}}\right]} = \sqrt{\frac{2\gamma}{\gamma-1} \frac{R_u}{\overline{m}} T_{0c} \left[1 - \left(\frac{p_e}{p_{0n}}\right)^{\frac{\gamma-1}{\gamma}}\right]} \tag{3.60}$$

$$u_e = \sqrt{\frac{2.5}{0.25} \frac{8314.51}{25} 3430 \left[1 - \left(\frac{1}{82.67}\right)^{\frac{0.25}{1.25}}\right]} = 2586.5 \text{m/s}$$

$$F = \dot{m} u_e + A_e (p_e - p_a) \tag{3.15}$$

$$F = 180.7 \times 2586.45 + 0.5758\,(58065.3 - 101325)$$
$$= 467372 - 24909 = 442463 \text{ N}$$

Under vacuum condition,

$$F_{vac} = 180.7 \times 2586.45 + 0.5758\,(58065.3 - 0)$$
$$= 467372 + 33434 = 500806 \text{ N}$$

Because of one-dimensional idealization the exit velocity and the corresponding thrust are over-estimated. However, we can calculate the thrust augmentation due to vacuum operation by taking the ratio of the calculated thrusts. Therefore the thrust at vacuum,

$$F_{vac} = \frac{500806}{442463} \times 426000 = 1.131 \times 426000 = 482172 \text{ N}$$

Specific impulse at vacuum

$$I_{sp_{vac}} = \frac{482172}{183.3} = 2631\ Ns/kg$$

Discussion
Due to the boundary layer and other real flow effects the actual nozzle dimensions will be more than the values calculated under the idealized assumptions. The maximum mass flux occurs at the nozzle throat and the material to be used in the throat region has to be of heat and erosion resistant. We will consider the nozzle throat materials to be adopted in Sect. 3.11.

3.5.5 Thrust Coefficient

We have seen that the thrust of a rocket engine is given by,

$$F = \dot{m}u_e + A_e\left(p_e - p_a\right) \tag{3.15}$$

The mass-flow rate of the choked nozzle is

$$\dot{m} = p_{0n}A_t\left(\frac{\Gamma}{\sqrt{\dfrac{R_u}{\overline{m}}T_{0c}}}\right) \tag{3.55}$$

And, the nozzle-exit velocity,

$$u_e = \sqrt{\frac{2\gamma}{\gamma-1}RT_{0c}\left[1-\left(\frac{p_e}{p_{0n}}\right)^{\frac{\gamma-1}{\gamma}}\right]} = \sqrt{\frac{2\gamma}{\gamma-1}\frac{R_u}{\overline{m}}T_{0c}\left[1-\left(\frac{p_e}{p_{0n}}\right)^{\frac{\gamma-1}{\gamma}}\right]} \tag{3.60}$$

Substituting Eqs. 3.55 and 3.60 into Eq. 3.15,

$$F = p_{0n}A_t\left\{\Gamma\sqrt{\frac{2\gamma}{\gamma-1}\left[1-\left(\frac{p_e}{p_{0n}}\right)^{\frac{\gamma-1}{\gamma}}\right]} + \frac{A_e}{A_t}\left(\frac{p_e}{p_{0n}} - \frac{p_a}{p_{0n}}\right)\right\} \tag{3.68}$$

The thrust coefficient is defined as

$$C_F \equiv \frac{F}{p_{0n}A_t} \tag{3.69}$$

Therefore, Eq. 3.68 can be written as

$$C_F = \Gamma \sqrt{\frac{2\gamma}{\gamma - 1}\left[1 - \left(\frac{p_e}{p_{0n}}\right)^{\frac{\gamma-1}{\gamma}}\right]} + \frac{A_e}{A_t}\left(\frac{p_e}{p_{0n}} - \frac{p_a}{p_{0n}}\right) \tag{3.70}$$

Under the adapted condition $p_e = p_a$, the pressure difference term in Eq. 3.70 vanishes. The thrust coefficient under adapted condition is known as **characteristic thrust coefficient** C_F^0. That is,

$$C_F^0 = \Gamma \sqrt{\frac{2\gamma}{\gamma - 1}\left[1 - \left(\frac{p_e}{p_{0n}}\right)^{\frac{\gamma-1}{\gamma}}\right]} \tag{3.71}$$

Then,

$$C_F = C_F^0 + \frac{A_e}{A_t}\left(\frac{p_e}{p_{0n}} - \frac{p_a}{p_{0n}}\right) \tag{3.72}$$

Under the adapted condition that the nozzle is flowing full, the nozzle pressure ratio p_{0n}/p_e is constant for the fixed nozzle area ratio A_e/A_t, see Eq. 3.62. Therefore, under the above conditions, C_F^0 is also a constant immaterial of changes in p_{0n} as well as altitude. The only variable in Eq. 3.70 or 3.72 is p_a/p_{0n} whose value may change due to the variation in chamber stagnation pressure and/or the altitude of operation.

When the rocket engine is operating in vacuum ($p_a = 0$), the corresponding thrust coefficient in known as **vacuum thrust coefficient** and is given by

$$C_{F_{vac}} = C_F^0 + \frac{A_e}{A_t}\left(\frac{p_e}{p_{0n}}\right) \tag{3.73}$$

For the fixed nozzle configuration, by Eq. 3.73, note that for any change in chamber pressure $C_{F_{vac}}$ remains constant.

3.5.6 Maximum Thrust

At a constant altitude of operation ($p_a = constant$) and with a fixed p_{0n}, the maximum thrust is obtained when the nozzle is adapted. In order to prove this point, let us consider a hypothetical rocket engine, the nozzle area ratio A_e/A_t of which can be increased continuously from an initial value of under-expanded condition. The change of A_e/A_t will bring about a corresponding change in p_{0n}/p_e [see Eq. 3.62]. This variable nozzle, that is initially under-expanded, will go into an over-expanded situation after passing through the adapted condition. From Eq. 3.68 or 3.70, we see that the hypothetical engine is going to deliver different thrusts at different values of

A_e/A_t. The question of interest here is when do we get the maximum thrust for this constant-altitude operation. Let us differentiate Eq. 3.70 with respect to p_e/p_{0n} and equate the result to zero.

$$\frac{dC_F}{d\left(\frac{p_e}{p_{0n}}\right)} = \frac{-\Gamma}{\left(\frac{p_e}{p_{0n}}\right)^{\frac{1}{\gamma}}\sqrt{\frac{2\gamma}{\gamma-1}\left[1-\left(\frac{p_e}{p_{0n}}\right)^{\frac{\gamma-1}{\gamma}}\right]}} + \frac{A_e}{A_t} + \left(\frac{p_e}{p_{0n}} - \frac{p_a}{p_{0n}}\right)\frac{d\frac{A_e}{A_t}}{d\left(\frac{p_e}{p_{0n}}\right)}$$

Recall A_e/A_t is given by Eq. 3.62. Therefore,

$$\frac{dC_F}{d\left(\frac{p_e}{p_{0n}}\right)} = \left(\frac{p_e}{p_{0n}} - \frac{p_a}{p_{0n}}\right)\frac{d\frac{A_e}{A_t}}{d\left(\frac{p_e}{p_{0n}}\right)}$$

In the above equation, we note that only the pressure difference term within the brackets can be zero. Therefore at the constant altitude of operation with a fixed p_{0n} the maximum thrust is obtained when

$$\left(\frac{dC_F}{d\left(p_e/p_{0n}\right)}\right)_{(p_a\&p_{0n})_{constant}} = 0 \Longrightarrow p_e = p_a \tag{3.74}$$

The condition of maximum thrust can also be proved by the application of the momentum equation to the thrust equation.

$$F = \dot{m}u_e + A_e\left(p_e - p_a\right) \tag{3.15}$$

For the constant p_{0n} and throat area A_t, the mass-flow rate is constant [Eq. 3.55]. Taking the differential of the thrust for the constant mass-flow rate and constant altitude ($p_a = constant$), Eq. 3.15 becomes

$$dF = \dot{m}du_e + \left(p_e - p_a\right)dA_e + A_e dp_e = 0 \tag{3.75}$$

By the momentum conservation of the isentropic flow through the elemental nozzle extension that resulted due to the elemental variation of nozzle area ratio, Fig. 3.8,

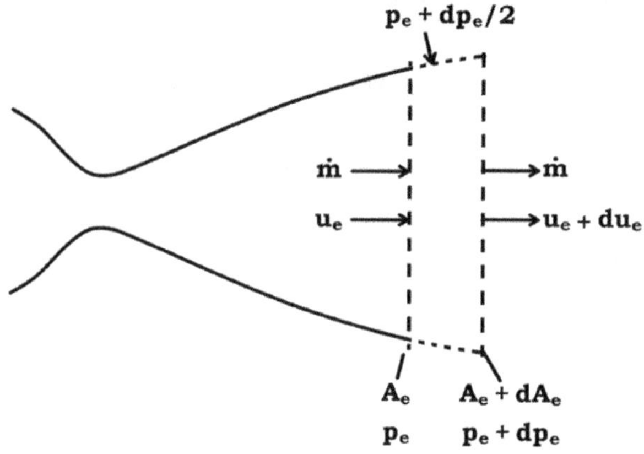

Fig. 3.8 Momentum conservation in the nozzle

$$\dot{m}u_e + p_e A_e + \left(p_e + \frac{dp_e}{2}\right) dA_e$$
$$= \dot{m}\,(u_e + du_e) + (p_e + dp_e)\,(A_e + dA_e) \tag{3.76}$$

Simplifying the above equation and neglecting the second-order terms,

$$\dot{m}du_e = -A_e dp_e \tag{3.77}$$

Substituting Eq. 3.77 in Eq. 3.75, we get

$$dF_{(p_a \& p_{0n})_{constant}} = (p_e - p_a)\,dA_e = 0$$

Therefore,

$$\left(\frac{dF}{dA_e}\right)_{(p_a \& p_{0n})_{constant}} = 0 \Rightarrow p_e = p_a \tag{3.78}$$

As a third proof for the maximum thrust condition, following [8], a very interesting and simple physical reasoning can be put forth to prove the condition for maximum thrust. Similar to the fundamental principle that the lift of an airfoil is simply the vertical component of the pressure integral over the surface of the airfoil, the thrust produced by a rocket engine is the resultant of static pressures acting on the internal and external surfaces of the engine. As shown in Fig. 3.9, suppose there is a certain nozzle-exit area for which $p_e = p_a$. If the nozzle is extended further, the static pressure on the extended *internal surface* will be less than the atmospheric pressure on the *external surface*. The net effect is that the extended surface will give a negative thrust component. Similarly, if the nozzle is cut off upstream of the exit plane where $p_e = p_a$, the engine loses an area which has been giving a positive thrust component. Hence we conclude that the thrust is maximum when $p_e = p_a$.

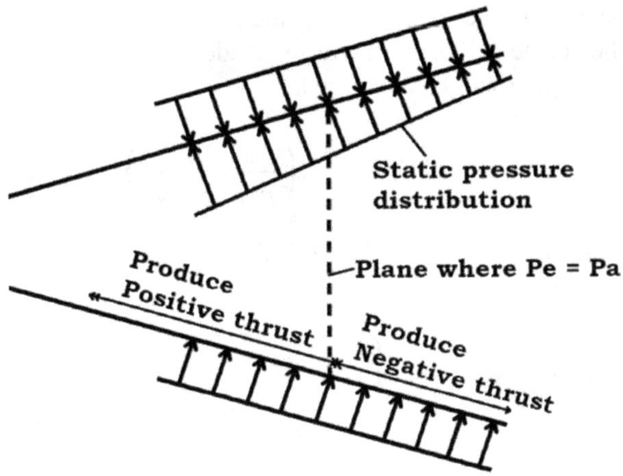

Fig. 3.9 Pressure distribution on internal and external surfaces of diverging portion of nozzle [8]

3.5.7 Thrust Variation

As a continuation of the previous discussion on maximum thrust, let us now consider a rocket engine operating at an altitude h with a fixed A_e/A_t and a constant nozzle stagnation pressure p_{0n}. Furthermore, the engine nozzle is adapted for the altitude h to give the maximum thrust at h. We will analyze two different maneuvers for the engine. As the first maneuver, let the engine climb to a higher altitude, say h^+ where the nozzle will become under-expanded with a lower ambient pressure and the C_F of the engine will be increased by the quantity

$$\Delta C_F = \frac{A_e}{A_t}\left(\frac{p_{e_h}}{p_{0n}} - \frac{p_{a_{h+}}}{p_{0n}}\right) = \frac{A_e}{A_t}\left(\frac{p_{a_h}}{p_{0n}} - \frac{p_{a_{h+}}}{p_{0n}}\right) \tag{3.79}$$

And, at the limit, that is the engine climbing to an altitude where vacuum prevails the ΔC_F becomes

$$\Delta C_{F_{vac}} = \frac{A_e}{A_t}\left(\frac{p_{e_h}}{p_{0n}}\right) = \frac{A_e}{A_t}\left(\frac{p_{a_h}}{p_{0n}}\right) \tag{3.80}$$

The C_F correspondingly becomes $C_{F_{vac}}$, Eq. 3.73. Likewise the thrust of the engine will be enhanced to the limit F_{vac}.

$$F_{vac} = p_{0n} A_t \left[C_F^0 + \frac{A_e}{A_t}\left(\frac{p_{e_h}}{p_{0n}}\right)\right] \tag{3.81}$$

Or,

$$F_{vac} = p_{0n} A_t \left\{\Gamma \sqrt{\frac{2\gamma}{\gamma - 1}\left[1 - \left(\frac{p_{e_h}}{p_{0n}}\right)^{\frac{\gamma-1}{\gamma}}\right]} + \frac{A_e}{A_t}\left(\frac{p_{e_h}}{p_{0n}}\right)\right\} \tag{3.82}$$

As the second maneuver, let the engine descend to a lower altitude, say h^- where the nozzle will become over-expanded with a higher ambient pressure and the C_F of the engine will be decreased.

$$\Delta C_F = \frac{A_e}{A_t}\left(\frac{p_{e_h}}{p_{0n}} - \frac{p_{a_{h^-}}}{p_{0n}}\right) = \frac{A_e}{A_t}\left(\frac{p_{a_h}}{p_{0n}} - \frac{p_{a_{h^-}}}{p_{0n}}\right) \tag{3.83}$$

Note $p_{a_{h^-}} > p_{e_h}$ and hence ΔC_F is negative. At the limit, the thrust at sea level becomes

$$F_{sea\ level} = p_{0n} A_t \left\{ \Gamma \sqrt{\frac{2\gamma}{\gamma - 1}\left[1 - \left(\frac{p_{e_h}}{p_{0n}}\right)^{\frac{\gamma-1}{\gamma}}\right]} + \frac{A_e}{A_t}\left(\frac{p_{e_h} - p_{a_0}}{p_{0n}}\right)\right\} \tag{3.84}$$

For conical nozzles, we have learned that the over-expanded condition may lead to nozzle-flow separation if the ambient pressure is equal to or greater than 2.5 times the nozzle-exit plane pressure [Eq. 3.66]. In writing Eq. 3.84 we have implicitly assumed that the extent of over-expansion at sea level is not up to the separation margin.

By the two altitude maneuvers that we considered previously we note that for a rocket engine operating with constant p_{0n} and A_e/A_t, C_F always increases from sea level to its absolute maximum in vacuum and in this process for a rocket engine adapted at an altitude h the nozzle flow may separate at a lower altitude if the limit of the over-expansion is reached.

3.5.8 *Thrust Control*

The required variation of the thrust with respect to time—thrust–time trace or thrust–time history—is the basic specification to a rocket propulsion system [Figs. 3.4 and 3.5]. As a ready reference to the present discussion, Eq. 3.68 is repeated below. It is clear from this equation that the thrust can be varied by varying p_{0n}, A_t, p_{0n}/p_e, and γ. Taking the last one first, the value of γ cannot be a variable under practical considerations. Being largely populated with poly-atomic gases, the combustion products of most propellants have the values of γ varying from about 1.2 to 1.3. And, once a propellant for the engine is selected, the value of γ is more or less fixed. Furthermore, even under theoretical considerations, the effect of γ on the thrust variation is very small, Eq. 3.68. Considering next p_{0n}/p_e, the variation of this quantity means the variation of A_e/A_t adopting an involved variable configuration nozzle, say with the nozzle extensions for the fixed A_t or throat-plug insertion for the fixed A_e, Fig. 3.10. However, for upper stage engines, once the rocket vehicle is in hard vacuum, the A_e/A_t is enhanced by adopting nozzle extensions, more so to enhance the thrust and the specific impulse than for the thrust control—see Fig. 2.19, the engine with the retracted nozzle extension for the ground display. Next, with respect to A_t variation, we see from Eq. 3.68, the thrust is linearly proportional to A_t. But, here again we end up with the variable configuration nozzle. However, A_t can be varied through a traversing nozzle plug or through a secondary fluid injection near the throat, both involving mechanical and control complications. Therefore, the thrust control through the variation of nozzle configuration, either by the variation of A_e for the fixed A_t or by the variation of A_t for the fixed A_e, is acceptable only under special applications. The schematic sketches for enhancing A_e/A_t through a nozzle extension, a nozzle plug insertion, or a secondary fluid injection are given in Fig. 3.10. We may note here that the control on A_t alters both A_e/A_t and p_{0n}. Finally as the fourth one on thrust control, let us consider p_{0n}, perhaps a powerful control variable. For a fixed-configuration engine, the thrust can be varied by changing the chamber stagnation pressure, which can be effected by the propellants injected into the chamber in the case of liquid propellant engines or by the burning area of the propellant grain in the case of solid propellant motors. In addition to all these, for the fixed nozzle and constant stagnation pressure p_{0n}, the thrust increases naturally with an increase in altitude (represented by the decrease in p_a).

$$F = p_{0n} A_t \left\{ \Gamma \sqrt{\frac{2\gamma}{\gamma - 1}\left[1 - \left(\frac{p_e}{p_{0n}}\right)^{\frac{\gamma-1}{\gamma}}\right]} + \frac{A_e}{A_t}\left(\frac{p_e}{p_{0n}} - \frac{p_a}{p_{0n}}\right)\right\} \tag{3.68}$$

In consideration of our previous discussions on the maximum thrust and the thrust variation, and the current treatment on the thrust control, three plots on the variations of thrust coefficient C_F (or C_T) with A_e/A_t with parametric values of p_{0n}/p_a are given in Figs. 3.11, 3.12, and 3.13 for three typical values of γ. The parametric contour of a p_{0n}/p_a represents the performance of a hypothetical rocket engine flying with a fixed p_{0n} at an altitude of pressure p_a with a widely varying A_e/A_t nozzle: through the under-expansion ($p_e > p_a$), the adapted condition ($p_e = p_a$), the over-expansion ($p_a < p_e$), the limit of nozzle-flow separation in conical nozzles ($p_a/p_e = 2.5$), and finally the limit of nozzle-flow separation in contoured nozzles ($p_a/p_e = 3.5$). The upward movement on a constant value of A_e/A_t represents the performance of an ascending engine of fixed configuration nozzle with a constant p_{0n}: through the complete nozzle-flow separation, the nozzle-flow separation limits, the safe over-expansion, the adapted condition, the under-expansion, and finally the limit of maximum thrust in vacuum. The plots are useful for quickly understanding the rocket performance at different operating conditions and sometimes become handy at arriving at the first-cut estimation of nozzle configuration.

Let us consider a general operating condition of a fixed-configuration rocket engine with respect to the reference altitude h^* where the ambient pressure is p_a^*. Under this operating condition, we give the provision that the nozzle may or may not be adapted at h^*. The engine has a constant A_e/A_t with a fixed A_t. Therefore, as long as its nozzle flows full, it has a fixed nozzle pressure ratio p_{0n}^*/p_e^* [Eq. 3.62]. The thrust coefficient under reference condition is given by

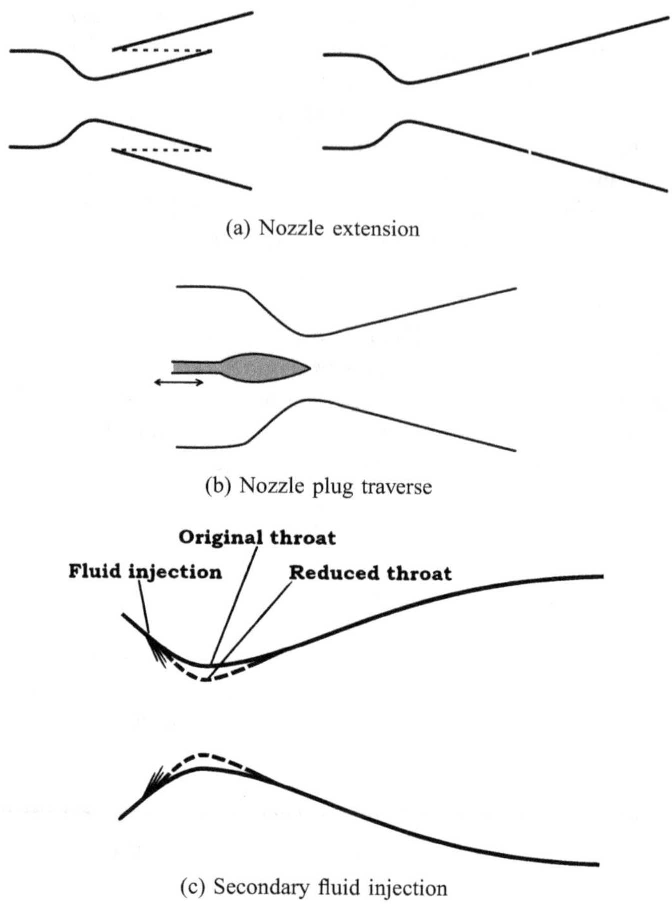

Fig. 3.10 Thrust controls through nozzle extension, nozzle plug traverse, and secondary fluid injection

$$
\begin{aligned}
C_F^* &= \Gamma \sqrt{\frac{2\gamma}{\gamma - 1}\left[1 - \left(\frac{p_e^*}{p_{0n}^*}\right)^{\frac{\gamma-1}{\gamma}}\right]} + \frac{A_e}{A_t}\left(\frac{p_e^*}{p_{0n}^*} - \frac{p_a^*}{p_{0n}^*}\right) \\
&= C_F^{0\,*} + \frac{A_e}{A_t}\left(\frac{p_e^*}{p_{0n}^*} - \frac{p_a^*}{p_{0n}^*}\right)
\end{aligned}
\tag{3.85}
$$

The thrust of the engine is being varied by the variation of nozzle stagnation pressure and altitude. At any nozzle stagnation pressure p_{0n} and arbitrary altitude (represented by the corresponding ambient pressure p_a), and under the condition of nozzle full flow,

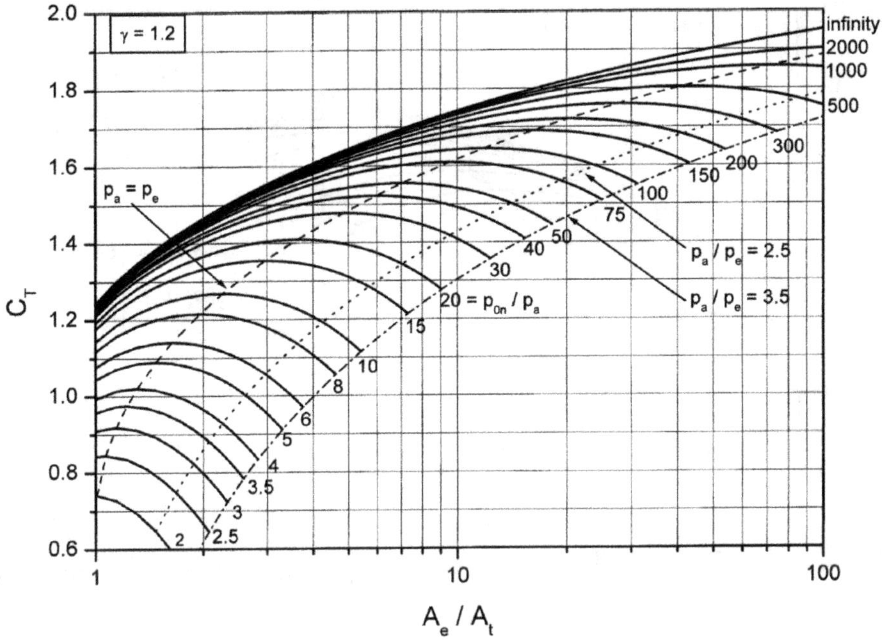

Fig. 3.11 Thrust coefficient versus nozzle area ratio; $\gamma = 1.2$

$$
\begin{aligned}
C_F &= \Gamma\sqrt{\frac{2\gamma}{\gamma-1}\left[1-\left(\frac{p_e^*}{p_{0n}^*}\right)^{\frac{\gamma-1}{\gamma}}\right]}+\frac{A_e}{A_t}\left(\frac{p_e^*}{p_{0n}^*}-\frac{p_a}{p_{0n}}\right) \\
&= C_F^{0\,*}+\frac{A_e}{A_t}\left(\frac{p_e^*}{p_{0n}^*}-\frac{p_a}{p_{0n}}\right)
\end{aligned}
\tag{3.86}
$$

Thus for the ratio,

$$
\frac{C_F}{C_F^*}=1+\frac{1}{C_F^*}\frac{A_e}{A_t}\left(\frac{p_a^*}{p_{0n}^*}-\frac{p_a}{p_{0n}}\right) \tag{3.87}
$$

Note that although nozzle stagnation pressure is changing, the nozzle pressure ratio remains the same at the reference value. Taking the ratio of the thrust at any altitude and any p_{0n} to the reference thrust (Eq. 3.69),

$$
\frac{F}{F^*}=\frac{p_{0n}}{p_{0n}^*}\frac{C_F}{C_F^*} \tag{3.88}
$$

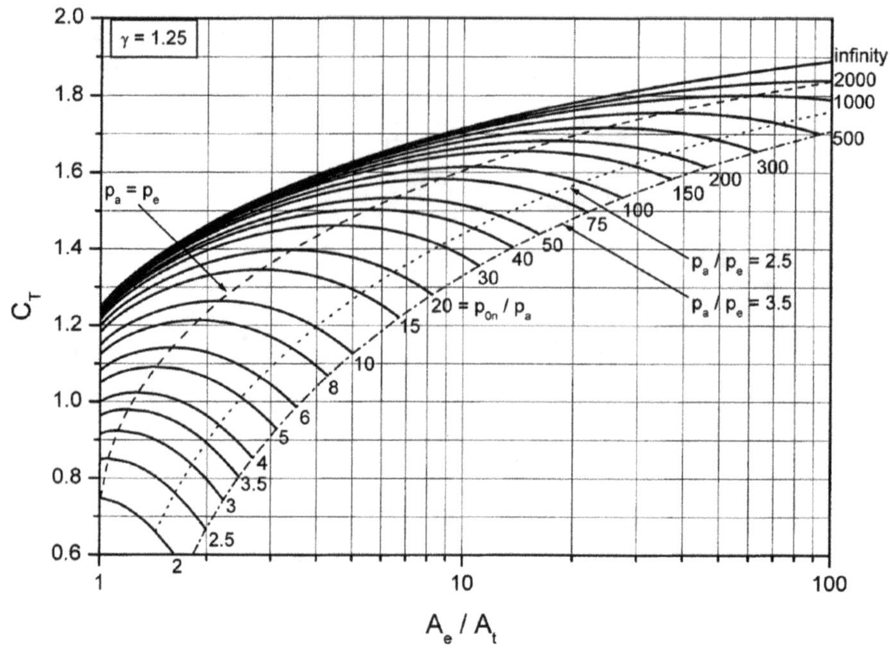

Fig. 3.12 Thrust coefficient versus nozzle area ratio; $\gamma = 1.25$

Substituting Eq. 3.87 into Eq. 3.88,

$$\frac{F}{F^*} = \frac{p_{0n}}{p_{0n}^*}\left[1 + \frac{1}{C_F^*}\frac{A_e}{A_t}\left(\frac{p_a^*}{p_{0n}^*} - \frac{p_a}{p_{0n}}\right)\right] \tag{3.89}$$

Or,

$$\frac{F}{F^*} = \frac{p_{0n}}{p_{0n}^*}\left[1 + \frac{1}{C_F^*}\frac{A_e}{A_t}\left(\frac{p_a^*}{p_{0n}^*}\right)\right] - \frac{1}{C_F^*}\frac{A_e}{A_t}\frac{p_a}{p_{0n}^*} \tag{3.90}$$

Equation 3.90 indicates that, for the engine with a fixed nozzle configuration, the thrust *at a particular altitude* (p_a = constant) varies linearly with respect to the nozzle stagnation pressure with a negative intercept due to

$$-\frac{1}{C_F^*}\frac{A_e}{A_t}\frac{p_a}{p_{0n}^*}$$

As the altitude increases (decreases), the magnitude of the negative intercept decreases (increases). And, in vacuum the intercept becomes zero. Therefore in vacuum, Eq. 3.90 becomes

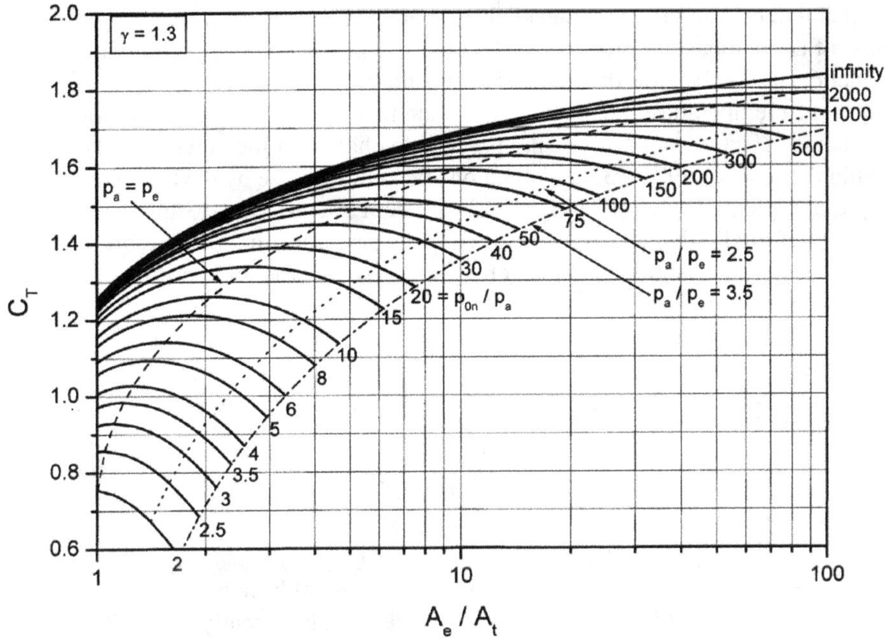

Fig. 3.13 Thrust coefficient versus nozzle area ratio; $\gamma = 1.3$

$$\frac{F_{vac}}{F^*} = \frac{p_{0n}}{p_{0n}^*}\left(1 + \frac{1}{C_F^*}\frac{A_e}{A_t}\frac{p_a^*}{p_{0n}^*}\right) \tag{3.91}$$

The thrust F_{vac} varies linearly with p_{0n} with no danger of nozzle-flow separation.

For the general thrust control on p_{0n}, there will be a limitation on the maximum possible p_{0n} on the points of view of the structural safety and mass enhancement, and the feed-system margins (applicable for liquid propellant engines). Similarly on the minimum possible p_{0n} there will be a limitation on the points of view of the nozzle-flow separation (not applicable in vacuum), the reduction in combustion efficiency, and the feed-system margins. In the case of liquid propellant engines, deep thrust controls to low levels are always followed with combustion instability problems. However, thrust controls to about 20 times the minimum values have been achieved for special applications in engines adopting cryogenic as well as the Earth-storable propellants [9].

We have considered previously the thrust variation with p_{0n} and altitude. Let us now look at the variation of specific impulse. Recalling that we have assumed the ratio of specific heats γ to be constant, by Eqs. 3.39 and 3.55,

$$\frac{I_{sp}}{I_{sp}^*} = \frac{F}{F^*}\frac{p_{0n}^*}{p_{0n}}\left(\frac{\sqrt{T_{0c}/\overline{m}}}{\sqrt{T_{0c}^*/\overline{m}^*}}\right) \tag{3.92}$$

With an increase in pressure, the adiabatic flame temperature T_{0c} and the molecular mass of the combustion products $\overline{m}$ increase slightly. Both the effects are due to the reduction in combustion dissociation. With the increase in pressure, the extent of increase in temperature is a shade higher than that of molecular mass. However the combined effect does not result in appreciable change in the value of the quantity within the brackets of Eq. 3.92—about one percent increase by the twofold increase in p_{0n} at typical engine operating pressures (from 5 to 10 MPa). Similar arguments can be put forth for the reduction in p_{0n} within the practical limits. Therefore the value of the bracketed quantity can be taken as unity for all practical purposes. Equation 3.92 thus becomes

$$\frac{I_{sp}}{I_{sp}^*} \approx \frac{F}{F^*} \frac{p_{0n}^*}{p_{0n}} \tag{3.93}$$

Substituting Eq. 3.89 in Eq. 3.93,

$$\frac{I_{sp}}{I_{sp}^*} \approx 1 + \frac{1}{C_F^*} \frac{A_e}{A_t} \left(\frac{p_a^*}{p_{0n}^*} - \frac{p_a}{p_{0n}} \right) \tag{3.94}$$

The I_{sp}/I_{sp}^* Eq. 3.94 and C_F/C_F^* Eq. 3.87 are identical because we are neglecting the change in T_{0c} and $\overline{m}$ in Eq. 3.92—we will learn subsequently that γ, T_{0c}, and $\overline{m}$ are combustion properties of a propellant, defined under the characteristic parameter c^* that is practically invariant with combustion chamber pressure and understand $I_{sp} = C_F c^*$ [Sec. 3.6]. For a fixed-configuration engine as p_{0n}/p_a is increased (by way of either p_{0n} increase for a fixed altitude, increased altitude for a fixed p_{0n}, or by the combinations) I_{sp} increases—look at any of the Figs. 3.11, 3.12 and 3.13 and track any A_e/A_t vertical. For the operation in vacuum, Eq. 3.94 becomes

$$I_{sp_{vac}} \approx I_{sp}^* \left(1 + \frac{1}{C_F^*} \frac{A_e}{A_t} \frac{p_a^*}{p_{0n}^*} \right) \tag{3.95}$$

For a fixed-configuration rocket engine, Eq. 3.95 indicates that the specific impulse in vacuum is enhanced with respect to the reference value and is held constant, irrespective of p_{0n}. But, this is not so for F_{vac}—inspect Eq. 3.91.

With reference to Eq. 3.89, let us now introduce a quantity known as nozzle separation parameter ψ, defined as,

$$\psi \equiv \frac{p_a}{p_e} \tag{3.96}$$

You may recall our discussion on nozzle-flow separation [Sect. 3.5.4]. With reference to the definition of ψ, when $\psi = 1$, the nozzle flow is adapted, when $\psi < 1$ it is under-expanded and when $\psi > 1$ it is over-expanded. Introducing ψ into Eq. 3.89,

$$\frac{F}{F^*} = \frac{p_{0n}}{p^*_{0n}}\frac{C_F}{C^*_F} = \frac{p_{0n}}{p^*_{0n}}\left[1 + \frac{1}{C^*_F}\frac{A_e}{A_t}\left(\frac{p^*_a}{p^*_{0n}} - \psi\frac{p^*_e}{p^*_{0n}}\right)\right]$$

$$\frac{C_F}{C^*_F} = 1 + \frac{1}{C^*_F}\frac{A_e}{A_t}\left(\frac{p^*_a}{p^*_{0n}} - \psi\frac{p^*_e}{p^*_{0n}}\right) \tag{3.97}$$

Rocket vehicles (or the stages) launched from sea level (or lower altitudes) may reach higher altitudes in a matter of a few seconds. We have seen previously that, given the altitude and p_{0n}, and for a fixed nozzle configuration, the maximum thrust at an altitude is obtained when the nozzle is adapted for that altitude. Noting that a launch vehicle or its stage operates mostly at higher altitudes, in order to take advantage of the maximum thrust condition, there is always a desire to adapt its nozzle for an optimum higher altitude. However, this should not endanger the nozzle-flow separation at lower altitudes. Going even a step further in arriving at the optimum higher altitude operation or towards achieving acceptable performance for a wider range of altitudes, there may be designs to have (1) the aforementioned nozzle extensions, Fig. 3.10; (2) a stepped nozzle or dual nozzle; (3) a trip ring attached to the inside of a nozzle; (4) an ablative or ejectable secondary-nozzle inserts in the nozzle divergence to give a lower A_e/A_t at launch; or (5) a secondary fluid injection system to enable a lower A_e/A_t at launch, Fig. 3.14. The most disturbing aspect of this is that the thrust oscillations that may be caused by the transitory nozzle-flow oscillations from the separated-flow condition to the full-flow condition. In addition, as a more important note, during the transients of establishing full flow from the separated flow, unsymmetrical nozzle flow may prevail to cause the so-called "side loads" to the ascending vehicle. In satellite launch vehicles with many strap-on propulsion units, it is frequently observed that a few among the strap-ons are designed to be of lower A_e/A_ts (to give higher p_e's to avoid nozzle-flow separation) for ignition at launch with the remaining ones of higher A_e/A_ts for a later ignition at a higher altitude. We may note another more serious situation for a launch vehicle having a cluster of strap-on units required to come into operation simultaneously within the operational margins. However, the units may have momentarily different operating conditions—separated flow, transitory flow, and full flow. The vehicle dynamics should take into account all these complex possibilities.

In view of the previous discussion, let us look more closely at the minimum thrust condition under thrust control with variation in altitudes. Let us assign the nozzle-flow separation constraint as

$$\frac{p_a}{p_e} < \psi_0 \tag{3.98}$$

where ψ_0 is the limit on the ratio of p_a/p_e below which there is no danger of nozzle-flow separation. Recall that in Sect. 3.5.4, we have noted that ψ_0 is about 2.5 for conical nozzles and about 3.5 for contoured nozzles. While controlling the thrust by varying p_{0n}, it should be the value for which $p_a/p_e < \psi_0$. This constraint for avoiding the nozzle-flow separation at the reference altitude h^* can be represented as

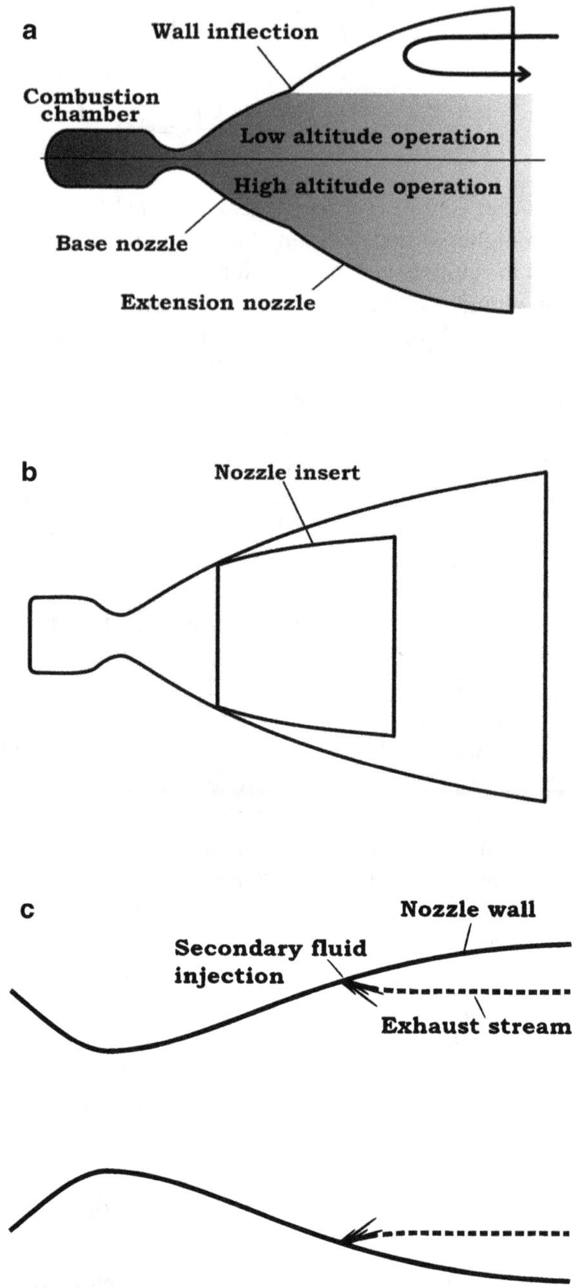

Fig. 3.14 Nozzle designs to have low area ratios at lower altitudes. **a** Stepped nozzle or dual nozzle (adopted from [10]). **b** Nozzle insertion to be ejected at high altitude. **c** Secondary fluid injection

$$p_{0n} \geq \frac{p_a}{\psi_0}\left(\frac{p_{0n}^*}{p_e^*}\right) \tag{3.99}$$

On applying this constraint to Eq. 3.97 we find that the minimum thrust that occurs at the limit of separation at a particular altitude of ambient pressure p_a is

$$\frac{F_{min}}{F^*} = \frac{p_a}{\psi_0}\frac{p_{0n}^*}{p_e^*}\frac{1}{p_{0n}^*}\left[1 + \frac{1}{C_F^*}\frac{A_e}{A_t}\left(\frac{p_a^*}{p_{0n}^*} - \psi_0\frac{p_e^*}{p_{0n}^*}\right)\right] \tag{3.100}$$

Equation 3.100 indicates that as the altitude of operation increases (p_a decreases), the possible minimum thrust decreases—more control is feasible. At the limit of vacuum operation, the possible minimum thrust becomes zero! Of course, this is not practical because there will be another constraint on F_{min} due to possible minimum p_{0n}. In liquid propellant engines, this constraint comes from the point of view of feed-system limitations and the issues caused by unsatisfactory propellant injection and combustion instabilities.

As discussed previously, the applicability of Eq. 3.94 is also subject to the constraint of Eq. 3.98. With the reduction in p_{0n} the I_{sp} slightly reduces. On applying Eq. 3.98 for an altitude of $p_a > 0$, Eq. 3.94 becomes

$$\frac{I_{sp_{min}}}{I_{sp}^*} \approx 1 + \frac{1}{C_F^*}\frac{A_e}{A_t}\left(\frac{p_a^*}{p_{0n}^*} - \psi_0\frac{p_e^*}{p_{0n}^*}\right) \tag{3.101}$$

Equation 3.101 indicates that the minimum specific impulse with respect to the reference specific impulse is always a constant.

Example 3.4 considers the aspects of thrust control limits with respect to nozzle-flow separation for a fixed-configuration engine by the variation of chamber pressure at different altitudes, from sea level to vacuum. Furthermore, it demonstrates the ease of understanding the engine operation by the use of Figs. 3.11, 3.12 and 3.13.

Example 3.4 A rocket engine of the nozzle area ratio $A_e/A_t = 10$ has the following operating conditions: (1) nozzle stagnation pressure = 5 MPa, and (2) reference altitude of operation = 11 km (22700 Pa). Ratio of specific heats $\gamma = 1.25$. Calculate the lowest relative change in thrust that is possible by the reduction in p_{0n}: (1) in vacuum, (2) at the reference altitude, (3) at the altitude where the nozzle becomes adapted, and (4) at sea level. Also calculate the relative change in the specific impulse values at the lowest thrust. Assume $\psi_0 = 2.5$.

Assumptions
As we have stated already the c_p and $\overline{m}$ are assumed to be constant by which γ is constant. Furthermore, because of the isentropic flow assumption, there is no stagnation pressure loss in the nozzle flow.

Analysis
We have

$$\frac{p_{0n}}{p_e} = \left\{ \frac{A_e}{A_t} \sqrt{\frac{2\gamma}{\gamma - 1}\left[1 - \left(\frac{p_{0n}}{p_e}\right)^{-(\gamma-1)/\gamma}\right]} / \Gamma \right\}^{\gamma} \tag{3.63}$$

$$\frac{F}{F^*} = \frac{p_{0n}}{p_{0n}^*}\left(1 + \frac{1}{C_F^*}\frac{A_e}{A_t}\frac{p_a^*}{p_{0n}^*}\right) - \frac{1}{C_F^*}\frac{A_e}{A_t}\frac{p_a}{p_{0n}^*} \tag{3.90}$$

$$\frac{I_{sp}}{I_{sp}^*} \approx \left(1 + \frac{1}{C_F^*}\frac{A_e}{A_t}\frac{p_a^*}{p_{0n}^*}\right) - \frac{A_e}{A_t}\frac{1}{C_F^*}\frac{p_a}{p_{0n}} \tag{3.94a}$$

$$\Gamma = \sqrt{1.25}\left(\frac{2}{2.25}\right)^{2.25/0.5} = 0.6581$$

On substituting the numbers in Eq. 3.63,

$$\frac{p_{0n}^*}{p_e^*} = 126.523\left[1 - \left(\frac{p_{0n}^*}{p_e^*}\right)^{-0.2}\right]^{1.25}$$

By iteration $p_{0n}^*/p_e^* = 91.433$

$$p_{0n}^* = 5 \times 10^6 \text{ Pa} \quad p_e^* = 54685 \text{ Pa}$$

$$C_F^* = 0.6581\sqrt{\frac{2.5}{0.25}\left[1 - \left(\frac{1}{91.433}\right)^{0.2}\right]} + 10\left(\frac{1}{91.433} - \frac{2.27 \times 10^4}{5 \times 10^6}\right)$$
$$= 1.6048 + 0.0640 = 1.6688$$

Therefore at the reference condition $p_a^*/p_e^* = 22700/54685 = 0.4151$. Note that the nozzle is under-expanded.

(1) Under vacuum conditions
$p_a = 0$
Equations 3.90 and 3.94 become

$$\frac{F_{vac}}{F^*} = \frac{p_{0n}}{p_{0n}^*}\left(1 + \frac{1}{C_F^*}\frac{A_e}{A_t}\frac{p_a^*}{p_{0n}^*}\right)$$

$$\frac{I_{sp_{vac}}}{I_{sp}^*} \approx \left(1 + \frac{1}{C_F^*}\frac{A_e}{A_t}\frac{p_a^*}{p_{0n}^*}\right)$$

Substituting the values,

$$\frac{F_{vac}}{F^*} = \frac{p_{0n}}{p^*_{0n}}\left(1 + \frac{1}{1.6688}10\frac{2.27 \times 10^4}{5 \times 10^6}\right) = 1.0272\frac{p_{0n}}{p^*_{0n}}$$

$$\frac{F_{vac}}{F^*} = 1.0272\frac{p_{0n}}{p^*_{0n}}$$

$$\frac{I_{sp_{vac}}}{I^*_{sp}} \approx 1 + \frac{1}{1.6688}10\left(\frac{2.27 \times 10^4}{5 \times 10^6}\right) = 1.0272$$

$$\frac{I_{sp_{vac}}}{I^*_{sp}} \approx 1.0272$$

From the above, it follows that the thrust varies linearly with respect to p_{0n}. Specific impulse is slightly more than the design value and it is constant. On the point of nozzle-flow separation, there is no danger of nozzle-flow separation in the case of operation in vacuum, because p_e is always greater than the ambient pressure that is equal to zero. Nozzle stagnation pressure can be reduced to any practical value without any danger of nozzle-flow separation. However, as discussed earlier, there will be constraints on the nozzle stagnation pressure on the point of view of feed-system limitations and combustion efficiency.

(2) At the reference altitude of 11 km (22700 Pa)

$p^*_a = p_a$

Substituting the values in Eq. 3.90,

$$\frac{F}{F^*} = \frac{p_{0n}}{p^*_{0n}}\left(1 + \frac{1}{C^*_F}\frac{A_e}{A_t}\frac{p^*_a}{p^*_{0n}}\right) - \frac{1}{C^*_F}\frac{A_e}{A_t}\frac{p_a}{p^*_{0n}}$$

$$\frac{F}{F^*} = 1.0272\frac{p_{0n}}{p^*_{0n}} - \frac{1}{1.6688} \times 10 \times \frac{2.27 \times 10^4}{5 \times 10^6}$$

$$\frac{F}{F^*} = 1.0272\frac{p_{0n}}{p^*_{0n}} - 0.0272$$

Here again we find that the thrust is linearly proportional to p_{0n} with the negative intercept of 0.0272.

$$\frac{I_{sp}}{I^*_{sp}} \approx \left(1 + \frac{1}{C^*_F}\frac{A_e}{A_t}\frac{p^*_a}{p^*_{0n}}\right) - \frac{A_e}{A_t}\frac{1}{C^*_F}\frac{p^*_a}{p_{0n}}$$

$$\frac{I_{sp}}{I^*_{sp}} \approx 1.0272 - \frac{10}{1.6688}\frac{2.27 \times 10^4}{p_{0n}} = 1.0272 - \frac{136026}{p_{0n}}$$

At the values of $p_{0n} < p^*_{on}$, $I_{sp} < I^*_{sp}$ and at the values of $p_{0n} > p^*_{on}$, $I_{sp} > I^*_{sp}$. As the reduction in p_{0n} will bring down p_e and if p_e falls below p_a beyond the allowable limit nozzle flow may separate. The limit on nozzle-flow separation determines the minimum p_{0n} that is possible. As long as the full flow is achieved for the nozzle of fixed area ratio,

$$\frac{p^*_{0n}}{p^*_e} = \frac{p_{0n}}{p_e} \Rightarrow \quad \frac{p_{0n}}{p^*_{0n}} = \frac{p_e}{p^*_e}$$

For the nozzle to flow full, $p_a/p_e < 2.5$. Therefore,

$$p_{0n_{min}} = \frac{p^*_{0n}}{p^*_e}\frac{p^*_a}{2.5} = 91.433 \times \frac{2.27 \times 10^4}{2.5} = 0.8302 \times 10^6\,Pa$$

$$\frac{p_{0n_{min}}}{p^*_{0n}} = \frac{0.8302 \times 10^6}{5 \times 10^6} = 0.166$$

However, $p_{0n_{min}}/p^*_{0n} = 0.166$ is only a theoretical value from the point of view of nozzle-flow separation. This appears to be too low and may override the limitations due to feed-system capability and acceptable combustion efficiency level.

$$\frac{F_{min}}{F^*} = 1.0272 \times 0.166 - 0.0272 = 0.1433$$

$$\left(\frac{F_{min}}{F^*}\right)_{h^*} = 0.1433$$

$$\frac{I_{sp_{min}}}{I^*_{sp}} = 1.0272 - \frac{136026}{p_{0n_{min}}} = 1.0272 - \frac{136026}{0.8302 \times 10^6} = 0.8634$$

(3) At the altitude where the nozzle is adapted for the reference conditions
The nozzle-exit plane pressure at the design conditions = 54685 Pa. This pressure exists around 4900 m altitude. As the nozzle is adapted to the reference conditions at this altitude, $p_a = p^*_e$. Equations 3.90 and 3.94 become

$$\frac{F}{F^*} = \frac{p_{0n}}{p_{0n}^*}\left[1 + \frac{1}{C_F^*}\frac{A_e}{A_t}\left(\frac{p_a^*}{p_{0n}^*}\right)\right] - \frac{1}{C_F^*}\frac{A_e}{A_t}\frac{p_a}{p_{0n}^*} \tag{3.90}$$

$$\frac{F}{F^*} = \frac{p_{0n}}{p_{0n}^*}\left[1 + \frac{1}{C_F^*}\frac{A_e}{A_t}\left(\frac{p_a^*}{p_{0n}^*}\right)\right] - \frac{1}{C_F^*}\frac{A_e}{A_t}\frac{p_e^*}{p_{0n}^*}$$

$$\frac{F}{F^*} = 1.0272\frac{p_{0n}}{p_{0n}^*} - \frac{1}{1.6688}10\frac{1}{91.433}$$

$$\left(\frac{F}{F^*}\right)_{ad} = 1.0272\frac{p_{0n}}{p_{0n}^*} - 0.0655$$

$$p_{0n_{min}} = \frac{p_{0n}^*}{p_e^*}p_{e_{min}} = \frac{p_{0n}^*}{p_e^*}\frac{p_a}{2.5} = \frac{p_{0n}^*}{p_e^*}\frac{p_e^*}{2.5} = \frac{5 \times 10^6}{2.5} = 2 \times 10^6 Pa$$

$$\left(\frac{F_{min}}{F^*}\right)_{ad} = 1.0272 \times \frac{2 \times 10^6}{5 \times 10^6} - 0.0655 = 0.3454$$

$$\left(\frac{F_{min}}{F^*}\right)_{ad} = 0.3454$$

$$\frac{I_{sp}}{I_{sp}^*} \approx \left(1 + \frac{1}{C_F^*}\frac{A_e}{A_t}\frac{p_a^*}{p_{0n}^*}\right) - \frac{A_e}{A_t}\frac{1}{C_F^*}\frac{p_a}{p_{0n}} \tag{3.94}$$

$$\left(\frac{I_{sp}}{I_{sp}^*}\right)_{ad} \approx \left(1 + \frac{1}{C_F^*}\frac{A_e}{A_t}\frac{p_a^*}{p_{0n}^*}\right) - \frac{A_e}{A_t}\frac{1}{C_F^*}\frac{p_e^*}{p_{0n}}$$

$$\left(\frac{I_{sp}}{I_{sp}^*}\right)_{ad} \approx 1.0272 - 10\frac{1}{1.6688}\frac{54685}{p_{0n}} = 1.0272 - \frac{327691}{p_{0n}}$$

$$\left(\frac{I_{sp_{min}}}{I_{sp}^*}\right)_{ad} \approx 1.0272 - \frac{327691}{2 \times 10^6} = 0.8634$$

$$\left(\frac{I_{sp_{min}}}{I_{sp}^*}\right)_{ad} = 0.8634$$

(4) At sea level

$$\frac{F}{F^*} = \frac{p_{0n}}{p_{0n}^*}\left(1 + \frac{1}{C_F^*}\frac{A_e}{A_t}\frac{p_a^*}{p_{0n}^*}\right) - \frac{1}{C_F^*}\frac{A_e}{A_t}\frac{p_a}{p_{0n}^*} \tag{3.90}$$

$$\frac{F}{F^*} = 1.0272\frac{p_{0n}}{p_{0n}^*} - \frac{1}{1.6688}10\frac{101325}{5 \times 10^6}$$

$$\frac{F}{F^*} = 1.0272\frac{p_{0n}}{p_{0n}^*} - 0.1214$$

$$p_{0n_{min}} = \frac{p_{0n}^*}{p_e^*}p_{e_{min}} = \frac{p_{0n}^*}{p_e^*}\frac{p_a}{2.5} = 91.433\frac{101325}{2.5}$$

$$p_{0n_{min}} = 3.7058 \times 10^6 \text{ Pa}$$

$$\left(\frac{F_{min}}{F^*}\right)_{sl} = 1.0272 \times \frac{3.7058 \times 10^6}{5 \times 10^6} - 0.1214$$

$$\left(\frac{F_{min}}{F^*}\right)_{sl} = 1.0272 \times 0.7412 - 0.1214 = 0.6399$$

$$\left(\frac{F_{min}}{F^*}\right)_{sl} = 0.6399$$

$$\frac{I_{sp}}{I_{sp}^*} \approx \left(1 + \frac{1}{C_F^*}\frac{A_e}{A_t}\frac{p_a^*}{p_{0n}^*}\right) - \frac{A_e}{A_t}\frac{1}{C_F^*}\frac{p_a}{p_{0n}} \tag{3.94}$$

$$\frac{I_{sp}}{I_{sp}^*} \approx 1.0272 - 10\frac{1}{1.6688}\frac{101325}{p_{0n}} = 1.0272 - \frac{607173}{p_{0n}}$$

$$\frac{I_{sp_{min}}}{I_{sp}^*} \approx 1.0272 - \frac{607173}{3.7058 \times 10^6} = 0.8634$$

$$\left(\frac{I_{sp_{min}}}{I_{sp}^*}\right)_{sl} \approx 0.8634$$

Note that for all operations except the one in vacuum, $I_{sp_{min}}/I_{sp}^* = 0.8634$. This is because of the nozzle separation constraint:

$$\frac{p_{0n_{min}}}{p_a} = \frac{1}{2.5}\frac{p_{0n}^*}{p_e^*} = \frac{1}{2.5} \times 91.433 = 36.573$$

In the case of vacuum operation, the nozzle separation constraint does not hold good. For the other operations, due to the nozzle separation constraint $p_{0n_{min}}/p_a = 36.573$. In Fig. 3.12, it is a fixed point at the intersection of the vertical $A_e/A_t = 10$ and $p_{0n}/p_a = 36.573$ on the nozzle separation contour of $p_a/p_e = 2.5$. You may track all the operating points of this example in Fig. 3.12 and appreciate the use of such a figure to understand the thrust variation with altitude and the thrust control.

Discussion

Several observations can be made from the results of this example. For an engine of fixed nozzle configuration:

1. Under vacuum conditions, without the danger of nozzle-flow separation, thrust can be reduced to any *practical limit* by reducing the nozzle stagnation pressure. Furthermore, there is no change in specific impulse.
2. At an altitude where the nozzle is under-expanded, the thrust reduction with reduction in p_{0n} is possible to wider limits than that is possible at an altitude where the nozzle is over-expanded.
3. At an altitude where the ambient condition is not vacuum, there is a loss in specific impulse in reducing the thrust through reduction in p_{0n}. This reduction becomes more significant as the ambient pressure increases.
4. In addition to the limit due to the nozzle-flow separation, there is a practical limit on reducing the nozzle stagnation pressure in liquid propellant engines. This constraint comes from the point of view of feed-system limitations and the issues caused by unsatisfactory propellant injection and combustion instabilities. In these engines, generally it may not be possible to reduce the p_{0n} below 40 percent of the design value.
5. Similar to the reduction in thrust by way of reducing p_{0n}, thrust can be increased by increasing p_{0n}. Under this control, as expected the specific impulse variation will be opposite to what is observed in the cases of pressure reduction.
6. There is again a practical limit on the increase in p_{0n}. This time it is the allowable increase in structural mass (to withstand the increased pressures) that can be permitted in addition to the feed-system constraint for liquid propellant engines.

3.6 Characteristic Parameters

There are three characteristic parameters, which describe the performance of rocket engines. Specific impulse and thrust coefficient are two of them and have already been introduced. The third one is **characteristic velocity** and we have considered this rather implicitly in Sect. 3.5.1, Eq. 3.55. Characteristic velocity is defined as

$$c^* \equiv \frac{p_{0n A_t}}{\dot{m}} \tag{3.102}$$

The characteristic velocity can also be written as the time integral

$$c^* \equiv \frac{\int_0^{t_b} p_{0n} A_t dt}{\int_0^{t_b} \dot{m} dt} = \frac{\int_0^{t_b} p_{0n} A_t dt}{m_p} \tag{3.103}$$

Note that the numerator and the denominator of Eq. 3.103 are easily measurable quantities. By the one-dimensional isentropic flow assumption, the mass-flow rate through the choked throat is given by

$$\dot{m} = p_{0n} A_t \left(\frac{\Gamma}{\sqrt{\frac{R_u}{\overline{m}} T_{0c}}} \right) \tag{3.55}$$

Substituting Eq. 3.55 for $\dot{m}$ in Eq. 3.102, we get

$$c^* = \frac{\sqrt{\frac{R_u}{\overline{m}} T_{0c}}}{\Gamma} \tag{3.104}$$

Thus we note from Eq. 3.104, being dependent on the properties of combustion—the adiabatic flame temperature T_{0c}, molecular mass $\overline{m}$, and ratio of specific heats γ—c^* theoretically represent the global characteristics of combustion in the engine. Furthermore, for the given propellant combination the value of c^* is essentially independent of pressure as the values of $\sqrt{T_{0c}/\overline{m}}$ and γ do not change appreciably with pressure [refer to our discussion on the change of specific impulse with the change in p_{0n} under Eqs. 3.93 and 3.94]. In view of this minimal dependence on chamber pressure, the combustion characteristics of different propellants can be compared independent of the respective pressures employed.

By the definition of c^*, Eq. 3.102, and the theoretical meaning to c^* by Eq. 3.104,

$$c^* \equiv \frac{p_{0n A_t}}{\dot{m}} = \frac{\sqrt{\frac{R_u}{\overline{m}} T_{0c}}}{\Gamma} \tag{3.105}$$

From Eq. 3.105, we observe that the characteristic velocity can be estimated through experimental methods and calculated as well using the principles of equilibrium thermodynamics in combustion systems. While the experimental value represents the c^* achieved by the constructed combustion chamber hardware, the theoretical one gives an estimate of the limit that an ideal engine can reach. This possibility to compare experimental value with the theoretical one is of great practical relevance in the engine development and it is common for all the three characteristic parameters.

For the sake of continuity, the other two characteristic parameters can be recalled under this section to stress the importance of the three c^*, C_F, and I_{sp}: *these relate and compare the experimental values achieved by the constructed engine with the theoretical upper limits that only an ideal engine can deliver.*

The thrust coefficient has been defined as

$$C_F \equiv \frac{F}{p_{0n} A_t} \tag{3.69}$$

By the isentropic flow through the convergent–divergent nozzle,

$$C_F = \Gamma \sqrt{\frac{2\gamma}{\gamma - 1}\left[1 - \left(\frac{p_e}{p_{0n}}\right)^{\frac{\gamma-1}{\gamma}}\right]} + \frac{A_e}{A_t}\left(\frac{p_e}{p_{0n}} - \frac{p_a}{p_{0n}}\right) \tag{3.70}$$

Equation 3.70 indicates that being dependent on the properties of nozzle flow—nozzle pressure ratio p_{0n}/p_e, molecular mass $\overline{m}$, and ratio of specific heats γ—C_F theoretically represents the overall characteristics of nozzle flow. Equating the experimental and the theoretical groups, Eqs. 3.69 and 3.70,

$$C_F \equiv \frac{F}{p_{0n} A_t} = \Gamma \sqrt{\frac{2\gamma}{\gamma - 1}\left[1 - \left(\frac{p_e}{p_{0n}}\right)^{\frac{\gamma-1}{\gamma}}\right]} + \frac{A_e}{A_t}\left(\frac{p_e}{p_{0n}} - \frac{p_a}{p_{0n}}\right) \tag{3.106}$$

We find again that the characteristic parameter C_F can be estimated through experimental methods as well as calculated via the theoretical route. If the quality of flow through the nozzle is not satisfactory (say due to nozzle-flow separation, high wall friction, eroding throat section, or incomplete combustion reflected by a value of γ substantially different from the theoretical one), the experimental C_F value can become unacceptably lower than the theoretical one, thus indicating the need for corrective design procedures.

The specific impulse has been defined as

$$I_{sp} = \frac{I}{m_p} = \frac{\int_0^{t_b} F dt}{\int_0^{t_b} \dot{m} dt} = \frac{F}{\dot{m}} = \overline{u}_e \tag{3.39}$$

Equation 3.39 can be written as

$$I_{sp} = \frac{p_{0n} A_t}{\dot{m}} \frac{F}{p_{0n} A_t} = c^* C_F \tag{3.107}$$

The rocket engine fundamentally has two major processes: the first, the combustion taking place in the combustion chamber and the second expansion in the nozzle. As c^* represents the global quality of combustion and C_F represents the overall characteristics of nozzle flow, I_{sp} represents the total performance of the engine. For the sake of uniformity of the algebraic treatment, let us write down the experimental and the theoretical groups.

$$I_{sp} = \frac{F}{\dot{m}} = \frac{\int F dt}{m_p} = \frac{p_{0n} A_t}{\dot{m}} \frac{F}{p_{0n} A_t}$$

$$= \left(\frac{\sqrt{\frac{R_u}{\overline{m}} T_{0c}}}{\Gamma} \right) \left\{ \Gamma \sqrt{\frac{2\gamma}{\gamma - 1} \left[1 - \left(\frac{p_e}{p_{0n}} \right)^{\frac{\gamma-1}{\gamma}} \right]} + \frac{A_e}{A_t} \left(\frac{p_e}{p_{0n}} - \frac{p_a}{p_{0n}} \right) \right\} \tag{3.108}$$

$$= c^* C_F$$

For assessing the experimental performance of rocket engines, the respective ratios of the experimental characteristic parameters and the theoretical ones are evaluated. These ratios are known as quality factors. The quality factor for c^*, namely, ξ_c is also known as c^* efficiency, η_{c^*}. And, the quality factor for specific impulse, namely, ξ_s is sometimes referred to as velocity correction factor. Definitions for the three quality factors ξ_c, ξ_n, and ξ_s can be written as follows:

$$\xi_c \equiv \frac{c^*_{expt}}{c^*_{theo}} \tag{3.109}$$

$$\xi_c = \frac{\left(\frac{p_{0n} A_t}{\dot{m}} \right)_{expt}}{\frac{\sqrt{\frac{R_u}{\overline{m}} T_{0c}}}{\Gamma}} = \left(\frac{\int_0^{t_b} p_{0n} A_t dt}{m_p} \right)_{expt} \frac{\Gamma}{\sqrt{\frac{R_u}{\overline{m}} T_{0c}}} \tag{3.110}$$

Note in the integral $\int_0^{t_b} p_{0n} A_t dt$, the throat area A_t can vary with time because of nozzle erosion. The quality factor for nozzle flow is defined as

$$\xi_n \equiv \frac{C_{F\,expt}}{C_{F\,theo}} \tag{3.111}$$

$$\xi_n = \frac{\left(\frac{F}{p_{0n} A_t}\right)_{expt}}{\Gamma \sqrt{\frac{2\gamma}{\gamma - 1}\left[1 - \left(\frac{p_e}{p_{0n}}\right)^{\frac{\gamma-1}{\gamma}}\right]} + \frac{A_e}{A_t}\left(\frac{p_e}{p_{0n}} - \frac{p_a}{p_{0n}}\right)} \tag{3.112}$$

$$= \frac{I}{\int_0^{t_b} p_{0n} A_t dt} \times \frac{1}{\Gamma \sqrt{\frac{2\gamma}{\gamma - 1}\left[1 - \left(\frac{p_e}{p_{0n}}\right)^{\frac{\gamma-1}{\gamma}}\right]} + \frac{A_e}{A_t}\left(\frac{p_e}{p_{0n}} - \frac{p_a}{p_{0n}}\right)}$$

$$\xi_s \equiv \frac{I_{sp\,expt}}{I_{sp\,theo}} \tag{3.113}$$

$$\xi_s = \frac{\left(\frac{F}{\dot{m}}\right)_{expt}}{\left(\frac{\sqrt{\frac{R_u}{\overline{m}} T_{0c}}}{\Gamma}\right)\left\{\Gamma \sqrt{\frac{2\gamma}{\gamma - 1}\left[1 - \left(\frac{p_e}{p_{0n}}\right)^{\frac{\gamma-1}{\gamma}}\right]} + \frac{A_e}{A_t}\left(\frac{p_e}{p_{0n}} - \frac{p_a}{p_{0n}}\right)\right\}}$$

$$= \left(\frac{I}{m_p}\right)_{expt} \frac{1}{\left(\frac{\sqrt{\frac{R_u}{\overline{m}} T_{0c}}}{\Gamma}\right)\left\{\Gamma \sqrt{\frac{2\gamma}{\gamma - 1}\left[1 - \left(\frac{p_e}{p_{0n}}\right)^{\frac{\gamma-1}{\gamma}}\right]} + \frac{A_e}{A_t}\left(\frac{p_e}{p_{0n}} - \frac{p_a}{p_{0n}}\right)\right\}} \tag{3.114}$$

$$\xi_s = \xi_c \xi_n \tag{3.115}$$

The typical variations for these quality factors are as per the following:

$$0.87 \le \xi_c \le 1.0 \tag{3.116}$$

$$0.90 \le \xi_n \le 0.99 \tag{3.117}$$

$$0.85 \le \xi_s \le 0.99 \tag{3.118}$$

Example 3.5 shows how to qualitatively fix the engine nozzle dimensions and estimate the propellant flow rate, using the experimentally known values of quality factors, ξ_c and ξ_n.

Example 3.5 A large cryogenic engine used in a satellite launch vehicle is to deliver a thrust of 1078 kN in vacuum. The engine is to use liquid oxygen and liquid hydrogen as propellants. The chamber pressure selected = 12.7 MPa. Nozzle pressure ratio = 783. From equilibrium combustion calculations for the propellant combination, the properties of the combustion products are the following. Adiabatic flame temperature $T_{0c} = 3546$ K. Molecular mass of combustion products $\overline{m} = 13.38$ kg/kgmol. Ratio of specific heats $\gamma = 1.2$. Estimate the following for the engine: (1) nozzle throat diameter, (2) nozzle-exit diameter, (3) actual specific impulse in vacuum, and (4) propellant flow rate. Assume $\xi_c = \xi_n = 0.96$.

Analysis

$$\Gamma = \sqrt{1.2}\left(\frac{2}{2.2}\right)^{2.2/0.4} = 0.6485$$

$$\frac{A_e}{A_t} = \frac{0.6485}{\left(\frac{1}{783}\right)^{1/1.2}\sqrt{\frac{2.4}{0.2}\left[1-\left(\frac{1}{783}\right)^{0.2/1.2}\right]}} = 58.96$$

$$\left(C_{F_{vac}}\right)_{theo} = 0.6485\sqrt{\frac{2.4}{0.2}\left[1-\left(\frac{1}{783}\right)^{0.2/1.2}\right]} + 58.96\times\frac{1}{783} = 1.915$$

$$\left(C_{F_{vac}}\right)_{expt} = 1.915\times 0.96 = 1.8384$$

$$A_t = \frac{1078\times 10^3}{1.8384\times 12.7\times 10^6} = 0.04617\ \text{m}^2 \quad \Rightarrow D_t = 0.243\ \text{m}$$

$$D_e = \sqrt{58.96}\times 0.243 = 1.866\ \text{m}$$

$$c^*_{theo} = \frac{\sqrt{\frac{8314.51}{13.38}\times 3546}}{0.6485} = 2289\ \text{m/s}$$

$$c^*_{expt} = 2289\times 0.96 = 2197.4\ \text{m/s}$$

$$\left(I_{sp_{vac}}\right)_{expt} = 2197.4 \times 1.8384 = 4040 \text{ Ns/kg}$$

$$\dot{m} = \frac{1078 \times 10^3}{4040} = 266.83 \text{ kg/s}$$

Discussion

From this example, we see that for a given propellant combination and chamber pressure one can get the properties of the combustion products by running a program such as CEC71, [11]. Assuming these properties to be constant and by adopting the values of quality factors that are known from experiments, we are able to qualitatively calculate the nozzle dimensions and propellant flow rate.

3.7 Frozen Flow and Shifting-Equilibrium Flow

So far we have been considering the operation of rocket engines under the assumptions of utmost simplicity. Under this, we have assumed that the specific heat at constant pressure c_p and the molecular mass $\overline{m}$ to be invariant. But in the environment of accelerating combustion products from the head end of the combustion chamber to the nozzle exit the equilibrium composition of the combustion products can change subject to the local static temperature and pressure. Furthermore, the change of composition is influenced by the related reaction rates and the local velocity—the inverse of the velocity gives the residence time per unit length of the flow field. Under low velocities but with "not-so-low" reaction rates (high reaction rates) the equilibrium composition may shift from one location to another. On the other hand, under high velocities but with "not-so-high" reaction rates (low reaction rates) the composition may remain fixed or frozen. Thus we have to account for the two extreme possibilities, namely, **frozen flow** assumption, which adopts infinitely slow reaction rates, and the **shifting-equilibrium flow** assumption that adopts infinitely fast reaction rates.

On the choice of the assumption of frozen flow or shifting-equilibrium flow through the nozzle, there are different considerations. The predicted specific impulse under the assumption of frozen flow is lower than that under the assumption of shifting-equilibrium flow, because the latter takes into account the energy recovery within the nozzle from the recombination of dissociated species such as H, O, OH, and CO into stable ones H_2O, CO_2, H_2, and O_2. Therefore, the calculation under frozen flow assumption is considered as a conservative one. However, such a conservative assumption of frozen flow may turn out to be erroneous when the local static temperature falls well below the condensation or the solidification temperature of one or more of the species in the flow. For example, the products of combustion of a solid propellant containing magnesium may have magnesium oxide MgO in gaseous form under combustion chamber conditions. When these products accelerate through the nozzle under high velocity, the local static temperature may fall well below the condensation temperature of MgO vapor. In such a situation, although delayed due to the thermal inertia, all the MgO vapor has to condense well below the condensation temperature under flash mode into $MgO\ (l)$, possibly through a **condensation shock** or **condensation front** in the nozzle divergence. Similar situation is possible for solid propellants containing aluminum. At the combustion chamber conditions liquid alumina $Al_2O_3\ (l)$ may be one of the species. During the nozzle flow, the local temperature may fall well below the solidification temperature of $Al_2O_3\ (l)$.

In continuation of our discussion on frozen flow and shifting-equilibrium flow, the composition in the former is of fixed one, while it in the latter is of reacting and variable one. This means that the molecular mass of the combustion products is constant in the frozen flow and it is a variable in the shifting-equilibrium flow. Furthermore, with a view to make the analysis simpler, all through the previous sections we did not take into account the dependence of specific heats c_p and c_v on the temperature and the composition of the combustion products and we assumed them to be invariant. But the specific heats are indeed functions of them. Therefore, the ratio of specific heats $c_p/c_v = \gamma$ can vary with respect to the temperature in the frozen flow and in addition to this it can vary with respect to the reacting composition in shifting-equilibrium flow. In view of this, with the variables γ and $\overline{m}$, we briefly reconsider here the equations that we derived for c^*, C_F, and I_{sp} in the previous sections.

For the reacting composition flow (shifting-equilibrium flow), the velocity of sound is calculated differently [11–13]. By fundamentals the velocity of sound is defined by

$$a \equiv \left(\frac{\partial p}{\partial \rho}\right)_s = \sqrt{\gamma_s\,(R_u/\overline{m})\,T} \tag{3.119}$$

where

$$\gamma_s = -\frac{\gamma}{(\partial ln\ v/\partial ln\ p)_T} \tag{3.120}$$

and γ_s is termed as **isentropic exponent**. Nevertheless, for the inert composition flow (frozen flow) the velocity of sound is given by the usual expression, $a = \sqrt{\gamma\,(R_u/\overline{m})\,T}$ with $(\partial ln\ v/\partial ln\ p)_T$ becoming -1. The so-called **key derivatives** $(\partial ln\ v/\partial ln\ p)_T$ and $(\partial ln\ v/\partial ln\ T)_p$ represent the potential for reaction. For the more reactive composition, the former tends to be more negative than -1 while the latter becomes more positive than $+1$. When the composition becomes the one containing the least dissociated species with the lowest possibility of further reactions both the key derivatives tend to be negative and positive unity, respectively, and hence $\gamma_s \to \gamma$. We will discuss more on the fundamentals of shifting-equilibrium flow, and the evaluation of key derivatives and isentropic exponent under Sect. 4.5.

The characteristic velocity c^* is defined as

$$c^* \equiv \frac{p_{0n} A_t}{\dot{m}} \tag{3. 102}$$

Noting $\dot{m}/A_t = \rho_t a_t$ with $a_t = \sqrt{\gamma_{s_t} (R_u/\overline{m}_t) T_t}$, the characteristic velocity becomes

$$c^* = \frac{(p_{0n}/p_t)\, a_t}{\gamma_{s_t}} \tag{3.121}$$

For the isentropic flow with invariant γ and $\overline{m}$, Eq. 3.121 becomes Eq. 3.104. Similarly, noting $u_e = \sqrt{2(h_0 - h_e)}$ and using the thrust equation Eq. 3.15 and the definition of thrust coefficient C_F Eq. 3.69, we can show that the thrust coefficient with varying γ and $\overline{m}$ as

$$C_F \equiv \frac{F}{p_{0n} A_t} = \frac{\gamma_{s_t} \sqrt{2(h_0 - h_e)}}{a_t\,(p_{0n}/p_t)} + \frac{A_e}{A_t}\left(\frac{p_e}{p_{0n}} - \frac{p_a}{p_{0n}}\right) = C_F^0 + \frac{A_e}{A_t}\left(\frac{p_e}{p_{0n}} - \frac{p_a}{p_{0n}}\right) \tag{3.122}$$

From these, it follows that the specific impulse I_{sp} for the varying γ and $\overline{m}$ is given as

$$I_{sp} \equiv \frac{F}{\dot{m}} = \sqrt{2(h_0 - h_e)} + \frac{A_e}{A_t}\frac{p_{0n}}{p_t}\frac{a_t}{\gamma_{s_t}}\left(\frac{p_e}{p_{0n}} - \frac{p_a}{p_{0n}}\right) \tag{3.123}$$

Or, in a more "revealing" form as

$$I_{sp} \equiv \frac{F}{\dot{m}} = \left[\frac{\gamma_{s_t} \sqrt{2(h_0 - h_e)}}{a_t\,(p_{0n}/p_t)} + \frac{A_e}{A_t}\left(\frac{p_e}{p_{0n}} - \frac{p_a}{p_{0n}}\right)\right] \frac{(p_{0n}/p_t)\, a_t}{\gamma_{s_t}} = C_F c^* \tag{3.124}$$

For the relatively rigorous calculation of rocket engine performance and internal ballistics design, we will discuss the adoption of these, Eqs. 3.121–3.124, and other performance-parameter equations under Sect. 4.7.

NASA's codes CEC71 (Chemical Equilibrium Calculations, 1971) [11] and its improved version CEA (Chemical Equilibrium and Applications) [12] are widely used by industries and research institutions for rocket and other applications. In order to demonstrate the difference between the results of shifting-equilibrium flow and that of frozen flow, the extracted outputs of CEC71 for a cryogenic rocket engine are given in Tables 3.1 and 3.2. Adopting liquid hydrogen (LH2) and liquid oxygen (LOX) as propellants, the cryogenic rocket engine operates at the oxidizer–fuel ratio of 6.2, the chamber pressure of 102 bar, and the nozzle area ratio of 45. Carefully compare the results of the shifting-equilibrium flow with those of the frozen flow to understand and appreciate the reasons for the differences, Tables 3.1 and 3.2. Let us first consider the characteristics of the shifting-equilibrium flow, Table 3.1. The composition is varying (shifting) with increasing $\overline{m}$, and at the nozzle exit it is mainly of the stable species of H_2O and H_2 with traces of dissociated species H and OH. The p_{0n}/p_e is 529.31 and the exit static temperature is 1453 K. And at the exit, because the

Table 3.1 Theoretical rocket performance assuming shifting-equilibrium flow with restricted combustion product species: H, H_2, H_2O, and OH (LOX/LH2, $O/F = 6.2$, $A_e/A_t = 45$, and $p_{0c} = 102\ bar$) [11]

	Chamber	Throat	Nozzle exit
p_0/p	1	1.737	529.31
p (bar)	102	58.722	0.1927
T (K)	3589	3383	1453
h (kJ/kg)	−970.2	−2117.4	−10016.7
s (kJ/(kgK))	17.3345	17.3345	17.33
$(\partial\ ln\ v/\partial\ ln\ p)_T$	−1.0213	−1.0169	−1.0000
$\overline{m}$ (kg/kgmol)	13.877	14.016	14.514
c_p(kJ/(kgK))	7.7265	7.2026	3.0057
γ_s	1.1435	1.1434	1.2355
M	0	1.000	4.195
A_e/A_t	–	1	45
c^* (m/s)	–	2301.1	2301.1
C_F^0	–	0.6583	1.848
$I_{sp_{vac}}$ (m/s)	–	2839.5	4449.2
I_{sp}^0 (m/s)	–	1514.7	4253.6

Table 3.2 Theoretical rocket performance assuming frozen flow with restricted combustion product species: H, H_2, H_2O, and OH (LOX/LH2, $O/F = 6.2$, $A_e/A_t = 45$, and $p_{0c} = 102\ bar$) [11]

	Chamber	Throat	Nozzle exit
p_0/p	1	1.7688	627.38
p (bar)	102	57.667	0.1626
T (K)	3589	3272	1110
h (kJ/kg)	−970.2	−2141.7	−9351.1
s (kJ/(kgK))	17.3345	17.3345	17.3345
$\overline{m}$ (kg/kgmol)	13.877	13.877	13.877
c_p(kJ/(kgK))	3.7228	3.6707	2.7880
γ	1.1918	1.1951	1.2737
M	0	1.000	4.448
A_e/A_t	–	1	45
c^* (m/s)	–	2265.5	2265.5
C_F^0	–	0.6756	1.8071
$I_{sp_{vac}}$ (m/s)	–	2811.6	4256.6
I_{sp}^0 (m/s)	–	1530.7	4094.1

composition is mainly of stable species, the key derivative $(\partial ln\ v/\partial ln\ p)_T \rightarrow -1$, and hence $\gamma_s \rightarrow \gamma$. Now consider the characteristics of the frozen flow, Table 3.2. The composition is invariant with the species H, H_2, H_2O, and OH as at the combustion chamber and hence $\overline{m}$ is constant throughout at 13.877 kg/kgmol. For the same A_e/A_t of 45, the p_{0n}/p_e is 627.38 and the exit static temperature is 1110 K. Since the composition is frozen $(\partial ln\ v/\partial ln\ p)_T = -1$ and $\gamma_s = \gamma$.

3.8 Two-Phase Flow

At the chamber conditions, combustion products of liquid propellants are mostly of gaseous mixtures. But, it is not so for solid propellants. Most **heterogeneous solid propellants** (popularly known as **composite propellants**) and also sometimes **homogeneous solid propellants** (popularly known as **double base propellants**) contain metal powders as one of the fuel components (usually powders of aluminum among the possible group of aluminum, boron, magnesium, and zirconium). For example, a typical heterogeneous propellant contains particles of the oxidizer ammonium perchlorate NH_4ClO_4 (AP) and the fuel aluminum. The particles of the both are randomly distributed by mixing them with a liquid pre-polymer fuel, say hydroxyl-terminated poly butadiene (HTPB, a pre-polymer compound of hydrogen, carbon, and oxygen), and the mix is cured into a solid by adding a curing agent and subsequent heating. On combustion, the added aluminum gets oxidized and forms alumina Al_2O_3 particulates. At the combustion chamber conditions, the alumina may be in liquid phase, $Al_2O_3\ (l)$. During the nozzle flow as the local temperature and pressure fall, $Al_2O_3\ (l)$ may get solidified into $Al_2O_3\ (s)$. In this process, for a significant time interval of the solidification there may be both $Al_2O_3\ (l)$ and $Al_2O_3\ (s)$ particulates resulting in a situation of multi-phase nozzle flow. Alternatively, even with the temperature of the gaseous mixtures falling well below the solidification point the solidification may not have been initiated due to thermal inertia, and when the temperature difference becomes unsustainable an instantaneous heat release into the nozzle flow may follow due to the flash solidification. And, as previously noted, in the case of solid propellants containing magnesium, one of the combustion products MgO vapor formed at the chamber conditions may condense during the nozzle flow, resulting in a condensation shock or condensation front with detectable pressure changes following the sudden heat addition.

The oxides of the metals added to solid propellants being thermally resistant, the particulates—more importantly the large ones—could assume a higher temperature than the gaseous mixture and due to the drag phenomena may assume velocity lags in the nozzle flow. The presence of condensed particulates at very high temperature considerably increases radiation losses to the surroundings with little absorption by the expanding gases. Combustion of a rocket-grade multimodal $AP/HTPB$ composite propellant with aluminum of particle size 33 μm indicated close to the burning surface the distribution of $Al_2O_3\ (s)$ particulates of modes 1 μm and 4 μm,

Table 3.3 Abridged results of shifting-equilibrium flow calculations using program CEC71 [11] for a typical heterogeneous propellant of composition $NH_4ClO_4 : HTPB : Al = 69 : 12 : 19$

	Chamber	Throat	Nozzle exit
p_{0c}/p	1.0000	1.7255	60.000
p (bar)	60.000	34.772	1.0000
T (K)	3609	3429	2336
$\overline{m}$ (kg/kgmol)	29.800	30.071	31.298
$\overline{m}_c$ (kg/kgmol)	27.257	27.390	28.194
$(\partial ln\ v/\partial ln\ p)_T$	−1.02833	−1.02327	−1.00250
γ_s	1.1233	1.1246	1.1521
M	0	1.000	3.035
A_e/A_t	–	1.0000	9.1568
c^* (m/s)	–	1584.5	1584.5
C_F^0	–	0.6517	1.6198
$I_{sp_{vac}}$ (m/s)	–	1951	2808
I_{sp}^0 (m/s)	–	1033	2566
Mole fractions			
$AlCl$	7.4977×10^{-3}	5.5245×10^{-3}	6.3809×10^{-5}
$AlOH$	7.6951×10^{-3}	5.1961×10^{-3}	3.6247×10^{-5}
$Al\,(OH)\,Cl$	1.1291×10^{-3}	6.9367×10^{-4}	3.6706×10^{-6}
CO	2.0658×10^{-1}	2.0760×10^{-1}	2.1000×10^{-1}
CO_2	1.3711×10^{-2}	1.3790×10^{-2}	1.7915×10^{-2}
Cl	1.5960×10^{-2}	1.4565×10^{-2}	2.4268×10^{-3}
H	4.6118×10^{-2}	4.0903×10^{-2}	5.8644×10^{-3}
HCl	1.3042×10^{-1}	1.3647×10^{-1}	1.6295×10^{-1}
H_2	2.4684×10^{-1}	2.5312×10^{-1}	2.8114×10^{-1}
H_2O	1.3819×10^{-1}	1.3740×10^{-1}	1.3701×10^{-1}
NO	1.0399×10^{-3}	7.2920×10^{-4}	1.4031×10^{-5}
N_2	7.9490×10^{-2}	8.0047×10^{-2}	8.2782×10^{-2}
O	1.3531×10^{-3}	9.2616×10^{-4}	6.8004×10^{-6}
OH	1.3464×10^{-2}	1.0534×10^{-2}	5.3639×10^{-4}
$Al_2O_3\,(l)$	8.5346×10^{-2}	8.9158×10^{-2}	9.9182×10^{-2}

and unburned Al particulates of mode $70\,\mu m$. However at the nozzle entry conditions downstream, the particulates transformed essentially to $Al_2O_3\,(s)$ of mode $4\,\mu m$ [14].

Let us consider the two-phase flow situation for a typical rocket motor adopting an aluminized composite propellant. The results of the shifting-equilibrium flow calculations are given in Table 3.3. Most symbols in the first column of Table 3.3 have been introduced or are self-explanatory except the difference between the molecular mass symbols $\overline{m}$ and $\overline{m}_c$. In a mixture of gaseous and condensed species, let there be a total of NS number of species in which the number of gaseous species is NG. Volume of the condensed species is negligible compared to the volume of the gaseous species. Furthermore, condensed species are incompressible. Per kg of the mixture, let n_j be the number of kgmoles of species j of molecular mass $\overline{m}_j$ with $j = 1, .., NS$. Thus, the molecular mass considering only the gaseous species,

$$\overline{m} = \frac{1}{\sum_{j=1}^{NG} n_j} = \frac{1}{n} \tag{3.125}$$

where n is the number kgmoles of gaseous species per kg of combustion products. Considering now all the species, the molecular mass $\overline{m}_c$ is given by

$$\overline{m}_c = \frac{1}{\sum_{j=1}^{NS} n_j} \tag{3.126}$$

It follows then

$$\overline{m}_c = \overline{m} \sum_{j=1}^{NG} x_j = \overline{m} \left(1 - \sum_{j=NG+1}^{NS} x_j\right) \tag{3.127}$$

where the mole fraction

$$x_j = \frac{n_j}{\sum_{j=1}^{NS} n_j} = n_j \overline{m}_c \tag{3.128}$$

Note that in a mixture of gaseous and condensed species $\overline{m}_c < \overline{m}$. In Table 3.3, examine the mole fractions of various species as the combustion products flow through the nozzle. Note that the dissociated species are getting recombined to form more stable species. We see that the only condensed species present is $Al_2O_3\,(l)$ and its mole fraction at the nozzle exit is 9.9182×10^{-02}. Let us look at the impact of such condensed species present on the rocket performance through Example 3.6.

Example 3.6 A solid propellant rocket motor, using a heterogeneous solid propellant of composition $NH_4ClO_4 : HTPB : Al = 69 : 12 : 19$, operates at a chamber pressure of 60 bar with a nozzle pressure ratio p_{0n}/p_e of 60. The theoretical rocket performance calculations (Table 3.3) indicate that the condensed species $Al_2O_3\,(l)$ is present in the combustion products and its mole fraction at the nozzle exit as 9.9182×10^{-02}. Considering all the gaseous and condensed species, the molecular mass of the nozzle flow at the nozzle exit $\overline{m}_c$ is 28.194 kg/kgmol. Nozzle-exit static temperature $T_e = 2336$ K. Calculate the mass and volume fraction of $Al_2O_3\,(l)$ at the nozzle exit. The density of $Al_2O_3\,(l)$ is 3800 kg/m^3.

Analysis

Mole fractions of the combustion products are given and all these add up to 1 kgmole. Of this, the mole fraction of $Al_2O_3\,(l)$ is 9.9182×10^{-02}.

Molecular mass of the mixture $\overline{m}_c = 28.194$ kg/kgmol.

In this 1 kgmole mixture, the number of kgmoles of gaseous species = $\left(1 - 9.9182 \times 10^{-02}\right) = 0.900818$ kgmol/kgmol.

Number of kgmoles of $Al_2O_3\,(l)$ per kg of mixture = $9.9182 \times 10^{-02}/28.194 = 3.518 \times 10^{-3}$ kgmol/kg.

Noting that the molecular mass of $Al_2O_3\,(l) = 101.96$ kg/kgmol, the mass of $Al_2O_3\,(l)$ per kg of mixture = $3.518 \times 10^{-3} \times 101.96 = 0.3587$ kg/kg.

Volume of $Al_2O_3\,(l) = 0.3587/3800 = 9.4389 \times 10^{-5}$ m^3/kg *of mixture*.

Ideal gas equation of state for the mixture is given by

$$pv = nR_uT = \frac{R_uT}{\overline{m}} = \frac{R_uT\left(1 - \sum_{j=NG+1}^{NS} x_j\right)}{\overline{m}_c}$$

Therefore,

$$v = \frac{8314.51 \times 2336 \times \left(1 - 9.9182 \times 10^{-02}\right)}{28.194 \times 10^5} = 6.2057 \text{ m}^3/\text{kg}$$

Therefore the volume fraction of $Al_2O_3\,(l)$ with respect to the volume of gaseous mixture = $9.4389 \times 10^{-5}/6.2057 = 1.521 \times 10^{-5}$.

Discussion

1. In a typical heterogeneous propellant containing a metal component, the condensed species in the combustion products are of mole fraction around 8–10%. However, its mass fraction exceeds 30%. Hence realize that for every kg of combustion products being ejected by the rocket nozzle the mass of condensed particulates exceeds 0.3 kg.
2. Environmental aspect on the ejection of abrasive particulates such as $Al_2O_3(s)$ apart, rocket scientists bother about the realities of the particulates flow with respect to gas flow. In analyzing this multi-phase flow they have to look into the aspects of heat transfer between gaseous species and condensed species and of velocity lag between them—recall that the rocket thrust is equal to the nozzle-exit flow rate multiplied by the effective nozzle-exit velocity.
3. Despite the high mass fraction, note that the volume fraction of the condensed species with respect to gaseous species is around 1.5×10^{-5}.

In Sect. 3.7, we have discussed the shifting-equilibrium flow, which is related to chemical analyses. Added to this now is the two-phase flow situation, the rigorous analysis of which needs the application of laws of physics: condensation and solidification processes, heat transfer between the gaseous mixture and the condensed particles, and drag characteristics of the particles. References [15–18] describe detailed flow calculation methods for two-phase flows in rocket nozzles. With reference to the ideal frozen flow I_{sp} values, experimental studies and flow analyses on typical solid motors have shown that the two-phase flow loss to be around 5% along with the other losses due to nozzle divergence and boundary layer around 3% and 2%, respectively [19]. Now, to understand the basic effects of the two-phase nozzle flow, let us next consider a simple model that is based on the one given in [1].

3.8.1 *Two-Phase Frozen Flow Model*

We follow [1] in analyzing the two-phase nozzle flow. It is assumed that the equilibrium composition formed at the chamber is frozen as the combustion products flow through the nozzle. The total mass-flow rate through the nozzle $\dot{m}$ comprises the gaseous mixture flow rate $\dot{m}_g$ and the condensed-matter flow rate $\dot{m}_c$ and their values are constant. In this adiabatic system, the stagnation enthalpy of the total combustion products is constant. Furthermore, specific heats c_c for the condensed matter and c_p for the gaseous mixture are assumed to be constant. Applying the momentum conservation for the frictionless flow in the elemental control volume of the nozzle, Fig. 3.15,

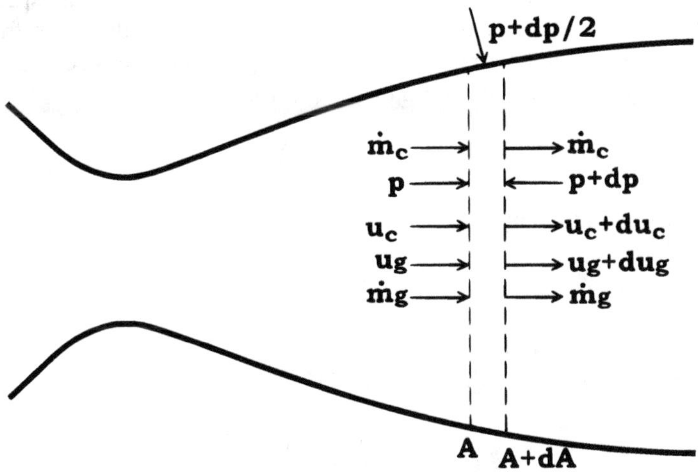

Fig. 3.15 Frozen two-phase flow in nozzle control volume

$$-Adp = \dot{m}_c (u_c + du_c) + \dot{m}_g (u_g + du_g) - \dot{m}_c u_c - \dot{m}_g u_g \tag{3.129}$$

where u_c is the velocity of condensed particles and u_g is the gas velocity. By continuity,

$$\dot{m}_c = \rho_c A u_c \tag{3.130a}$$

$$\dot{m}_g = \rho_g A u_g \tag{3.130b}$$

where ρ_c is the total mass of condensed particles per unit volume of the total mixture, and ρ_g is the density of the gas. Note that the definition of ρ_c is different from the usually adopted one for density. Combining Eqs. 3.129 and 3.130, we get the mass and momentum conservation as

$$-dp = \rho_c u_c du_c + \rho_g u_g du_g \tag{3.131}$$

We assume that the velocity of combustion products in the chamber is negligibly small. By the conservation of energy for the adiabatic nozzle flow,

$$\dot{m}_c h_{0c} + \dot{m}_g h_{0g} = \dot{m}_c h_c + \frac{u_c^2}{2} + \dot{m}_g h_g + \frac{u_g^2}{2} \tag{3.132}$$

Taking differentials with constant specific heats c_c and c_p,

$$\dot{m}_c (c_c dT_c + u_c du_c) + \dot{m}_g (c_p dT_g + u_g du_g) = 0 \tag{3.133}$$

where c_c and T_c are, respectively, the specific heat and the temperature of the condensed particles, and c_p and T_g are, respectively, the constant pressure specific heat and the static temperature of the gaseous mixture. Let χ be the mass fraction of the condensed matter in the total flow and following this $(1-\chi)$ becomes the mass fraction of the gaseous mixture. With this, Eq. 3.133 becomes

$$\chi c_c dT_c + (1-\chi) c_p dT_g + \chi u_c du_c + (1-\chi) u_g du_g = 0 \tag{3.134}$$

From the mass and momentum conservation Eq. 3.131, we get

$$u_g du_g = -\frac{dp}{\rho_g} - \frac{\rho_c}{\rho_g} u_c du_c \tag{3.135}$$

Noting $\chi = \rho_c u_c / (\rho_g u_g + \rho_c u_c)$, we obtain

$$u_g du_g = -\frac{dp}{\rho_g} - \frac{\chi}{1-\chi} u_g du_c \tag{3.136}$$

Substituting for $u_g du_g$ from Eq. 3.136 into Eq. 3.134, we have

$$\frac{\chi}{1-\chi} c_c dT_c + c_p dT_g + \frac{\chi}{1-\chi}\left(u_c - u_g\right) du_c = \frac{dp}{\rho_g} \tag{3.137}$$

Equation 3.137 is the governing equation having into it momentum, mass, and energy conservation. Solution of this governing equation requires knowledge of T_c variation with respect to T_g and of u_c variation with respect to u_g. The heat transfer characteristics between the gaseous mixture and the condensed particles are required to determine T_c versus T_g while drag characteristics between them are needed to fix u_c versus u_g. However, four limiting cases can be applied to understand the effects of the two-phase frozen flow. These are

1. Very high heat transfer rate between the gaseous mixture and the condensed particles, and extreme drag experienced by the condensed particles: $T_g = T_c$ & $u_g = u_c$.
2. Very high heat transfer rate between the gaseous mixture and the condensed particles, and not so significant drag experienced by the condensed particles that a velocity lag may be experienced by the particles: $T_g = T_c$ and $u_c = \phi u_g$, where $\phi < 1$.
3. A low heat transfer rate between the gaseous mixture and the condensed particles so that the particles assume higher temperature than the gaseous mixture, and extreme drag experienced by the condensed particles: $T_c = \tau T_g$, where $\tau > 1$ and $u_g = u_c$.
4. A low heat transfer between the gaseous mixture and the condensed particles, and not so significant drag experienced by the condensed particles: $T_c = \tau T_g$, where $\tau > 1$ and $u_c = \phi u_g$, where $\phi < 1$.

By the introduction of the temperature difference parameter τ and the velocity lag parameter ϕ, the energy conservation Eq. 3.134 becomes

$$\left[\chi c_c \tau + (1-\chi) c_p\right] dT_g + \left[\chi \phi^2 + (1-\chi)\right] u_g du_g = 0 \tag{3.138}$$

Similarly, the equation of momentum conservation Eq. 3.136 becomes

$$u_g du_g = -\frac{(1-\chi)}{[1-(1-\phi)\chi]} \frac{dp}{\rho_g} \tag{3.139}$$

Combining the energy conservation Eq. 3.138 and the momentum conservation Eq. 3.139, and assuming that the gaseous mixture behaves as an ideal gas $\rho = p/(RT)$, we get the revised governing equation with the parameters τ and ϕ.

$$\frac{dT_g}{T_g} = \frac{R}{\left[\frac{\chi}{(1-\chi)} c_c \tau + c_p\right]} \frac{\left[1-\chi\left(1-\phi^2\right)\right]}{[1-\chi(1-\phi)]} \frac{dp}{p} \tag{3.140}$$

Let us consider the first case, $T_g = T_c$ and $u_g = u_c$, that is, $\tau = 1$ & $\phi = 1$. Then the governing Eq. 3.140 reduces to

$$\frac{dT_g}{T_g} = \frac{R}{\left[\left(\frac{\chi}{1-\chi}c_c + c_p\right)\right]}\frac{dp}{p} \tag{3.141}$$

Integrating between the combustion chamber to the nozzle exit with the limits p_{0c} & p_e and T_{0c} & T_e, we get

$$\frac{T_e}{T_{0c}} = \left(\frac{p_e}{p_{0c}}\right)^n \tag{3.142a}$$

$$n = \frac{R}{[\chi/(1-\chi)]\,c_c + c_p} \tag{3.142b}$$

Under the condition $T_g = T_c$ and $u_g = u_c$, we get from the energy balance equation Eq. 3.138

$$I_{sp}^0 = u_e = \sqrt{2\left[\chi c_c + (1-\chi)\,c_p\right] T_{0c}\left[1 - (p_e/p_{0c})^n\right]} \tag{3.143}$$

in which n is given by Eq. 3.142b. Note I_{sp}^0, the specific impulse under adapted condition $p_e = p_a$, is also given by Eq. 3.143. Note $(p_e/p_{0c})^n = T_e/T_{0c}$. For the given condensed-matter fraction χ and $c_c < c_p$, $n < \left(R/c_p\right) = (\gamma - 1)/\gamma$. Therefore, for the chosen nozzle pressure ratio, the nozzle-exit temperature will be hotter, and more importantly, u_e the specific impulse will be less. This loss in the specific impulse happens even under the "liberal" condition of $T_g = T_c$ and $u_g = u_c$.

Proceeding similarly for the other three limiting cases, we can get the equations for the temperature ratio T_e/T_{0c}, nozzle-exit velocity, and specific impulse under adapted condition I_{sp}^0. The results are summarized below:

$\underline{T_g = T_c \rightarrow (\tau = 1);\ u_g = u_c \rightarrow (\phi = 1)}$

$$\frac{T_e}{T_{0c}} = \left(\frac{p_e}{p_{0c}}\right)^n \tag{3.142a}$$

$$n = \frac{R}{[\chi/(1-\chi)]\,c_c + c_p} \tag{3.142b}$$

$$I_{sp}^0 = u_e = \sqrt{2\left[\chi c_c + (1-\chi)\,c_p\right] T_{0c}\left[1 - (p_e/p_{0c})^n\right]} \tag{3.143}$$

$$\underline{T_g = T_c \to (\tau = 1);\; u_g > u_c \to (\phi < 1)}$$

$$\frac{T_e}{T_{0c}} = \left(\frac{p_e}{p_{0c}}\right)^{nm} \tag{3.144a}$$

$$nm = \frac{R}{[\chi/(1-\chi)]\,c_c + c_p}\,\frac{\left[1-\chi\left(1-\phi^2\right)\right]}{[1-\chi(1-\phi)]} \tag{3.144b}$$

$$u_{ge} = \sqrt{2\frac{\left[\chi c_c + (1-\chi)\,c_p\right]}{\left[1-\chi\left(1-\phi^2\right)\right]}T_{0c}\left[1-\left(p_e/p_{0c}\right)^{nm}\right]} \tag{3.145}$$

$$I_{sp}^0 = [1-\chi(1-\phi)]\,u_{ge} \tag{3.146}$$

$$\underline{T_g < T_c \to (\tau > 1);\; u_g = u_c \to (\phi = 1)}$$

$$\frac{T_e}{T_{0c}} = \left(\frac{p_e}{p_{0c}}\right)^{k} \tag{3.147a}$$

$$k = \frac{R}{[\chi/(1-\chi)]\,c_c\tau + c_p} \tag{3.147b}$$

$$I_{sp}^0 = u_e = \sqrt{2\left[\chi c_c\tau + (1-\chi)\,c_p\right]T_{0c}\left[1-\left(p_e/p_{0c}\right)^{k}\right]} \tag{3.148}$$

$$\underline{T_g < T_c \to (\tau > 1);\; u_g > u_c \to (\phi < 1)}$$

$$\frac{T_e}{T_{0c}} = \left(\frac{p_e}{p_{0c}}\right)^{km} \tag{3.149a}$$

$$km = \frac{R}{[\chi/(1-\chi)]\,c_c\tau + c_p}\,\frac{\left[1-\chi\left(1-\phi^2\right)\right]}{[1-\chi(1-\phi)]} \tag{3.149b}$$

$$u_{ge} = \sqrt{2\frac{\left[\chi c_c\tau + (1-\chi)\,c_p\right]}{\left[1-\chi\left(1-\phi^2\right)\right]}T_{0c}\left[1-\left(p_e/p_{0c}\right)^{km}\right]} \tag{3.150}$$

$$I_{sp}^0 = [1-\chi(1-\phi)]\,u_{ge} \tag{3.146}$$

Let us consider the general situation of the two-phase flow, $\tau > 1$ & $\phi < 1$, Eqs. 3.149 and 3.150. In the nozzle-exit velocity expression, the "usual c_p" of the homogeneous gaseous flow is replaced by the "effective c_p" of the two-phase flow

$$\frac{\left[\chi c_c \tau + (1-\chi)\, c_p\right]}{\left[1-\chi\left(1-\phi^2\right)\right]}$$

The pressure ratio index "R/c_p" now comprises two components, the inverse of the "effective c_p" and momentum deficiency contribution

$$\frac{R\,(1-\chi)}{[1-\chi\,(1-\phi)]}$$

All other expressions for the cases of τ and ϕ and also the ideal situation where $\chi = 0$ can be obtained by the respective substitutions in Eqs. 3.149 and 3.150.

Generally the values of c_p are marginally higher than that of c_c. For example, typical values of c_p for the non-reacting (frozen) combustion products of unmetallized composite propellants vary from about 1.85 (~3000 K)–1.61 kJ/(kgK)(~1500 K) while $Al_2O_3\ (l)$ has c_c of 1.60 $kJ/(kgK)$ (>2300 K) and $Al_2O_3\ (s)$ has c_c of 1.37 (~2300 K)–1.33 kJ/(kgK) (~1500 K). Despite the fact that this simple model over-predicts the two-phase flow losses, the basic effects are clearly brought out in Example 3.7.

Example 3.7 A solid propellant rocket motor, adopting a heterogeneous propellant of AP, HTPB, and aluminum, operates at a chamber pressure of 70 bar with a nozzle pressure ratio p_{0n}/p_e of 70. The chamber stagnation temperature and the mass fraction of $Al_2O_3\ (s)$ particles present in the combustion products are, respectively, 3600 K and 0.32. The specific heat c_c of $Al_2O_3\ (s)$ is 1370 $J/(kgK)$. For the gaseous mixture part of the combustion products, the specific heat c_p is 1853 $J/(kgK)$ and the molecular mass is 24.414 $kg/kgmol$. Under the assumption of the condensed particles being absent, calculate the nozzle-exit temperature and adapted specific impulse I^0_{sp} for the motor operating conditions: $T_{0c} = 3600\ K$, $p_{0c} = 70\ bar$, and $p_{0c}/p_e = 70$. With these two values as the reference, calculate and compare I^0_{sp} for the four limiting conditions with the condensed particles being present: $T_g = T_c\ \&\ u_g = u_c$, $T_g = T_c\ \&\ u_c = 0.9u_g$, $T_c = 1.1T_g\ \&\ u_g = u_c$, and $T_c = 1.1T_g\ \&\ u_c = 0.9u_g$.

Analysis

$p_{0c} = 70\ bar$

$p_{0c}/p_e = 70$

$T_{0c} = 3600$ K

$\chi = 0.32$

$c_c = 1370$ J/(kgK)

For the gaseous mixture part,
$c_p = 1853$ J/(kgK)

$$\gamma = \frac{c_p}{c_p - R_u/\overline{m}}$$

$$\gamma = \frac{1853}{1853 - 8314.51/24.414} = 1.2252$$

$$T_e = T_{0c}\left(\frac{p_e}{p_{0c}}\right)^{\frac{\gamma-1}{\gamma}} = 3600 \times \left(\frac{1}{70}\right)^{(0.2252/1.2252)} = 1648.77 \text{ K}$$

$$R = \frac{8314.51}{24.414} = 340.563 \text{ J/(kgK)}$$

$$u_e = \sqrt{2c_p T_{0c}\left[1 - \left(\frac{p_e}{p_{0c}}\right)^{(\gamma-1)/\gamma}\right]}$$

$$= \sqrt{2 \times 1853 \times 3600 \times \left[1 - \left(\frac{1}{70}\right)^{0.18381}\right]} = 2689.1 \text{ m/s}$$

$$I_{sp}^0 = u_e = 2689.1 \text{ Ns/kg}$$

$T_g = T_c$ & $u_g = u_c \rightarrow \tau = 1$ & $\phi = 1$

$$\frac{T_e}{T_{0c}} = \left(\frac{p_e}{p_{0c}}\right)^n \tag{3.142a}$$

$$n = \frac{R}{[\chi/(1-\chi)]\, c_c + c_p} \tag{3.142b}$$

$$n = \frac{340.563}{[0.32/(0.68)] \times 1370 + 1853} = 0.13635$$

$$T_e = 3600 \times \left(\frac{1}{70}\right)^{0.13635} = 2017.1 \text{ K}$$

$$I_{sp}^0 = u_e = \sqrt{2\left[\chi c_c + (1-\chi)\, c_p\right] T_{0c}\left[1 - (p_e/p_{0c})^n\right]} \tag{3.143}$$

$$I_{sp}^{0} = u_e = \sqrt{2\left[0.32 \times 1370 + 0.68 \times 1853\right] \times 3600 \times \left[1 - (1/70)^{0.13635}\right]}$$

$$I_{sp}^{0} = u_e = \sqrt{2\left[1698.44\right] \times 3600 \times \left[1 - (1/70)^{0.13635}\right]} = 2318.8 \text{ m/s}$$

$T_g = T_c \;\&\; u_c = 0.9 u_g \rightarrow \tau = 1 \;\&\; \phi = 0.9$

$$\frac{T_e}{T_{0c}} = \left(\frac{p_e}{p_{0c}}\right)^{nm} \tag{3.144a}$$

$$nm = \frac{R}{\left[\chi/(1-\chi)\right] c_c + c_p} \frac{\left[1 - \chi\left(1 - \phi^2\right)\right]}{\left[1 - \chi(1 - \phi)\right]} \tag{3.144b}$$

$$nm = \frac{340.563}{\left[0.32/0.68\right] \times 1370 + 1853} \frac{\left[1 - 0.32 \times (1 - 0.81)\right]}{\left[1 - 0.32 \times (1 - 0.9)\right]} = 0.132294$$

Therefore,

$$T_e = 3600 \times \left(\frac{1}{70}\right)^{0.132294} = 2052.1 \text{ K}$$

$$u_{ge} = \sqrt{2 \frac{\left[\chi c_c + (1-\chi) c_p\right]}{\left[1 - \chi\left(1 - \phi^2\right)\right]} T_{0c} \left[1 - \left(p_e/p_{0c}\right)^{nm}\right]} \tag{3.145}$$

$$u_{ge} = \sqrt{2 \frac{\left[0.32 \times 1370 + 0.68 \times 1853\right]}{\left[1 - 0.32\,(1 - 0.81)\right]} \times 3600 \times \left[1 - (1/70)^{0.132294}\right]}$$

$$u_{ge} = \sqrt{2 \times \left[1808.39\right] \times 3600 \times \left[1 - (1/70)^{0.132294}\right]} = 2366.1 \,\frac{\text{m}}{\text{s}}$$

$$I_{sp}^{0} = \left[1 - \chi(1 - \phi)\right] u_{ge} \tag{3.146}$$

$$I_{sp}^{0} = \left[1 - 0.32 \times 0.1\right] \times 2366.1 = 2290.4 \,\frac{\text{m}}{\text{s}}$$

$T_c = 1.1 T_g \;\&\; u_g = u_c \rightarrow \tau = 1.1 \;\&\; \phi = 1$

$$\frac{T_e}{T_{0c}} = \left(\frac{p_e}{p_{0c}}\right)^{k} \tag{3.147a}$$

$$k = \frac{R}{[\chi/(1-\chi)]\,c_c\tau + c_p} \tag{3.147b}$$

$$k = \frac{340.563}{[0.32/0.68]\times 1370\times 1.1 + 1853} = 0.132919$$

$$T_e = 3600\times\left(\frac{1}{70}\right)^{0.132919} = 2046.7\ \text{K}$$

$$I_{sp}^0 = u_e = \sqrt{2\left[\chi c_c\tau + (1-\chi)\,c_p\right]T_{0c}\left[1-(p_e/p_{0c})^k\right]} \tag{3.148}$$

$$I_{sp}^0 = u_e = \sqrt{2\times[0.32\times 1370\times 1.1 + 0.68\times 1853]\times 3600\times\left[1-(1/70)^{0.132919}\right]}$$

$$I_{sp}^0 = u_e = \sqrt{2\times[1742.28]\times 3600\times\left[1-(1/70)^{0.132919}\right]} = 2326.5\ \tfrac{\text{m}}{\text{s}}$$

$T_c = 1.1T_g$ & $u_c = 0.9u_g \rightarrow \tau = 1.1$ & $\phi = 0.9$

$$\frac{T_e}{T_{0c}} = \left(\frac{p_e}{p_{0c}}\right)^{\text{km}} \tag{3.149a}$$

$$km = \frac{R}{[\chi/(1-\chi)]\,c_c\tau + c_p}\,\frac{\left[1-\chi\left(1-\phi^2\right)\right]}{[1-\chi(1-\phi)]} \tag{3.149b}$$

$$km = \frac{340.563}{[0.32/0.68]\times 1370\times 1.1 + 1853}\,\frac{[1-0.32\times 0.19]}{[1-0.32\times 0.1]} = 0.128965$$

$$T_e = 3600\times(1/70)^{0.128965} = 2081.4\ \text{K}$$

$$u_{ge} = \sqrt{2\frac{\left[\chi c_c\tau + (1-\chi)\,c_p\right]}{\left[1-\chi\left(1-\phi^2\right)\right]}T_{0c}\left[1-(p_e/p_{0c})^{\text{km}}\right]} \tag{3.150}$$

$$I_{sp}^0 = [1-\chi(1-\phi)]\,u_{ge} \tag{3.146}$$

$$u_{ge} = \sqrt{2 \times \frac{[0.32 \times 1370 \times 1.1 + 0.68 \times 1853]}{[1 - 0.32 \times 0.19]} \times 3600 \left[1 - (1/70)^{0.128965}\right]}$$

$$u_{ge} = \sqrt{2 \times [1855.07] \times 3600 \left[1 - (1/70)^{0.128965}\right]} = 2373.7 \frac{\mathrm{m}}{\mathrm{s}}$$

$$I^0_{sp} = [1 - 0.32 \times 0.1] \times 2373.7 = 2297.7 \frac{\mathrm{m}}{\mathrm{s}}$$

Discussion

1. The model assumes frozen flow. Even in the frozen flow the specific heats of the gaseous mixture and the condensed matter reduce as the static temperature falls during the nozzle flow. Furthermore, under the possible shifting-equilibrium flow, the molecular mass of the combustion products increases due to the recombination of the dissociated species during the flow. Therefore, the frozen two-phase flow model being sensitive to the assumed values of c_c, c_p, τ, ϕ, and $\overline{m}$, these values should be carefully chosen to reflect the realities.
2. As the condensed particles of metal oxides are heat-resistant ones, they can assume temperatures higher than the temperature of the gaseous mixture. As per the model, this results in a higher "effective c_p" that renders relatively higher specific impulse. However, it is observed that the high-temperature particles lose the heat by way of radiation to the surroundings rather than contributing to the flow.
3. The addition of combustible metal powders in solid propellants gives higher adiabatic flame temperature rendering higher specific impulse values. But, there are challenges in accurately predicting the nozzle-flow performance considering the heat transfer effects between the gaseous species and the condensed ones and the drag characteristics of particulates in the flow.

3.9 Nozzle Contours

The nozzle design has three main aspects: the gas dynamic design, the thermal design, and the structural design. Here, we focus on the gas dynamic design aspects by which the nozzle contour is determined. Conical nozzles are easier to fabricate and are adopted for low nozzle area ratios A_e/A_t. However for engines operating in high altitudes or in a vacuum environment, taking advantage of the least or negligible external drag experienced by the vehicle, nozzles of high area ratios are designed to achieve maximum specific impulse, Eq. 3.108. But the conical nozzles of such high area ratios will be very long and consequently heavy. Added to this, the conical nozzles suffer from the specific impulse loss due to the nozzle-exit velocities not being parallel, Eq. 3.34. To circumvent this issue, the nozzle contour to produce the desired uniform parallel exit flow can be accomplished by the method of characteristics as done in wind tunnel design. But here again, the nozzle length is expected to be excessively long and unacceptably heavy. Therefore, **contoured nozzles** are designed through optimization routes by adopting either the classical method of characteristics or the modern computational fluid dynamics (considering real flow effects) to give the best possible specific impulse under specified constraints, say given the nozzle length and the area ratio A_e/A_t. Such optimized contoured nozzles are practically far better than the conical ones by way of length and mass and more importantly on specific impulse. On yet another aspect, even with low area ratios, high-performance boosters of large rocket vehicles are designed to have contoured nozzles to offer wider full-flow nozzle condition with over-expansion [see Sect. 3.5.4].

A rocket nozzle is basically of three sections, the convergent portion where the subsonic acceleration takes place, the throat region of transonic flow, and the divergent portion with supersonic acceleration. The nozzle being fundamentally a system of supporting pressure gradient, the characteristics of the flow are not that very sensitive to its geometrical design. Therefore, its geometry is frequently selected for reasons of ease of manufacture, more particularly for the convergent region. Usually about 75% or more of the total thrust is developed by the combustion products accelerating in the convergent portion to sonic velocity at the nozzle throat and the balance is by the expansion in the divergent portion. Depending on the velocity at the nozzle entrance, the nozzle **contraction ratio** or **convergence ratio** (the ratio of the combustion chamber area at the nozzle entrance to the nozzle throat area) A_n/A_t is around 2 or more. For higher values of this ratio, the conical convergence half angle can be higher. This half angle varies from 30 to 45°. Steeper convergence angle increases the erosion in the convergence section because of a thinner boundary layer and a greater number of condensed particle impacts in the case of two-phase flows. Generally the conical convergent portion of the nozzle is merged with the chamber with a small radius of curvature. Sacrificing to some extent the ease of fabrication, the convergent portion and the throat region are sometimes bi-sinusoidal or of two circular arcs. In two-phase flows, a large convergence radius of curvature provides more uniform acceleration of the gas-particle mixture from the low velocities in the chamber to approximately sonic velocity at the throat [18].

Near the throat, the radius of curvature must ensure a smooth velocity transition from the subsonic to supersonic values through the transonic regime. For liquid and cryogenic engines, the convergent and divergent portions are joined by double toroidal sections, say 1.5–2 times the throat radius for the convergent side and about 0.5 times or more for the divergent portion. Smaller the radius of curvature for the convergent side, shorter is the nozzle length but it with a faster flow acceleration results in a higher distortion of choking contour (the so-called **sonic line** of $M = 1$ at the throat region) from the nozzle wall to axis. For solid motors, the joining is simply by a single toroidal section of radius 0.5–2 times the throat radius. Smaller values are adopted for larger motors, although the radius of curvature being equal to

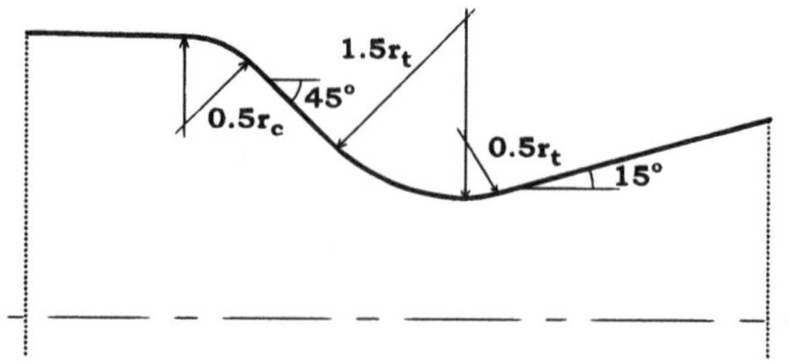

Fig. 3.16 Nozzle profile with conical convergent and divergent portions joined by double toroidal sections

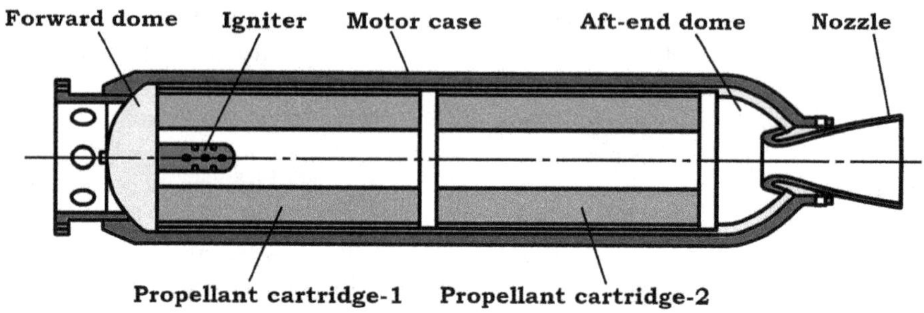

Fig. 3.17 A large solid propellant rocket motor of 1.22 m diameter with its submerged nozzle (adopted from [21])

the throat radius is common [20]. For large particle-laden flows in solid motors, the throat radius of curvature plays a vital role on the specific impulse quality factor or velocity correction factor ξ_s (Eq. 3.113). The factor ξ_s increases with the increase in throat radius of curvature.

For the conical divergent portion, as mentioned previously, a half cone angle α_d of 15–20° is adopted. Note that a higher α_d gives shorter nozzle length but with higher divergence loss. A schematic of nozzle profile for a conical nozzle with a nozzle contraction ratio around 3 is shown in Fig. 3.16. In large solid motors of advanced designs, **submerged nozzles** are often used to improve motor performance on the point of view of enhanced propellant loading and longer nozzle divergence within the constraint of overall motor length, Fig. 3.17. In addition, submerged nozzle designs facilitate to some extent the thrust vector control capability.

The classical method to design optimized nozzle contours is that of Rao [6], [22–24] and it was adopted by rocket industries for more than two decades since the mid-1950s. This design is based on the method of characteristics [8], which is applicable only to two-dimensional inviscid isentropic supersonic flows. To initiate the supersonic flow analysis, one requires a contour of transonic Mach number slightly greater than unity, a little downstream of the throat. This is done by the transonic flow analysis by applying small perturbation techniques [24]. The nozzle contour thus computed being applicable only for inviscid isentropic flows, in order

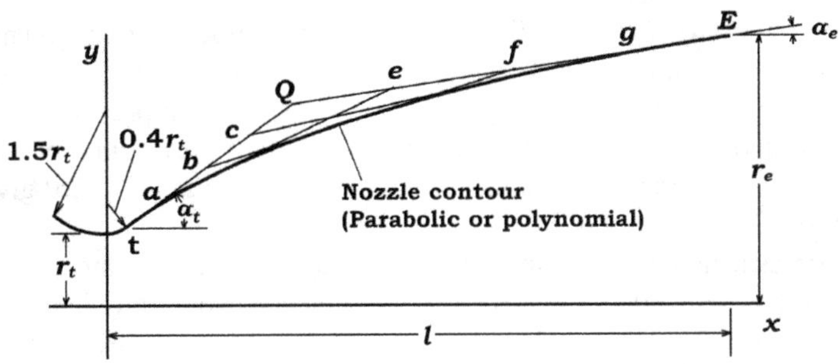

Fig. 3.18 A schematic of a contoured nozzle

to compensate to some extent the real flow effects, the displacement thickness of the boundary layer along the nozzle contour is calculated and the computed nozzle contour is shifted accordingly to arrive at the final design. However, the rocket nozzle flows being viscous, turbulent, of two phases (in the case of solid motors), and reacting (more applicable for cryogenic and also other liquid engines), full justification is not possible by Rao's method [22]. Therefore adopting real flow effects, an integrated nozzle contour design that includes the portions of convergence, throat, and divergence is currently adopted for high-performance rocket engines. For the optimization of nozzle contours in an integrated way, the thrust and the specific impulse can be accurately calculated by numerically solving the governing equations representing the real flows. However, detailing here either the method of characteristics or the computational fluid mechanics is beyond the scope of this book. Therefore, a brief treatment on the current methods being followed by rocket industries to arrive at the optimized nozzle contours for high-performance engines is given here. There may be some procedural variations from one industry to another. A schematic of a contoured nozzle is given in Fig. 3.18.

For the optimization of nozzle contour in solid motors, the objective function to be maximized is usually the ratio of thrust to nozzle surface area as the surface area can be easily related to the mass from the nozzle structural configuration. However, for liquid or cryogenic engines, as a first step the objective function to be maximized is thrust or specific impulse as the related nozzle mass with its involved cooling system is not easy to be arrived at. Subsequently, if needed, adopting the optimized contour reached in the first step along with the associated structural mass with its cooling system can form a second step optimization. Let us now look at the optimization procedure. Initially, choosing a suitable radius or radii of curvature at the throat for the specified throat radius r_t and area ratio A_e/A_t, the conical nozzle length with $\alpha_d = 15°$ is calculated as $l_{15°}$. For each of the different fractional lengths of $l_{15°}$, say 0.7, 0.75, and 0.8 of $l_{15°}$, choosing α_t and α_e (Fig. 3.18), contours of a polynomial, say $y = a + bx + cx^2 + dx^3$ (where y is the radial distance from the nozzle axis and x is the axial distance from the throat section) are generated. The coefficients a, b, c, and d are derived by the coordinates and the tangent conditions at α_t and α_e. Taking into account the major losses due to the nozzle divergence, viscosity, two-phase flow (for solid motors) or reacting flow (for liquid and cryogenic engines) numerical analyses for the axisymmetric turbulent flow field are performed. By systematically varying α_t and α_e, a series of contours are generated and the procedure described in [25] is adopted to select the optimum nozzle contour. For contours, instead of polynomials, canted parabolas can also be adopted. References [17, 26] give details of the numerical procedures.

One major issue in the case of solid motors is that the maximum thrust nozzle contours have a high **turn-back angle** ($\alpha_t - \alpha_e$), which can cause particle impingement at the nozzle exit [27]. The usual design practice is to avoid particle impingement by sacrificing some amount of I_{sp}, as particle impingement can severely damage the nozzle exit apart from the momentum loss (thrust loss). Hence, a sub-optimum contour is selected to avoid particle impingement by conducting an Eulerian–Lagrangian gas-particle flow simulation for the nozzle. The length reduction of a contoured nozzle with particle-laden flows in solid motors is typically 80–90% of the length of an equivalent conical nozzle with 15° half angle. Generally the impingement can be minimized with α_t between 20 and 26° with a turn-back angle of 10–15°. Typically $\alpha_t = 25°$ with $\alpha_e = 13°$ is a representative of a common contour nozzle for solid rocket motors. As the two-phase losses are not there in liquid engines, particle impingement also is not there and one can select the maximum thrust contour.

In situations where the demand for quick design overshadows the advanced calculations to arrive at the optimum nozzle contour, a single canted parabola for the divergence can very closely approximate the optimum. It must be realized that although the optimization calculations are mathematically correct, input property values of the combustion product flow are only approximate unless substantiated through extensive experimental verification. Therefore, for engineering purposes, near-optimum parabolic contours are suitable for many applications [28]. Figure 3.19 shows the initial and final nozzle-wall angles α_t and α_e for a range of r_e/r_t and l/r_t. The parabola has to fit α_t at the throat region and α_e at the nozzle exit. The values of α_t and α_e are the results of calculations adopting Rao's method [22] using $\gamma = 1.23$ and the throat region curvatures shown in Fig. 3.18. A different value of γ does not appear to change the nozzle contour appreciably when the area ratio and the length are prescribed. Similarly it can be expected that minor changes in the throat region curvatures will not cause large changes in the contour. Hence, the parabolic contours with the values shown in Fig. 3.19 can be used to describe approximately the optimum-thrust nozzle contours. This however is subject to the small shift in the contour that can be effected for boundary layer correction. Furthermore, it must be noted that the results of Fig. 3.19 are more applicable for the liquid engine conditions, where two-phase flow issues are absent. For classroom exercise, the drafting procedure shown in Fig. 3.18 can be followed.

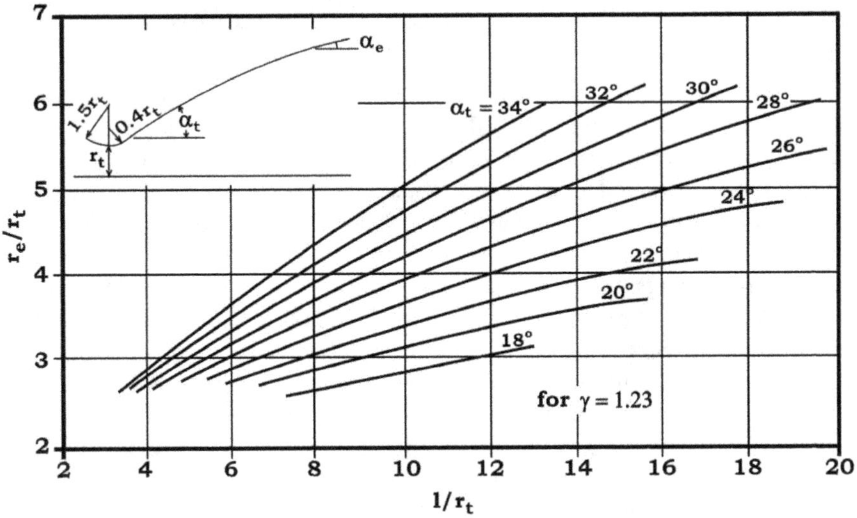

(a) Parametric values of α_t corresponding to various exit-radius ratios and nozzle-length ratios

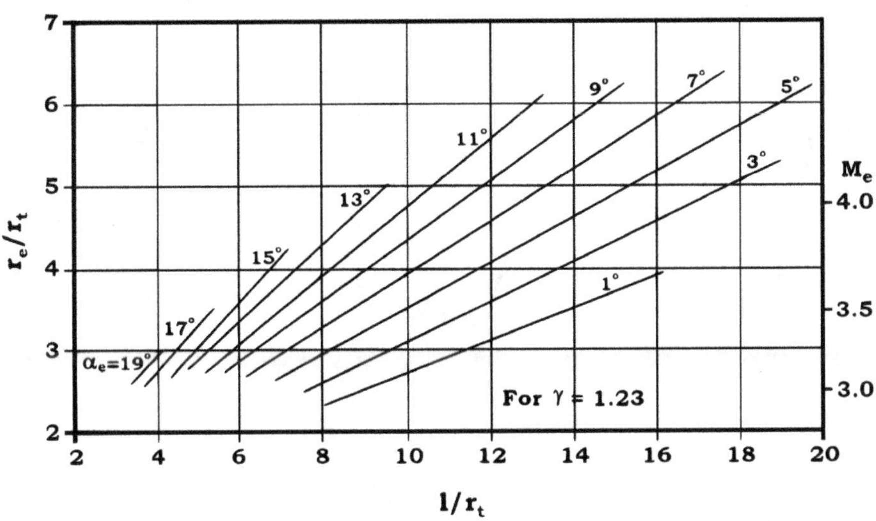

(b) Parametric values of α_e corresponding to various exit-radius ratios and nozzle-length ratios

Fig. 3.19 GVR Rao's contoured nozzle charts (adopted from [6])

3.10 Advanced Concepts in Nozzle Configurations

Although the conventional conical nozzles of simple designs and the bell or contoured nozzles of high-performance designs have all along been adopted by rocket industries, advanced concepts known as **plug nozzles** (also known as **spike nozzles**) and **expansion–deflection nozzles** or in short **ED nozzles** have been under research and development since the 1950s. These advanced concepts are expected to give acceptable nozzle performance practically for an unlimited range of altitudes with attractive savings on nozzle mass and length. In the conventional nozzles, the flow directions in the throat region as well as at the nozzle exit are essentially axial, whereas in the advanced concepts the flow direction at the throat is inward for the plug type and outward for the expansion-deflection one, as shown in Fig. 3.20. Furthermore, the supersonic expansion process in the conventional nozzles is solely governed by the enclosed divergent wall. And, the ambient pressure does not affect the internal expansion process until the separation limit due to the over-expansion occurs. On the other hand, the supersonic expansion process in the advanced nozzle concepts is directly or indirectly governed by the ambient pressure, which results in a very wide range of altitudes with acceptable performance. Therefore, these nozzles are also known as self-adjusting [29] or altitude-adaptive ones [30].

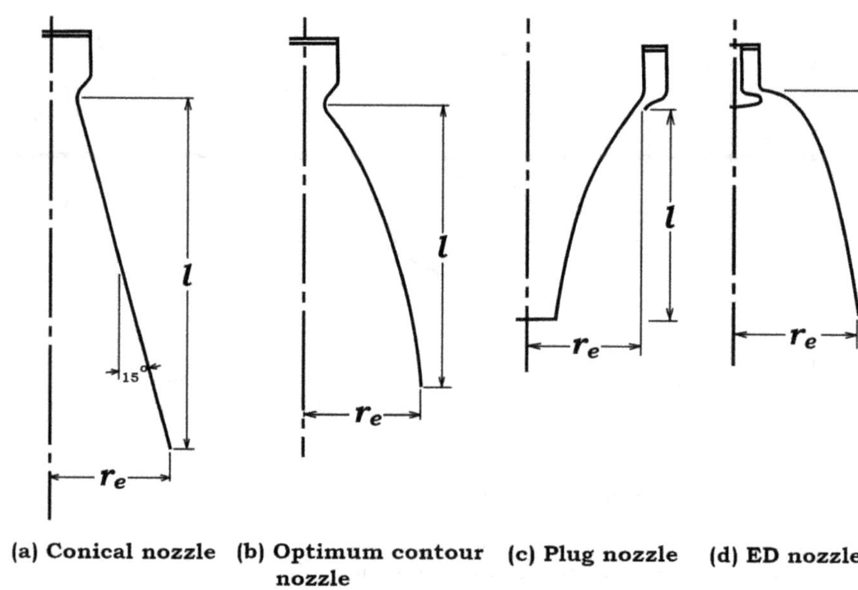

Fig. 3.20 Main features of various nozzles

3.10.1 Plug Nozzles

The nozzle with the straight conical central body was initially known as plug nozzle and the one with the contoured conical central body with or without truncation bore the name spike nozzle. However, the recent nozzle configurations being studied under this category are always contoured and mostly truncated ones, and referred to as plug nozzles. The nozzle without truncation is known as a full-length plug nozzle while the one with truncation is identified as a truncated plug nozzle. As the pointed end of the central body in a full-length plug nozzle poses structural and thermal issues, the plug nozzles being studied are mostly truncated. The plug-type nozzle is schematically shown in Fig. 3.20. If the central body is conical, the nozzle is known as circular, annular, or toroidal plug nozzle. In this type, the gas-ejection exit annulus can be fully open with a few supporting struts or can be formed by a cluster of circular contoured nozzle modules or of quasi-rectangular modules. If the central body is a wedge-shaped rectangular one, the nozzle is known as linear plug nozzle. This type was adopted and extensively ground-tested for the aerospike engine of Single-Stage-to-Orbit (SSTO) vehicle X-33, the project of which was prematurely closed in 2001, apparently for the want of lighter propellant tanks. The image of the test in progress of the Linear Aerospike XRS-2200 engine of X-33 is shown in Fig. 3.21. In the case of a linear plug, the nozzle exit can be of a cluster of rectangular nozzle modules.

In a plug nozzle, at the outer edge of the exit, the cowl lip, the gases expand abruptly to the ambient pressure p_a. The external diameter with reference to the cowl lip of a full-length plug nozzle can be considered as the exit diameter of a

Fig. 3.21 The test of Linear Aerospike XRS-2200 engine of X-33 program on August 6, 2001 at NASA's Stennis Space Center (Courtesy NASA Marshall Space Flight Center)

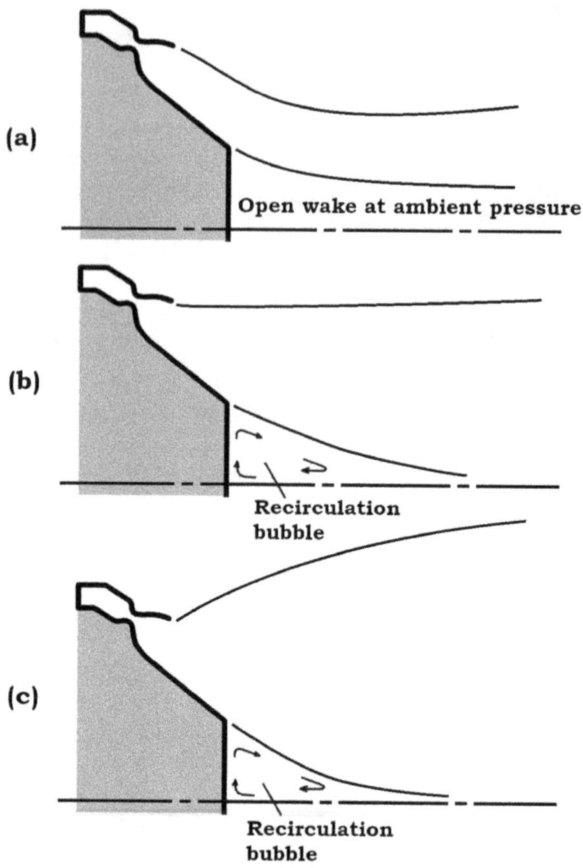

Fig. 3.22 Plug nozzle flows with constant chamber pressure at different altitudes: **a** sub-design altitude, **b** design altitude, and **c** above design altitude (adopted from [30])

uniform flow convergent–divergent contoured nozzle expanding the gases to the same ambient pressure, Fig. 3.20. We may consider the behavior of the exhaust either with the variation in chamber pressure at a constant altitude (constant ambient pressure) or with the variation in altitude (varying ambient pressure) at a constant chamber pressure. As the self-adjusting advanced concepts give acceptable nozzle performance for a much wider range of altitudes, for better clarity let us consider their performance with the variation in altitude for a constant chamber pressure. Under the design condition, the exhaust flow, controlled by the expansion waves emanating from the cowl lip, glides over the contoured surface with its exposed outer boundary parallel to the nozzle axis, Fig. 3.22. While the outer boundary will be parallel to the axis at the design altitude (design ambient pressure), the ballooning out of the exhaust occurs similar to the conventional nozzle plume as the operational altitude is further increased.

Let us briefly discuss the flow field performance of the circular plug nozzle [30]. In the case of conventional conical or contoured nozzles, if the over-expansion limit for full flow is not satisfied, as discussed previously in Sect. 3.5.4, the flow separates.

But in the case of a circular plug nozzle, even beyond the over-expansion limit, with a system of compression and expansion waves the flow expands without separation with its exposed outer boundary more or less parallel to the contoured surface, diverging a little with several inflections, as shown in part (a) of Fig. 3.22. And, the wake of the truncated plug is open and exposed to the ambient pressure. As the operational altitude is increased towards the design value of a full-length plug nozzle, the exposed boundary tends to be parallel to the nozzle axis, part (b) of Fig. 3.22. The exiting boundary from the contoured surface of the truncated plug closes forming a **recirculation bubble** at the wake with a pressure less than the ambient value causing a negative thrust component. With further increase in the operational altitude above the design value, the ballooning of the exposed boundary as in the conventional nozzle happens. And, advantageously the recirculation pressure, being more or less constant and less than the ambient pressure at the design altitude, becomes greater than the ambient pressure at higher altitudes, and it renders a positive thrust component from the truncated wake. However, this positive thrust component is only minor and it cannot compensate for the higher pressure-surface integral (the major thrust component) that is possible in a full-length plug nozzle. While the full-length plug nozzle, without the wake flow issues, is able to operate without separation and over-expansion losses up to its design altitude and thereafter behaves like a conventional contoured nozzle, the truncated one repeats this performance only up to the design altitude but thereafter suffers due to the wake losses and renders specific impulse lower than that of the former. Larger the truncation higher is this penalty. The flow field performance of a linear plug nozzle is more or less similar to that of a circular plug except the flow deviation that can occur due to the two end surfaces open to the atmospheric air and beyond, Fig. 3.21. The main expansion can spread sideways causing thrust losses. And, the surroundings will influence the wake recirculation zone.

All in all, the altitude compensation capability of circular and linear plug nozzles at launch and lower altitudes is indisputable. But, beyond design altitudes the full-length plug performs similar to the conventional contoured nozzle and the truncated one exhibits lower performance, nonetheless with its intrinsic advantage of attractive saving on nozzle length and mass. Therefore, the designers would always try to enhance the design altitudes as much as possible for the truncated plug nozzles. In spite of the annulus exit in the circular plug or the rectangular exit in the linear plug being far removed from the engine axis causing ducting difficulties and heat transfer issues, the facility that these exit areas can be formed by a cluster of quasi-rectangular or of rectangular modules enables wide possibility of thrust control.

3.10.2 Expansion–Deflection Nozzles

In an expansion–deflection (ED) nozzle, the throat annulus is located around a central plug close to the engine axis, and hence the combustion chamber is compact, Fig. 3.23. Consequently, in contrast to plug nozzles, the small engine envelope due

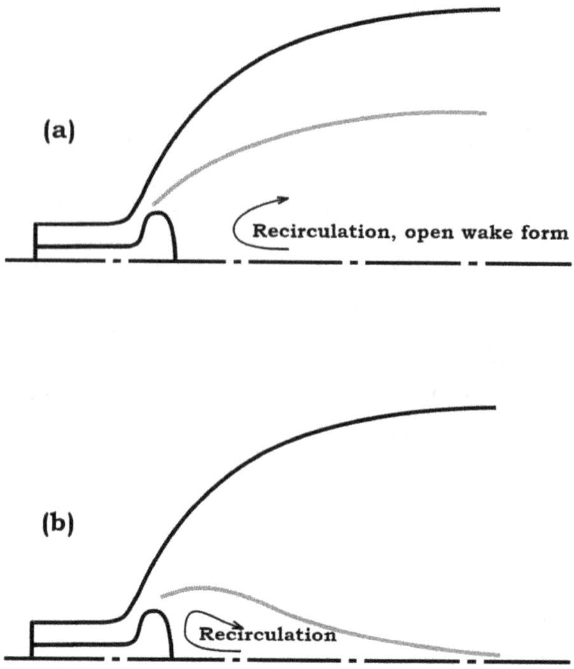

Fig. 3.23 Expansion–deflection nozzle flows with constant chamber pressure at different altitudes: **a** sub-design altitude and **b** above design altitude (adopted from [30])

to the compact combustion chamber is a major advantage of ED nozzles. Because of this inherent compactness, the ED nozzles could become the preferred ones against the plug nozzles for the engines operating in space with very high nozzle area ratios (A_e/A_t). In an ED nozzle, the gases exit the throat outwardly and expand around the central plug shoulder, bounded at the outer by the contoured nozzle-wall shroud. And, the wake pressure at the base of the central plug controls the expansion of the exposed inner boundary, making the nozzle altitude adaptive. At sub-design altitudes, the gases expand in an annular form at the nozzle exit with an enlarging open wake behind the central plug, Fig. 3.23. Due to the ejector action of the expanding supersonic stream, the wake pressure that controls the expansion is lower than the ambient value. At these altitudes, similar to the performance of plug nozzles, with a system of compression and expansion waves the flow expands with its exposed inner boundary more or less parallel to the contoured shroud, diverging a little with several inflections. At sub-design altitudes, in the case of a plug nozzle, the exposed outer boundary of the expanding gases is directly controlled by the ambient pressure, whereas in the case of an ED nozzle, the exposed inner boundary is controlled by the open-wake pressure which is lower than the ambient pressure due to the ejector action. This results in some over-expansion losses at sub-design altitudes. Therefore, under similar sub-design conditions, the thrust performance of an ED nozzle is expected to be lower than that of a plug nozzle. As the altitude is increased beyond the design value, the inner boundary exposed to the wake pressure closes forming a full flow at the nozzle exit, Fig. 3.23. And, the enclosed wake becomes a recirculation bubble with a pressure essentially independent of ambient value. The length of an ED nozzle is nearly the same as that of a plug nozzle for the equivalent thrust performance. Like plug nozzles, the ED nozzles suffer from higher heat fluxes at the throat region relative to the conventional contoured nozzles of equal throat areas [30].

3.11 Nozzle Construction

Under the topic of nozzle construction, let us first consider the common construction aspects for liquid- as well as solid-propulsion units. As we noted previously, the three basic phases of nozzle design are: (1) the gas dynamic design to give within the constrained envelope the maximum possible nozzle-exit velocities parallel to the nozzle axis; (2) the thermal design to select the liners that form the gas dynamically contoured inner surface and the insulators or the coolant systems behind the liners such that the nozzle is protected against erosion and keep it within the permissible temperature limits; and (3) the structural design to select suitable materials to support the thermal components and sustain the predicted loads, Fig. 3.24. And in addition if necessary, the structural design involves the development of components that will facilitate thrust vector control, be it a movable nozzle thrust vector control or a liquid injection one. Near the nozzle throat region the mass flux is maximum and hence the heat transfer flux is also the maximum. Therefore, a superior heat- and erosion-resistant material as **throat insert** is provided to limit the increase in throat area. As an example of these common aspects, Fig. 3.24 shows the construction features of solid motor nozzles. In liquid engines of short-duration thrusting, for cooling and retention of nozzle contour, the liner can be of ablative material with an insulator behind. But for the engines of long duration thrusting, to protect the liner, the insulator is replaced by a cooling system, wherein one of the propellants, generally the fuel, is circulated around the nozzle and chamber. Depending on the complexity of nozzle design in liquid- as well as solid-propulsion units, the **liner**, the **insulator** or **coolant system**, and the **structure** can be separate, two in one, or all in one.

As composites are extensively adopted in rockets, a brief introduction to polymer materials and composites is appropriate. Polymeric materials can be broadly classified into **elastomers** and **plastics**. Elastomers are the polymers, having high elongation within the elastic range. Natural rubber, polyurethane elastomer, silicone rubber, and polybutadiene are some examples. Plastics, the second category, are broadly of two kinds, **thermoplastics** and **thermosets**. The former requires heat to make it formable and after cooling can be reheated and reformed into new shapes without significant changes in properties. Cellulosics (celluloid), fluoroplastics (teflon), nylons (polyamides), and polyimide are examples. A thermoset uses heat to make its shape permanent. Epoxy, phenolic, and polyester (Mylar) are some examples.

A **composite** is a matrix material reinforced in random or oriented form with a special fiber, tape, fabric, etc. For example, in the term for the composite carbon cloth /phenolic, the first part of the term identifies the reinforcement material as carbon cloth and the second one identifies the thermoset phenolic-resin matrix that binds

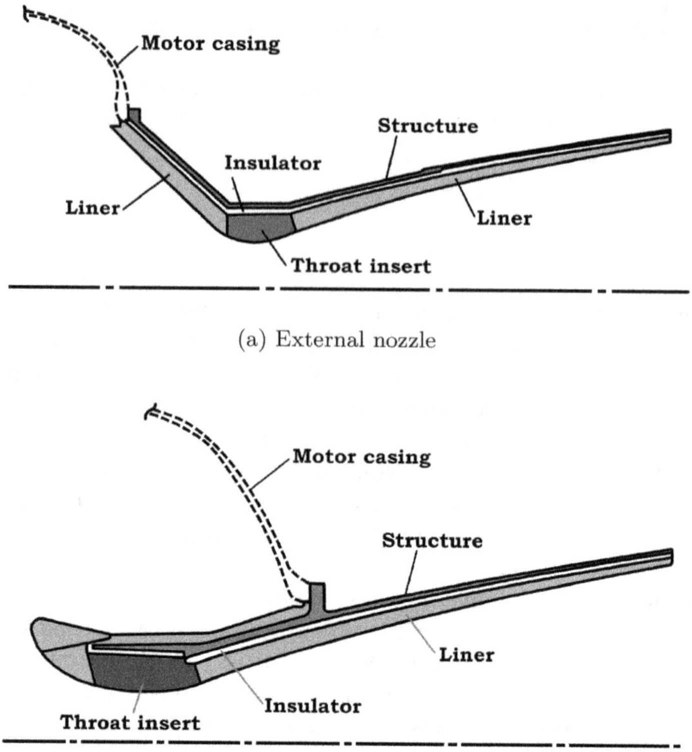

(b) Submerged nozzle

Fig. 3.24 Basic solid rocket motor nozzle configurations (adopted from [31])

the reinforcement together. Composites are classified into **polymer matrix composites**, **ceramic matrix composites**, and **metal matrix composites**. A polymer matrix composite is the one composed of a reinforcement and a binding polymer matrix of an elastomer or a plastic. The previously mentioned carbon cloth/phenolic is a polymer matrix composite; some more examples are graphite/phenolic, glass /epoxy, and silica-asbestos/butadiene acrylonitrile rubber (NBR) elastomer. Apart from the conventional polymer matrix composites, ceramic matrix composites and metal matrix composites find applications in rocket construction. A ceramic matrix composite consists of a ceramic fiber embedded in a ceramic matrix. The composite carbon/carbon is a ceramic matrix composite. The first carbon refers to the fibrous carbon reinforcement and the second one is some type of amorphous-carbon matrix material that binds the reinforcement. Some more examples of ceramic matrix composites are carbon/silicon carbide, graphite/carbon, and alumina/alumina. A metal matrix composite is the one with at least two constituent parts, one belonging to a metal necessarily, the other material may be a different metal or another material type, say a ceramic or a polymer; some examples are carbon/aluminum, carbon/titanium, graphite/aluminum, and silicon carbide/titanium. While a ceramic matrix composite is used as a superior heat- and erosion-resistant material, a metal matrix composite is used mainly for its high strength–mass ratio as a structural member.

Orientation of reinforcements plays a vital role on the composite-material properties, such as erosion, thermal resistance, and strength. More particularly in the development of polymer matrix composite components various methods under tape wrap, layup, and die molding are adopted [31]. Typical examples are graphite/phenolic tape for throat insert, glass cloth/epoxy lamination layup for structure, and silica/phenolic die molding for liner cum insulator.

Apart from the composites, special metal alloys are widely used in rocket construction. In comparison between composites and metal alloys, the strength of metals generally drops less rapidly with the increase in temperature than that of most composites. Furthermore, the strength reproducibility of composite materials is not as good as that of metals. However, the attractive higher strength–mass ratio or better heat resistance of composites in special applications is indisputable.

3.11.1 Solid Rocket Nozzles

As we know, the nozzle configuration in solid motors is of two types, one the conventional external nozzle and the other the submerged nozzle, Fig. 3.24. In the submerged nozzle configuration, the entry, throat, and a part or whole of the nozzle-exit section are cantilevered into the combustion chamber. The design of a submerged nozzle is more complex as the submerged inner as well as outer surfaces are exposed to high-temperature combustion products in addition to the structural load to be withstood against the related external and internal pressures. However, the submerged nozzle design more efficiently uses the constrained envelope allotted to the propulsion system along with the thrust vector control that can be incorporated by the movement of the entire nozzle.

With the maximum heat flux in the throat region, the throat insert should be of a material having high resistance to heat and erosion. Erosion is much more significant in small throats than in large ones since the relative change in the throat area is greater in the former. Thermoset-based polymer matrix composites (carbon/phenolic rosette layup, carbon/phenolic tape, graphite/phenolic tape, silica/phenolic die molding, etc.), polycrystalline graphite, pyrolytic graphite, refractory metals tungsten and molybdenum, and ceramic matrix composites (carbon/carbon, graphite/graphite, silicon carbide/graphite, silicon carbide/silicon-nitride) are used for throat inserts. **Polycrystalline graphite** is erosion resistant and has the characteristic of increase in strength with the rise in temperature. However, because of its relatively low strength, thick sections with good structural backups have to be provided. Although polycrystalline graphite finds application in flight qualified nozzles as liner parts, it as a regular throat insert is used generally in ground test motors. **Pyrolytic graphite** is an anisotropic material formed by the vapor deposition of carbon layer by layer on a substrate. It is denser, stronger, and more erosion resistant than polycrystalline graphite

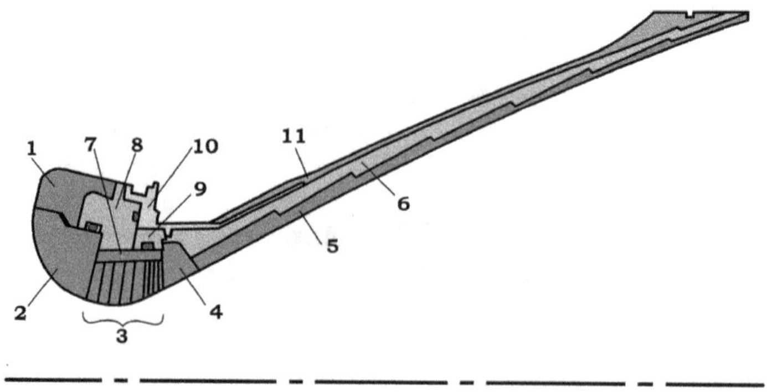

Fig. 3.25 Nozzle details of solid motor X-260 used in the second stage of Polaris A3 missile. (1) graphite/phenolic tape to enhance thermal liner; (2) polycrystalline graphite—throat approach thermal liner; (3) pyrolytic graphite washers—throat insert; (4) polycrystalline graphite—throat extension thermal liner; (5) graphite/phenolic tape - exit thermal liner; (6) asbestos/phenolic tape—exit insulation; (7) polycrystalline graphite—support ring for washers; (8) asbestos/phenolic die molding—entrance and forward throat insulation; (9) asbestos/phenolic die molding—aft throat insulation; (10) steel—entrance, throat, and forward exit structure; and (11) glass/epoxy—aft exit structure (adopted from [31])

and exhibits similar increase in strength with the rise in temperature. It has a thermal conductivity two orders of magnitudes higher along the layer than the one across [31]. In order to provide adequate thermal expansion, usually washers of pyrolytic graphite are stacked together to form the required throat insert with the layers parallel to the nozzle axis. A typical throat insert arrangement using pyrolytic graphite washers is shown in Fig. 3.25. Among refractory metals, the highly erosion-resistant tungsten, its alloys, and its metal matrix composites (known as infiltrated tungsten) such as silver/tungsten find application as throat inserts in large long duration high-performance motors. Also, molybdenum may find application for throat insert in low-temperature short-duration motors. Ceramic matrix composites carbon/carbon and the previously mentioned ceramic composites are adopted for throat inserts. Like polycrystalline and pyrolytic graphite, the structural properties of carbon/carbon and its variants improve with the rise in temperature. Carbon/carbon composites possess good erosion resistance like polycrystalline graphite and in addition they have better strength. Therefore, these as throat inserts do not require thick sections and elaborate structural backups.

Figure 3.25 is to be closely studied also for our ensuing discussions on the liner, insulator, and structure in solid motor nozzle construction. In general, thermoset plastic-based polymer matrix composites and the previously noted polycrystalline graphite find extensive application as nozzle liners and liner parts. And, elastomer-based polymer matrix composites find application as liners in low Mach number zones, say the entry region of the nozzle convergence. Thermoset-phenolic composites reinforced with graphite cloth, carbon cloth, silica cloth, or glass cloth are adopted as nozzle liners. Of these, the first two reinforcements are most often preferred in that order where erosion concern is more in the regions just upstream and downstream of the throat. The other two silica cloth/phenolic and glass cloth/phenolic are for the liners in the lower nozzle divergence, where erosion concern is less. Some of the composites and their layup patterns adopted for liners in flight qualified nozzles are as follows: asbestos/phenolic die molding, asbestos/phenolic tape, carbon/phenolic die molding, carbon/phenolic rosette, carbon/phenolic tape, graphite/phenolic die molding, graphite/phenolic rosette, graphite/phenolic tape, silica- and asbestos-filled butadiene acrylonitrile rubber (NBR) elastomer, silica/phenolic die molding, and silica/phenolic tape.

Glass- and silica-reinforced phenolic or epoxy matrix are commonly used for insulators. The filled elastomers (NBR and polysulphide epoxy), the other group of insulators, are frequently adopted as insulator cum liners in nozzle-convergence entry zones where low Mach numbers prevail. The thermoset-epoxy matrix reinforced with glass (glass/epoxy) has been extensively used for insulators and structural components. Some of the composites and their layup patterns adopted for insulators in flight qualified nozzles are as follows: asbestos/phenolic die molding, asbestos/phenolic tape, carbon/phenolic die molding, carbon/phenolic rosette, carbon/phenolic tape, glass/phenolic die molding, glass/phenolic tape, graphite/phenolic die molding, silica and asbestos-filled NBR elastomer, silica/phenolic die molding, silica/phenolic rosette layup, silica/phenolic tape, and silicon rubber (flexible insulator). Also cork, polycrystalline graphite, and zirconium oxide are found used as insulators.

Composites as well as metals are adopted for the structural components in nozzles. Frequently, if a composite is used as a structural member it may also be chosen to serve as an insulator under the two-in-one concept. Thermoset resins, phenolic and epoxy, and thermoplastic polyimide are used for the structural composites—epoxy for low-temperature applications while phenolics and polyimides are for high-temperature applications. Glass/epoxy and glass/phenolic composites are frequently adopted. In advanced designs, graphite/epoxy is adopted to achieve lighter structures. Some of the composites and their layup patterns adopted for structural members in flight qualified nozzles are as follows: asbestos/phenolic tape, carbon/phenolic rosette, glass cloth/epoxy lamination, glass-filament/epoxy, glass/phenolic layup, glass/phenolic tape, silica/phenolic die molding, and silica/phenolic tape. Adopted metals for nozzle structures include 4130 steel (normalized or heat treated), 18% nickel maraging steel, 7075 aluminum alloy, and Ti64 titanium alloy. Steel is the most common structural material where the strength–mass ratio is not a primary consideration. Aluminum alloy is often adopted for small rockets with short burning times. Titanium has higher strength–mass ratio and retains its strength better at elevated temperatures and hence adopted for nozzle structure when the mass saving is vital and/or the operation at elevated temperature is the requirement. Sometimes molybdenum is used as the throat structure. Niobium (also known as columbium) is sometimes adopted as the three-in-one nozzle material for small motors and such nozzles are essentially radiation cooled. Honeycomb exit structure is used for large nozzles. Although asbestos-based composites have been adopted as nozzle materials, their use is being avoided due to health and environmental issues. Table 3.4 summarizes the materials adopted for the nozzle components of solid rocket motors.

Table 3.4 Materials for solid propellant rocket nozzles

Component	Materials
Throat region	Polymer composites: carbon/ phenolic, graphite/ phenolic, silica/ phenolic Ceramic composites: carbon/ carbon, graphite/ graphite, silicon carbide/ graphite, silicon carbide/ silicon carbide Metal composites: silver/ tungsten Refractory metals and alloys: tungsten, molybdenum Pyrolytic graphite Polycrystalline graphite
Liner	Polymer composites: graphite/ phenolic, carbon/ phenolic, silica/ phenolic, glass/ phenolic Polycrystalline graphite Silica and asbestos-filled NBR elastomer Elastomer-based polymer matrix composites (for entrance and convergence)
Insulator	Polymer composites: glass/ phenolic, silica/ phenolic, glass/ epoxy, silica/ epoxy Filled elastomer (NBR and polysulphide epoxy)
Structure	Polymer composites: glass/ epoxy, graphite/ epoxy, glass/ phenolic, carbon/ phenolic, silica/ phenolic, asbestos/ phenolic 4130 steel, 18% nickel maraging steel, Ti64 titanium alloy, 7075 aluminum alloy

3.11.2 Liquid Rocket Nozzles

Short duration and low-pressure liquid engines mostly adopt **ablative cooling**. Such units adopt simple pressure-fed liquid injection since a very high propellant-tank pressure is needed to negotiate a high pressure drop in the complex **regenerative cooling**. In these engines, the nozzle and chamber are ablatively or radiatively cooled and hence the details that we previously discussed in the construction of solid rocket nozzles are all applicable. Generally an ablative liner, a thin layer of insulation, and a high-strength outer shell are adopted. A typical nozzle construction of a short-duration and low-pressure liquid thruster is shown in Fig. 3.26—small liquid engines or solid motors are traditionally called as thrusters. As simplicity is the criteria for such small engines/ thrusters, the two-in-one concept of liner cum insulator is frequently adopted by providing a thick ablative liner material. Mostly, the ablative liner is fabricated from the high-temperature phenolic-based (say silica cloth/ phenolic and carbon/ phenolic) or advanced carbon/ carbon composites. Asbestos/ phenolic or rubber may form the insulator over the liner. The high-strength outer shell structure may be of glass/ epoxy or carbon/ carbon composites, titanium alloy, 7075 aluminum, or stainless steel of grade 304 or 316. In such an ablative cooled nozzle, the throat insert may be of quartz/phenolic composite, tungsten/molybdenum (98/2) alloy, or pyrolytic graphite. The throat insert is usually installed with heavy graphite backups for better structural strength. Attitude control and orbit-adjust thrusters,

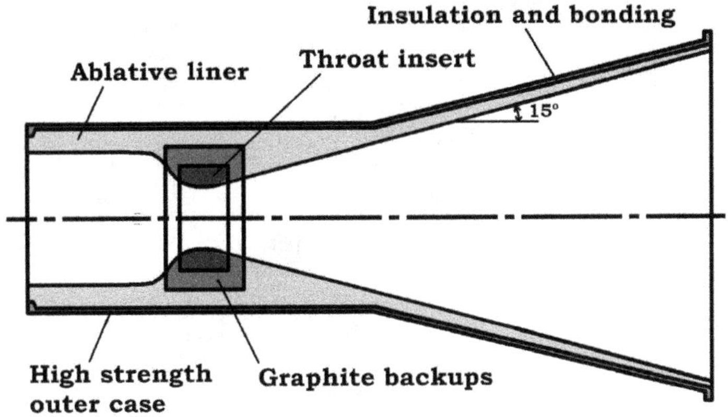

Fig. 3.26 Ablatively cooled nozzle and chamber (adopted from [32])

which are designed to give impulse pulses to continuous thrust, adopt a single metal for the radiatively cooled chamber-nozzle unit. Control thrusters of low temperatures are radiatively cooled adopting alloys of titanium, molybdenum or cobalt, or stainless steel. Also austenitic nickel–chromium-based alloys (Inconel) are used. Cobalt–nickel–chromium–tungsten-based Haynes 25 alloy is used in many hydrazine monopropellant thrusters. On the other hand, the high-temperature control units are of niobium. However, in some advanced designs of such thrusters, the parts of the chamber-nozzle unit can be constructed out of different high-temperature metals, say iridium-lined rhenium for chamber and throat region, silicide-coated niobium for upper nozzle, and titanium for lower nozzle [33]. As an additional support to the ablative as well as **radiative cooling**, injecting surplus fuel or a more fuel-rich layer close to the chamber-nozzle wall and thus keeping the liner at safe thermal level is invariably adopted. This is known as **film cooling**. For example, the Aerojet Rocketdyne built control thrusters R-4D (490 N thrust) and R-1E (110 N thrust) have integral chamber-nozzle units that are radiatively film cooled. The former one is of iridium-lined rhenium and the latter is of silcide-coated niobium and both use the hypergolic propellant combination, nitrogen tetroxide, and monomethyl hydrazine. Table 3.5 summarizes the materials adopted for the nozzles of ablatively and/or radiatively cooled liquid propellant thrusters.

Long-duration high-thrust liquid engines are pump fed and adopt the method of regenerative cooling along with film cooling. As noted previously, in a regeneratively cooled liquid engine, one of the propellants, usually the fuel, is used as a coolant. For this purpose, the chamber-nozzle unit is made up of the liner with a bundle of thin-walled tubes braced around or with the rectangular coolant passages milled and enclosed, Fig. 3.27. For the cooling process, the coolant is pumped at high velocities through the passages starting from the nozzle end or around the midway to pick up the heat transferred from the nozzle and chamber and is then injected into the combustion chamber with an enhanced temperature. As maximum heat flux

Table 3.5 Materials for nozzles of ablatively and/or radiatively cooled liquid propellant thrusters, [32, 33]

Component	Materials
Throat region	Polymer composite: quartz/ phenolic Tungsten/ molybdenum (98/2) alloy Pyrolytic graphite
Liner cum insulator	Polymer composites: carbon/ phenolic, silica/ phenolic Ceramic composite: carbon/ carbon
Structure	Polymer composite: glass/ epoxy Ceramic composite: carbon/ carbon, Titanium alloy, 7075 aluminum, 304/316 stainless steel
Single chamber-nozzle unit (low temperature)	Inconel, Haynes 25 Alloys of titanium, molybdenum or cobalt 304/316 stainless steel
Single chamber-nozzle unit (high temperature)	Niobium Iridium-lined rhenium Silicide-coated niobium Iridium-lined rhenium (chamber and throat), silicide-coated niobium (upper nozzle), and titanium (lower nozzle)

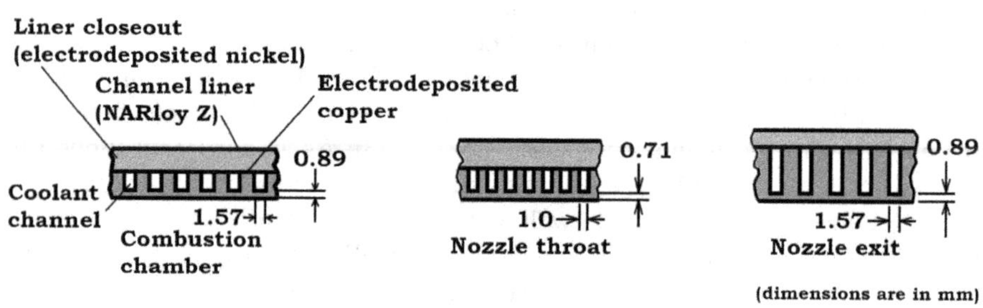

Fig. 3.27 Milled rectangular grooves for cooling nozzle and chamber of Space Shuttle Main Engine; dimensions are in mm (adopted from [28])

exists at the throat and its vicinity, in order to absorb the excessive heat flow, the coolant velocity in this region has to be the highest. Therefore, the cross-sectional area of the coolant tube or the milled coolant passages is designed to be minimal. The basic nozzle liner either with the bundle of tubes brazed around or with the milled and enclosed passages cannot withstand the external loads arising from side loads during startup, flow separation, or other mechanical forces from attachments. Therefore, either a continuous structural shell or intermittent structural bands are needed over the basic nozzle. Generally the continuous structural shell would stop just downstream of the throat followed by the intermittent rigid bands down to the nozzle exit, and evidently the latter arrangement instead of the continuous shell all around will save some engine mass. The rigid bands on the cryogenic engine nozzles can be seen in Figs. 2.19 and 3.28. The retaining rigid bands can be of a composite material or a suitable metal.

Most information presented here on materials is drawn from [32, 33]. In engines using Earth-storable propellants, especially the corrosive oxidizers such as nitric oxide or nitrogen tetroxide, the chamber-nozzle liners are usually of stainless steel, say A-286 or its variant JBK-75 with a bundle of thin-walled coolant tubes of stainless steel brazed around the liner. Coolant tubes of titanium alloy are for monomethyl hydrazine and kerosene. In LOX/kerosene semi-cryogenic engines, stainless steel 347 (a niobium-stabilized austenitic stainless steel) or Inconel-X750 (a nickel–chromium alloy) is used as the material for chamber-nozzle liner and coolant tubes. Stainless steel 347 and Inconel-X750 were adopted, respectively, in RS-27A and F-1 LOX/kerosene engines of Rocketdyne. While the chamber and upper nozzle of F-1 were regeneratively cooled, the lower nozzle was cooled by injecting close to the liner wall the highly fuel-rich combustion exhaust from the turbine—note in Fig. 3.28 the turbine exhaust manifold positioned at the beginning of the lower nozzle. In a cryogenic environment, **metal embrittlement** is of concern. Copper has high electrical and thermal conductivity, excellent corrosion resistance and ease of fabrication, plus immunity to hydrogen-environment embrittlement. Alloying it improves its strength at the expense of good thermal and electrical conductivity. The major application of copper alloys in LOX/LH2 cryogenic engines is for high heat flux applications such as combustion chambers and nozzles. Copper–silver (Narloy A) and copper–silver–zirconium (Narloy Z) alloys are for these applications. Nickel-based alloys (Inconel, Rene 41, Waspaloy, etc.) find extensive application in cryogenic engines. These are oxygen compatible but susceptible to hydrogen-environment embrittlement. The Space Shuttle Main Engine adopted the alloy Narloy Z for its milled liner. The milled passages were enclosed by the electrodeposited copper and nickel, Fig. 3.27. Iron-based superalloys (iron–nickel–cobalt alloys: Incoloy) have found widespread use in cryogenic hydrogen environments because these alloys possess good resistance for hydrogen-environment embrittlement. Aluminum alloys, with the relatively high strength–mass ratio, have excellent cryogenic properties and are resistant to hydrogen embrittlement. Evidently aluminum alloys cannot be used in an oxygen environment due to oxidation issues. Cobalt-based alloys (Alloy 188 and Elgiloy) are similar in strength and corrosion resistance to nickel-based alloys but advantageously compatible to both oxygen and hydrogen environments. Nozzle extensions of upper stage engines are generally radiation cooled and are of advanced carbon/carbon composite variants. By nozzle extension, we refer here to the detached part of the nozzle that is slid and attached to the parent nozzle at higher altitudes or hard vacuum to effect increased area ratio A_e/A_t to render higher specific impulse, Sect. 3.5.8 and Fig. 3.10a. This detached nozzle part is also referred to as extendible nozzle. Furthermore, the lower nozzle (our reference) is sometimes referred to as nozzle extension as mentioned in the original, Fig. 3.28.

Generally in all types of rocket engines/motors as the combustion products expand through the nozzle, high static temperatures prevail in the chamber and the nozzle convergence. High heat flux zone exists just upstream and downstream of the throat

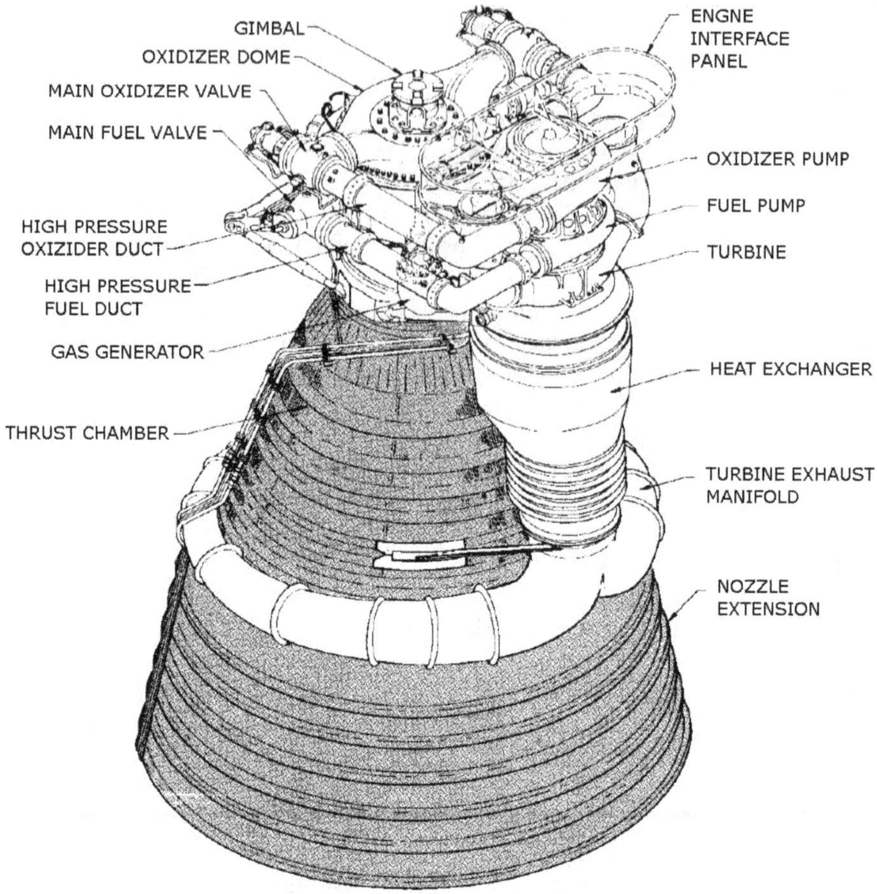

Fig. 3.28 F-1 LOX/kerosene rocket engine components (Courtesy NASA: NASA, Public domain via Wikimedia Commons)

where the erosion concern is more. Therefore, these areas of chamber and upper nozzle require effective thermal protection by way of employing highly heat-resistant materials, regenerative cooling, ablative along with film cooling or radiative cooling or selective combinations of these. In large engines, however, for the lower nozzle the required cooling effort is less because of the diminished heat flux and crosion concern due to the reduced static temperature and pressure of the expanding combustion products. Therefore, in many large pump-fed liquid engines, lower nozzle divergence is not regeneratively cooled but cooled by other means: injecting fuel-rich combustion exhaust from turbine, adopting ablative liner, or cooling through radiation by fabricating the lower nozzle out of graphite/ graphite or any other advanced composites. As a striking example, the large regeneratively cooled LOX/LH2 engine RS-68

(3312 kN thrust), built by Rocketdyne, has its lower nozzle out ablative lined material [34]. Invariably nozzle extensions of space engines (Fig. 3.10) are constructed out of high-temperature composite materials enabling radiation cooling. Also, in large solid motors, the lower nozzle is lined with low-temperature ablative composites or out of low-temperature metal alloys.

Problems

3.1 Briefly discuss the thrust production in the general thrust producer, namely, the basic turbojet. How does this thrust production get modified in the cases of propeller thrust and rocket engine thrust?

3.2 Briefly discuss the operating conditions under $p_e = p_a$, $p_e > p_a$, and $p_e < p_a$.

3.3 What are the limiting conditions for full nozzle-flow under over-expanded condition?

3.4 Distinguish between momentum thrust and pressure thrust.

3.5 Briefly compare and discuss the characteristics between air-breathing engines and rocket engines.

3.6 Flight vehicles fitted with air-breathing engines cannot fly at velocities equal to or greater than the nozzle exhaust velocities. But those fitted with rocket engines can do it. How?

3.7 Derive the following propulsive efficiency equations for air-breathing engines and rocket engines.

$$(\eta_p)_{ab} = \frac{2\nu}{1+\nu} \tag{3.25}$$

$$(\eta_p)_r = \frac{2\nu}{1+\nu^2} \tag{3.26}$$

3.8 Considering the divergence of nozzle-exit velocities in conical nozzles derive the thrust equation,

$$F = \left(\frac{1+\cos\alpha_d}{2}\right)\dot{m}u_e + A_e(p_e - p_a) \tag{3.33}$$

3.9 Define total impulse, specific impulse, and density specific impulse. What are their dimensions?

3.10 Define specific thrust for air-breathing engines and specific impulse for rocket engines. What are their typical values?

3.11 Adopting high-energy propellant ingredients to get high T_{0c} in order to get a high specific impulse is not a good solution. Discuss.

3.12 Discuss the importance of chamber stagnation temperature, molecular mass of combustion products, and nozzle pressure ratio in order to get high specific impulse. What are the limitations?

3.13 With a sketch, discuss the phenomenon of flow separation in an over-expanded nozzle.

3.14 Given the altitude and the chamber pressure, prove that the maximum thrust of a rocket engine is reached when the nozzle is adapted.

3.15 What are the different thrust control methods adopted for rocket engines? Briefly discuss the challenges in adopting such methods.

3.16 Prove

$$\frac{C_F}{C_F^*} = 1 + \frac{1}{C_F^*}\frac{A_e}{A_t}\left(\frac{p_a^*}{p_{0n}^*} - \frac{p_a}{p_{0n}}\right) \tag{3.87}$$

3.17 What are the limitations on the maximum or minimum thrust control by way of changing chamber pressure of a rocket engine?

3.18 Define the three characteristic parameters: c^*, C_F, and I_{sp}. Show how these are interconnected and how their experimental and theoretical values can be compared in the development of rocket engines.

3.19 Define c^*. Show how this theoretically represents the global characteristics of combustion and can be used to compare the combustion characteristics of different propellants independent of the chamber pressure employed.

3.20 Define the three quality factors: ξ_c or η_{c^*}, ξ_n, and ξ_s.

3.21 Discuss the results of rocket performance calculated assuming frozen flow against those assuming shifting-equilibrium flow.

3.22 The rocket performance values calculated through frozen flow assumption are conservative. However, with the nozzle flows of combustion products containing oxides of metallic fuels such values may be unacceptably conservative under certain conditions. Explain.

3.23 What is the condition under which condensation shock or condensation front can exist in rocket nozzle flows?

3.24 Is the ratio of specific heats constant under the frozen flow assumption? Explain.

3.25 Contoured nozzles are adopted for high-performance rocket engines operating in high altitudes and in vacuum. Discuss.

3.26 What is the turn-back angle? What is its maximum that is normally adopted in particle-laden nozzle flows? What happens when such a limit is exceeded?

3.27 List the basic differences between the conventional conical or contoured nozzles and the plug or ED nozzles. Give sketches of plug and ED nozzles.

3.28 Compare qualitatively the performance of plug nozzles with ED nozzles.

3.29 Classify polymer materials and list their basic characteristics. Give at least two examples under each category.

3.30 Define composite materials and classify them. Give at least two examples under each category.

3.31 What are the different materials used for throat inserts?

3.32 An intercontinental ballistic missile (ICBM) is at its terminal phase with flight Mach number of 22 at 30 km altitude (226.5 K). At this phase, the missile is powered by a solid propellant motor of specific impulse 2770 Ns/kg. The propellant mass-flow rate is estimated to be 30 kg/s. Calculate the thrust and the propulsive efficiency.
Answers:
$F = 83.1\ kN, \eta_p = 0.711.$

3.33 A cryogenic liquid rocket engine adopting liquid oxygen (LOX) and liquid hydrogen (LH2) is to be designed to develop a thrust of 500 kN at a chamber pressure of 80 bar. The nozzle is to be adapted to sea level pressure of 1 bar. The ratio of specific heats $\gamma = 1.2$. (1) Calculate the nozzle throat and nozzle-exit diameter. The rocket is ascending to an altitude where the pressure is 0.4 bar. (2) Calculate the engine thrusts at this altitude and also in vacuum. With no other changes, assume that the nozzle is redesigned for its adaptation to 0.4 bar ambient condition. (3) What is the corresponding nozzle-exit diameter? (4) Also calculate for the redesigned nozzle the engine thrusts at sea level pressure of 1 bar, at the altitude of 0.4 bar ambient pressure, and in vacuum. (5) Is there any danger of nozzle-flow separation for the second engine at sea level?
Answers:
(1) $D_t = 221.8$ mm; $D_e = 702.2$ mm. (2) $F_{p_a=0.4bar} = 523.24$ kN; $F_{vac} = 538.73\ kN$
(3)$D_e = 997.4$ mm. (4) $F_{p_a=1\ bar} = 485.02$ kN; $F_{p_a=0.4\ bar} = 531.90$ kN; $F_{vac} = 563.15\ kN$. (5) If the engine adopts conical nozzle, to avoid nozzle-flow separation p_a/p_e should be < 2.5. In the present case $p_a/p_e = 2.5$ and therefore it is a borderline situation.

3.34 For the propellant combination and the engines mentioned in Problem 3.33, the combustion products are of molecular mass of 12.87 kg/kgmol and adiabatic flame temperature of 3600 K. (1) Calculate the c^* value and the engine mass-flow rate. (2) Also calculate the specific impulse values for both the engines at the sea level pressure of 1 bar, at the altitude of ambient pressure 0.4 bar, and in vacuum.
Answers:
(1) $c^* = 2351.5$ m/s; $\dot{m} = 131.47$ kg/s. (2a) Engine with $p_e = 1\ bar$: $I_{sp_{sealevel}} = 3803$ m/s; $I_{sp_{pa=0.4\ bar}} = 3980$ m/s; and $I_{sp_{vac}} = 4098$ m/s. (2b) Engine with $p_e = 0.4$ bar: $I_{sp_{sealevel}} = 3689$ m/s; $I_{sp_{pa=0.4\ bar}} = 4046$ m/s; and $I_{sp_{vac}} = 4283$ m/s.

3.35 Using an aluminized composite propellant, a solid rocket motor is to be designed to develop 6 MN thrust at the sea level pressure of 1 bar with a chamber pressure of 70 bar. The calculated combustion properties of the propellant at this pressure are the following. Adiabatic flame temperature is 3500 K, the molar mass is 27 kg/kgmol. The ratio of specific heats is 1.18. Calculate (1) the specific impulse, (2) the mass-flow rate of combustion products, and (3) the nozzle throat and exit diameters. (4) At sea level, if the chamber pressure is reduced to 50 bar, what will be the thrust, specific impulse, and nozzle-exit pressure? Neglect the changes in the combustion properties at this reduced pressure.

Next, consider the operation of this motor with the chamber pressure of 70 bar at an altitude of 30 km (226.5 K and 1185.5 Pa) and in vacuum. (5) What will be the thrust and specific impulse at this altitude and in vacuum?

Answers:

(1) $I_{sp} = 2632.95$ Ns/kg; (2) $\dot{m} = 2278.809\ kg/s$; (3) $D_t = 0.8228\ m$ and $D_e =$ 2.5278 m; (4) $F = 4.1423$ MN, $I_{sp} = 2544.86$ Ns/kg, $p_e = 0.7143\ bar$; and (5) At 30 km altitude $F = 6.4959$ MN and $I_{sp} = 2850.57$ Ns/kg. In vacuum $F =$ 6.5019 MN and $I_{sp} = 2853.18$ Ns/kg.

3.36 A nuclear rocket engine using liquid hydrogen is to deliver a thrust of 73 kN for 1200 s. Selected nozzle area ratio is 100. The nozzle entry pressure and temperature are 31 bar and 2700 K, respectively. Accounting for some dissociation, the ratio of specific heats and molar mass are 1.31 and 2.008 kg/kgmol, respectively. And, these for simplicity can be assumed to be constant. Calculate (1) the nozzle pressure ratio, (2) the vacuum specific impulse, (3) the propellant flow rate and the total propellant mass required, and (4) the nozzle throat and exit diameters.

Answers:

(1) $p_0/p_e = 2555.5$; (2) $C_{F_{vac}} = 1.8259$; (3) $\dot{m}_p = 8$ kg/s and $m_p = 9600$ kg; and (4) $D_t = 128.1$ mm and $D_e = 1281$ mm.

3.37 A large cryogenic rocket engine adopting LOX/LH2 propellants delivers a thrust of 1075 kN in vacuum at a chamber pressure of 102 bar with a nozzle pressure ratio of 530. An average ratio of specific heats can be taken as 1.143. Molar mass of combustion products is 14 kg/kgmol. Chamber stagnation temperature is 3560 K. (1) Calculate the nozzle throat and exit diameters. (2) Only with the choked nozzle convergence, what thrust will be delivered in a vacuum? (3) Estimate the differential axial load to cause an axial stress at the throat wall section when the engine operates with the full convergent–divergent nozzle. (4) Estimate the nozzle-exit Mach number and velocity. (5) From this calculate the equivalent exhaust velocity and check it with the specified thrust.

Answers:

(1) $d_t = 260.5$ mm and $d_e = 1883.2$ mm; (2) 670.831 kN; (3) 404.169 kN; (4) $M_e =$ 4.083 and $u_e = 4287$ m/s: and $I_{sp} = \overline{u}_e = 4512.1$ m/s.

3.38 The available data for the Space Shuttle Main Engine (SSME) are the following. Total injection mass-flow rate = 469 kg/s. Chamber pressure = 205 bar. Oxidizer:fuel ratio = 6. Nozzle area ratio = 77.5. Nozzle-exit diameter = 2.39 m. Vacuum thrust

Table 3.6 Problem 3.38

Assumption	p_{0n}/p_e	$c^*(m/s)$	$C_{F_{vac}}$	$I_{sp_{vac}}(m/s)$	γ_n	γ_e
Equilibrium	1124.11	2322.7	1.9623	4557.9	1.149	1.257
Frozen	1301.78	2290.2	1.9144	4384.5	1.192	1.289
Data	–	2530.3	1.7612	4456.3	–	–

= 2090 kN. Vacuum specific impulse = 4464 Ns/kg. Running the program CEC71 for the given chamber pressure, oxidizer:fuel ratio, and nozzle area ratio, we get the results given in Table 3.6. Also the derived data from the SSME data are given in the row Data. Assuming isentropic one-dimensional nozzle flow and the constant ratio of specific heats, we have derived various governing equations in this chapter. Deduce the ratio of specific heats γ that would be needed to account for the measured thrust if the nozzle expansion had been isentropic and one dimensional with constant γ. What are the corresponding nozzle pressure ratio and vacuum specific impulse?
Answers:
$\gamma = 1.375$; $p_{0n}/p_e = 2424.44$; and $I_{sp_{vac}} = 4375$ Ns/kg.

3.39 An experimental rocket motor is to contain a composite propellant of composition AP:HTPB:Al= 0.80:0.15:0.05. Its chamber pressure and nozzle pressure ratios are, respectively, 40 bar and 40. Ambient pressure is 1 bar. Adopting the program CEC71, the calculated values are the following. Adiabatic flame temperature in the chamber is 3074 K. Molar mass of the gaseous mixture is 25 kg/kgmol. Specific heat at constant pressure of the gaseous mixture is 1840 J/(kgK). The specific heat of $Al_2O_3(s)$ is 1350 J/(kgK). The exhaust products have the $Al_2O_3(s)$ mass fraction of 0.094. Assuming the specific heats and molar mass of gaseous mixture to be constant, estimate the nozzle-exit temperature and specific impulse for the assumed conditions of (1) $T_c = T_g$ and $u_c = u_g$; (2) $T_c = T_g$ and $u_c = 0$; (3) $T_c = constant\ at 3074$K and $u_c = u_g$; and (4) $T_c = constant\ at 3074$K and $u_c = 0$. (5) Also estimate the nozzle-exit temperature and specific impulse for the homogeneous gaseous flow.
Answers:
(1) $T_e = 1654.3$ K and $I_{sp}^0 = 2256.9$ Ns/kg. (2) $T_e = 1654.3$ K and $I_{sp}^0 = 2148.2$ Ns/kg. (3) $T_e = 1578.1$ K and $I_{sp}^0 = 2233.3$ Ns/kg. (4) $T_e = 1578.1$ K and $I_{sp}^0 = 2125.7$ Ns/kg. (5) $T_e = 1578.1$ K and $I_{sp}^0 = 2346.3$ Ns/kg.

3.40 A large LOX/LH2 cryogenic engine is to operate in space to develop a thrust of 1371 kN. Its chamber pressure, oxidizer–fuel ratio, and nozzle area ratio are, respectively, 119 bar, 6.0, and 36. The theoretical performance calculations under frozen flow assumption give an adiabatic flame temperature of 3547 K and molar mass of 13.585 kg/kgmol. Although the c_p varies from 3722.8 J/(kgK) (throat) to 2868.6 J/(kgK) (exit), let us assume for simplicity an average c_p value of 3296 J/(kgK). Assume $\xi_c = \xi_n = 0.98$. (1) Calculate the vacuum specific impulse, propellant flow rate, and throat diameter. (2) Estimate the length of the conical nozzle divergence

for an $\alpha_d = 15°$. (3) If the length of contoured nozzle divergence is to be 75% of the conical one, estimate the nozzle angles adopting GVR Rao's contoured nozzle graphs. Nozzle convergence ratio A_c/A_t and contraction angle are, respectively, 1.5 and 40°. Choosing appropriate radii of curvature at the nozzle entry and throat region, sketch the nozzle configuration with parabolic nozzle divergence.
Answers:
(1) $I_{sp_{vac}} = 3998$ m/s; $\dot{m}_p = 342.9$ kg/s; and $D_t = 285$ mm.
(2) $l_{conical} = 2656$ mm. And, (3) $l_{contour} = 1992$ mm; $\alpha_t = 33.6°$; and $\alpha_e = 9.5°$.

3.41 A toy water rocket with a convergent nozzle is of dry mass 250 g and an empty internal volume of chamber and nozzle 1500 cc. (Why should it be only a convergent nozzle, not a convergent–divergent one?) The toy initially contains 1 liter of water pressurized to 9 bar. (1) What is its nozzle-exit diameter if it is required to vertically take-off with an acceleration of 0.15 g? (2) What is the mass-flow rate and the specific impulse at the take-off? (3) If the rocket is ground tested what is the final mass-flow rate and the specific impulse? At the end of water ejection, the entrapped air has to expand through the nozzle. (4) What is the thrust and specific impulse for this mode? Assume that the entire system is adiabatic and the expansion of the pressurized air is isentropic with $\gamma = 1.4$. Molar mass of air is 28.967 kg/kgmol.

Let us next consider this problem in a generalized way. Let the empty volume of the toy rocket be V. Initial mass of the liquid is m_l with its density ρ_l. p_0 is the initial pressure of the compressed inert gas with its ratio of specific heats γ. A_e is the exit area of nozzle convergence. The ambient pressure is p_e. (5) Derive the governing differential equation to show how the exhaust velocity u_e depends on time t if the toy is ground tested.
Answers:
(1) $d_t = 3.35$ mm. (2) $\dot{m}_{w_i} = 0.3524$ kg/s; $I_{sp_i} = 40.0$ m/s.
(3) $\dot{m}_{w_f} = 0.1204$ kg/s; $I_{sp_f} = 13.66$ m/s.
(4) $F_{air} = 1.278$ N; $I_{sp_{air}} = 257.4$ m/s.
(5) $du_e = -(p_e + \frac{\rho_l u_e^2}{2})^{(\gamma+1)/\gamma} \frac{A_e \gamma p_e}{V \rho_l (p_0/p_e)^{1/\gamma}} dt.$

References

Cited

1. Hill PG, Peterson CR (1999) Mechanics and thermodynamics of propulsion, 2nd edn. Addition-Wesley, Reading
2. Schmucker RH (1974) Status of flow separation prediction in liquid propellant rocket nozzles. NASA TM X-64890
3. Summerfield M, Foster C, Swan W (1954) Flow separation in overexpanded supersonic exhaust nozzles. Jet Propuls 24:319–321
4. Barrere M, Jaumotte A, De Veubuke BF, Vandenkerckhove J (1960) Rocket propulsion. Elsevier Publishing Company, New York
5. Green L Jr (1953) Flow separation in rocket nozzles. J ARS 23:34–35

6. Rao GVR (1961) Nozzle contours. In: Koelle HH (ed) Handbook of astronautical engineering. McGraw-Hill, New York
7. Ralf S (2013) Flow separation in rocket nozzles - an overview. In: AIAA joint propulsion conference, AIAA 2013-3840
8. Shapiro AH (1953) The dynamics and thermodynamics of compressible fluid flow, vols I & II. The Arnold Press, New York
9. Casiano MJ, Hulka JR, Yang V (2010) Liquid-propellant rocket engine throttling: a comprehensive review. J Propuls Power 26:897–923
10. Tomita T, Takahashi M, Sasaki M, Tamura H (2006) Investigation on characteristics of conventional-nozzle-based altitude compensating nozzles by cold flow tests. AIAA paper 2006-4375
11. Gordon S, McBride BJ (1976) Computer program for calculating complex chemical equilibrium composition, rocket performance, incident and reflected shocks and Chapman-Jouguet detonations. NASA SP-273
12. Gordon S, McBride BJ (1994) Computer program for calculation of complex chemical equilibrium compositions and applications: part I. Analysis, NASA RP, p 1311
13. Krishnan S, Raghavan J (2019) Chemical rockets: performance prediction and internal ballistics design. Springer, Switzerland
14. Jeenu R, Kiran P, Deepak D (2010) Size distribution of particles in combustion products of aluminized composite propellant. J Propuls Power 26:715–723
15. Chang IS (1980) One- and two-phase nozzle flows. AIAA J 18:1455–1461
16. Chang IS (1990) Three dimensional two-phase transonic canted nozzle flows. AIAA J 28:790–797
17. Mehta RC, Jayachandran T (1998) A fast algorithm to solve viscous two-phase flow in an axisymmetric rocket nozzle. Int J Numer Methods Fluids 26:501–517
18. Daniel E (2000) Eulerian approach for unsteady two-phase solid rocket flows with aluminum particles. J Propuls Power 16:309–317
19. Anon. (1971) Solid rocket motor performance analysis and prediction. NASA SP-8039
20. Wong EY (1968) Solid rocket nozzle design summary. In: AIAA 4th propulsion joint specialist conference, AIAA paper 68-655
21. Edward JW (1991) ASRM nozzle design and development. In: AIAA/SAE/ASME 27th joint propulsion conference, AIAA paper 91-2070
22. Rao GVR (1958) Exhaust nozzle contour for optimum thrust. Jet Propuls 28:377–382
23. Rao GVR (1965) One-dimensional approach to optimum rocket nozzles. In: Miele A (ed) Theory of optimum aerodynamic shapes. Academic Press, New York
24. Zucrow MJ, Hoffman JD (1977) Gas dynamics, vol II. Wiley, New York
25. Allman JG, Hoffman JD (1978) Design of maximum thrust nozzle contours by direct optimization methods. AIAA J 19:750–751
26. Doi T (1981) Gas-particle nozzle flows and optimum nozzle shape. The Institute of Space and Astronautical Science Report No. 596
27. Akiba R, Kohno M (1986) Experiments with sold rocket technology in the development of M-3SII. Acta Astronaut 13:349–361
28. Anon. (1976) Liquid rocket engine nozzles. NASA SP-8120
29. Rao GVR (1961) Recent developments in rocket nozzle configurations. ARS J 1488-1494
30. Hagemann G, Immich H, Nguyen TV, Dumnov GE (1998) Advanced rocket nozzles. J Propuls Power 14:620–634
31. Anon. (1975) Solid rocket motor nozzles. NASA SP-8115
32. Huzel DK, Huang DH (1992) Modern engineering for design of liquid-propellant rocket engines. Progress in Astronautics and Aeronautics Vol. 147. AIAA, Washington DC
33. Sutton GP, Biblarz O (2017) Rocket propulsion elements, 9th edn. Wiley, New York
34. Wood BK (2002) Propulsion for the 21st century—RS-68. In: AIAA/SAE/ASME 38th joint propulsion conference, AIAA paper 2002-4324

Further Reading

35. Heister SD, Anderson WE, Pourpoint TL, Cassady RJ (2019) Rocket propulsion. Cambridge University Press, Cambridge
36. Mukunda HS (2012) Understanding aerospace chemical propulsion. International Publishing House, New Delhi, I. K
37. Ramamurthi K (2010) Rocket propulsion. Macmillan Publishers India, Delhi
38. Bard TA (2010) Aerospace propulsion systems. Wiley, Singapore
39. Brown CD (1996) Spacecraft propulsion. AIAA education series, AIAA, Washington DC
40. Timnat YM (1987) Advanced chemical rocket propulsion. Academic, London
41. Williams FA, Barrere M, Huang NC (1969) Fundamental aspects of solid propellant rockets, AGARDograph 116, Neuilly sur Seine (France)
42. Shorr M, Zaerhinger AJ (1967) Solid rocket technology. Wiley, New York

Chapter 4
Quantitative Internal-Ballistics Prediction and Design

Synopsis

Composition selection for liquid and solid propellants is discussed. Theoretical performance values for different oxidizer–fuel ratios of LOX/LH2 and LOX/RP1 are presented. Besides, detailed data of flight proven engines using such propellants are tabled. The classification of homogeneous and heterogeneous solid propellants and the propellant ingredients are discussed. The contour mapping of specific impulse in a triangular diagram to arrive at a suitable composition for a heterogeneous propellant is presented.

Related thermodynamic fundamentals (equation of state, entropy principle, and chemical equilibrium conditions) are briefly discussed. That an isolated system attaining a maximum entropy for equilibrium is equivalent to it reaching a minimum energy is explained. The chemical potential, the potent energy of a species to react in a chemical product mixture, is introduced. The generalized definition of chemical potential is derived. Since the chemical potential does not change in its value when the method of its calculation differs, it is an intensive property. Chemical reactions occur under two distinct operating conditions, either at constant pressure or at constant volume. For the former, the chemical potential is evaluated with respect to Gibbs energy and for the latter with respect to Helmholtz energy. For an operating condition given by a pair of constant state functions (temperature & pressure, enthalpy & pressure, and entropy & pressure, etc.), the chemical equilibrium is reached for a product mixture when the sum of chemical potentials of all species reaches a minimum. Reactions under given pressures are of interest in rocket propulsion systems. Although many routes are available to evaluate the chemical potential, it is convenient to evaluate it based on Gibbs energy and it is shown that the chemical potential of a species is its unit molar Gibbs energy with respect to product-mixture's temperature and pressure.

S. Krishnan and J. Raghavan, *Rocket Propulsion Primer*,
https://doi.org/10.1007/978-981-97-5644-5_4

For equilibrium, in addition to energy minimization, the simultaneous satisfaction of the conservation of elements between a propellant and its products or between two operating conditions is mandatory. That is, the total kgatoms of an element between two operating conditions must be equal. For this mass conservation, a necessary governing equation is given for a multi-element system involving many product compositions.

Adopting Lagrange multipliers, the governing equations, encompassing in tandem the energy minimization and the conservation of elements, are derived. Since rocket propulsion units are analyzed under parametric values of pressure, applicable Newton–Raphson iteration equations are derived from the governing equations. Following this, in order to lessen the iteration load, reduced iteration equations are obtained for computations.

After the determination of equilibrium composition, to proceed further in rocket performance calculations we require the values of three key derivatives: *change of volume with respect to temperature at constant pressure, change of volume with respect to pressure at constant temperature, and specific heat at constant pressure, which is the change of enthalpy with respect to temperature at constant pressure.* These key derivatives assume different values for the reacting composition and the frozen composition and hence lead to different values of rocket performance parameters. Linked to this, against the conventional ratio of specific heats γ, the concept of the isentropic exponent γ_s for reacting composition is considered. Also a brief treatment on the use of Bridgman Table is presented.

The primary inputs required for the equilibrium composition calculations are the energies contained in propellant ingredients and additives and the molar standard state enthalpies. Thermodynamic data of many propellant ingredients and additives are given. Adopting fourth-order polynomials, the calculation of molar specific heat at constant pressure, the standard state enthalpy, and the standard state entropy of species is explained. For selected condensed and gaseous species, the least square coefficients of the fourth-order polynomials are presented.

Internal gas dynamics of liquid rocket engines and solid rocket motors with equilibrium flow and frozen flow is discussed. In evaluating the rocket performance under equilibrium flow, the application of the three key derivatives [$\left(\partial \ln v / \partial \ln T\right)_p$, $\left(\partial \ln v / \partial \ln p\right)_T$, and c_p] is explained. The performance parameters (characteristic velocity c^*, thrust coefficient C_F, and specific impulse I_{sp}) under the equilibrium flow and the frozen flow are discussed to show that the parameters assume different values for the two flows. To demonstrate the application of the knowledge gained, the design analysis on the internal ballistics and engine envelope determination of proven engines, such as Space Shuttle Main Engine (SSME), is carried out.

Worked examples are introduced under the Sections. A large number of concept questions and numerical problems with answers are added at the chapter end.

4.1 Introduction

The calculation of the equilibrium composition of a reacting mixture of gaseous and condensed species is of interest in many areas of science and engineering. Its important engineering applications include rocket propulsion, air-breathing jet propulsion, reciprocating internal-combustion engines, gun ballistics, system of shocks, and explosions and detonations. Discussion of this calculation and its extension applied to the performance prediction of rocket propulsion systems are of interest here. Furthermore, as an alert on the limitations in the calculation of equilibrium composition, we must note that we are not looking at the rate at which the related combustion reactions are taking place. This rate aspect comes under the topic of chemical kinetics, which is beyond the scope of this text.

4.1.1 Propellant Selection

Be it a solid propellant or a liquid propellant, the propellant combination is nothing but a chemical system, which undergoes chemical reactions to produce products of combustion, principally of gaseous mixture by volume. These products expand through the nozzle to produce the required thrust. Our ability to calculate theoretically the equilibrium composition of the combustion products for a given condition is the first step towards the performance prediction of a rocket motor (or -engine) that leads to the final system design. By the way, we recall at this point that by tradition, a rocket propulsion system containing a solid propellant is termed as a rocket motor and the one containing a liquid propellant combination as a rocket engine.

There are two methods available for the calculation of equilibrium composition, first the **method of equilibrium constants** and the other the **method of free energy minimization** or simply the **method of energy minimization**. Both the methods should undoubtedly lead to the same answers. For the calculation of equilibrium composition, first and foremost, both the methods maximize (or attain the specified) global entropy, the entropy of an isolated system. An isolated system is the one in which there is no interaction between the system and its surroundings. It is of fixed mass and energy, and there is no mass or energy transfer across the system boundary. While maximizing the global entropy or maintaining it at the specified level, the calculation by either method has two more conditions: (1) the conservation of elements (an avatar of mass conservation appropriate for chemical reactions, but not for nuclear reactions!) and (2) depending on the type of reactions, the assigned state of the system—constancy on any two-state functions: constant enthalpy h and pressure p, constant entropy s and pressure p, or constant temperature T and pressure p. The other constant volume states (constant internal energy u and constant volume v, constant entropy s and volume v, or constant temperature T and volume v) are not of interest to rocket propulsion.

The method of equilibrium constants is a classical one. In this method, one has to impose à priori a set of equilibrium equations containing the possible product species. And, the calculation proceeds with the tabulated or the calculated values of equilibrium constants with respect to the equilibrium equations. It has been observed however that the method of equilibrium constants is less amenable for generalized treatment with more difficulty in testing for the presence of condensed product species (liquid and solid species). Invariably, most heterogeneous (composite) propellants and some homogeneous (double base) propellants contain metals as fuel components and their oxides in condensed phase do exist in significant proportion in the combustion products by mass although not by volume [see Sect. 3.8]. Nonetheless, for a detailed treatment of the method of equilibrium constants [1–4] may be referred to. But, here we discuss the second method, namely the method of energy minimization for finding equilibrium composition.

In the method of energy minimization also, one has to identify à priori the possible product species, not along with the equilibrium equations but individually. The difference here is that each identified-species is treated individually such that the composition at a specified system state (constant $h\&p$, $s\&p$, or $T\&p$) attains global-minimum energy with all the species and their coefficients. For example, let us assume that one kgmole of hydrogen is burning with 0.35 kgmole of oxygen. For simplicity we restrict the possible product species to only H, H_2, H_2O, O, OH, and O_2. We may write the combustion reaction to attain equilibrium as

$$H_2 + 0.35 O_2 \rightarrow a_1 H + a_2 H_2 + a_3 H_2O + a_4 O + a_5 OH + a_6 O_2$$

At the specified system state, as the equilibrium composition assumes the minimum global energy, the values of the coefficients a_1–a_6 become known.

Adopting the method of energy minimization, NASA Lewis Research Center has been developing an elaborate computer code since 1950s. The code originally had the name CEC71 (Chemical Equilibrium Calculations, 1971) [5] and in 1994 the code with many developments has the name CEA (Chemical Equilibrium and Applications) [6, 7]. The program has been widely used for many applications and can be downloaded through a request from [8]. Also one can find extensions to this code in public web sites. The treatment here closely follows the two related reports by NASA [6, 7] but, as previously indicated, restricts its focus on the application of the method to the performance prediction of rocket propulsion systems. The detailed treatment on the topic of theoretical rocket performance prediction by the method of energy minimization can be found in Part I of [9]. Here, however, we give a simplified and concise treatment of the topic.

In order to select a suitable propellant combination for a rocket propulsion system we have to first determine the equilibrium compositions of different propellant blends at a specified combustion pressure and then calculate the related theoretical rocket performance values. For this it is appropriate to divide a propellant blend into two groups, namely, oxidizer and fuel. In each group we include the relevant ingredients and their proportions and states. For every blend characterized by the oxidizer–fuel ratio or equivalence ratio, rocket performance values have to be calculated for a

Table 4.1 Summary of theoretical rocket performance calculation for the selection of a suitable propellant combination

Select a rocket chamber pressure	
Oxidizer group: specify ingredients and their proportions and states	Fuel group: specify ingredients and their proportions and states
Choose blends of different oxidizer–fuel ratios	
Calculate equilibrium composition in the combustion chamber	
Select a nozzle pressure ratio or nozzle area ratio	
Adopt shifting-equilibrium nozzle flow	Adopt frozen-equilibrium nozzle flow
Calculate theoretical rocket performance for shifting-equilibrium flow	Calculate theoretical rocket performance for frozen-equilibrium flow

typical geometrical parametric-ratio: nozzle entry stagnation pressure to nozzle-exit static pressure (p_{0n}/p_e)—**nozzle pressure ratio** that fixes nozzle-exit area to throat area (A_e/A_t)—**nozzle area ratio** or vice versa. Furthermore, in calculating rocket performance values we have the option of choosing either **shifting-equilibrium nozzle flow** (equilibrium composition shifts according to local static temperature and pressure along the nozzle axis) or **frozen-equilibrium nozzle flow** (equilibrium composition as at the combustion chamber exit is constant along the nozzle axis) [see Sect. 3.7]. In all these exercises, the assignment of constancy on any two-state functions has to be employed: $h\&p$ for the process in the combustion chamber and $s\&p$ for the nozzle flow. The constancy on $T\&p$ is adopted at the places where heat transfer losses are to be accounted for. Thus the selection of a suitable propellant combination can be summarized as given in Table 4.1.

4.1.1.1 Liquid Propellants

In the case of liquid propellants, the optimum propellant combination (say optimum oxidizer–fuel ratio) is primarily based on thermodynamic calculations for the maximum possible thrust per unit mass flow rate of propellant, that is, on the maximum possible **specific impulse**. But in the case of solid propellants, the propellant blend of the maximum possible specific impulse cannot be based on thermodynamic calculations alone. Because such a blend may be impracticable from the point of view of the propellant manufacturing process. In addition, such a blend may not normally give the required mechanical strength for the cured propellant to withstand a steep pressure rise during the ignition as well as a high chamber pressure during the motor operation. Hence the blend of a solid propellant is chosen by the simultaneous considerations of the ability to process the propellant mix (mix, pour, and cure), acceptable mechanical strength of cured propellant, and possible maximum specific impulse.

Previously through simplifying assumptions [$p_{0c} = p_{0n}$, adiabatic engine, one-dimensional isentropic nozzle flow, ideal gas equation of state, and c_p & $\overline{m}$ constant] we have derived the equations for the characteristic parameters [see Sect. 3.5]. The derived equation for the specific impulse is repeated here.

$$I_{sp} = \frac{F}{\dot{m}} = \frac{p_{0n A_t}}{\dot{m}} \frac{F}{p_{0n} A_t}$$

$$= \left(\frac{\sqrt{\frac{R_u}{\overline{m}} T_{0c}}}{\Gamma} \right) \left\{ \Gamma \sqrt{\frac{2\gamma}{\gamma - 1} \left[1 - \left(\frac{p_e}{p_{0n}} \right)^{\frac{\gamma-1}{\gamma}} \right]} + \frac{A_e}{A_t} \left(\frac{p_e}{p_{0n}} - \frac{p_a}{p_{0n}} \right) \right\} \tag{3.108}$$

$$= c^* C_F$$

But under the shifting-equilibrium flow as the composition of the reacting combustion products changes as per the local temperature and pressure the values of the **isentropic exponent** γ_s using which the sound velocity is determined is not simply the ratio of specific heats γ but,

$$\gamma_s = -\frac{\gamma}{\left(\frac{\partial \ln v}{\partial \ln p} \right)_T} \tag{4.1}$$

The velocity of sound a is

$$a = \sqrt{\gamma_s R T} \tag{4.2}$$

Under the shifting-equilibrium flow assumption the isentropic exponent γ_s, the ratio of specific heats γ, and the molar mass of combustion products $\overline{m}$ all change as per the local nozzle-flow conditions. But, under the frozen-equilibrium flow assumption the difference between γ_s and γ vanishes as the denominator in the expression for γ_s becomes -1. Nonetheless, although the composition is constant under this assumption, γ changes, differently though when compared under shifting-equilibrium flow—the specific heats c_p & c_v are dependent on temperature. Evidently under the frozen-equilibrium assumption, $\overline{m}$ remains constant. Thus, in variance to the simplified equation for the specific impulse Eq. 3.108, the general equation for specific impulse can be written as [9]

$$I_{sp} = \frac{F}{\dot{m}} = \frac{p_{0n A_t}}{\dot{m}} \frac{F}{p_{0n} A_t}$$

$$= \frac{(p_{0n}/p_t)\, a_t}{\gamma_{st}} \left[\frac{\gamma_{st} \sqrt{2 (h_0 - h_e)}}{a_t (p_{0n}/p_t)} + \frac{A_e}{A_t} \left(\frac{p_e}{p_{0n}} - \frac{p_a}{p_{0n}} \right) \right] \tag{4.3}$$

$$= c^* C_F$$

For the specified propellant blend, chamber pressure (nozzle entry total pressure p_{0n} assumed to be equal to chamber pressure p_{0c}), nozzle pressure ratio (p_{0n}/p_e) or nozzle area ratio (A_e/A_t), and ambient pressure (p_a) the theoretical performance calculations enable us to find all the remaining quantities in Eq. 4.3.

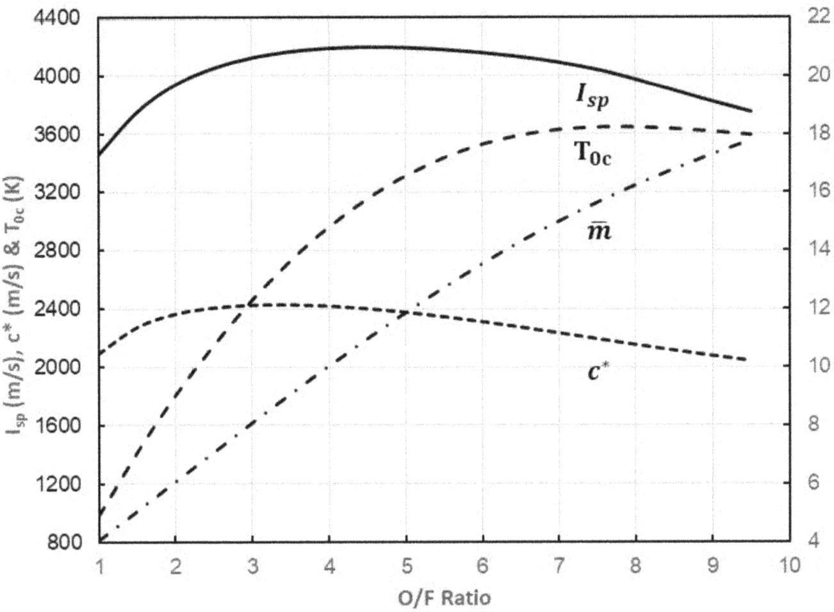

Fig. 4.1 LOX/LH2 rocket engine shifting-equilibrium performance: chamber pressure = 100 bar and nozzle pressure ratio = 100. T_{0c} & $\overline{m}$ are respectively the total temperature and molar mass of combustion products in the chamber. I_{sp} & c^* are respectively the vacuum specific impulse and characteristic velocity

Let us now consider the theoretical performance of the cryogenic propellant blends, liquid oxygen (LOX, boiling point 90.17 K, and density 1149 kg/m^3) and liquid hydrogen (LH2, boiling point 20.27 K, and density 70.9 kg/m^3). Figure 4.1 presents the theoretical rocket performance of a LOX/LH2 engine with the chamber pressure of 100 bar and nozzle pressure ratio of 100. We know that the **stoichiometric oxidizer–fuel ratio** for the oxygen and hydrogen mixture is 7.937 that corresponds to the equivalence ratio of unity. As the oxidizer–fuel ratio O/F increases from its lower values the total temperature and the molar mass of the combustion products in the chamber continue to increase, Fig. 4.1. Because of the negative contribution to the temperature rise by the enhanced dissociated species close to the **equivalence ratio** of unity, the maximum chamber temperature occurs not at the stoichiometric O/F of 7.937 but at a slightly fuel rich condition, say $O/F \approx 7.77$. With the increase in O/F the value of isentropic exponent at a point, say at the throat, reduces but can be taken as more or less constant. The vacuum specific impulse $I_{sp_{vac}}$ theoretically peaks around $O/F \approx 4.62$. Thereafter however, the $I_{sp_{vac}}$ is essentially flat up to O/F of 6. Therefore the rated O/F's of most flight proven LOX/LH2 engines are practically in the range of 5–6, Table 4.2. Here m_e and Δt are respectively dry engine mass and rated thrusting time. In the operating range of $5 \leq O/F \leq 6$ the difference between the I_{sp} frozen and the I_{sp} shifting-equilibrium is about 2–4% for the specified engine

Table 4.2 Some flight proven LOX/LH2 cryogenic engines[a]

Engine	F_{vac} (kN)	m_e (kg)	Isp_{vac} (m/s)	Δt (s)	p_c (bar)	A_e/A_t	F_{vac}/m_e (kN/kg)	O/F	Typ.[b]	Cycl.[c]	Orgn.
RS68A	3137	6740	4040	–	102.6	21.5	0.465	–	B	GG	USA
SSME	2278	3177	4442	480	204	77.5	0.717	6	B	SC	USA
RD0120	1961	3450	4462	600	218	85.7	0.568	6	B	SC	Russia
Vulcain2	1359	1800	4207	–	117.3	58.2	0.755	6.1	B	GG	France
LE7A	1098	1800	4295	390	121	51.9	0.61	5.9	B	SC	Japan
YF77	700	2700	4218	525	102	49	0.259	5.5	B	GG	China
J2	1033	1438	4129	475	30	28	0.718	5.5	U	GG	USA
LE5B	137	269	4384	534	36	110	0.509	5	U	E	Japan
RL10B2	110.1	277	4565	1125	44.12	280	0.397	5.88	U	E	USA
YF75D	88.36	265	4340	780	41	80	0.333	6	U	E	China
HM7A	61.7	149	4344	563	30	62.5	0.414	5.15	U	GG	France

[a] Values are either from [10] or from different web sites, more particularly from Wikipedia, the free encyclopedia

[b] B: booster/first stage. U: upper stage

[c] GG: gas generator cycle. SC: staged combustion cycle. E: expander cycle

conditions, Fig. 4.1. However, this difference will increase as the nozzle pressure ratio is increased. Note that upper stage engines are designed to have very high nozzle pressure ratios. Incidentally, on a note of related aspect, the developments of medium-performing version of cryogenic engines using liquid methane CH_4 (LCH4, boiling point 111.6 K, and density 421.1 kg/m^3) with LOX have been initiated in recent times by spacefaring nations. In comparison with LOX/LH2 engines, LOX/LCH4 engines have a reduced rigor of dealing with the cryogenic fuel LCH4 relatively of high-temperature and -density as against the hyper-cooled low-temperature and -density of LH2. However this is at the expense of specific impulse and higher carbon print in the atmosphere. Yet, LOX/LCH4 engines are expected to have higher specific impulse with lesser carbon print in the atmosphere when compared to LOX/Kerosene engines.

Rocket engines using liquid oxygen and the Earth storable kerosene are increasingly being used for boosters and first-stage main engines. In these the kerosene fuel replaces the hyper-cooled low-density liquid hydrogen at the expense of specific impulse but with the advantage of higher thrust–mass ratios and reduced rigor of the need to handle only one cryogenic propellant. In addition, when compared to the solid motors that are preferred for their simplicity as well as the capacity to deliver practically unlimited thrust power, these so-called semi-cryogenic engines are increasingly being used although with lower **thrust–mass ratios** but advantageously with higher specific impulse values. Figure 4.2 presents the theoretical rocket performance of a LOX/Kerosene engine with the chamber pressure of 100 bar, and nozzle pressure ratio of 100 using the rocket grade kerosene RP-1. The stoichiometric O/F ratio for the oxygen and the assumed RP-1 composition is 3.4 that corresponds to the equivalence ratio of unity. As the oxidizer–fuel ratio O/F increases from its lower values the total

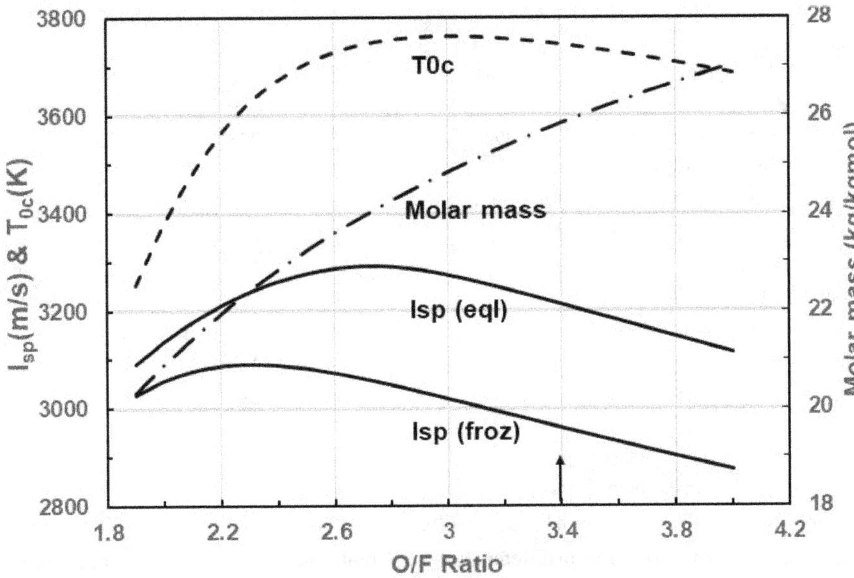

Fig. 4.2 LOX/Kerosene rocket engine performance: chamber pressure = 100 bar and nozzle pressure ratio = 100. T_{0c} & Molar mass are respectively the total temperature and molar mass of combustion products in the chamber. I_{sp} is the vacuum specific impulse

temperature and the molar mass of the combustion products in the chamber continue to increase, Fig. 4.2. For the reasons explained previously, the maximum chamber temperature occurs not at the stoichiometric O/F of 3.4 but at a fuel rich condition of $O/F \approx 3$. The vacuum specific impulse $I_{sp_{vac}}$ theoretically peaks around $O/F \approx 2.3$ for the frozen flow and around $O/F \approx 2.725$ for shifting-equilibrium flow. And, the rated O/F's of most flight proven LOX/Kerosene engines are practically in the range of 2.25–2.85, Table 4.3. Here m_e and Δt are respectively dry engine mass and rated thrusting time. In the operating range of $2.25 \leq O/F \leq 2.85$ the difference between the I_{sp} frozen and the I_{sp} shifting-equilibrium is about 4–7%. However, this difference will increase as the nozzle pressure ratio is increased. The RP-1 used for the present calculations is assumed to be of chemical composition $CH_{1.9423}$ and the enthalpy of formation −22723 kJ/kgmol [7]. Rocket grade kerosene RP-1 and its variants can be of different physical properties and chemical compositions. Their densities are found to vary from 810 to 1020 kg/m^3. Russian rocket grade kerosene has the designation of RG-1. However, Russian LOX/Kerosene engines developed using RG-1 have been reported to have satisfactorily performed with the use of RP-1.

There is a third class of liquid propellant combinations known as **Earth storable hypergolics**. In these, the oxidizer and the fuel when injected into the combustion chamber get automatically ignited without any necessity for a separate igniter. This renders the propulsion system more reliable, particularly desirable in upper stage engines and the thrusters in space with multiple starts capability. The mostly used

Table 4.3 Some flight proven LOX/Kerosene engines[a]

Engine	F_{vac} (kN)	m_e (kg)	I_{spvac} (m/s)	Δt (s)	p_c (bar)	A_e/A_t	F_{vac}/m_e (kN/kg)	O/F	Typ.[b]	Cycl.[c]	Orgn.
RD171	7903	9500	3305	150	245	36.9	0.832	2.63	B	SC	Russia
F1	7783	8409	2981	165	70	16	0.926	2.27	B	GG	USA
RS56	2100	1610	2903	167	49	8	1.304	2.25	B	GG	USA
RD191	2090	2290	3305	325	258	37	0.913	2.6	B	SC	Russia
NK33	1638	1222	3246	600	145.7	27	1.34	2.8	B	SC	Russia
YF100	1340	–	3285	155	180	35	–	2.6	B	SC	China
RD108	977	1230	3099	290	53.6	–	0.712	2.39	B	GG	Russia
LR89-5	822.5	720	2844	120	41	8	1.142	2.25	B	GG	USA
RS56OSA	386.4	460	3099	283	48	25	0.84	2.25	B	GG	USA
RD120	912	1125	3432	290	178	114.5	0.811	2.6	U	SC	Russia
RD0124	294.3	572	3521	300	157	–	0.515	2.6	U	SC	Russia
YF115	176.5	–	3349	341	120	–	–	2.5	U	SC	China
RD58M	85	340	3540	1200	79	280	0.25	2.82	U	SC	Russia
RD0109	54.52	121	3177	430	50	79.4	0.451	–	U	GG	Russia

[a] Values are either from [10] or from different web sites, more particularly from Wikipedia, the free encyclopedia
[b] B: booster/first stage. U: upper stage
[c] GG: gas generator cycle. SC: staged combustion cycle

oxidizer under this class is nitrogen tetroxide N_2O_4. The fuels used with N_2O_4 are hydrazine N_2H_4, monomethyl hydrazine (MMH) CH_6N_2, and unsymmetrical dimethyl hydrazine (UDMH) $C_2H_8N_2$. Many flight proven engines, boosters as well as upper stage ones, use N_2O_4 with MMH or UDMH. N_2O_4–MMH combination was used in the Space Shuttle orbiter. Also engines with N_2O_4 use variants of UDMH such as Aerozine 50 (50% UDMH + 50% Hydrazine) and UH 25 (75% UDMH +25% hydrazine hydrate ($N_2H_4.xH_2O$)). As previously discussed, the theoretical rocket performance calculations for different blends of N_2O_4 and the related fuels form the precursor to choose the suitable O/F ratios. Incidentally, the oxidizer N_2O_4 and the adopted fuels are all toxic and therefore there have been efforts to develop green propellant combinations.

4.1.1.2 Solid Propellants

As indicated previously the suitable blend of a solid propellant is chosen principally by the simultaneous considerations of the ability to process the propellant blend, the acceptable mechanical strength of the processed propellant, and the possible maximum specific impulse. Solid propellants are broadly classified into homogeneous propellants (or popularly known as double base propellants) and heterogeneous propellants (or popularly known as composite propellants). It is termed as homogeneous in the sense, the mixing of the principal ingredients, the oxidizer nitrocellulose (NC) and the fuel, say nitroglycerin (NG), is at a molecular- or colloidal-level in the finished grain. On the other hand, a heterogeneous propellant contains fine oxidizer crystals, say of ammonium perchlorate (AP, NH_4ClO_4), mechanically dispersed into a continuous binder matrix of a liquid fuel, say hydroxyl terminated polybutadiene (HTPB), which is subsequently cured into a solid grain. Also fine powders of a fuel metal, mostly of aluminum among the possible group of aluminum, beryllium, boron, magnesium, and zirconium, are dispersed into the fuel matrix along with the oxidizer crystals. During the last few decades, however, many hybridization steps have been adopted towards developing new solid propellant compositions. This has made the classification of solid propellants into homogeneous- and heterogeneous-ones as classical. For example, new compositions developed from the common homogeneous propellant have become heterogeneous. AP and aluminum are added to the homogeneous propellant and this is known as a **composite modified double base** (CMDB) one. Nitroamine high explosives RDX (cyclotrimethylene-trinitramine, $C_3H_6N_6O_6$) and HMX (cyclotetramethylene-tetranitramine, $C_4H_8N_8O_8$) are effectively stoichiometrically balanced chemicals. NG, RDX, or HMX is added into the plain heterogeneous propellant and this is known as **modified composite propellant**.

The most common ingredients of a homogeneous propellant are NC and NG. NC is an oxygen deficient fibrous solid chemical of varying molecular form depending on the process adopted. It is generally classified by the nitrogen content, for example as 13.25% N_2 content ($C_{21.25}H_{25.96}N_{9.46}O_{36.63}$) and 12.6% N_2 content ($C_{600}H_{755}N_{245}O_{990}$). NG is an oxygen rich plasticizer liquid $C_3H_5N_3O_9$. Sometimes along with NG or replacing it, nitroguanidine ($CH_4N_4O_2$) or diethylene glycol dinitrate ($C_4H_8N_2O_7$) is adopted. The stoichiometric blend of NC and NG is about 1 : 8.5 and this is not practical for a propellant. With plasticizers such as diethylphthalate and stabilizers, say diphenylamine, only a maximum of about 43% of NG can be added to NC. And, the propellant is extruded by solvent and solventless processes or cast. **Solvent process** is adopted to get grains of small web dimensions below 10 mm. In this, the propellant blend is mechanically mixed or gelatinized in the presence of a solvent, say acetone, and extruded allowing the excess solvent to evaporate. In the **solventless process**, the colloiding or plasticization is accomplished by hot working of the blend through differential rollers and the mix thus obtained is extruded as grains, relatively of large web thicknesses. Apart from these two extrusion processes, casting procedure is adopted for large and complicated grain configurations. In this process an evacuated mold is loaded with small pellets of NC. NG and the other ingredients are added into the mold and the mixture is heated for several days at a temperature of about 70 °C to get the grains to the finished form. JPN Ballistite, Russian Cordite, and Cordite are typical examples of homogeneous propellants. Table 4.4 gives the composition of JPN Ballistite, along with the functions of the various ingredients [4]. Notwithstanding their lower specific impulse compared to the heterogeneous propellants, homogeneous propellants are preferred in warfare applications, mainly because of their smokeless exhaust and longer shelf life.

A heterogeneous solid propellant is basically a mechanical mixture of a liquid polymer fuel (frequently termed as binder) with graded multimodal particles of an oxidizer and powders of a metallic fuel. In this mixture other principal ingredients added are a plasticizer, a curing agent (or crosslinker), and if necessary a burning

Table 4.4 Typical homogeneous propellant [4]

Ingredients	Mass%	Purpose
Nitrocellulose (13.25% N_2)	51.40	Polymer
Nitroglycerine	42.93	Explosive plasticizer
Diethylphthalate	3.20	Nonexplosive plasticizer
Ethyl centralite	1.00	Stabilizer
Potassium sulfate	1.20	Flash suppressor
Carbon black	0.20	Opasifying agent
Candelilla wax	0.07	Die lubricant

rate modifier. Aided by the crosslinker the mixture is solidified or cured at a suitable temperature. Although the oxidizer for a heterogeneous propellant can be nitrate or perchlorate of potassium, sodium, lithium, or ammonium, the ammonium perchlorate (AP, NH_4ClO_4) has become the workhorse oxidizer for the last sixty to seventy years. However, in special applications such as igniters, potassium perchlorate is used. As a new entrant to the list of oxidizers, ammonium dinitramide (ADN, $H_4N_4O_4$), first synthesized in 1971 in the USSR, is being adopted in small experimental motors. Presently, the oxidizer is quite expensive and its use in large rocket motors is yet to be realized. The most significant advantage of this chemical is that it gives a chlorine free exhaust with low signature. In Sect. 3.11, we have discussed the classification of polymers. As you recall polymers are classified into plastics and elastomers. Plastics in turn are subdivided into thermoplastics and thermosets. During the development of heterogeneous propellants many polymers under each of these categories were tried as binder fuels. Asphalt, cellulosic resin, polyamide, polyethylene, polystyrene, and polyvinyl under thermoplastics were studied as possible fuel candidates. Under thermosets, each of polyester, epoxy, and phenolic was used in the development of heterogeneous propellants. Polyacrylate, polybutadiene, polysiloxane, polysulphide, polyurethane, and synthetic rubber all formed the trial polymer fuels under elastomers. In recent years copolymers and terpolymers under the butadiene base have become preferred. Of these, polybutadiene acrylic acid (PBAA), polybutadiene acrylic acid acrylonitrile (PBAN), carboxyl terminated polybutadiene (CTPB), and hydroxyl terminated polybutadiene (HTPB) are to be mentioned. In the last few decades, HTPB has become the most preferred polymer fuel. PBAN was used as the fuel in the solid rocket booster (SRB) of the Space Shuttle. In recent years quite a few energetic binder fuels such as glycidyl azide polymer (GAP), bis-azidomethyloxetane (BAMO), and azidomethyl-methyloxetane (AMMO) are being studied and the prime one among these is GAP.

To arrive at a suitable propellant composition, the first step is to calculate the theoretical rocket performance of the propellant with different O/F ratios. In a typical metallized heterogeneous propellant, we can divide the ingredients into three groups: oxidizer, polymer-fuel binder and plasticizer, and metallic fuel. And, selecting different propellant compositions the specific impulse values for the specified

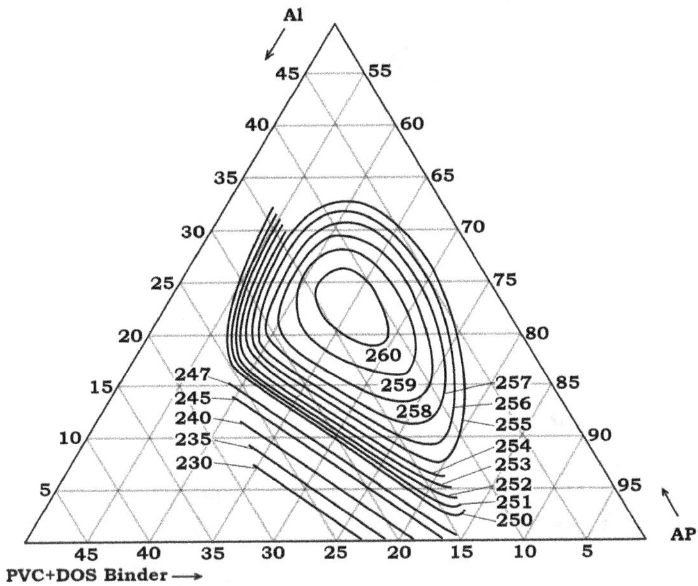

Fig. 4.3 Theoretical specific impulse for AP+Al+(PVC+Dioctyl sebacate) propellant. Chamber pressure = 70 bar and nozzle pressure ratio = 70. Shifting-equilibrium nozzle flow assumed. Specific impulse values are in kgf-s/kgm

chamber pressure and nozzle pressure ratio can be calculated. From these values we can arrive at the contour mapping of specific impulse in a **triangular composition diagram**. Although straightforward, the calculations are rather laborious. One such contour mapping of specific impulse for a polyvinyl chloride plastisol propellant is given in Fig. 4.3. PVC plastisol propellants were in use in control motors, sounding rockets, and ship and shoulder fired missiles. The triangular composition mapping of Fig. 4.3 is from one of our earlier studies in 1970s. Equal mass proportions of polyvinyl chloride powder and plasticizer dioctyl sebacate were adopted for the calculations. The specific impulse quoted is for the adapted condition in the units of kgf-s/kgm. The maximum specific impulse of 260.6 kgf-s/kgm (2555.6 m/s) is obtained at the composition of $AP : Al : (PVC + DOS) = 64.7 : 22.8 : 12.5$. Let us consider the specific impulse contour of 260 kgf-s/kgm in the triangular composition diagram. This gives multiple propellant compositions of binder+plasticizer percentages from 10 to 15% along with AP:Al proportions. With this range of compositions, the provisional propellant composition can be chosen considering the ability to process the propellant blend. The other properties to be considered to arrive at the final composition are the mechanical strength, burning rate, temperature sensitivity, density, and aging. Table 4.5 gives the composition of a typical AP/Al/HTPB heterogeneous propellant.

Table 4.5 Typical AP/Al/HTPB heterogeneous propellant

Ingredients	Mass%
Ammonium perchlorate (bimodal coarse and fine particles)	67.7
Aluminum powder	18.0
Hydroxyl terminated polybutadiene—HTPB (fuel binder)	10.0
Dioctyl adipate (plasticizer)	3.0
Toluene diisocyanate (crosslinker)	0.7
Ferric oxide (burning rate modifier)	0.3
Anti oxidant and other additives	0.3

4.1.2 *Equilibrium*

In order to understand the method to verify equilibrium conditions, let us briefly review the related thermodynamic principles.

4.1.2.1 Entropy Principle

The **entropy principle** is one of the important statements of the second law and it is the basis for the determination of equilibrium composition. The principle is stated as follows. *The entropy of an isolated system can never decrease; for all real processes, it always increases and becomes a maximum at the state of equilibrium; and at the equilibrium, any conceivable differential change in entropy would be zero.* By the definition, the isolated system cannot exchange either mass or energy outside its boundaries. In rocket engine operation three states are of interest. One, at the specified energy and pressure (constant enthalpy and pressure: h & p). Second, at the specified entropy and pressure (constant entropy and pressure: s & p). Third, at the specified temperature and pressure (constant temperature and pressure: T & p). At the first state ($h\&p$) we calculate the products of combustion at equilibrium for the energy contained in the propellant ingredients in the adiabatic combustion chamber of specified pressure. And, the temperature thus calculated for the stagnant condition is the **adiabatic flame temperature** at the specified pressure. At the second state ($s\&p$), we make an idealization that the nozzle flow is isentropic and calculate the equilibrium composition at a pressure along the nozzle (convergent portion, throat section, and divergent portion). At the third state ($T\&p$), we consider the possibility of the products of combustion to reach a particular temperature and pressure and calculate the equilibrium composition for that specified state. We may note here that the equilibrium composition is dependent only on the static values of temperature and pressure. In the combustion chamber since the velocities there are generally low, the properties of equilibrium composition are calculated with respect to the chamber total pressure. In advanced designs, however, the velocities at the nozzle entry are expected to be high with certain total pressure loss with respect to the total value at the head end of the combustion chamber. Therefore, one may calculate the equilibrium composition at the specified static values at the entry to the nozzle (constant $T\&p$). Evidently, further downstream in the nozzle the velocities are high from subsonic to supersonic values. And hence, the equilibrium composition there is calculated with respect to the static values. Now, to understand the equilibrium composition, let us take a simple example of the combustion reaction between methane CH_4 and oxygen O_2 at the stoichiometric proportion. With sufficient time and cooling the combustion products will have only stable species, Eq. 4.4.

$$CH_4 + 2O_2 \rightarrow CO_2 + 2H_2O \tag{4.4}$$

But, the equilibrium composition, corresponding to any of the previously specified three states, may occur even before the completion of reactions to have only stable species. At the equilibrium we may have additionally dissociated species, Eq. 4.5.

$$CH_4 + 2O_2 \rightarrow a_1CO_2 + a_2H_2O + a_3CO + a_4H + a_5O + a_6OH \tag{4.5}$$

The calculation of equilibrium composition at a specified state involves evaluating the coefficients a_1–a_6 such that the system entropy is maximized for $h\&p$ and $T\&p$ or reaches the specified value for $s\&p$. At the attained equilibrium state, any differential change in any of the coefficients a_1–a_6 will only decrease the entropy of the system that is against the entropy principle.

The entropy principle (the entropy of an isolated system can never decrease; for all real processes, it always increases and becomes a maximum at the state of equilibrium) can be written as

$$dS_u \geq 0 \tag{4.6}$$

where S_u is the entropy of the isolated system—the isolated system is also known as the universe. For our analysis, we can assume the isolated system that does not exchange either mass or energy outside its boundaries to contain two subsystems within: one, the system of interest and the other, its infinite surroundings, Fig. 4.4. We are concerned in finding the equilibrium of the system of interest. We may write the entropy of the isolated system S_u as the sum of the entropy of the system of interest S and the entropy of the surroundings S_r. And, the path towards equilibrium can be written as in Eq. 4.7.

$$dS_u = dS + dS_r \geq 0 \tag{4.7}$$

We see here that the entropy of the isolated system follows the path of positive slope to reach the maximum entropy at the equilibrium. Following the path of negative slope would mean the decrease in entropy that is against the entropy principle. In the expression of Eq. 4.7, the calculation of the differential change of entropy of the surroundings dS_r is an arduous task. Therefore we have to eliminate dS_r by expressing it in terms of the properties of the system of interest. For this, let us assume an elemental irreversible heat transfer δQ from the system of interest that is assumed

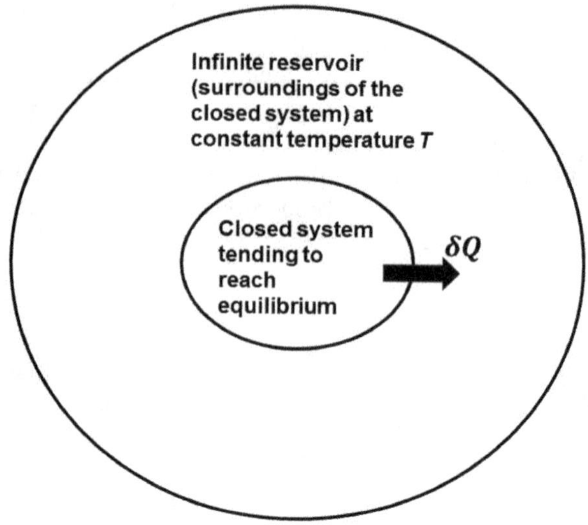

Fig. 4.4 Isolated system consisting the system of interest and the infinite surroundings

to be closed to the infinite surroundings of constant temperature T, Fig. 4.4. Note by the convention that we follow in thermodynamics, for the system of interest the algebraic quantity δQ is a negative number but for the surroundings it is a positive one of equal magnitude by conservation. Therefore we get $dS_r = -\delta Q/T$. Substituting this into Eq. 4.7, we get the entropy principle as in Eq. 4.8.

$$\delta Q \le TdS \tag{4.8}$$

Only with a reversible pressure–volume work and in the absence of any other energy and work transfer from the system of interest to its surroundings, we note by the first law $\delta Q = dU + pdV$. Substituting this into Eq. 4.8, we get

$$dU + pdV - TdS \le 0 \tag{4.9}$$

The equilibrium condition given by Eq. 4.9 is for an arbitrary mass of the system. It can be written for unit mass as

$$du + pdv - Tds \le 0 \tag{4.10}$$

With the "restrictive" conditions, (the system of interest is a closed one and except δQ no other energy and work transfer occurs from it to its surroundings) we have transformed the equilibrium inequality of Eq. 4.7 into the one of Eq. 4.10. Although the conditions are restrictive, the expression in Eq. 4.10 contains only the properties u, p, v, T, and s of the system, which are readily measurable or calculable. Furthermore, more importantly note that despite the restrictive conditions, since the inequality Eq. 4.10 contains only properties (point functions and not path functions) it is applicable for any change, reversible or irreversible. Moreover, since the path that we follow to calculate these properties is immaterial, the inequality holds good for both closed- as well as open-system. Note that rocket engines come under the class of open systems. It is interesting to observe that in variance to the path of positive slope followed by the expression in Eq. 4.7 to reach a maximum value of entropy to attain equilibrium, equivalently the expression in Eq. 4.10 containing the properties of the system of interest follows the path of negative slope to reach a minimum value to attain equilibrium.

Example 4.1 demonstrates numerically the values of $dS_u \geq 0$ (Eq. 4.6) and $dU + pdV - TdS \leq 0$ (Eq. 4.9) when a cylinder-piston system is far way from equilibrium (highly non-equilibrium) and very close to equilibrium.

Example 4.1 A mechanical actuator using a cylinder–piston system stores 1 g of air at 10 bar at 300 K. On triggering, the cylinder expands to the atmospheric pressure of 1 bar. The expansion can be assumed to be an isentropic one. After the actuation and expansion, the actuator comes to equilibrium with respect to the atmospheric condition of 1 bar and 300 K. By considering the heating of the actuator at the condition immediately after the actuation and at the condition close to equilibrium, discuss the equilibrium process with respect to Eqs. 4.7, 4.8, and 4.9. Assume for air the molar mass as 28.971 kg/kgmol, the ratio of specific heats as 1.4, and the specific heat at constant pressure c_p as 1005 J/(kgK).

Analysis
Specific gas constant,

$$R = \frac{R_u}{\overline{m}} = \frac{8314.51}{28.971} = 286.994\ \frac{\text{J}}{\text{kgK}}$$

$$V_i = \frac{0.001 \times 286.994 \times 300}{10 \times 10^5} = 8.6098 \times 10^{-5}\ \text{m}^3$$

As the air expands under the isentropic condition, the final volume can be found.

$$p_i V_i^{\gamma} = p_f V_f^{\gamma} \Rightarrow 10 \times \left(8.6098 \times 10^{-5}\right)^{1.4} = 1 \times V_f^{1.4}$$
$$V_f = 4.4594 \times 10^{-4}\ \text{m}^3$$

$$1 \times 10^5 \times 4.4594 \times 10^{-4} = 0.001 \times 286.994 \times T_f$$
$$T_f = 155.38\ \text{K}$$

After expansion, the actuator volume of 4.4594E-04 m^3 is at 1 bar and 155.38 K. The major constituents of air, nitrogen (boiling point at 77.352 K) and oxygen (boiling point at 90.17 K) will not condense! As the ambient pressure and temperature are 1 bar and 300 K respectively, the heat transfer will take place from the surrounding atmosphere (the infinite surroundings) to the actuator (the system) and under constant pressure. In this process, the system will be doing work on the surroundings by volume expansion under constant pressure until the temperature of the system is equal to the ambient temperature 300 K. Let us consider a temperature rise of 1 K by way of heat transfer at (1) the condition immediately after the actuation 1 bar, 155.38 K and (2) the system condition close to equilibrium 1 bar and 299 K.

For the heating by 1 K temperature difference,

$$\delta Q = 0.001 \times 1005 \times 1 = 1.005 \text{ J}$$

$$c_v = \frac{c_p}{\gamma} = \frac{1005}{1.4} = 717.857 \ \frac{\text{J}}{\text{kgK}}$$

(1) Heating at constant pressure of 1 bar from 155.38 to 156.38 K

$$1 \times 10^5 \times V_a = 0.001 \times 286.994 \times 156.38$$
$$V_a = 4.4880 \times 10^{-4} \text{ m}^3$$

The heat transfer of 1.005 J with respect to the surroundings is negative but with respect to the system it is positive. The system's entropy change,

$$\Delta S = \int_{155.38\,\text{K}}^{156.38\,\text{K}} dS = \int_{155.38\,\text{K}}^{156.38\,\text{K}} \frac{dQ}{T} = mc_p \ln \frac{T_2}{T_1}$$
$$= 0.001 \times 1005 \times \ln \frac{156.38}{155.38} = 6.4473 \times 10^{-3} \ \frac{\text{J}}{\text{K}}$$

The entropy change of the surroundings,

$$\Delta S_r = \frac{-1.005}{300} = -3.35 \times 10^{-3} \ \frac{\text{J}}{\text{K}}$$

As per entropy principle $\Delta S_u = \Delta S + \Delta S_r \geq 0$. Therefore the entropy change of the isolated system,

$$\Delta S_u = \Delta S + \Delta S_r = 6.4473 \times 10^{-3} - 3.35 \times 10^{-3}$$
$$= 3.0973\ \frac{\text{J}}{\text{K}} > 0$$

$$T\Delta S = 300 \times 6.4473 \times 10^{-3} = 1.9342\ \text{J}$$

As per entropy principle $\Delta Q \le T\Delta S$. We see that in our calculations this is satisfied.

$$\Delta Q \le T\Delta S$$
$$1.005\ \text{J} \le 1.9342\ \text{J}$$

The work delivered by the system on its surroundings,

$$pdV = 1 \times 10^5 \times \left(4.4880 \times 10^{-4} - 4.4594 \times 10^{-4}\right) = 0.286\ \text{J}$$

The change in internal energy of the system,

$$\Delta U = m \times c_v \times \Delta T = 0.001 \times 717.857 \times 1 = 0.717857\ \text{J}$$

By entropy principle $\Delta U + p\Delta V - T\Delta S \le 0$. By our calculations,

$$\Delta U + p\Delta V - T\Delta S \le 0$$
$$0.717857 + 0.286 - 1.9342 = -0.930343\ \text{J}$$

(2) Heating at constant pressure of 1 bar from 299 to 300 K

$$1 \times 10^5 \times V_b = 0.001 \times 286.994 \times 299$$
$$V_b = 8.5811 \times 10^{-4}\ \text{m}^3$$

At the equilibrium condition,

$$1 \times 10^5 \times V = 0.001 \times 286.994 \times 300$$
$$V = 8.6098 \times 10^{-4}\ \text{m}^3$$

The system's entropy change,

$$\Delta S = \int_{299\,\mathrm{K}}^{300\,\mathrm{K}} dS = \int_{299\,\mathrm{K}}^{300\,\mathrm{K}} \frac{dQ}{T} = mc_p \ln \frac{T_2}{T_1}$$
$$= 0.001 \times 1005 \times \ln \frac{300}{299} = 3.3556 \times 10^{-3}\ \frac{\mathrm{J}}{\mathrm{K}}$$

The entropy change of the surroundings remains the same as calculated previously,

$$\Delta S_r = \frac{-1.005}{300} = -3.35 \times 10^{-3}\ \frac{\mathrm{J}}{\mathrm{K}}$$

As per entropy principle $\Delta S_u = \Delta S + \Delta S_r \geq 0$. Therefore the entropy change of the isolated system,

$$\Delta S_u = \Delta S + \Delta S_r = 3.3556 \times 10^{-3} - 3.35 \times 10^{-3}$$
$$= 5.6 \times 10^{-6}\ \frac{\mathrm{J}}{\mathrm{K}} > 0$$

$$T\Delta S = 300 \times 3.3556 \times 10^{-3} = 1.00668\ \mathrm{J}$$

As per entropy principle $\Delta Q \leq T\Delta S$. We see that in our calculations this is satisfied.

$$\Delta Q \leq T\Delta S$$
$$1.005\ \mathrm{J} \leq 1.00668\ \mathrm{J}$$

The work delivered by the system on its surroundings,

$$pdV = 1 \times 10^5 \times \left(8.6098 \times 10^{-4} - 8.5811 \times 10^{-4}\right) = 0.287\ \mathrm{J}$$

The change of internal energy remains the same at 0.717857 J. By entropy principle $\Delta U + p\Delta V - T\Delta S \leq 0$. By our calculations,

$$\Delta U + p\Delta V - T\Delta S \leq 0$$
$$0.717857 + 0.287 - 1.00668 = -1.823 \times 10^{-3}\ \mathrm{J}$$

Discussion

In the example, we have considered the heat flow from the infinite surroundings to the system. Hence the heat flow from the surroundings is a negative number but it is equal in magnitude but positive for the system. Also we see here the system is doing work on the surroundings—a positive quantity for the system.

Secondly, we observe that when the system is at 155.38 K with the surroundings at 300 K, the system is far from equilibrium and hence $\Delta U + p\Delta V - T\Delta S = -0.930343$ J (relatively a highly negative quantity). But close to the equilibrium with the system temperature at 299 K and the surrounding temperature at 300 K, $\Delta U + p\Delta V - T\Delta S = -1.823 \times 10^{-}3$ J (close to a 0 value). Similar processes occur regarding the entropy change of the isolated system ΔS_u. With the system temperature at 155.38 K against the surrounding temperature of 300 K, ΔS_u is having a highly positive value at 3.0973 J/K. But, close to the equilibrium it is close to zero at 5.6×10^{-6} J/K. Also notice the inequality values of $\Delta Q \leq T\Delta S$ far away from the equilibrium and very close to the equilibrium.

4.1.2.2 Equilibrium Conditions

Equilibrium conditions for many particular cases can be drawn from the general condition given by Eq. 4.10. As a particular case, let us consider equilibrium reactions at constant temperature and pressure ($T\&p$). Writing the Gibbs energy with respect to the system.

$$g \equiv h - Ts = u + pv - Ts \tag{4.11}$$

Taking differentials,

$$dg = du + pdv + vdp - Tds - sdT \tag{4.12}$$

At constant temperature and pressure ($dT = 0$ & $dp = 0$),

$$(dg)_{T,p} = du + pdv - Tds \tag{4.13}$$

By comparing Eq. 4.13 with the general equilibrium condition given by Eq. 4.10, we deduce that for the reacting system that operates at constant temperature and pressure, the path towards equilibrium condition is given by

$$(dg)_{T,p} = du + pdv - Tds \leq 0 \tag{4.14}$$

And, at the equilibrium,

$$(dg)_{T,p} = du + pdv - Tds = 0 \tag{4.15}$$

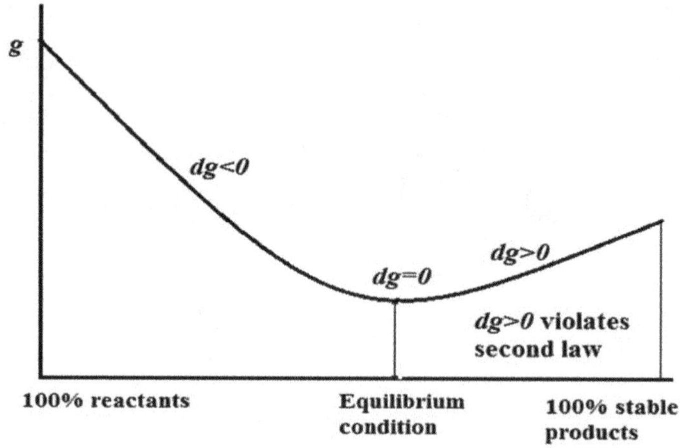

Fig. 4.5 Equilibrium criteria for constant temperature and pressure reaction

Thus we conclude that all irreversible processes at constant temperature and pressure ($T\&p$) have to proceed in the direction of a decreasing Gibbs energy and never in the direction of an increasing value, Fig. 4.5. The second particular case applicable for a rocket engine is a reacting system at constant entropy and pressure. Following similar procedure, we can show the criteria for equilibrium for the constant entropy and pressure ($s\&p$) as

$$(dh)_{s,p} = du + pdv = 0 \tag{4.16}$$

Two other particular cases of reactions are the one at constant temperature and volume ($T\&v$) and the other at constant entropy and volume. Although these two are not of interest for rocket engines, for the sake of completion let us consider these cases. Writing Helmholtz energy with respect to the system,

$$f \equiv u - Ts \tag{4.17}$$

Taking differentials and applying constant temperature and volume condition,

$$(df)_{T,v} = du - Tds \tag{4.18}$$

For the condition of constant volume, the general equilibrium condition Eq. 4.10 with $pdv = 0$ becomes, $du - Tds \leq 0$. Comparing this with Eq. 4.18, we deduce that for the reacting system that operates at constant temperature and volume, the path towards equilibrium condition is

$$(df)_{T,v} = du - Tds \leq 0 \tag{4.19}$$

And, at the equilibrium,

$$(df)_{T,v} = du - Tds = 0 \tag{4.20}$$

Following similar procedure, we can show the criteria for the equilibrium for the constant entropy and volume ($s\&v$) as

$$(du)_{s,v} = 0 \tag{4.21}$$

Have a re-look at the equilibrium conditions for the four particular cases, Eqs. 4.15, 4.16, 4.20, and 4.21 and recognize that we have arrived at these conditions by minimizing respectively Gibbs energy, enthalpy, Helmholtz energy, and internal energy. Hence the method of arriving at the equilibrium is known as **method of energy minimization**. Many reacting systems in engineering have the application of minimizing Gibbs energy or Helmholtz energy for arriving at their equilibrium conditions and because of their popular names as Gibbs free energy and Helmholtz free energy, the method came to be known as the method of free energy minimization. However, most recent compilations have avoided the use of "free" and adopted the terminologies as Gibbs Energy and Helmholtz energy.

It is convenient to proceed with the equilibrium analyses by considering unit mass of the reacting mixture. Example 4.2 demonstrates the simple calculations involved.

Example 4.2 By considering unit mass of the reactants mixture, it is easier to proceed with the calculation of equilibrium composition at the specified state. In the stoichiometric composition of reaction between CH_4 and O_2 calculate the reactants mixture composition for one kg of mixture.

Analysis

Molar mass of CH_4, $\overline{m}_{CH_4} = 16.04246$ kg/kgmol

Molar mass of O_2, $\overline{m}_{O_2} = 31.9988$ kg/kgmol

For the stoichiometric composition of $CH_4 + 2O_2$, the total mass of mixture = 80.0401 kg.

Therefore for 1 kg of the reactants mixture, the composition is

$12.494 \times 10^{-3} CH_4 + 24.987 \times 10^{-3} O_2$

Therefore, we can write the equilibrium composition that we have to consider equivalent to Eq. 4.5 as

$$\begin{gathered} 12.494 \times 10^{-3} CH_4 + 24.987 \times 10^{-3} O_2 \rightarrow \\ n_{CO_2} CO_2 + n_{H_2O} H_2O + n_{CO} CO + n_H H + n_O O + n_{OH} OH \end{gathered} \tag{4.5a}$$

Discussion

As previously stated, the calculation of equilibrium composition at a specified state involves evaluating the coefficients n_{CO_2}–n_{OH} such that the system entropy is maximized for $h\&p$ and $T\&p$ or reaches the specified value for $s\&p$. At the attained equilibrium state, any differential change in any of the coefficients n_{CO_2}–n_{OH} will only decrease the entropy of the system that is against the entropy principle. Equivalently at the equilibrium, the expression in Eq. 4.10 of the system is minimized. We will learn later that apart from the maximum or minimum condition, the equilibrium composition has to satisfy element balance (number of kgatoms of each element in the reactants side should be equal to that in the products side) and energy balance (total energy of the reactants should be equal to that of all the products for the specified states of $h\&p$ and $s\&p$). On the point of element balance, we see that there are 12.494×10^{-3} kgatoms of carbon, 49.975×10^{-3} kgatoms of hydrogen, and 49.975×10^{-3} kgatoms of oxygen in the reactants mixture and these have to be balanced by the respective kgatoms in the products at the equilibrium.

4.2 Chemical Potential

We will learn in this section an important intensive property known as chemical potential that is linked to the thermodynamic equilibrium of reacting systems. It governs the general equilibrium condition given by Eq. 4.9 or 4.10 and in consequence the particular cases given by Eqs. 4.15, 4.16, 4.20, and 4.21.

4.2.1 Reacting System and Chemical Potential

Let us consider a non-reacting system of a fixed composition with negligible change in kinetic- and potential-energy. Combining the first and second laws of thermodynamics and in the absence of any other energies, the system delivering only reversible pressure–volume work is governed by

$$du = Tds - pdv \tag{4.22}$$

Noting the definitions of enthalpy $h \equiv u + pv$, Gibbs energy $g \equiv h - Ts$, and Helmholtz energy $f \equiv u - Ts$, and adopting Eq. 4.22, we have

$$dg = vdp - sdT \tag{4.23}$$

$$dh = Tds + vdp \tag{4.24}$$

$$df = -sdT - pdv \tag{4.25}$$

Equations 4.22–4.25 are known as **Gibbs equations**. Involving only properties (point functions and not path functions), these equations are valid for both closed and open systems as well as reversible and irreversible processes. But, their applicability only to non-reacting systems is the common limitation. Therefore, for reacting systems that are characterized by variable composition, we have to revise these equations taking into account the energy contributions by the reacting species to the energies u, g, h, and f.

Note that for a non-reacting system the specific entropy s (through ds) and specific volume v (through dv) are appearing as the natural variables of the internal energy u, Eq. 4.22. Therefore, for a unit mass system of a reacting composition, let us consider internal energy u as a function of specific entropy s, specific volume v, and additionally now the number of kgmoles of different reacting species comprised in different phases $n_1,\ n_2, \ldots,\ n_{NG},\ n_{NG+1}, \ldots,$ & n_{NS}. Here $n_1,\ n_2, \ldots,$ & n_{NG} represent kgmoles of gaseous species and $n_{NG+1}, \ldots,$ & n_{NS} represent kgmoles of condensed species (liquid- as well as solid species).

$$u = u\left(s, v, n_1, n_2, \ldots, n_{NS}\right) \tag{4.26}$$

For small changes in u,

$$\begin{aligned} du = &\left(\frac{\partial u}{\partial s}\right)_{v,n_1,n_2,\ldots,n_{NS}} ds + \left(\frac{\partial u}{\partial v}\right)_{s,n_1,n_2,\ldots,n_{NS}} dv \\ &+ \left(\frac{\partial u}{\partial n_1}\right)_{s,v,n_2,n_3,\ldots,n_{NS}} dn_1 + \left(\frac{\partial u}{\partial n_2}\right)_{s,v,n_1,n_3,\ldots,n_{NS}} dn_2 + \cdots\cdots \\ &+ \left(\frac{\partial u}{\partial n_{NS}}\right)_{s,v,n_1,n_2,\ldots,n_{NS-1}} dn_{NS} \end{aligned} \tag{4.27}$$

Notice that the first two gradients in the right-hand side of the expression are for the fixed composition of unit mass comprising kgmoles of all species $n_1,\ n_2, \ldots,$ & n_{NS}. Also observe that the differential du in Eq. 4.22 is the change in the internal energy for a fixed composition involving only heat transfer and pressure–volume work, while the differential du in Eq. 4.27 involves in addition the energy contribution by the reacting species. Comparing Eq. 4.27 with Eq. 4.22, we rewrite Eq. 4.27 as

$$du = Tds - pdv + \left(\frac{\partial u}{\partial n_1}\right)_{s,v,n_2,n_3,\ldots,n_{NS}} dn_1 + \left(\frac{\partial u}{\partial n_2}\right)_{s,v,n_1,n_3,\ldots,n_{NS}} dn_2 + \cdots\cdots + \left(\frac{\partial u}{\partial n_{NS}}\right)_{s,v,n_1,n_2,\ldots,n_{NS-1}} dn_{NS} \tag{4.28}$$

Notice that in the right-hand side of the above equation, in the third term the number of kgmoles n_1 is missing in the subscripts of the gradient $(\partial u/\partial n_1)$. Therefore, the third term $(\partial u/\partial n_1)_{s,v,n_2,n_3,\ldots,n_{NS}}$ represents the change in internal energy of the unit mass system per unit kgmole of just the species 1 when all other quantities $s, v, n_2, \ldots, n_{NS}$ are held constant. This meaning of the partial derivatives continues for all the remaining terms. Therefore we can write the Eq. 4.28 in a short form as

$$du = Tds - pdv + \sum_{j=1}^{NS}\left(\frac{\partial u}{\partial n_j}\right)_{s,v,n_{i\neq j}} dn_j \tag{4.29}$$

where the sub-subscript $i \neq j$ implies that the number of kgmoles of all the species is kept constant except that of the species j under consideration. The term $\sum_{j=1}^{NS}\left(\frac{\partial u}{\partial n_j}\right)_{s,v,n_{i\neq j}} dn_j$ represents the energy contribution of the reacting species to the du.

For a unit mass system of reacting composition, let us next consider Gibbs energy as a function of its natural variables temperature T and pressure p, Eq. 4.23, and additionally the number of kgmoles of different reacting species in different phases $n_1, n_2, \ldots, n_{NS}$.

$$g = g\,(T, p, n_1, n_2, \ldots, n_{NS}) \tag{4.30}$$

Proceeding in similar lines we get

$$dg = vdp - sdT + \sum_{j=1}^{NS}\left(\frac{\partial g}{\partial n_j}\right)_{T,p,n_{i\neq j}} dn_j \tag{4.31}$$

Similar procedures for enthalpy h and Helmholtz energy f yield

$$dh = Tds + vdp + \sum_{j=1}^{NS}\left(\frac{\partial h}{\partial n_j}\right)_{s,p,n_{i\neq j}} dn_j \tag{4.32}$$

$$df = -sdT - pdv + \sum_{j=1}^{NS}\left(\frac{\partial f}{\partial n_j}\right)_{T,v,n_{i\neq j}} dn_j \tag{4.33}$$

Thus for a unit mass system, Eqs. 4.22–4.25 applicable for fixed compositions get modified respectively as Eqs. 4.29, and 4.31–4.33 for reacting compositions.

Now we are interested in knowing whether these energy contributions by the reacting species to the energy differentials $du, dg, dh,$ and df are different from one another. Recalling $g = h - Ts$ and $h = u + pv$, note

$$dg = du + pdv + vdp - Tds - sdT$$

Substituting $du + pdv + vdp - Tds - sdT$ for dg in Eq. 4.31 and canceling the corresponding terms, we get

$$du = Tds - pdv + \sum_{j=1}^{NS} \left(\frac{\partial g}{\partial n_j}\right)_{T,p,n_{i\neq j}} dn_j \tag{4.34}$$

Compare Eq. 4.34 with Eq. 4.29. As the interactions among species are negligible and as the differentials in the summations are only with respect to individual species we find for the reacting system of unit mass,

$$\left(\frac{\partial u}{\partial n_j}\right)_{s,v,n_{i\neq j}} = \left(\frac{\partial g}{\partial n_j}\right)_{T,p,n_{i\neq j}} \tag{4.35}$$

Following similar procedures for other energy differentials dh and df, we can show

$$\begin{aligned} \left(\frac{\partial u}{\partial n_j}\right)_{s,v,n_{i\neq j}} &= \left(\frac{\partial g}{\partial n_j}\right)_{T,p,n_{i\neq j}} = \\ \left(\frac{\partial h}{\partial n_j}\right)_{s,p,n_{i\neq j}} &= \left(\frac{\partial f}{\partial n_j}\right)_{T,v,n_{i\neq j}} \equiv \mu_j \end{aligned} \tag{4.36}$$

Thus we understand that the energy contributions to the differentials $du, dg, dh,$ and df by the reacting species j, with the constancy constraints on the respective natural variables and $n_{i\neq j}$, are all one and the same. And, it is the intensive property known as **chemical potential** or **partial molar energy**. Recall that an intensive property does not change in value when the method of its calculation differs as demonstrated by Eq. 4.36 or when the related system size varies. Examples of intensive properties are pressure, temperature, density, and specific properties such as specific enthalpy and specific impulse. μ_j can be calculated with reference to either internal energy at constant entropy and volume, Gibbs energy at constant temperature and pressure, enthalpy at constant entropy and pressure, or Helmholtz energy at constant temperature and volume, Eq.4.36.

Fundamentally, chemical potential of a species in a mixture of gaseous and condensed species is the slope of the system energy with respect to a change in the number of kgmoles of just that species when kgmoles of all other species along with two other thermodynamic system-properties are held constant. The dimension of chemical potential of species j is J/kgmol of species j. Or in other words μ_j represents the unit molar contribution of energy by species j under the *system specified-conditions* to the unit mass of the reacting system. The chemical potential μ_j represents the driving energy (driving potential) that tends to make a chemical reaction take place.

4.2.2 Chemical Potential Evaluation

Now the link between the particular cases of equilibrium condition based on energy minimization (Eqs. 4.15, 4.16, 4.20, and 4.21) and the chemical potential μ_j can be addressed. Let n_j, the number of kgmoles of the species j per kg of the system, change by a differential amount dn_j. Then the energy of the system changes by $\mu_j dn_j$. Realize μ_j is an algebraic quantity. Given that μ_j is negative, if dn_j is negative the system energy increases and if dn_j is positive it decreases. During reactions such changes can take place for all the species. Therefore the energy change for such changes can be $\sum \mu_j dn_j$. Therefore Eq. 4.29 can be written as

$$du = Tds - pdv + \sum_{j=1}^{NS} \mu_j dn_j \tag{4.37}$$

Similarly we can rewrite Eqs. 4.31–4.33 as

$$dg = vdp - sdT + \sum_{j=1}^{NS} \mu_j dn_j \tag{4.38}$$

$$dh = Tds + vdp + \sum_{j=1}^{NS} \mu_j dn_j \tag{4.39}$$

$$df = -sdT - pdv + \sum_{j=1}^{NS} \mu_j dn_j \tag{4.40}$$

Based on the fundamental equilibrium condition given by Eq. 4.10, $du + pdv - Tds \leq 0$, where the expression containing the properties of the system follows the path of negative slope to reach a minimum value to attain equilibrium, we have derived the particular cases of equilibrium conditions under various constraints, Eqs. 4.15, 4.16, 4.20 and 4.21. The corresponding equations are repeated below for ready reference.

$$(dg)_{T,p} = du + pdv - Tds = 0 \tag{4.15}$$

$$(dh)_{s,p} = du + pdv = 0 \tag{4.16}$$

$$(df)_{T,v} = du - Tds = 0 \tag{4.20}$$

$$(du)_{s,v} = 0 \tag{4.21}$$

Either by substituting du of Eq. 4.37 with applicable constraints in the above equations of equilibrium conditions or by applying directly the respective constraints into Eqs. 4.37–4.40, we get the revised conditions of equilibrium, which are easy to work with and more importantly identical.

$$(du)_{s,v} = \sum_{j=1}^{NS} \mu_j dn_j = 0 \tag{4.41}$$

$$(dg)_{T,p} = \sum_{j=1}^{NS} \mu_j dn_j = 0 \tag{4.42}$$

$$(dh)_{s,p} = \sum_{j=1}^{NS} \mu_j dn_j = 0 \tag{4.43}$$

$$(df)_{T,v} = \sum_{j=1}^{NS} \mu_j dn_j = 0 \tag{4.44}$$

We note from the above that under the method of energy minimization, independent of the pair of state functions at their constant values, the common requirement for chemical reactions to reach equilibrium is

$$\sum_{j=1}^{NS} \mu_j dn_j = 0 \tag{4.45}$$

Thus the equilibrium point is the one at which the decrease in the system energy by certain reacting species is exactly countered by the increase in the system energy by the remaining reacting species. And, more importantly the values of μ_j's are independent of the methods of calculation, Eq. 4.36.

Chemical reactions can be grouped under two distinct operating conditions, those at constant pressure (reactions in open systems or control volumes as in combustion chambers of rockets and gas turbines) and others at constant volume (reactions in closed systems as in combustion chambers of reciprocating engines). Hence for the equilibrium requirement Eq. 4.45, as a matter of convenience the chemical potential for constant pressure reactions is evaluated with respect to g while that for constant volume reactions is calculated with respect to f, Eq. 4.36.

In order to apply the equilibrium condition Eq. 4.45, $\sum_{j=1}^{NS} \mu_j dn_j = 0$, we have to evaluate μ_j's. For constant pressure reactions, as previously indicated, let us consider the definition of μ_j with respect to Gibbs energy g. From Eq. 4.36, $(\partial g/\partial n_j)_{T,p,n_{i\neq j}} \equiv \mu_j$. Therefore,

$$\mu_j dn_j = \left(\frac{\partial g}{\partial n_j}\right)_{T,p,n_{i\neq j}} dn_j = (dg_j)_{T,p,n_{i\neq j}} \tag{4.46}$$

is the system's differential Gibbs energy variation by the species j, the phase of which can be gaseous or condensed. Recall that the chemical potential of a species j based on Gibbs energy is the change in the system's Gibbs energy per unit kgmole of just that species j when all other species' kgmoles along with the system's temperature and pressure are held constant, Eq. 4.36. Therefore the chemical potential of the constituent species j is nothing but its unit molar Gibbs energy imposed with the system's constraints: $(\overline{g}_j)_{T,p} = \mu_j$. *It is not the standalone unit molar energy of the species j at $T\&p$ but it is the partial unit molar energy of the species j in the mixture system, which is at $T\&p$*—hence the other name for chemical potential is **partial molar energy**. Therefore $(d\overline{g}_j)_{T,p} \equiv d\mu_j$. As we are interested only in changes in chemical potential, similar to enthalpy h or internal energy u, we should get the values of chemical potential with reference to a standard state. Standard state pressure for chemical potential is 1 bar. The temperature-dependent part of chemical potential has the standard state temperature of 298.15 K. For the constituent species j at its *standalone condition* Eq. 4.23 can be written on molar basis as

$$d\overline{g}_j = \overline{v}_j dp_j - \overline{s}_j dT_j \tag{4.47}$$

where $\overline{g}_j$, $\overline{v}_j$ and $\overline{s}_j$ are respectively specific molar Gibbs energy, volume, and entropy.

As the system under consideration is heterogeneous, the species j can be gaseous or condensed (liquid or solid). That all the gaseous species are ideal and the condensed species are incompressible (specific volume is independent of pressure) and of negligible volume are our fundamental assumptions. Realize that the specific molar volume of condensed species is negligibly small compared to that of gaseous species. Be it gaseous or condensed, every constituent species senses the system temperature and hence $T_j = T$. In contrast, while a constituent condensed species senses the system pressure p, a constituent gaseous one senses only related partial pressure p_j. A unit kgmole of gaseous species at the partial pressure p_j occupies the system's specific volume v, that is $(\overline{v}_j = v)$ and temperature T. Considering Eq. 4.47, at $T_j = T$ with constant system temperature $dT = dT_j = 0$, the differential chemical potential for gaseous species can be written as

$$\left(d\overline{g}_j\right)_{T,p} = d\mu_j = v\,dp_j = R_u T \frac{dp_j}{p_j} \qquad (j = 1, 2, \ldots, NG) \tag{4.48}$$

Integrating between the limits of the assigned temperature T and standard pressure p_{std}, and the assigned temperature T and partial pressure of the species p_j, we get

$$\int_{T,p_{std}}^{T,p_j} d\mu_j = R_u T \int_{p_{std}}^{p_j} \frac{dp_j}{p_j}$$

$$\mu_j\left(T, p_j\right) - \mu_j\left(T, p_{std}\right) = R_u T \ln \frac{p_j}{p_{std}}$$

Note $p_j = \left(n_j/n\right) p$ (n_j and n being the already existing kgmoles of the unit mass system).

$$\mu_j\left(T, p_j\right) = \mu_j^0\left(T\right) + R_u T \ln \frac{n_j}{n} + R_u T \ln \frac{p}{p_{std}} \qquad (j = 1, 2, \ldots, NG) \tag{4.49}$$

$$\mu_j\left(T, p_j\right) = \overline{g}_j^0\left(T\right) + R_u T \ln \frac{n_j}{n} + R_u T \ln \frac{p}{p_{std}} \qquad (j = 1, 2, \ldots, NG) \tag{4.49a}$$

where $\mu_j^0\left(T\right) = \mu_j\left(T, p_{std}\right) = \overline{g}_j^0\left(T\right)$ is the standard chemical potential of the species j at the assigned temperature T and the standard pressure $p_{std} = 1$ bar. Also note $\mu_j^0\left(T\right)$ is the unit molar Gibbs energy of the species j at the assigned temperature T and the standard pressure $p_{std} = 1$ bar, $\overline{g}_j^0\left(T\right)$. Noting $\overline{g}_j^0 = \overline{h}_j^0 - T\overline{s}_j^0$ and $\overline{s}_j = \overline{s}_j^0 - R_u\left(\ln \frac{n_j}{n} + \ln \frac{p}{p_{std}}\right)$ it follows immediately,

$$\begin{aligned}\mu_j\left(T, p_j\right) &= \overline{g}_j(T, p_j) = \overline{h}_j^0(T) - T\overline{s}_j(T, p_j)\\ &= \overline{h}_j^0(T) - T\left[\overline{s}_j^0(T) - R_u\left(\ln \frac{n_j}{n} + \ln \frac{p}{p_{std}}\right)\right] \qquad (j = 1, 2, \ldots, NG)\end{aligned} \tag{4.49b}$$

Since the standard state pressure is 1 bar Eq. 4.49 can be modified to Eq. 4.50 *noting with caution* that the system pressure p in Eq. 4.50 has to be in the unit bar.

$$\mu_j(T, p_j) = \mu_j^0\left(T\right) + R_u T \ln \frac{n_j}{n} + R_u T \ln p \qquad (j = 1, 2, \ldots, NG);\ p = [\text{bar}] \tag{4.50}$$

$$\begin{aligned}\mu_j\left(T, p_j\right) &= \overline{g}_j(T, p_j) = \overline{h}_j^0(T) - T\overline{s}_j(T, p_j)\\ &= \overline{h}_j^0(T) - T\left[\overline{s}_j^0(T) - R_u\left(\ln \frac{n_j}{n} + \ln p\right)\right] \qquad (j = 1, 2, \ldots, NG)\ p = [\text{bar}]\end{aligned} \tag{4.50a}$$

Using either Eq. 4.49 or Eq. 4.50, we can calculate μ_j, the chemical potential of the gaseous species j.

If the species j is in a condensed phase, it senses the system pressure p and temperature T. Noting that the condensed species are incompressible and of negligible volume and as per constraints $dT = dT_j = 0$ and $dp = dp_j = 0$. Considering Eq. 4.47, the differential chemical potential for condensed species can be written as

$$\left(d\overline{g}_j\right)_{T,p} = d\mu_j = 0 \qquad (j = NG + 1, \ldots, NS) \tag{4.51}$$

Integrating,

$$\mu_j\,(T) = \mu_j^0\,(T) = \overline{g}_j^0\,(T) \qquad (j = NG + 1, \ldots, NS) \tag{4.52}$$

As per the above discussion,

$$\mu_j\,(T) = \overline{g}_j^0\,(T) = \overline{h}_j^0(T) - T\overline{s}_j^0(T) \qquad (j = NG + 1, \ldots, NS) \tag{4.52a}$$

Recognize that μ_j for gaseous species is a function of species mole fraction and system pressure (that is, species partial pressure) and system temperature while μ_j for condensed species is dependent only on system temperature. Standard chemical potential $\mu_j^0\,(T)$ is nothing but the standard molar Gibbs energy, $\overline{g}_j^0\,(T) = \overline{h}_j^0\,(T) - T\overline{s}_j^0\,(T)$.

In Example 4.3, we apply the equations that we have derived to calculate the properties of a mixture in equilibrium, more importantly species partial pressure, chemical potential, and entropy.

Example 4.3 Combustion products of aluminum powder and oxygen in a certain proportion are kept at equilibrium at a temperature and pressure of 4000 K and 50 bar respectively. Mixture molar mass taking into account only the gaseous moles, $\overline{m} = 90.263$ kg/kgmol. The mole fractions of the product species are $x_O = 0.1183$, $x_{O_2} = 0.5058$, and $x_{Al_2O_3(l)} = 0.3759$. Calculate (1) molar mass of combustion products taking into account also the moles of condensed species, $\overline{m}_c$; (2) number of kgmoles of the individual species per kg of the combustion product mixture; (3) partial pressures of the gaseous species; (4) chemical potentials of the individual species in the product mixture, μ_j's in kJ/kgmol; (5) entropies of the individual species in the equilibrium mixture, $\overline{s}_j$'s in kJ/(kgmolK) and the entropy of the mixture kJ/(kgK); (6) Gibbs energy of the mixture, kJ/kg; (7) enthalpy of the mixture, in kJ/kg; and finally (8) density of the mixture.

Enthalpy and entropy for O and O_2 may be read from a standard text book. The enthalpy and entropy of $Al_2O_3\,(l)$ can be calculated from the following equations.

$$\frac{\overline{h}^0_{Al_2O_3(l)}(T)}{R_u T} = 1.959225499E+01 - \frac{2.027701571E+05}{T}$$

$$\frac{\overline{s}^0_{Al_2O_3(l)}(T)}{R_u} = 1.959225499E+01 \times \ln\ T - 1.108590952E+02$$

Solution

This example is given to understand the calculation of various properties required for the equilibrium mixture. On calculating the equilibrium product mixture by the program CEA for the conditions given in this example, we may have 10 species of mole fractions greater than 5E-6. However, in order to understand the calculation procedure we have simplified the product mixture contents to three major species. Polynomial functions with the values of the coefficients for the calculation of specific heat at constant pressure, enthalpy, and standard entropy can be found in the *thermo.inp*, which can be downloaded from [8]. The functions that are given above for Al_2O_3 (l) are from [8]. The solution of the required quantities is through straightforward application of the equations that we have learned. More discussion on the thermodynamic data on propellant constituents and products can be found in Sect. 4.6.

Analysis

(1) The molar mass of combustion products taking into account also the moles of condensed species, $\overline{m}_c$ is given by Eq. 3.127,

$$\overline{m}_c = \overline{m}\sum_{j=1}^{NG} x_j = \overline{m}\left(1 - \sum_{j=NG+1}^{NS} x_j\right) \tag{3.127}$$

Therefore

$$\overline{m}_c = 90.263\,(1 - 0.3759) = 56.3331\ \frac{\text{kg}}{\text{kgmol}}$$

(2) The number of gaseous moles per kg of mixture can be calculated from,

$$\overline{m} = \frac{1}{n} = \frac{\sum_{j=1}^{NS} n_j \overline{m}_j}{\sum_{j=1}^{NG} n_j} \tag{3.125}$$

Therefore,

$$n = \frac{1}{\overline{m}} = \frac{1}{90.263} = 0.011079 \ \frac{\text{kgmol}}{\text{kg}}$$

The number of kgmoles of individual species can be calculated from,

$$n_j = \frac{x_j}{\overline{m}_c} \tag{3.128}$$

Therefore,

$$n_O = \frac{0.1183}{56.3331} = 2.1 \times 10^{-3} \ \frac{\text{kgmol}}{\text{kg}}$$

$$n_{O_2} = \frac{0.5058}{56.3331} = 8.9787 \times 10^{-3} \ \frac{\text{kgmol}}{\text{kg}}$$

$$n_{Al_2O_3}(l) = \frac{0.3759}{56.3331} = 6.6728 \times 10^{-3} \ \frac{\text{kgmol}}{\text{kg}}$$

(3) Partial pressure of gaseous species is given by

$$p_j = \frac{n_j}{n} \times p$$

Therefore,

$$p_O = \frac{2.1 \times 10^{-3}}{0.011079} \times 50 = 9.4774 \text{ bar}$$

$$p_{O_2} = \frac{8.9787 \times 10^{-3}}{0.011079} \times 50 = 40.5213 \text{ bar}$$

(4) Chemical potentials of the individual gaseous species, μ_j is given by

$$\mu_j(T, p_j) = \overline{g}_j(T, p_j) = \overline{h}_j^0(T) - T\overline{s}_j(T, p_j)$$
$$= \overline{h}_j^0(T) - T\left[\overline{s}_j^0(T) - R_u\left(\ln\frac{n_j}{n} + \ln p\right)\right] \quad (j = 1, 2, \ldots, NG)\ p = [\text{bar}] \tag{4.50a}$$

Notice $\overline{g}_j^0 = \overline{h}_j^0 - T\overline{s}_j^0$. $\overline{h}_j^0(T)$ is the sum of the enthalpy of formation at the standard temperature of 298.15 K $\overline{h}_{fj}^0$ (298.15 K) (and, when applicable, heat of transition) and sensible heat at the assigned temperature T. And, $\overline{s}_j^0(T) = \int_{T_{std}}^{T} \overline{c}_{p,j}^0 dT/T$ is the standard state molar entropy for species j at temperature T. The values of $\overline{h}_j^0(T)$ and $\overline{s}_j^0(T)$ for common species are given in standard text books. However, for species generally not considered in standard text books, we have to calculate these values through the values of the coefficients of the respective polynomials given in literature, as we have drawn for $Al_2O_3\,(l)$ from *thermo.inp* of CEA code. Detailed information on $\overline{h}_j^0(T)$ and $\overline{s}_j^0(T)$ is given in Sect. 4.6.

$$\mu_O\,(4000\text{ K, } 9.4774\text{ bar}) = (249190 + 77676) - 4000 \times 215.778$$
$$+8.31451 \times 4000 \times \left(\ln\frac{2.1 \times 10^{-3}}{0.011079} + \ln 50\right)$$
$$= (326866 - 863112) + 74794.303$$
$$= -461451.697\ \frac{\text{kJ}}{\text{kgmol } O}$$

$$\mu_{O_2}\,(4000\text{ K, } 40.5213\text{ bar}) = 138913 - 4000 \times 296.236$$
$$+8.31451 \times 4000\,(\ln 0.810425 + \ln 50)$$
$$= -1046031 + 123115.495 = -922915.505\ \frac{\text{kJ}}{\text{kgmol } O_2}$$

$$\mu_{Al_2O_3(l)}\,(4000\text{ K}) = \overline{h}^0_{Al_2O_3(l)}\,(4000\text{ K}) - 4000 \times \overline{s}^0_{Al_2O_3(l)}\,(4000\text{ K})$$

$$\overline{h}^0_{Al_2O_3(l)}\,(4000\ \text{K}) = 8.31451 \times 4000 \times \left(1.959225499E+01 - \frac{2.027701571E+05}{4000}\right) = -1034334.5\ \frac{\text{kJ}}{\text{kgmol } Al_2O_3\ (l)}$$

$$\overline{s}^0_{Al_2O_3(l)}\,(4000\ \text{K}) = 8.31451 \times (1.959225499E+01 \times \ln\ 4000 - 1.108590952E+02) = 429.3616\ \frac{\text{kJ}}{\text{kgmol } Al_2O_3\ (l)}$$

Therefore,

$$\mu_{Al_2O_3(l)}\,(4000\ \text{K}) = -1034334.5 - 4000 \times 429.3616 = -2751780.9\ \frac{\text{kJ}}{\text{kgmol } Al_2O_3\ (l)}$$

(5) Entropy of individual species in the mixture is given for gaseous species as

$$\overline{s}_j = \overline{s}^0_j\,(T) - R_u \ln\ \frac{n_j}{n} - R_u \ln\ p$$

and for condensed species,

$$\overline{s}_j = \overline{s}^0_j\,(T)$$

Therefore,

$$\overline{s}_O\,(4000\ \text{K},\ 9.4774\ \text{bar}) = 215.778 - 8.31451 \times \left(\ln\ \frac{2.1 \times 10^{-3}}{0.011079} + \ln\ 50\right) = 197.079\ \frac{\text{kJ}}{\text{kgmol } O - K}$$

$$\bar{s}_{O_2}\,(4000\text{ K},\ 40.5213\text{ bar})$$
$$= 296.236 - 8.31451 \times \left(\ln \frac{8.9787 \times 10^{-3}}{0.011079} + \ln 50\right)$$
$$= 265.457\ \frac{\text{kJ}}{\text{kgmol}\ O_2 - K}$$

Entropy of the condensed species $Al_2O_3\ (l)$ has been calculated previously.

$$\bar{s}_{Al_2O_3(l)}\,(4000\text{ K}) = \bar{s}^0_{Al_2O_3(l)}\,(4000\text{ K})$$
$$= 429.3616\ \frac{\text{kJ}}{\text{kgmol}\ Al_2O_3\ (l) - K}$$

Therefore the entropy per kg of the mixture,

$$s\,(4000\text{ K},\ 50\text{ bar}) = n_O \times \bar{s}_o\,(4000\text{ K},\ 9.4774\text{ bar})$$
$$+ n_{O_2} \times \bar{s}_{O_2}\,(4000\text{ K},\ 40.5213\text{ bar})$$
$$+ n_{Al_2O_3(l)} \times \bar{s}^0_{Al_2O_3(l)}\,(4000\text{ K})$$
$$= 2.1 \times 10^{-3} \times 197.079 + 8.9787 \times 10^{-3} \times 265.457$$
$$+ 429.3616 \times 6.6728 \times 10^{-3}$$
$$= 5.6624\ \frac{\text{kJ}}{\text{kgK}}$$

(6) Gibbs energy of the mixture is given by

$$g = \sum_{j=1}^{NS} (\bar{g}_j)_{T,p} n_j = \sum_{j=1}^{NS} \mu_j n_j$$

Therefore,

$$g = -461451.697 \times 2.1 \times 10^{-3} - 922915.505 \times 8.9787 \times 10^{-3}$$
$$-2751780.9 \times 6.6728 \times 10^{-3} = -27617.71\ \frac{\text{kJ}}{\text{kg}}$$

(7) The enthalpy of the mixture is given by

$$h = \sum_{j=1}^{NS} \bar{h}^0_j\,(T)\, n_j$$

Therefore,

$$h = 326866 \times 2.1 \times 10^{-3} + 138913 \times 8.9787 \times 10^{-3}$$
$$-1034334.5 \times 6.6728 \times 10^{-3} = -4968.23 \ \frac{\text{kJ}}{\text{kg}}$$

(8) The density of the mixture is

$$\rho = \frac{p}{nR_uT} = \frac{\overline{m}p}{R_uT} = \frac{50 \times 10^5}{0.011079 \times 8314.51 \times 4000} = 13.570 \ \frac{\text{kg}}{\text{m}^3}$$

Discussion
Recall that the chemical potential of a species is nothing but its partial molar energy, here the partial molar Gibbs energy for gaseous species:

$$\begin{aligned}\mu_j &= \left(\overline{g}_j\right)_{T,p} \\ &= \overline{h}_j^0(T) - T\left[\overline{s}_j^0(T) - R_u\left(\ln\frac{n_j}{n} + \ln\frac{p}{p_{std}}\right)\right] \rightarrow (j = 1, \ldots, NG)\end{aligned}$$

But, for condensed species, the chemical potential is a pure function of temperature:

$$\mu_j = \left(\overline{g}_j\right)_{T,p} = \overline{h}_j^0(T) - T\overline{s}_j^0(T) \rightarrow \quad (j = NG + 1, \ldots, NS)$$

The enthalpy of formation of the element O_2 at its naturally occurring condition at the standard reference state is zero. However, as the sensible heat has gone into it from 298.15 to 4000 K, its enthalpy is +138913 kJ/kgmol. The species O being a dissociated one (heat has been absorbed), its enthalpy of formation at 298.15 K is +249190 kJ/kgmol. Sensible heat has gone into it to heat it from 298.15 to 4000 K by +77676 kJ/kgmol. Hence its enthalpy at 4000 K is +326866 kJ/kgmol. As $Al_2O_3\,(s)$ is formed by the combustion of Al with O_2 at the standard state by the release of −1675700 kJ/kgmol, hence it is its enthalpy of formation. Heating $Al_2O_3\,(s)$ and melting it at its melting point of 2327 K, which involves the sensible heat plus heat of fusion, and further heating the liquid to 4000 K requires a heat supply. Hence the enthalpy of $Al_2O_3\,(l)$ at 4000 K is −1034334.5 kJ/kgmol.

We have learned in Sect. 4.1.2.1 that to reach equilibrium all irreversible processes at constant temperature and pressure have to proceed in the direction of a decreasing Gibbs energy to attain the minimum Gibbs energy level, Eqs. 4.14 and 4.15, and Fig. 4.5. With reference to the present Example, at the equilibrium the minimum Gibbs energy level attained is −27617.71 kJ/kg that corresponds to the maximum entropy of 5.6624 kJ/(kgK).

4.3 Mass Balance

In a chemical reaction, all the elements of reactants must be conserved and distributed in the products. Suppose that we have liquid oxygen and liquid hydrogen as the propellant components in a certain proportion. As far as the number of elements is concerned here we have only two elements: hydrogen and oxygen. Similarly, a solid propellant composition having the oxidizer, ammonium perchlorate NH_4ClO_4, the widely used polymer fuel, hydroxyl terminated polybutadiene of acronym HTPB and of a possible molecular formula $C_{73}H_{110}O_6$, and the metallic fuel aluminum has six elements: carbon, hydrogen, nitrogen, oxygen, chlorine, and aluminum. The combustion products of these propellant compositions should have the respective elements of the propellants distributed in them. Furthermore, on the conservation point of view, the mass contribution of each element in the propellant should be equal to the mass contribution of the respective element in the products. The classical statement of mass conservation is that mass can neither be created nor be destroyed. This classical statement is modified for chemical reactions (propellant combustion) as follows. *The total kgatoms of an element in the reactants must be equal to the total kgatoms of that element in the products.*

In this conservation analysis, it is convenient to consider the total reactants (propellant ingredients) to be of unit mass. As a first step in this element conservation, we have to determine kgatoms of each element i per kg of total reactants. Let b_i^o represent the kgatoms of element i per unit mass of the total reactants. By mass conservation, it is mandatory to have an equal number of kgatoms of element i in the *total* products. Therefore we can write the mass balance for the element i as

$$\sum_{j=1}^{NS} a_{ij} n_j = b_i^o \tag{4.53}$$

where a_{ij} is the stoichiometric coefficient of element i of the species j of n_j kgmoles per unit mass of the products of NS species. With a total of l elements in the reactants we can write the general mass conservation equation as

$$\sum_{j=1}^{NS} a_{ij} n_j - b_i^o = 0 \qquad (i = 1, \ldots, l) \tag{4.54}$$

Denoting for brevity $\sum_{j=1}^{NS} a_{ij} n_j = b_i$, Eq. 4.54 is written as

$$b_i - b_i^o = 0 \qquad (i = 1, \ldots, l) \tag{4.55}$$

Thus, in addition to the energy minimization condition Eq. 4.45, the equilibrium composition has to satisfy the mass conservation condition, Eq. 4.54.

Of the four conditions required for calculating equilibrium composition, as mentioned previously, we have derived two: the first, energy minimization condition Eq. 4.45 and the second, the mass conservation condition, Eq. 4.54. The remaining two conditions are based on the type of reactions, that is on any one applicable-pair of state functions at their constant values ($T\&p$, $s\&p$, $T\&v$, or $s\&v$). For example, for a reacting system with a pair of state functions temperature T and pressure p at their constant values, the two equations are $T = T^*$ and $p = p^*$. The two are called "trivial" in the sense that during our search for equilibrium composition at constant temperature and pressure we should keep in mind to evaluate chemical potentials and other quantities dependent on temperature and pressure with respect to the assigned T^* and p^*. Example 4.4 details the methods of calculating kgatoms of elements per unit mass of the total reactants (propellant composition) and of the total products (combustion products). Also the example explains the procedure to check the mass balance condition.

Example 4.4 A propellant combination adopts hydrazine (N_2H_4) and beryllium (Be) as fuels and hydrogen peroxide (H_2O_2) as an oxidizer. Total-fuel percentage is 67% of propellant mass. The mass proportion of fuels are 20% of Be and 80% of N_2H_4. The combustion chamber pressure is 200 bar. The calculated values of equilibrium composition through a computer program are the following. Adiabatic flame temperature = 3018.68 K. Molecular mass counting also the moles of condensed species ($\overline{m}_c$) = 13.331 kg/kgmol.

Mole fractions of the species at equilibrium calculated by the computer program are given in the following table.

$$
\begin{array}{lclclcl}
x_{BeOH} & = & 0.00025 & \quad & x_{NH_3} & = & 0.00029 \\
x_H & = & 0.00758 & & x_{N_2} & = & 0.22283 \\
x_{H_2} & = & 0.51064 & & x_{OH} & = & 0.00033 \\
x_{H_2O} & = & 0.06012 & & x_{BeO(l)} & = & 0.19796
\end{array}
$$

Calculate (1) the assigned number of kgatoms of each element (b_i^o) per kg of total propellant combination, (2) molar mass of the combustion products considering only gaseous species $\overline{m}$, and (3) the number of kgatoms of each element per kg of combustion products (b_i). (4) By comparing the values obtained by (1) and (3), show that the results satisfy $b_i - b_i^o = 0 \quad (i = 1, \ldots\ldots, l)$. (5) Also calculate the density of the combustion products.

Solution

Results of complex chemical equilibrium calculation for a liquid propellant combination at an assigned pressure are given. We have to calculate the kgatoms of each element per kg of total propellant combination. Also, we have to distinguish between the two molecular masses resulting from the presence and absence of condensed species. One of the constraints of complex chemical equilibrium calculation is the satisfaction of the mass balance of elements. This has to be verified.

Analysis

(1) The fuels form 67% of total propellant combination. Of the fuels, 20% is by beryllium and the balance is by hydrazine. Therefore the propellant combination for 1 kg of propellant is

$$0.33\,\text{kg}\ H_2O_2 + 0.536\,\text{kg}\ N_2H_4 + 0.134\,\text{kg}\ Be$$

Molecular mass of H_2O_2	=	34 kg/kgmol
Molecular mass of N_2H_4	=	32 kg/kgmol
Atomic mass of Beryllium	=	9.01 kg/kgatom

Therefore the above propellant combination can be written in terms of kgmoles for 1 kg of propellant combination.

$$9.7059 \times 10^{-3}\,\text{kgmol}\ H_2O_2 + 16.75 \times 10^{-3}\,\text{kgmol}\ N_2H_4$$

$$+14.8724 \times 10^{-3}\,\text{kgatom}\ Be$$

$$b_H^o = 9.7059 \times 10^{-3} \times 2 + 16.75 \times 10^{-3} \times 4 = 0.086412\ \frac{\text{kgatom of } H}{\text{kg propellant}}$$

$$b_N^o = 16.75 \times 10^{-3} \times 2 = 0.0335\ \frac{\text{kgatom of } N}{\text{kg propellant}}$$

$$b_O^o = 9.7059 \times 10^{-3} \times 2 = 0.019412\ \frac{\text{kgatom of } O}{\text{kg propellant}}$$

Thus the propellant molecular form for 1 kg of propellant becomes

$$H_{0.086412}N_{0.0335}O_{0.019412}Be_{0.014872}$$

(2) Molecular mass $\overline{m}$ is given by

$$\overline{m} = \frac{1}{n} = \frac{\sum_{j=1}^{NS} n_j \overline{m}_j}{\sum_{j=1}^{NG} n_j} \tag{3.125}$$

Table 4.6 Calculated results for Example 4.4

Species	x_j	$\overline{m}_j$	$x_j\overline{m}_j$	$n_j = x_j/\overline{m}_c$ (kgmol/kg)
$BeOH$	0.00025	26.01	0.006503	1.87712E-05
H	0.00758	1	0.00758	0.000569142
H_2	0.51064	2	1.02128	0.038341281
H_2O	0.06012	18	1.08216	0.004514096
NH_3	0.00029	17	0.00493	2.17746E-05
N_2	0.22283	28	6.23924	0.016731137
OH	0.00033	17	0.00561	2.4778E-05
$BeO(l)$	0.19796	25.01	4.95098	0.014863779
Total	$\sum_{j=1}^{NS} x_j=1$	–	$\overline{m}_c = 13.31828$	$\sum_{j=1}^{NS} n_j = 0.075084759$

As per the results of the calculations given in Table 4.6,

$$n = \sum_{j=1}^{NS} n_j - n_{BeO(l)} = 0.075084759 - 0.014863779$$

$$= 0.06022098 \frac{\text{kgmol gas}}{\text{kg}}$$

$$\overline{m} = \frac{1}{n} = \frac{1}{0.06022098} = 16.60551 \frac{\text{kg}}{\text{kgmol}}$$

$$\overline{m} = \frac{\sum_{j=1}^{NS} x_j\overline{m}_j}{\sum_{j=1}^{NG} x_j} = 16.60551 \text{ kg/kgmol}$$

$$\overline{m}_c = \sum_{j=1}^{NS} x_j\overline{m}_j = 13.31828 \text{ kg/kgmol}$$

(3) Number of kgatoms of each element i per kg of combustion products is given by

$$b_i = \sum_{j=1}^{NS} a_{ij} n_j \qquad (i = 1, \ldots\ldots, l) \tag{4.53}$$

Therefore

$$b_H = 1.87712 \times 10^{-5} + 0.000569142 + 2 \times (0.038341281 + 0.004514096)$$

$$+3 \times 2.17746 \times 10^{-5} + 2.4778 \times 10^{-5}$$

$$= 0.086389 \text{ kgatom of } H/\text{kg propellant}$$

$$b_N = 2.17746 \times 10^{-5} + 2 \times 0.016731137$$

$$= 0.033484 \text{ kgatom of } N/\text{kg propellant}$$

$$b_O = 1.87712 \times 10^{-5} + 0.004514096$$

$$+2.4778 \times 10^{-5} + 0.014863779$$

$$= 0.019421 \text{ kgatom of } O/\text{kg propellant}$$

$$b_{Be} = 1.87712 \times 10^{-5} + 0.014863779$$

$$= 0.014883\text{kgatom of } Be/\text{kg propellant}$$

(4) Now we can check the satisfaction of the mass balance

$$b_i - b_i^o = 0 \qquad (i = 1, \ldots\ldots, l) \tag{4.55}$$

$$b_H - b_H^o = 0.086412 - 0.086389 = 23 \times 10^{-6} \approx 0$$

$$b_N - b_N^o = 0.0335 - 0.033484 = 16 \times 10^{-6} \approx 0$$

$$b_O - b_O^o = 0.019412 - 0.01942 = -8 \times 10^{-6} \approx 0$$

$$b_{Be} - b_{Be}^o = 0.014872 - 0.014883 = -11 \times 10^{-6} \approx 0$$

The small differences that we get are due to the convergence error band and also possibly due to the round-off errors resulting from hand calculations.

(5) The density can now be calculated

$$\sum_{j=1}^{NG} n_j = 1.87712 \times 10^{-5} + 0.000569142 + 0.03834128$$

$$+0.004514096 + 2.17746 \times 10^{-5} + 0.016731137$$

$$+2.4778 \times 10^{-5} = 0.060221 \text{ kgmol gas/kg propellant}$$

$$\rho = \frac{p}{nR_uT} = \frac{200 \times 10^5}{0.06022 \times 8314.51 \times 3018.68} = 13.2323\ \frac{\text{kg}}{\text{m}^3}$$

Discussion
We have checked the elemental mass balance condition. Furthermore, we note that when the combustion products contain condensed species we have two molecular mass values, $\overline{m}$ and $\overline{m}_c$, and $\overline{m}_c < \overline{m}$.

4.4 Iteration Equations

Our task is to calculate the n_j's of NS species at equilibrium through a suitable iteration procedure. Although not in forms suitable for iterative operations, we have obtained the four necessary and sufficient conditions to determine equilibrium composition. They are:

1. the energy minimization equation for equilibrium,

$$\sum_{j=1}^{NS} \mu_j dn_j = 0 \tag{4.45}$$

2. the mass balance equation

$$\sum_{j=1}^{NS} a_{ij}n_j - b_i^o = 0 \qquad (i = 1, \ldots, l) \tag{4.54}$$

and

Table 4.7 Pairs of state functions held constant and their application areas

Assigned pair of state functions	Applications
$h\&p$	Equilibrium composition for constant pressure combustion in adiabatic systems: combustion chamber performance in rocket engines, gas turbines, and ramjets
$T\&p$	Equilibrium composition at specified temperature and pressure
$s\&p$	Isentropic expansion/compression to a specified pressure: compression and expansion in turbo machineries; nozzle/diffuser flow characteristics
$u\&v$	Equilibrium composition for constant volume combustion in adiabatic systems: combustion in reciprocating engines at constant volume
$T\&v$	Equilibrium composition at specified temperature and volume
$s\&v$	Isentropic expansion/compression to a specified volume: compression and expansion in reciprocating engines and compressors

3. depending on the type of reaction, a pair of assigned state functions at their constant values ($h = h^*\&p = p^*$ for $h\&p$, $T = T^*\&p = p^*$ for $T\&p$, $s = s^*\&p = p^*$ for $s\&p$, $u = u^*\&v = v^*$ for $u\&v$, $T = T^*\&v = v^*$ for $T\&v$, or $s = s^*\&v = v^*$ for $s\&v$).

Note that in addition to the already considered four pairs of assigned state functions, we have now two more such pairs, namely constant $h\&p$, and constant $u\&v$. While the constant $h\&p$ represents the constant pressure combustion in adiabatic systems, as in ideal rocket propulsion systems and gas turbines, the constant $u\&v$ characterizes the constant volume combustion in adiabatic systems, such as in ideal reciprocating engines of spark-ignition type. Table 4.7 details the application areas for these six pairs of constant state functions. Of the six pairs, the first three correspond to the reactions at constant pressure and the others correspond to the ones at constant volume. We mentioned previously that our focus here is on the calculation of equilibrium composition applicable to the performance prediction of rocket propulsion systems. Since constant pressure reactions come under this focus, *hereafter we consider only the analyses related to the first three pairs of constant state functions, namely $h\&p$, $T\&p$, and $s\&p$.*

Energy minimization condition is given in general by Eq. 4.45. For our analyses of constant pressure reactions, let us consider the energy minimization through Gibbs energy minimization condition at assigned temperature and pressure, Eq. 4.42.

$$(dg)_{T,p} = \sum_{j=1}^{NS} \mu_j dn_j = 0 \tag{4.42}$$

Equilibrium condition is identified by finding a set of n_j's that satisfies Eq. 4.42. Basically this is to be done by starting a *trial set* of n_j's, calculating its μ_j's subject to the specified pair of state functions and mass balance (Eq. 4.54), getting the total energy $\sum_{j=1}^{NS} \mu_j n_j$, and checking whether the total energy has reached its minimum or not. If not reached, mathematically proceed to estimate the next trial set to continue the process of checking until the total energy reaches its minimum. Thus the objective function of energy to be minimized for equilibrium is

$$g_{T,p} = \sum_{j=1}^{NS} \mu_j n_j \tag{4.56}$$

where n_j's become the variables for analyses.

The mass balance condition Eq. 4.54 $\sum_{j=1}^{NS} a_{ij} n_j - b_i^o = 0 \quad (i = 1, \ldots, l)$—kgatoms of elements in the reactants should be equal to the kgatoms of respective elements in the products—becomes the effective *constraint* under which we have to find the minimum of the $g_{T,p}$ function. For the given values of the chosen state function pair (Table 4.7), the minimum of $g_{T,p}$ along with the mass balance constraint determines the equilibrium composition, n_j's and thus n.

4.4.1 Newton–Raphson (NR) Equations

Subject to the l constraints given by Eq. 4.54, we have to minimize $g_{T,p}$ given by Eq. 4.56. For this we can adopt the method of **Lagrange multipliers**. This method finds the extremum of a function subject to the specified constraints. The method introduces a new unknown scalar variable known as **Lagrange multiplier** for each constraint and forms a linear combination involving the Lagrange multipliers as coefficients. This reduces the constrained problem to an unconstrained one. Detailed information on Lagrange Multipliers and related procedures to find extrema can be found in text books dealing with Calculus of Variations or Nonlinear Programming.

On adopting the method of Lagrange multipliers for the minimization of Gibbs energy $g_{T,p}$ (Eq. 4.56) with l mass balance constraints equal to the number of elements in the system (Eq. 4.54), we can write the combined expression for optimization as

$$G = \sum_{j=1}^{NS} \mu_j n_j + \sum_{i=1}^{l} \lambda_i \left(\sum_{j=1}^{NS} a_{ij} n_j - b_i^o \right) \tag{4.57}$$

where λ_i's $(i = 1, \ldots, l)$ are the Lagrange multipliers. With n_j's and λ_i's as variables, the condition of extremum becomes

$$dG = \sum_{j=1}^{NS} \left(\frac{\partial G}{\partial n_j} \right) dn_j + \sum_{i=1}^{l} \left(\frac{\partial G}{\partial \lambda_i} \right) d\lambda_i = 0 \tag{4.58}$$

On differentiation, Eq. 4.58 can be shown to be

$$dG = \sum_{j=1}^{NS} \left(\mu_j + \sum_{i=1}^{l} \lambda_i a_{ij} \right) dn_j + \sum_{i=1}^{l} \left(\sum_{j=1}^{NS} a_{ij} n_j - b_i^o \right) d\lambda_i = 0 \tag{4.59}$$

Treating the differentials dn_j and $d\lambda_i$ as independent, it immediately follows for equilibrium,

$$\mu_j + \sum_{i=1}^{l} \lambda_i a_{ij} = 0 \qquad (j = 1, \ldots, NS) \tag{4.60}$$

and

$$\sum_{j=1}^{NS} a_{ij} n_j - b_i^o = b_i - b_i^o = 0 \qquad (i = 1, \ldots, l) \tag{4.54}$$

The above two equations (totaling to $NS + l$ simultaneous individual equations corresponding to the species and elements) encompass energy minimization with the constraint on mass conservation. These are required for all the three reactions with the pairs of constant pressure state functions ($h\&p$, $T\&p$, and $s\&p$). These $NS + l$ equations along with an assigned pair of state functions have to be iterated to determine the equilibrium composition. For this, we can use a descent **Newton–Raphson method**, which adopts **Taylor series** expansion of the appropriate equations, truncating all terms with derivatives of second and higher orders.

4.4.1.1 NR Equations for Single Variable

To find the root of a function, the Newton–Raphson (NR) method can be applied, which can be briefly explained as follows. For the application of this method, the function must be continuous. For simplicity, let us consider a continuous function of a single variable $f(x)$. And, we have to find its root—for what value of x, $f(x) = 0$. If we know the value of the function at an arbitrary point x_1, the value of the function in the immediate neighborhood of x_1, say $x_1 + \delta x = x_2$ is given by Taylor series as

$$f(x_2) = f(x_1) + \frac{df}{dx}(x_1)\,\delta x + \frac{d^2 f}{dx^2}(x_1)\,\delta x^2 + \cdots \tag{4.61}$$

Noting $\delta x = (x_2 - x_1)$, and by truncating higher order terms,

$$f(x_2) \approx f(x_1) + \left[\frac{df}{dx}(x_1)\right](x_2 - x_1) \tag{4.62}$$

If (x_2) happens to be the root or closer to the root within an acceptable error band, then $f(x_2) \approx 0$. Then,

$$f(x_1) + \left[\frac{df}{dx}(x_1)\right]\Delta x \approx 0 \tag{4.63}$$

Or,

$$x_2 = x_1 - \frac{f(x_1)}{\left[\dfrac{df}{dx}(x_1)\right]} \tag{4.64}$$

Equations 4.63 and 4.64 are known as **Newton–Raphson equations** (NR equations) and Δx is known as **correction**. Although Eqs. 4.63 and 4.64 are derived assuming the root to be x_2, even in the absence of it, the equations help us to obtain corrections to approach the root by successive substitutions—just obtained x_2 being substituted for x_1 to find the next x_2 and so on until we are closer to the root within an acceptable error band, $(x_2 - x_1) = \Delta x \approx 0$ and $f(x_2) \approx 0$. And, thus the last found x_2 becomes the root. For simplicity, we have considered here a single variable function. For multi-variable function the above procedure has to be suitably adopted, as we do it in our subsequent analyses.

4.4.1.2 NR Equations for Equilibrium

As Eqs. 4.54 and 4.60, and the equations for enthalpy h and entropy s are continuous, we can adopt the NR method for the solution. The three pairs of constant state functions under our consideration $h\&p$, $T\&p$, and $s\&p$ have in common the constant system pressure. Therefore, presently we can assume the system pressure only to be constant while deriving the NR equations and thereafter suitably adopt the constancy on h, T, or s in the equations.

For a single variable function $f(x) = 0$ we obtained the NR equation along with the correction Δx, Eq. 4.63. Now we have to obtain NR equations along with applicable corrections for the multi-variable functions in Eqs. 4.54 and 4.60. These two equations have in them the variables μ_j's, λ_i's, and n_j's. The gaseous species' μ_j's in turn are functions of system temperature (T) and pressure (p), n_j's, and n, Eq. 4.49. Whereas, the condensed species' μ_j's are functions of only system temperature (T), Eq. 4.52. At constant system pressure $dp = 0$. Therefore, in order to iterate Eqs. 4.54 and 4.60 we have to first obtain corrections to (1) temperature T, (2) n_j's, the initial estimates of kgmoles of various species per unit mass of products, (3) n, the total number of kgmoles of gaseous species per unit mass of products, and (4) λ_i's, the Lagrange multipliers. Notice that the expression $\mu_j + \sum_{i=1}^{l} \lambda_i a_{ij}$ in Eq. 4.60 has been obtained through partial differentiation keeping λ_i's constant. Hence, there is no restriction to start each iteration with the estimate for the λ_i's equal to zero because they appear linearly in Eq. 4.60 [6]. In order to normalize the magnitudes of the corrections, the corrections for T, and gaseous n_j's and n are logarithmic ones. The respective correction variables therefore are $\Delta ln\ T$, $\Delta \ln\ n_j\quad (j = 1, \ldots, NG)$, $\Delta n_j\quad (j = NG + 1, \ldots, NS)$, $\Delta ln\ n$, and $\pi_i = -\lambda_i / R_u T\quad (i = 1, \ldots, l)$. Equations containing thermodynamic functions are made dimensionless by suitably dividing them by $R_u T$ or R_u.

First, let us get the NR equation for Eq. 4.60 of the form Eq. 4.63, that is

$$\left(\mu_j + \sum_{i=1}^{l} \lambda_i a_{ij}\right) + d\left(\mu_j + \sum_{i=1}^{l} \lambda_i a_{ij}\right) = 0 \qquad (j = 1, \ldots, NS)$$

Dividing Eq. 4.60 by $R_u T$, and noting that $\pi_i = -\lambda_i / R_u T$ is assumed zero for each iteration, μ_j's for gaseous species are given by Eq. 4.49, and at constant pressure $dln(p) = 0$,

$$d\left(\frac{\mu_j}{R_u T} - \sum_{i=1}^{l} a_{ij}\pi_i\right) = -\frac{\overline{h}_j^0}{R_u T} d\ln\ T + d\ln\ n_j - d\ln\ n \quad (j = 1, \ldots, NG) \tag{4.65}$$

Adopting the form of Eq. 4.63 and using Eq. 4.65, the NR equation for Eq. 4.60, corresponding to gaseous species, becomes

$$\Delta \ln\ n_j - \sum_{i=1}^{l} a_{ij}\pi_i - \Delta \ln\ n - \frac{\overline{h}_j^0}{R_u T}\Delta \ln\ T = -\frac{\mu_j}{R_u T} \quad (j = 1, \ldots, NG) \tag{4.66}$$

where $\overline{h}_j^0$ is the standard state molar enthalpy for species j at temperature T, which is the sum of molar heat of formation at the standard temperature of 298.15 K, $\overline{h}_{f,j}^0$ (and, when applicable, heat of transition) and sensible heat at the assigned temperature T (see Sect. 4.6). Similarly, noting that μ_j's for condensed species are given by Eq. 4.52, the NR equation for Eq. 4.60 corresponding to condensed species becomes

$$\sum_{i=1}^{l} a_{ij}\pi_i + \frac{\overline{h}_j^0}{R_u T}\Delta \ln\ T = \frac{\mu_j^0}{R_u T} \qquad (j = NG + 1, \ldots, NS) \tag{4.67}$$

Considering now the mass balance equation Eq. 4.54, which is indeed the second condition for extremum under Eq. 4.59, the NR equation for it becomes

$$\sum_{j=1}^{NG} a_{kj} n_j \Delta \ln\ n_j + \sum_{j=NG+1}^{NS} a_{kj}\Delta\ n_j = b_k^o - b_k \qquad (k = 1, \ldots, l) \tag{4.68}$$

In addition we have to keep track of the variation in kgmoles of gaseous species per unit mass. Changes in n_j's and n in $NG + l$ individual equations of Eqs. 4.66 and 4.68 should satisfy the condition that the total number of gaseous moles n is equal to the sum of the individual gaseous moles $\sum_{j=1}^{NG} n_j$. That is

$$\left(n - \sum_{j=1}^{NG} n_j\right) + d\left(n - \sum_{j=1}^{NG} n_j\right) = 0$$

Hence,

$$\sum_{j=1}^{NG} n_j \Delta \ln\, n_j - n\Delta \ln\, n = n - \sum_{j=1}^{NG} n_j \tag{4.69}$$

Equations 4.66–4.69 (totaling to $NS + l + 1$ individual equations) are the NR equations that encompass the energy minimization with the constraint on mass conservation for reactions *only under constant pressure*. Now let us look at the NR equations required for the pair of constant state functions $T\&p$, where both $T\&p$ are constants. Notice that $\Delta ln\ T$, the correction to temperature T, is *not required* for constant state functions $T\&p$ evidently because the equilibrium composition is to be calculated for the *assigned temperature:* T^* ($dT = 0$). By imposing selectively the constant temperature condition ($\Delta \ln\ T = 0$), the equations Eqs. 4.66 and 4.67 (with $\Delta \ln\ T = 0$), and Eqs. 4.68 and 4.69 become applicable for reactions under constant state functions $T\&p$. But the correction to temperature T is very much required for constant state functions $h\&p$ or $s\&p$ because we need to find *that* temperature T that satisfies the assigned enthalpy h^* or the assigned entropy s^*.

The second pair of constant state functions to be considered under constant pressure is $h\&p$. Combustion of propellants under constant pressure is governed by the pair of state functions $h\&p$. The pair is given by

$$h = h^* \tag{4.70}$$

$$p = p^* \tag{4.71}$$

where h is the specific enthalpy of the products and h^* a constant equal to the specific enthalpy of the reactants. The expression for h is

$$h = \sum_{j=1}^{NS} n_j \overline{h}_j^0 (T) \tag{4.72}$$

where $\overline{h}_j^0(T)$ is the standard state molar enthalpy for species j at temperature T. Remember that $\overline{h}_j^0(T)$ is the sum of molar heat of formation at the standard temperature of 298.15 K, $\overline{h}^0_{f,j}(298.15)$ (and, when applicable, heat of transition) and sensible heat at the assigned temperature T (see Sect. 4.6). While taking the differential on $\overline{h}^0_j(T)$, note that both the heat of formation and the heat of transition are constants. For the condition of constant enthalpy, Eq. 4.70, we have to get the NR equation. Equation 4.72 indicates the effective variables to be n_j and T.

$$dh = \sum_{j=1}^{NG} n_j \overline{h}^0_j d\ln n_j + \sum_{j=NG+1}^{NS} \overline{h}^0_j dn_j + \sum_{j=1}^{NS} n_j \overline{c}^0_p dT \tag{4.73}$$

With the above expression for dh and dividing throughout by $R_u T$, the NR equation of the form Eq. 4.63 for constant enthalpy condition becomes

$$\begin{aligned} &\sum_{j=1}^{NG} \frac{n_j \overline{h}^0_j}{R_u T} \Delta \ln n_j + \sum_{j=NG+1}^{NS} \frac{\overline{h}^0_j}{R_u T} \Delta n_j \\ &+ \left(\sum_{j=1}^{NS} \frac{n_j \overline{c}^0_{p,j}}{R_u} \right) \Delta \ln T = \frac{h^* - h}{R_u T} \end{aligned} \tag{4.74}$$

The third and final pair of constant state functions to be considered under constant pressure is $s\&p$. For isentropic expansion in a nozzle or isentropic diffusion in a diffuser to an assigned pressure, the thermodynamic state is specified by

$$s = s^* \tag{4.75}$$

$$p = p^* \tag{4.76}$$

where s is the specific entropy of the mixture and s^* the assigned specific entropy. The expression for s is

$$s = \sum_{j=1}^{NS} n_j \overline{s}_j \tag{4.77}$$

From Eq. 4.24, $d\overline{s} = \dfrac{\overline{c}^0_p dT}{T} - \dfrac{R_u}{p} dp$

$$d\overline{s}_j = \frac{\overline{c}^0_{p,j} dT}{T} - R_u \frac{dp_j}{p_j} \quad (j = 1, \ldots, NG) \tag{4.78}$$

$$d\overline{s}_j = \frac{\overline{c}^0_{p,j} dT}{T} \quad (j = NG + 1, \ldots, NS) \tag{4.79}$$

On integration

$$\bar{s}_j = \bar{s}_j^0(T) - R_u \ln \frac{n_j}{n} - R_u \ln p \quad (j = 1, \ldots, NG);\; p = [\text{bar}] \tag{4.80}$$

$$\bar{s}_j = \bar{s}_j^0(T) \qquad (j = NG + 1, \ldots, NS) \tag{4.81}$$

$\bar{s}_j^0 = \int_{T_{std}}^{T} \bar{c}_{p,j}^0 dT/T$ is the standard state molar entropy for species j at temperature T. Equations 4.80 and 4.81 are similar to Eqs. 4.50 and 4.52, and the same discussion concerning standard state pressure 1 bar that is applicable to Eq. 4.50 also applies to Eq. 4.80, that is, the unit of p in Eq. 4.80 is in bar.

Considering now the constant entropy condition and Eqs. 4.77–4.81, we find the variables are n_j, n, and T. At constant system pressure, $(dp = 0)$.

$$\begin{aligned} ds = &\sum_{j=1}^{NG} n_j \bar{s}_j d\ln n_j + \sum_{j=NG+1}^{NS} \bar{s}_j^0 dn_j \\ &+ \sum_{j=1}^{NS} n_j \bar{c}_{p,j}^0 d\ln T - R_u \left(\sum_{j=1}^{NG} n_j d\ln n_j - n d\ln n \right) \end{aligned} \tag{4.82}$$

Substituting for the last expression within the parenthesis of Eq. 4.82 from Eq. 4.69,

$$\begin{aligned} ds = &\sum_{j=1}^{NG} n_j \bar{s}_j d\ln n_j + \sum_{j=NG+1}^{NS} \bar{s}_j^0 dn_j \\ &+ \sum_{j=1}^{NS} n_j \bar{c}_{p,j}^0 d\ln T - R_u \left(n - \sum_{j=1}^{NG} n_j \right) \end{aligned} \tag{4.83}$$

With the above expression for ds and dividing throughout by R_u, the NR equation of the form Eq. 4.63 for constant entropy condition becomes

$$\begin{aligned} &\sum_{j=1}^{NG} \frac{n_j \bar{s}_j}{R_u} \Delta \ln n_j + \sum_{j=NG+1}^{NS} \frac{\bar{s}_j^0}{R_u} \Delta n_j + \left(\sum_{j=1}^{NS} \frac{n_j \bar{c}_{p,j}^0}{R_u} \right) \Delta \ln T \\ &= \frac{s^* - s}{R_u} + n - \sum_{j=1}^{NG} n_j \end{aligned} \tag{4.84}$$

Thus, for constant pressure systems we have derived NR equations Eqs. 4.66–4.69, 4.74, and 4.84 for all the three pairs of constant state functions ($h\&p$, $T\&p$, and $s\&p$). Note that the Eqs. 4.66–4.69 are the ones that resulted from the energy minimization condition for equilibrium with the constraint on mass balance, Eqs. 4.54, and 4.58–4.60. Constant state function systems $h\&p$ and $s\&p$ are being governed by Eqs. 4.74 and 4.84 respectively.

4.4.1.3 Reduced Correction Equations for Equilibrium

For all the three constant pressure reactions, we have derived NR equations for (1) energy minimization, Eqs. 4.66 and 4.67 totaling to NS equations, (2) mass conservation, Eqs. 4.68 and 4.69 totaling to $l + 1$ equations, (3) constant enthalpy, Eq. 4.74, and (4) constant entropy, Eq. 4.84. For problems with assigned thermodynamic states $h\&p$, $T\&p$, and $s\&p$, various combinations of Eqs. 4.66–4.69, 4.74, and 4.84 could be used to obtain corrections to estimates. However, for reacting systems containing quite a few elements numbering to l and thus a lot more species numbering to NS, it would be necessary to solve per se a very large number of simultaneous equations for the processes in propulsion units. This large number of equations can be reduced quite simply to a much smaller number by algebraic substitution. Actually Eq. 4.66 represents NG simultaneous equations corresponding to gaseous species. And, in propellant combustion, the gaseous species constitute a very large number when compared to the number of condensed species. If the expression for $\Delta ln\ n_j$ obtained from Eq. 4.66 is substituted into Eqs. 4.68, 4.69, 4.74, and 4.84, the total number of simultaneous equations to be solved is largely reduced by NG. Thus the resulting reduced equations are

$$\sum_{i=1}^{l}\sum_{j=1}^{NG} a_{kj}a_{ij}n_j\pi_i + \sum_{j=NG+1}^{NS} a_{kj}\Delta n_j + \left(\sum_{j=1}^{NG} a_{kj}n_j\right)\Delta \ln\ n$$
$$+\left(\sum_{j=1}^{NG}\frac{a_{kj}n_j\overline{h}_j^0}{R_uT}\right)\Delta \ln\ T = b_k^o - b_k + \sum_{j=1}^{NG}\frac{a_{kj}n_j\mu_j}{R_uT} \quad (k = 1, \ldots, l) \tag{4.85}$$

Equation 4.85 is a combination of Eqs. 4.66 and 4.68.

$$\sum_{i=1}^{l} a_{ij}\pi_i + \frac{\overline{h}_j^0}{R_uT}\Delta \ln\ T = \frac{\mu_j^0}{R_uT} \qquad (j = NG + 1, \ldots, NS) \tag{4.67}$$

$$\sum_{i=1}^{l}\sum_{j=1}^{NG} a_{ij}n_j\pi_i + \left(\sum_{j=1}^{NG} n_j - n\right)\Delta \ln\ n + \left(\sum_{j=1}^{NG}\frac{n_j\overline{h}_j^0}{R_uT}\right)\Delta \ln\ T$$
$$= n - \sum_{j=1}^{NG} n_j + \sum_{j=1}^{NG}\frac{n_j\mu_j}{R_uT} \tag{4.86}$$

Equation 4.86 is a combination of Eqs. 4.66 and 4.69.

Table 4.8 Applicable equations and correction variables for constant pressure reactions

Type of problem	Equations required	Correction variables
Assigned temperature and pressure, $T\&p$	Equations 4.85, 4.67, and 4.86; $\Delta\ lnT = 0$	$\pi_i,\ \Delta n_j,\ \&\ \Delta \ln\ n$
Assigned enthalpy and pressure, $h\&p$	Equations 4.85, 4.67, 4.86, and 4.87	$\pi_i,\ \Delta n_j,\ \Delta \ln\ n,$ $\&\ \Delta \ln\ T$
Assigned entropy and pressure, $s\&p$	Equations 4.85, 4.67, 4.86, and 4.88	$\pi_i,\ \Delta n_j,\ \Delta \ln\ n,$ $\&\ \Delta \ln\ T$

$$\sum_{i=1}^{l}\left(\sum_{j=1}^{NG}\frac{a_{ij}n_j\overline{h}_j^0}{R_uT}\right)\pi_i+\sum_{j=NG+1}^{NS}\frac{\overline{h}_j^0}{R_uT}\Delta n_j+\left(\sum_{j=1}^{NG}\frac{n_j\overline{h}_j^0}{R_uT}\right)\Delta \ln\ n$$
$$+\left(\sum_{j=1}^{NS}\frac{n_j\overline{c}_{p,j}^0}{R_u}+\sum_{j=1}^{NG}\frac{n_j\overline{h}_j^{0\,2}}{R_u^2T^2}\right)\Delta \ln\ T=\frac{h^*-h}{R_uT}+\sum_{j=1}^{NG}\frac{n_j\overline{h}_j^0\mu_j}{R_u^2T^2} \tag{4.87}$$

Equation 4.87 is the combination of Eqs. 4.66 and 4.74.

$$\sum_{i=1}^{l}\left(\sum_{j=1}^{NG}\frac{a_{ij}n_j\overline{s}_j}{R_u}\right)\pi_i+\sum_{j=NG+1}^{NS}\frac{\overline{s}_j^0}{R_u}\Delta n_j+\left(\sum_{j=1}^{NG}\frac{n_j\overline{s}_j}{R_u}\right)\Delta \ln\ n$$
$$+\left(\sum_{j=1}^{NS}\frac{n_j\overline{c}_{p,j}^0}{R_u}+\sum_{j=1}^{NG}\frac{n_j\overline{h}_j^0\overline{s}_j}{R_u^2T}\right)\Delta \ln\ T=\frac{s^*-s}{R_u}+n \tag{4.88}$$
$$-\sum_{j=1}^{NG}n_j+\sum_{j=1}^{NG}\frac{n_j\overline{s}_j\mu_j}{R_u^2T}$$

Equation 4.88 is a combination of Eqs. 4.66 and 4.84.

Equations 4.85, 4.67, 4.86–4.88 are known as **reduced Gibbs correction equations** or simply **reduced correction equations**. The correction equations required for several types of constant pressure problems are summarized in Table 4.8, where $i = 1, \ldots, l$ and $j = NG + 1, \ldots, NS$. For energy minimization and mass conservation, Eqs. 4.85, 4.67, and 4.86 are required for all the three constant pressure reactions $h\&p$, $T\&p$, and $s\&p$. But for constant temperature and pressure reactions these equations should be imposed with constant temperature condition, $\Delta \ln\ T = 0$. For constant enthalpy and pressure reactions, additionally we require Eq. 4.87. Similarly Eqs. 4.85, 4.67, 4.86 along with Eq. 4.88 are required for constant entropy and pressure reactions. After obtaining the correction variables π_i's, Δn_j's for condensed species, $\Delta ln\ n$, and (if applicable) $\Delta ln\ T$ from these correction equations the corrections for gaseous species $\Delta ln\ n_j$'s can be obtained from Eq. 4.66.

Example 4.5 shows the calculation procedure to arrive at the equilibrium composition of combustion products for a typical 5.6 oxidizer–fuel ratio of LOX and LH2 at a chamber pressure of 70 bar and temperature of 3200 K.

Example 4.5 Liquid hydrogen and liquid oxygen are burned in a rocket combustion chamber. The oxidizer–fuel ratio is 5.6. Possible equilibrium species above 5.0×10^{-6} mole fractions are many, say H, HO_2, H_2, H_2O, H_2O_2, O, OH, and O_2. However, to make the numerical example less cumbersome and easier to understand the method of equilibrium calculation through energy minimization technique, let us consider only three major species: H_2, H_2O, and OH. Calculate the equilibrium composition at the specified pressure of 70 bar and the temperature of 3200 K. Standard state enthalpies and entropies for the considered species may be taken from a standard text book.

Solution

Equilibrium composition at a specified temperature and pressure has to be calculated. This is a case of the specified constant pair of state functions $T\&p$, hence $\Delta ln\ (T) = 0$. Standard state enthalpies and entropies at 3200 K are the following:

$$\overline{h}^{0}_{H_2}\ (3200\,\text{K}) = 96199\ \text{kJ/kgmol}$$
$$\overline{s}^{0}_{H_2}\ (3200\,\text{K}) = 205.293\ \text{KJ}/(\text{kgmolK})$$

$$\overline{h}^{0}_{H_2O}\ (3200\,\text{K}) = -104267\ \text{kJ/kgmol}$$
$$\overline{s}^{0}_{H_2O}\ (3200\,\text{K}) = 289.994\ \text{KJ}/(\text{kgmolK})$$

$$\overline{h}^{0}_{OH}\ (3200\,\text{K}) = 136420\ \text{kJ/kgmol}$$
$$\overline{s}^{0}_{OH}\ (3200\,\text{K}) = 259.199\ \text{KJ}/(\text{kgmolK})$$

There are two elements H and O (hence, $l = 2$), three gaseous species H_2, H_2O, and OH (hence, $NG = 3$), and no condensed species (hence, $NG = NS$). For the specified oxidizer–fuel ratio, the molar form of the propellant combination for unit mass has to be calculated to obtain b^o_H and b^o_O. On forming the a_{ij} matrix, the numerical values corresponding to the Eqs. 4.85 and 4.86 have to be obtained. With the resulting three simultaneous equations and with suitable trial values for the number of moles of H_2, H_2O, and OH, iteration has to be initiated to calculate n_{H_2}, n_{H_2O}, and n_{OH} within the acceptable convergence error. For convergence let us consider the ratio of the change in the number of moles of a species to the total number of moles of species $\delta_{n_j} = |\Delta n_j|/n$ as the convergence error. Although the acceptable convergence error while using computer programs can be finer, let us have the acceptable convergence error for this example to be $\delta_{n_j} = |\Delta n_j|/n \leq 1.0E - 04$.

Table 4.9 a_{ij} matrix: Example 4.5

Atom	n_{H_2}	n_{H_2O}	n_{OH}
H	$a_{11} = 2$	$a_{12} = 2$	$a_{13} = 1$
O	$a_{21} = 0$	a_{22}=1	$a_{23} = 1$

Analysis

Let us first get the a_{ij} matrix for the considered species n_{H_2}, n_{H_2O}, and n_{OH}, Table 4.9.

The reaction equation is

$$H_2 + xO_2 \rightarrow products\ of\ H_2,\ H_2O,\ \&\ OH$$

The oxidizer–fuel ratio is 5.6. Therefore,

$$x = \frac{5.6 \times 2.016}{31.999} = 3.528110E - 01$$

$$H_2 + 3.528110E - 01O_2 \rightarrow products\ of\ H_2,\ H_2O,\ \&OH$$

For unit mass of propellant combination, the kgmoles of H_2 and O_2 are to be determined. Total mass of the above molar form of propellant combination is given by

$$2.016 + 3.528110E - 01 \times 31.999 = 1.330560E + 01 \text{ kg}$$

Therefore for unit mass of propellant combination, the molar form of propellant combination is given as

$$\frac{H_2}{1.330560E + 01} + \frac{3.528110E - 01O_2}{1.330560E + 01} = n_{H_2}H_2, + n_{H_2O}H_2O + n_{OH}OH$$

$$7.515633E - 02H_2 + 2.651598E - 02O_2 = n_{H_2}H_2, + n_{H_2O}H_2O + n_{OH}OH$$

Let us now calculate b_H^o and b_O^o, the assigned number of kgatoms of elements H and O per kg of total propellant combination.

$$b_H^o = 7.515633E - 02 \times 2 = 1.503127E - 01 \text{ kgatoms}$$

$$b^o_O = 2.651598E - 02 \times 2 = 5.303196E - 02 \text{ kgatoms}$$

Initial estimates of n_j***'s*** The propellant combination is a fuel rich one. Therefore, after the formation of the species H_2O with the available O_2, the remaining H_2 should be the next abundant species. Therefore the stoichiometric equation can be written as

$$\begin{aligned} &7.515633E - 02H_2 + 2.651598E - 02O_2 \\ &= 2.212437E - 02H_2 + 5.303196E - 02H_2O \end{aligned}$$

The above two product species add up to 0.075156 kgmol/kg. Since the species OH is to be also considered for the equilibrium composition calculation, arbitrarily the sum of the product species n is taken as 0.08 kgmol/kg with trial mole fraction 0.31, 0.66, and 0.03 for H_2, H_2O, and OH respectively. Noting $n_j = x_j \times n$ the trial kgmoles of product species per kg of propellant are the following.

$$\begin{aligned} n_{H_2} = 2.480000E - 02 \text{ kgmol}; \quad n_{H_2O} = 5.280000E - 02 \text{ kgmol}; \\ n_{OH} = 2.400000E - 03 \text{ kgmol} \end{aligned}$$

There is no condensed species being considered. From Eq. 4.50,

$$\mu_j(T, p_j) = \overline{g}^0_j(T) + R_u T \ln \frac{n_j}{n} + R_u T \ln p \qquad (j = 1, 2, .\&3); \; p = [\text{bar}]$$

$$\mu_j(T, p_j) = \overline{h}^0_j(T) - T\overline{s}^0_j(T) + R_u T \ln p + R_u T \ln \frac{n_j}{n} \quad p = [\text{bar}]$$

Therefore,

$$\begin{aligned} &\mu_{H_2}(3200\,\text{K}, p_{H_2}) = 96199 - 3200 \times 205.293 \\ &+8.31451 \times 3200 \times \ln(70) + 8.31451 \times 3200 \times \ln\left(\frac{n_{H_2}}{n}\right) \\ &= -4.477013E + 05 + 26606.432 \times \ln\left(\frac{n_{H_2}}{n}\right) \frac{\text{kJ}}{\text{kgmol}} \end{aligned}$$

$$\mu_{H_2O}\left(3200\,\mathrm{K},\,p_{H_2O}\right) = -104267 - 3200 \times 289.994$$
$$+8.31451 \times 3200 \times \ln(70) + 8.31451 \times 3200 \times \ln\left(\frac{n_{H_2O}}{n}\right)$$
$$= -9.192105E + 05 + 26606.432 \times \ln\left(\frac{n_{H_2O}}{n}\right) \frac{\mathrm{kJ}}{\mathrm{kgmol}}$$

$$\mu_{OH}\left(3200\,\mathrm{K},\,p_{OH}\right) = 136420 - 3200 \times 259.199$$
$$+8.31451 \times 3200 \times \ln(70) + 8.31451 \times 3200 \times \ln\left(\frac{n_{OH}}{n}\right)$$
$$= -5.799795E + 05 + 26606.432 \times \ln\left(\frac{n_{OH}}{n}\right) \frac{\mathrm{kJ}}{\mathrm{kgmol}}$$

From Eq. 4.80

$$\bar{s}_j = \bar{s}_j^0(T) - R_u \ln\, p - R_u \ln\, \frac{n_j}{n} \quad (j = 1, \ldots, NG) \quad p = [\mathrm{bar}]$$

Therefore,

$$\bar{s}_{H_2}\left(3200,\, p_{H_2}\right) = 205.293 - 8.31451 \times \ln(70) - 8.31451 \ln\, \frac{n_{H_2}}{n}$$
$$= 169.9688 - 8.31451 \ln\, \frac{n_{H_2}}{n} \frac{\mathrm{kJ}}{\mathrm{kgmolK}}$$

$$\bar{s}_{H_2O}\left(3200,\, p_{H_2O}\right) = 289.994 - 8.31451 \times \ln(70) - 8.31451 \ln\, \frac{n_{H_2O}}{n}$$
$$= 254.66984 - 8.31451 \ln\, \frac{n_{H_2O}}{n} \frac{\mathrm{kJ}}{\mathrm{kgmolK}}$$

$$\bar{s}_{OH}\left(3200,\, p_{OH}\right) = 259.199 - 8.31451 \times \ln(70) - 8.31451 \ln\, \frac{n_{OH}}{n}$$
$$= 223.87484 - 8.31451 \ln\, \frac{n_{OH}}{n} \frac{\mathrm{kJ}}{\mathrm{kgmolK}}$$

From the above results for the trial values of n and mole fractions x_j's we can list the calculated chemical potential values before we start the iterations, Table 4.10.

Table 4.10 Chemical potentials for the first trial n_j's: Example 4.5

Species	$n_j/n = x_j$	n_j (kgmol)	$\mu_j \left(\frac{\text{kJ}}{\text{kgmol}}\right)$	$n_j\mu_j$ (kJ)
H_2	0.31	2.480000E-02	−4.788623E+05	−1.187579E+04
H_2O	0.66	5.280000E-02	−9.302659E+05	−4.911804E+04
OH	0.03	2.400000E-03	−6.732765E+05	−1.615864E+03
	$\sum n_j = n$ (kgmol/kg) =	0.08	$\sum n_j\mu_j = g$ (kJ/kg) =	−6.260969E+04

Recall,

$$b_i = \sum_{j=1}^{NS} a_{ij} n_j \qquad (i = 1, \ldots\ldots, l)$$

Hence with the trial values of n_j's and using the a_{ij} matrix we calculate the trial b_H and b_O as

$$\begin{aligned} b_H &= 2n_{H_2} + 2n_{H_2O} + n_{OH} \\ &= 2 \times 2.48E-02 + 2 \times 5.28E-02 + 2.4E-03 \\ &= 0.1576 \text{ kgatom} \end{aligned}$$

$$\begin{aligned} b_O &= n_{H_2O} + n_{OH} \\ &= 5.28E-02 + 2.4E-03 \\ &= 0.0552 \text{ kgatom} \end{aligned}$$

Therefore,

$$\begin{aligned} b_H^o - b_H &= 1.503127E-01 - 0.1576 = -7.2873E-03 \\ b_O^o - b_O &= 5.303196E-02 - 0.0552 = -2.16804E-03 \end{aligned}$$

We have to get the numbers for the reduced iteration equations Eqs. 4.85 and 4.86. The equations are reproduced here for ready reference.

$$\sum_{i=1}^{l}\sum_{j=1}^{NG} a_{kj}a_{ij}n_j\pi_i + \sum_{j=NG+1}^{NS} a_{kj}\Delta n_j + \left(\sum_{j=1}^{NG} a_{kj}n_j\right)\Delta \ln n + \left(\sum_{j=1}^{NG}\frac{a_{kj}n_j\bar{h}_j^0}{R_uT}\right)\Delta \ln T = b_k^o - b_k + \sum_{j=1}^{NG}\frac{a_{kj}n_j\mu_j}{R_uT} \qquad (k = 1, \ldots, l) \tag{4.85}$$

Table 4.11 Matrix numbers for the three simultaneous correction equations for the first trial values: Example 4.5

Atom	π_1	π_2	$\Delta ln\ (n)$	RHS
H	3.1280E-01	1.0800E-01	1.5760E-01	−4.6529E+00
O	1.0800E-01	5.5200E-02	5.5200E-02	−1.9090E+00
	1.5760E-01	5.5200E-02	0.0000E+00	−2.3532E+00

$$\sum_{i=1}^{l}\sum_{j=1}^{NG} a_{ij} n_j \pi_i + \left(\sum_{j=1}^{NG} n_j - n\right)\Delta \ln\ n + \left(\sum_{j=1}^{NG} \frac{n_j \overline{h}_j^0}{R_u T}\right)\Delta \ln\ T = n - \sum_{j=1}^{NG} n_j + \sum_{j=1}^{NG} \frac{n_j \mu_j}{R_u T} \tag{4.86}$$

$\Delta \ln\ T = 0$ for the specified constant pair of state functions $T\&p$. The two equations covering Eq. 4.85 are

$$\begin{aligned}&\left(4n_{H_2} + 4n_{H_2O} + n_{OH}\right)\pi_1 + \left(2n_{H_2O} + n_{OH}\right)\pi_2 \\ &+ \left(2n_{H_2} + 2n_{H_2O} + n_{OH}\right)\Delta \ln (n) = \\ &= b_H^o - b_H + \frac{\left(2n_{H_2}\mu_{H_2} + 2n_{H_2O}\mu_{H_2O} + n_{OH}\mu_{OH}\right)}{R_u T}\end{aligned}$$

$$\begin{aligned}&\left(2n_{H_2O} + n_{OH}\right)\pi_1 + \left(n_{H_2O} + n_{OH}\right)\pi_2 \\ &+ \left(n_{H_2O} + n_{OH}\right)\Delta \ln (n) = \\ &= b_O^o - b_O + \frac{\left(n_{H_2O}\mu_{H_2O} + n_{OH}\mu_{OH}\right)}{R_u T}\end{aligned}$$

The single equation covering Eq. 4.86 is

$$\begin{aligned}&\left(2n_{H_2} + 2n_{H_2O} + n_{OH}\right)\pi_1 + \left(n_{H_2O} + n_{OH}\right)\pi_2 \\ &+ \left(n_{H_2} + n_{H_2O} + n_{OH} - n\right)\Delta \ln (n) = \\ &= n - \left(n_{H_2} + n_{H_2O} + n_{OH}\right) + \frac{\left(n_{H_2}\mu_{H_2} + n_{H_2O}\mu_{H_2O} + n_{OH}\mu_{OH}\right)}{R_u T}\end{aligned}$$

Now, we can calculate the coefficients for the reduced iteration equations for our first iteration, Table 4.11.

Thus we have three simultaneous equations to be solved as shown below.

$$\begin{bmatrix} 3.128000E-01 & 1.080000E-01 & 1.576000E-01 \\ 1.576000E-01 & 5.520000E-02 & 0.000000E+00 \\ 1.080000E-01 & 5.520000E-02 & 5.520000E-02 \end{bmatrix} \begin{bmatrix} \pi_1 \\ \pi_2 \\ \Delta \ln(n) \end{bmatrix} = \begin{bmatrix} -4.652913E+00 \\ -2.353179E+00 \\ -1.908997E+00 \end{bmatrix}$$

Iteration Let us adopt the method of Gaussian elimination with partial pivoting. The result after forward elimination is

$$\begin{bmatrix} 3.128000E-01 & 1.080000E-01 & 1.576000E-01 \\ 0.000000E+00 & 1.791100E-02 & 7.856777E-04 \\ 0.000000E+00 & 0.000000E+00 & -7.943907E-02 \end{bmatrix} \begin{bmatrix} \pi_1 \\ \pi_2 \\ \Delta \ln(n) \end{bmatrix} = \begin{bmatrix} -4.652913E+00 \\ -3.024922E-01 \\ 4.396710E-03 \end{bmatrix}$$

By back substitution, solutions on the first trial values are,

$$\begin{aligned} \pi_1 &= -9.016880E+00 \\ \pi_2 &= -1.688620E+01 \\ \Delta \ln(n) &= -5.534694E-02 \end{aligned}$$

Using Eq. 4.66 and noting again $\Delta ln\ T = 0$,

$$\Delta \ln\ n_j - \sum_{i=1}^{l} a_{ij}\pi_i - \Delta \ln\ n - \frac{\overline{h}_j^0}{R_u T}\Delta \ln\ T = -\frac{\mu_j}{R_u T} \quad (j = 1, \ldots, NG) \tag{4.66}$$

$$\begin{aligned} \Delta \ln\ n_{H_2} &= -\frac{\mu_{H_2}}{R_u T} + (a_{11}\pi_1 + a_{21}\pi_2) + \Delta \ln(n) \\ &= -\frac{\mu_{H_2}}{R_u T} + 2\pi_1 + \Delta \ln(n) \end{aligned}$$

$$\begin{aligned} \Delta \ln\ n_{H_2O} &= -\frac{\mu_{H_2O}}{R_u T} + (a_{12}\pi_1 + a_{22}\pi_2) + \Delta \ln(n) \\ &= -\frac{\mu_{H_2O}}{R_u T} + 2\pi_1 + \pi_2 + \Delta \ln(n) \end{aligned}$$

Table 4.12 n_j's after the first iteration: Example 4.5

Species	$n_j^{(-)}$	$\Delta ln\,(n_j)$	Δn_j	n_j	$\delta_{n_j} = \lvert\Delta n_j\rvert/n$
H_2	2.480000E-02	−9.111757E-02	−2.259716E-03	2.254028E-02	2.990140E-02
H_2O	5.280000E-02	−1.136133E-02	−5.998783E-04	5.220012E-02	7.937812E-03
OH	2.400000E-03	−6.534006E-01	−1.568161E-03	8.318385E-04	2.075049E-02

$\sum n_j = n$ kgmol/kg = 7.557224E-02

Table 4.13 Chemical potentials for the second iteration: Example 4.5

Species	$n_j/n = x_j$	$n_j\ (kgmol)$	$\mu_j\ \left(\frac{\text{kJ}}{\text{kgmol}}\right)$	$n_j\mu_j$ (kJ)
H_2	2.982614E-01	2.254028E-02	−4.798894E+05	−1.081684E+04
H_2O	6.907314E-01	5.220012E-02	−9.290550E+05	−4.849678E+04
OH	1.100720E-02	8.318385E-04	−6.999534E+05	−5.822482E+02
	$\sum n_j = n$ (kgmol/kg) =	7.557224E-02	$\sum n_j\mu_j = g$ (kJ/kg) =	−5.989587E+04
			$\delta_g \equiv (g - g^{(-)})/g =$	−4.530884E-02

Table 4.14 Matrix numbers for the three simultaneous correction equations for the second trial values: Example 4.5

Atom	b	$b^o - b$	π_1	π_2	$\Delta ln\,(n)$	RHS
H	1.503127E-01	0.000000E+00	2.997935E-01	1.052321E-01	1.503127E-01	−4.480477E+00
O	5.303196E-02	1.387779E-16	1.052321E-01	5.303196E-02	5.303196E-02	−1.844630E+00
			1.503127E-01	5.303196E-02	0.000000E+00	−2.251180E+00

$$\begin{aligned}\Delta \ln\, n_{OH} &= -\frac{\mu_{OH}}{R_u T} + (a_{13}\pi_1 + a_{23}\pi_2) + \Delta \ln\,(n)\\ &= -\frac{\mu_{OH}}{R_u T} + \pi_1 + \pi_2 + \Delta \ln\,(n)\end{aligned}$$

Note $\Delta n_j = n_j \Delta \ln\, n_j$. From the above three equations, using the solutions on the first trial values, the calculated n_j after the first iteration are given in Table 4.12. The numerical values in the last column of Table 4.12 represent δn_j the convergence errors on n_j's. With the new values of n_j's and n we initiate the next iteration by calculating chemical potentials, Table 4.13.

Table 4.15 Converged values: Example 4.5

Species	n_j	$x_j = n_j/n$	$n_j\overline{g}_j$	$n_j\overline{s}_j$	$\delta_{n_j} = \lvert\Delta n_j\rvert/n$
H_2	2.274275E-02	3.001364E-01	−1.091021E+04	4.093137E+00	1.418488E-05
H_2O	5.179520E-02	6.835421E-01	−4.813501E+04	1.335452E+01	2.836975E-05
OH	1.236762E-03	1.632157E-02	−8.527128E+02	3.191975E-01	2.836975E-05
		g (kJ/kg) & s (kJ/(kgmolK))	−5.989793E+04	1.776686E+01	

The numerical value in the last row of Table 4.13 represents the convergence error on Gibbs energy δ_g where $g^{(-)}$ is the Gibbs energy of the previous iteration. With the new values of n_j's we are ready for the next iteration.

The calculated coefficients of the reduced iteration equations for the second iteration are given in Table 4.14. The iterative solution procedure is carried for two more times and the converged solutions are listed in Table 4.15. The converged values of n and $\overline{m}$ are,

$$n = 7.577471E-02\ \frac{\text{kgmol}}{\text{kg}} \qquad \overline{m} = \frac{1}{n} = 13.197\ \frac{\text{kg}}{\text{kgmol}}$$

And, the other convergence errors are listed below.

$$\delta_g = \left(g - g^{(-)}\right)/g = 8.595193E-10$$

$$\delta_s = \left(s - s^{(-)}\right)/s = 1.092024E-05$$

$$b_H^o - b = -1.332268E-15\ \text{kgatom}$$
$$b_O^o - b = -5.551115E-17\ \text{kgatom}$$

Discussion

We have calculated the equilibrium composition at the assigned temperature and pressure. Although the calculation of entropy per kg of propellant combination is not required for the solution, it has been done in parallel to "feel" how the entropy is getting maximized towards equilibrium while the Gibbs energy is getting minimized. Only the converged solution value of entropy is given in Table 4.15. The adiabatic flame temperature for the specified propellant combination and pressure may be higher than the temperature, for which calculations are done in a typical $T\&p$ problem. After accounting for the combustion efficiency and other heat transfer losses, equilibrium composition calculations are carried out for a lower temperature in the combustion chamber for the onward expansion in rocket nozzle [for c^* efficiency see Sect. 3.6]. For the specified propellant combination and pressure of the present Example,

Table 4.16 Bridgman Table written for unit mass of the system

$$(\partial T)_p = -(\partial p)_T = 1$$
$$(\partial v)_p = -(\partial p)_v = \left(\partial v/\partial T\right)_p$$
$$(\partial s)_p = -(\partial p)_s = \left(c_p/T\right)$$
$$(\partial u)_p = -(\partial p)_u = c_p - p\left(\partial v/\partial T\right)_p$$
$$(\partial h)_p = -(\partial p)_h = c_p$$
$$(\partial g)_p = -(\partial p)_g = -s$$
$$(\partial f)_p = -(\partial p)_f = -s - p\left(\partial v/\partial T\right)_p$$

$$(\partial v)_T = -(\partial T)_v = -\left(\partial v/\partial p\right)_T$$
$$(\partial s)_T = -(\partial T)_s = \left(\partial v/(\partial T\right)_p$$
$$(\partial u)_T = -(\partial T)_u = T\left(\partial v/\partial T\right)_p + p\left(\partial v/\partial p\right)_v$$
$$(\partial h)_T = -(\partial T)_h = -v + T\left(\partial v/\partial T\right)_p$$
$$(\partial g)_T = -(\partial T)_g = -v$$
$$(\partial f)_T = -(\partial T)_f = p\left(\partial v/\partial p\right)_T$$

$$(\partial s)_v = -(\partial v)_s = \left(1/T\right)\left[c_p\left(\partial v/\partial p\right)_T + T\left(\partial v/\partial T\right)_p^2\right]$$
$$(\partial u)_v = -(\partial v)_u = c_p\left(\partial v/\partial p\right)_T + T\left(\partial v/\partial T\right)_p^2$$
$$(\partial h)_v = -(\partial v)_h = c_p\left(\partial v/\partial p\right)_T + T\left(\partial v/\partial T\right)_p^2 - v\left(\partial v/\partial T\right)_p$$
$$(\partial g)_v = -(\partial v)_g = -v\left(\partial v/\partial T\right)_p - s\left(\partial v/\partial p\right)_T$$
$$(\partial f)_v = -(\partial v)_f = -s\left(\partial v/\partial p\right)_T$$

$$(\partial u)_s = -(\partial s)_u = \left(p/T\right)\left[c_p\left(\partial v/\partial p\right)_T + T\left(\partial v/\partial T\right)_p^2\right]$$
$$(\partial h)_s = -(\partial s)_h = -\left(vc_p/T\right)$$
$$(\partial g)_s = -(\partial s)_g = -\left(1/T\right)\left[vc_p - sT\left(\partial v/\partial T\right)_p\right]$$
$$(\partial f)_s = -(\partial s)_f = \left(1/T\right)\left\{p\left[c_p\left(\partial v/\partial p\right)_T + T\left(\partial v/\partial T\right)_p^2\right] + sT\left(\partial v/\partial T\right)_p\right\}$$

$$(\partial h)_u = -(\partial u)_h = -v\left[c_p - p\left(\partial v/\partial T\right)_p\right] - p\left[c_p\left(\partial v/\partial p\right)_T + T\left(\partial v/\partial T\right)_p^2\right]$$
$$(\partial g)_u = -(\partial u)_g = -v\left[c_p - p\left(\partial v/\partial T\right)_p\right] + s\left[T\left(\partial v/\partial T\right)_p + p\left(\partial v/\partial p\right)_T\right]$$
$$(\partial f)_u = -(\partial u)_f = p\left[c_p\left(\partial v/\partial p\right)_T + T\left(\partial v/\partial T\right)_p^2\right] + s\left[T\left(\partial v/\partial T\right)_p + p\left(\partial v/\partial p\right)_T\right]$$

$$(\partial g)_h = -(\partial h)_g = -v\left(c_p + s\right) + Ts\left(\partial v/\partial T\right)_p$$
$$(\partial f)_h = -(\partial h)_f = -\left[s + p\left(\partial v/\partial T\right)_p\right]\left[v - T\left(\partial v/\partial T\right)_p\right] + c_p p\left(\partial v/\partial p\right)_T$$
$$(\partial f)_g = -(\partial g)_f = -s\left[v + p\left(\partial v/\partial p\right)_T\right] - pv\left(\partial v/\partial T\right)_p$$

the adiabatic flame temperature is higher than 3200 K. Recall that the adiabatic flame temperature and the corresponding equilibrium composition are calculated under the specified constant pair of state functions $h\&p$.

4.5 Three Key Derivatives

Equilibrium composition of n_j's totaling to n for an assigned condition $T\&p$, $h\&p$, or $s\&p$ can be calculated as detailed in Sect. 4.4. As our next step, we have to evaluate the rocket performance. For this, we require the values of the **derived thermodynamic properties** such as the ratio of specific heats γ, isentropic exponent γ_s, and velocity of sound a. In Sect. 3.7, we briefly considered these derived properties and noted how these properties are dependent on the condition of the equilibrium composition, whether the composition is inert (frozen) or reactive (shifting-equilibrium). A derived thermodynamic property is usually defined by a partial derivative of **basic thermodynamic properties**. Leaving the path-dependent heat and work, there are eight basic point-dependent thermodynamic properties: p, T, v or ρ, s, u, h, g, and f. And, the partial derivatives of all these basic properties are of the form $\left(\partial a/\partial b\right)_c$ where a, b, and c are any three of the above eight basic properties. For example, the derived thermodynamic property velocity of sound is defined as $a \equiv (\partial p/\partial \rho)_s$. Expressions for these partial derivatives can be obtained from the Bridgman Table. This Table adopts a simple shorthand notation,

$$\left(\frac{\partial a}{\partial b}\right)_c = \frac{(\partial a)_c}{(\partial b)_c} \tag{4.89}$$

where a, b, and c are any three of the above eight basic properties. Table 4.16 is a part of Bridgman Table for the unit mass system. And, it lists expressions for the differentials of the basic thermodynamic properties principally in terms of **three key derivatives**. The three key derivatives are *change of volume with respect to temperature at constant pressure* $(\partial v/\partial T)_p$, *change of volume with respect to pressure at constant temperature* $(\partial v/\partial p)_T$, *and the change of enthalpy with respect to temperature at constant pressure* $(\partial h/\partial T)_p$, *which is specific heat at constant pressure* c_p:

$$\left(\frac{\partial v}{\partial T}\right)_p, \ \left(\frac{\partial v}{\partial p}\right)_T, \ \textit{and} \ \left(\frac{\partial h}{\partial T}\right)_p \equiv c_p$$

The three key derivatives are considered as primary variables as values of these are directly obtainable from experiments or from assumed ideal conditions. On adopting the expressions for $(\partial a)_c$ and $(\partial b)_c$ from the Bridgman Table, one can obtain the relation for any derived thermodynamic property $\left(\partial a/\partial b\right)_c$ in terms of the key derivatives. Obviously, bypassing the Bridgman Table such relations can be derived by adopting selectively the first and second laws, the definitions of h, g, f, c_p, and c_v and the Maxwell relations, although such derivations could sometimes be quite involved and lengthy. And as expected the key derivatives in the relations, being the variables for the derived thermodynamic properties, are dependent on whether the equilibrium composition is inert (frozen) or reactive (shifting-equilibrium). Accordingly we derive here the equations for the key derivatives and evaluate thereafter the derived thermodynamic properties for rocket performance calculations. Example 4.6 details the use of Bridgman Table.

Example 4.6 Adopting regular thermodynamic relations as well as Bridgman Table, prove for a gaseous system containing species j

$$\left(\frac{\partial \frac{\mu_j^0}{R_u T}}{\partial T}\right)_p = \frac{-\overline{h}_j^0}{R_u T^2}$$

Solution

Note $\mu_j^0 = \overline{g}_j^0$ and $\overline{h}_j^0$. All these are pure functions of temperature, Eq. 4.49. The proof is drawn by the straightforward differentiation. The proof through the use of Bridgman Table is by differentiating the given quantity and substituting for the numerator and denominator from Table 4.16.

Analysis

For a gaseous system containing species j

$$\mu_j^0 = \overline{g}_j^0 = \overline{h}_j^0 - T\overline{s}_j^0$$

Remember $\mu_j^0(T) = \mu_j(T, p_{std})$ is the standard chemical potential of the species j at the assigned temperature T and the standard pressure $p_{std} = 1$ bar, Eq. 4.49. Therefore it is a pure function of temperature.

$$\left(\frac{\partial \frac{\mu_j^0}{R_u T}}{\partial T}\right)_p = \left(\frac{\partial\left(\frac{\overline{h}_j^0}{R_u T}\right)}{\partial T}\right)_p - \left(\frac{\partial\left(\frac{\overline{s}_j^0}{R_u}\right)}{\partial T}\right)_p$$

$$= \frac{1}{R_u T}\left(\frac{\partial \overline{h}_j^0}{\partial T}\right)_p - \frac{\overline{h}_j^0}{R_u T^2} - \frac{1}{R_u}\left(\frac{\partial \overline{s}_j^0}{\partial T}\right)_p$$

Adopting the general thermodynamic relation $dh = Tds + vdp, d\overline{h}_j^0 = Td\overline{s}_j^0$ as $\overline{h}_j^0$ and hence $\overline{s}_j^0$ are pure functions of temperature and calculated with respect to the standard pressure of one bar that is constant,

$$\frac{1}{R_u T}\left(\frac{\partial \overline{h}_j^0}{\partial T}\right)_p = \frac{1}{R_u}\left(\frac{\partial \overline{s}_j^0}{\partial T}\right)_p$$

Substituting this in the previous expression

$$\left(\frac{\partial \dfrac{\mu_j^0}{R_u T}}{\partial T}\right)_p = -\frac{\overline{h}_j^0}{R_u T^2}$$

We can also prove the above result by using the Bridgman Table.

$$\left(\frac{\partial \dfrac{\mu_j^0}{R_u T}}{\partial T}\right)_p = \frac{1}{R_u T}\left(\frac{\partial \mu_j^0}{\partial T}\right)_p - \frac{\mu_j^0}{R_u T^2} = \frac{1}{R_u T}\frac{(\partial \overline{g}_j^0)_p}{(\partial T)_p} - \frac{\overline{h}_j^0}{R_u T^2} + \frac{\overline{s}_j^0}{R_u T}$$

From the Bridgman Table

$$(\partial \overline{g}_j^0)_p = -\overline{s}_j^0$$

$$(\partial T)_p = 1$$

Substituting the above,

$$\left(\frac{\partial \dfrac{\mu_j^0}{R_u T}}{\partial T}\right)_p = -\frac{\overline{s}_j^0}{R_u T} - \frac{\overline{h}_j^0}{R_u T^2} + \frac{\overline{s}_j^0}{R_u T} = -\frac{\overline{h}_j^0}{R_u T^2}$$

Thus proved.

Discussion
Note that we have previously adopted the result of this Example in deriving Eq. 4.66. By adopting Bridgman Table, relations for derived thermodynamic properties can be easily obtained, bypassing the involved algebraic manipulations.

4.5.1 Key Derivatives for Inert- and Reactive-Systems

From Bridgman Table, we can obtain a relation in terms of the key derivatives $(\partial v/\partial T)_p$, $(\partial v/\partial p)_T$, and c_p, for any chosen derived thermodynamic property, take for example the just cited derived thermodynamic property velocity of sound for which the derivative is $(\partial p/\partial \rho)_s$. The key derivatives in the obtained expression in turn *require equations with respect to the system specific quantities, here in the example the terms involving the calculated quantities n_j's and n* of the equilibrium composition, Sect. 4.4. From such equations, we evaluate the chosen derived thermodynamic property.

For a system of *fixed (inert) composition*, it is a straightforward matter to obtain equations for the key derivatives with respect to the system specific quantities, n_j's and n of the equilibrium composition. But for a system of *variable (reactive) composition*, obtaining the equations for the key derivatives depends on how the n_j's and hence n are affected by the applicable reaction. If the reaction rates are assumed to be infinitely slow, the composition remains fixed (frozen) and the key derivatives are referred to as "frozen" ones. If, on the other hand, the rates are assumed to be infinitely fast, the composition becomes a variable one. And, as the composition reaches its equilibrium value instantaneously from one condition to another, the key derivatives are referred to as "equilibrium" ones consisting of "frozen" contribution and "reaction" contribution. Therefore we have two types of key derivatives: **frozen key derivatives** and **equilibrium key derivatives**, and we have to obtain related equations for these. Adopting such equations any chosen derived thermodynamic property of a system $(\partial a/\partial b)_c$ can be evaluated, Eq. 4.89.

For example for an ideal gas-mixture of inert (frozen) composition $pv = nR_uT$, where n is a constant, the two frozen key derivatives have the equations: $(\partial v/\partial T)_p = nR_u/p = (v/T)$ [or $(\partial \ln v/\partial \ln T)_p = 1$] and $(\partial v/\partial p)_T = -nR_uT/p^2 = -(v/p)$ [or $(\partial \ln v/\partial \ln p)_T = -1$]. But, for an ideal gas-mixture of variable (reacting) composition, where n is a variable, the two equilibrium key derivatives are

$$\left(\frac{\partial v}{\partial T}\right)_p = \left(\frac{v}{T}\right) + \left(\frac{v}{n}\right)\left(\frac{\partial n}{\partial T}\right)_p$$

or

$$\left(\frac{\partial \ln\ v}{\partial \ln\ T}\right)_p = 1 + \left(\frac{\partial \ln\ n}{\partial \ln\ T}\right)_p \tag{4.90}$$

$$\left(\frac{\partial v}{\partial p}\right)_T = -\left(\frac{v}{p}\right) + \left(\frac{v}{n}\right)\left(\frac{\partial n}{\partial p}\right)_T$$

or

$$\left(\frac{\partial \ln\ v}{\partial \ln\ p}\right)_T = -1 + \left(\frac{\partial \ln\ n}{\partial \ln\ p}\right)_T \tag{4.91}$$

Observe that in the right-hand sides of the above equations, the first term represents the frozen component and the second one represents the reaction component. Furthermore, as may have been noticed from the above equations, another specific that is to be adopted for the key derivatives in reacting systems is the use of logarithmic forms namely, $(\partial \ln\ v/\partial \ln\ T)_p$ and $(\partial \ln\ v/\partial \ln\ p)_T$ instead of $(\partial v/\partial T)_p$ and $(\partial v/\partial p)_T$. This is for a better indication of the extent of chemical reaction occurring among the reacting species.

We now have the equations for the first two key derivatives, Eqs. 4.90 and 4.91. Next let us consider the equation for the third key derivative, specific heat at constant pressure c_p for a reacting ideal gas mixture. Remembering that the equilibrium value of a key derivative is the sum of a frozen contribution and a reaction contribution,

$$c_{p,e} = c_{p,f} + c_{p,r} \tag{4.92}$$

By Eq. 4.72, $h = \sum_{j=1}^{NS} n_j \overline{h}_j^0$, $c_{p,e} = \sum_{j=1}^{NS} \left(\partial\left(n_j \overline{h}_j^0\right)/\partial T\right)_p$, and differentiating,

$$c_{p,e} = \sum_{j=1}^{NS} \left(\frac{\partial\left(n_j \overline{h}_j^0\right)}{\partial T}\right)_p = \sum_{j=1}^{NS} n_j \left(\frac{\partial \overline{h}_j^0}{\partial T}\right)_p$$
$$+ \sum_{j=1}^{NG} n_j \frac{\overline{h}_j^0}{T}\left(\frac{\partial \ln n_j}{\partial \ln T}\right)_p + \sum_{j=NG+1}^{NS} \frac{\overline{h}_j^0}{T}\left(\frac{\partial n_j}{\partial \ln T}\right)_p$$

Here the frozen component,

$$c_{p,f} = \sum_{j=1}^{NS} n_j \left(\frac{\partial \overline{h}_j^0}{\partial T}\right)_p = \sum_{j=1}^{NS} n_j \overline{c}_{p,j}^0 \tag{4.93}$$

And the reaction component, where n_j's are variables,

$$c_{p,r} = \sum_{j=1}^{NG} n_j \frac{\overline{h}_j^0}{T} \left(\frac{\partial \ln n_j}{\partial \ln T} \right)_p + \sum_{j=NG+1}^{NS} \frac{\overline{h}_j^0}{T} \left(\frac{\partial n_j}{\partial \ln T} \right)_p \tag{4.94}$$

Substituting Eqs. 4.93 and 4.94 into Eq. 4.92, we get the third key derivative at equilibrium as

$$c_{p,e} = \sum_{j=1}^{NS} n_j \overline{c}_{p,j}^0 + \sum_{j=1}^{NG} n_j \frac{\overline{h}_j^0}{T} \left(\frac{\partial \ln n_j}{\partial \ln T} \right)_p + \sum_{j=NG+1}^{NS} \frac{\overline{h}_j^0}{T} \left(\frac{\partial n_j}{\partial \ln T} \right)_p \tag{4.95}$$

Notice that in the right-hand side of the above equation, the first term represents the frozen component and the remaining two terms characterize contribution by reaction.

4.5.1.1 Temperature Dependent Key Derivatives

Of the three key derivatives, $(\partial \ln v / \partial \ln T)_p$ and $c_{p,e}$ are the ones with respect to temperature, Eqs. 4.90 and 4.95. And to evaluate the two, the derivatives of n and n_j with respect to temperature are required. These may be obtained by differentiating Eqs. 4.54, 4.60, and 4.96 that govern the equilibrium condition.

$$\mu_j + \sum_{i=1}^{l} \lambda_i a_{ij} = 0 \qquad (j = 1, \ldots, NS) \tag{4.60}$$

$$\sum_{j=1}^{NS} a_{ij} n_j - b_i^o = 0 \qquad (i = 1, \ldots, l) \tag{4.54}$$

$$n = \sum_{j=1}^{NG} n_j \tag{4.96}$$

Adopting Eqs. 4.49 and 4.52 for μ_j in Eq. 4.60, and noting $\pi_i = -\lambda_i / R_u T$ and $\left(\partial \frac{\mu_j^0}{R_u T} \middle/ \partial T \right)_p = \frac{-\overline{h}_j^0}{R_u T^2}$ (Example 4.6) the corresponding differentiation with respect to temperature at constant pressure leads to the following equations.

$$\left(\frac{\partial \ln n_j}{\partial \ln T} \right)_p = \sum_{i=1}^{l} a_{ij} \left(\frac{\partial \pi_i}{\partial \ln T} \right)_p + \left(\frac{\partial \ln n}{\partial \ln T} \right)_p + \frac{\overline{h}_j^0}{R_u T} \qquad (j = 1, \ldots, NG) \tag{4.97}$$

$$\sum_{i=1}^{l} a_{ij}\left(\frac{\partial \pi_i}{\partial \ln\ T}\right)_p = -\frac{\overline{h}_j^0}{R_u T} \qquad (j = NG+1, \ldots, NS) \tag{4.98}$$

$$\sum_{j=1}^{NG} a_{kj} n_j \left(\frac{\partial \ln\ n_j}{\partial \ln\ T}\right)_p + \sum_{j=NG+1}^{NS} a_{kj}\left(\frac{\partial n_j}{\partial \ln\ T}\right)_p = 0 \qquad (k = 1, \ldots, l) \tag{4.99}$$

$$\sum_{j=1}^{NG} n_j \left(\frac{\partial \ln\ n_j}{\partial \ln\ T}\right)_p - n\left(\frac{\partial \ln\ n}{\partial \ln\ T}\right)_p = 0 \tag{4.100}$$

Equations 4.97–4.100 are known as **temperature-derivative equations**. Using Eq. 4.97, we can eliminate $\left(\partial \ln\ n_j / \partial \ln\ T\right)_p$ from Eq. 4.95. On dividing by R_u, the resulting equation for the third key derivative $c_{p,e}$ becomes

$$\begin{aligned} \frac{c_{p,e}}{R_u} &= \sum_{j=1}^{NS} \frac{n_j \overline{c}_{p,j}^0}{R_u} + \sum_{i=1}^{l}\left(\sum_{j=1}^{NG}\frac{a_{ij} n_j \overline{h}_j^0}{R_u T}\right)\left(\frac{\partial \pi_i}{\partial \ln\ T}\right)_p \\ &+ \sum_{j=NG+1}^{NS} \frac{\overline{h}_j^0}{R_u T}\left(\frac{\partial\ n_j}{\partial \ln\ T}\right)_p + \left(\sum_{j=1}^{NG}\frac{n_j \overline{h}_j^0}{R_u T}\right)\left(\frac{\partial \ln\ n}{\partial \ln\ T}\right)_p + \sum_{j=1}^{NG}\frac{n_j \overline{h}_j^{0\,2}}{R_u^2 T^2} \end{aligned} \tag{4.101}$$

The term $\sum_{j=1}^{NS} n_j \overline{c}_{p,j}^0 / R_u$ in Eq. 4.101 is the frozen contribution to the specific heat; the remainder of the terms is the reaction contribution.

To evaluate the two key derivatives $\left(\partial \ln\ v / \partial \ln\ T\right)_p$ through Eq. 4.90 and $c_{p,e}$ through Eq. 4.101, we have to now solve the dependent temperature-derivatives found in the $(NS + l + 1)$ temperature-derivative equations, Eqs. 4.97–4.100. The $(NS + l + 1)$ "unknowns" are

$$\left(\frac{\partial \ln n_j}{\partial \ln T}\right)_p \qquad (j = 1, \ldots, NG)$$

$$\left(\frac{\partial n_j}{\partial \ln T}\right)_p \qquad (j = NG+1, \ldots, NS)$$

$$\left(\frac{\partial \pi_i}{\partial \ln T}\right)_p \qquad (i = 1, \ldots, l)$$

$$\left(\frac{\partial \ln\ n}{\partial \ln\ T}\right)_p$$

Similar to what we did in reducing the number of NR equations to get correction equations [Eqs. 4.85–4.88], the temperature-derivative equations can be reduced to a much smaller number of simultaneous equations by eliminating $\left(\partial \ln n_j/\partial \ln T\right)_p$, obtained from Eq. 4.97, from Eqs. 4.99 and 4.100. The resulting reduced number of temperature-derivative equations are

$$\begin{aligned}&\sum_{i=1}^{l}\sum_{j=1}^{NG} a_{kj}a_{ij}n_j\left(\frac{\partial \pi_i}{\partial \ln T}\right)_p + \sum_{j=NG+1}^{NS} a_{kj}\left(\frac{\partial n_j}{\partial \ln T}\right)_p \\ &+ \sum_{j=1}^{NG} a_{kj}n_j\left(\frac{\partial \ln n}{\partial \ln T}\right)_p = -\sum_{j=1}^{NG} a_{kj}n_j\frac{\overline{h}_j^0}{R_u T} \qquad (k=1,\ldots,l)\end{aligned} \tag{4.102}$$

$$\sum_{i=1}^{l} a_{ij}\left(\frac{\partial \pi_i}{\partial \ln T}\right)_p = -\frac{\overline{h}_j^0}{R_u T} \qquad (j = NG+1,\ldots,NS) \tag{4.98}$$

$$\sum_{i=1}^{l}\sum_{j=1}^{NG} a_{ij}n_j\left(\frac{\partial \pi_i}{\partial \ln T}\right)_p = -\sum_{j=1}^{NG}\frac{n_j\overline{h}_j^0}{R_u T} \tag{4.103}$$

Thus we have reduced the number of temperature-derivative equations from $(NS+l+1)$ to $[(NS-NG)+l+1]$.

The reduced number of effective $[(NS-NG)+l+1]$ "unknowns" are

$$\left(\frac{\partial n_j}{\partial \ln T}\right)_p \qquad (j = NG+1,\ldots,NS)$$

$$\left(\frac{\partial \pi_i}{\partial \ln T}\right)_p \qquad (i = 1,\ldots,l)$$

$$\left(\frac{\partial \ln n}{\partial \ln T}\right)_p$$

These temperature-derivatives are solved from Eqs. 4.98, 4.102, and 4.103. Using the obtained $\left(\partial \ln n/\partial \ln T\right)_p$ in Eq. 4.90 we determine the first key derivative $\left(\partial \ln v/\partial \ln T\right)_p$. And, substituting the solved temperature-derivatives in Eq. 4.101, the third key derivative $c_{p,e}$ can be evaluated.

4.5.1.2 Pressure Dependent Key Derivative

The derivative of n with respect to pressure $(\partial \ln n/\partial \ln p)_T$ is required to evaluate the pressure-dependent second key derivative, $(\partial \ln v/\partial \ln p)_T$, Eq. 4.91. This can be obtained in a manner similar to that described for obtaining the first and the third key derivatives $(\partial \ln\ v/\partial \ln\ T)_p$ and $c_{p,e}$. Differentiating Eqs. 4.54, 4.60, and 4.96 with respect to pressure at constant temperature and noting μ_j^0 is a pure function of temperature, we get

$$\left(\frac{\partial \ln n_j}{\partial \ln p}\right)_T - \sum_{i=1}^{l} a_{ij}\left(\frac{\partial \pi_i}{\partial \ln\ p}\right)_T - \left(\frac{\partial \ln n}{\partial \ln p}\right)_T = -1 \qquad (j = 1, \ldots, NG) \tag{4.104}$$

$$\sum_{i=1}^{l} a_{ij}\left(\frac{\partial \pi_i}{\partial \ln\ p}\right)_T = 0 \qquad (j = NG + 1, \ldots, NS) \tag{4.105}$$

$$\sum_{j=1}^{NG} a_{kj} n_j \left(\frac{\partial \ln\ n_j}{\partial \ln\ p}\right)_T + \sum_{j=NG+1}^{NS} a_{kj}\left(\frac{\partial n_j}{\partial \ln\ p}\right)_T = 0 \qquad (k = 1, \ldots, l) \tag{4.106}$$

$$\sum_{j=1}^{NG} n_j \left(\frac{\partial \ln\ n_j}{\partial \ln\ p}\right)_T - n\left(\frac{\partial \ln n}{\partial \ln p}\right)_T = 0 \tag{4.107}$$

As previously done we reduce the number of equations to be solved by eliminating $(\partial \ln n_j/\partial \ln p)_T$, obtained from Eq. 4.104, from Eqs. 4.106 and 4.107. Writing down the results,

$$\begin{aligned}\sum_{i=1}^{l}\sum_{j=1}^{NG} a_{kj} a_{ij} n_j \left(\frac{\partial \pi_i}{\partial \ln\ p}\right)_T + \sum_{j=NG+1}^{NS} a_{kj}\left(\frac{\partial n_j}{\partial \ln\ p}\right)_T \\ + \sum_{j=1}^{NG} a_{kj} n_j \left(\frac{\partial \ln\ n}{\partial \ln\ p}\right)_T = \sum_{j=1}^{NG} a_{kj} n_j \qquad (k = 1, \ldots l)\end{aligned} \tag{4.108}$$

$$\sum_{i=1}^{l} a_{ij}\left(\frac{\partial \pi_i}{\partial \ln\ p}\right)_T = 0 \qquad (j = NG + 1, \ldots, NS) \tag{4.105}$$

$$\sum_{i=1}^{l}\sum_{j=1}^{NG} a_{ij} n_j \left(\frac{\partial \pi_i}{\partial \ln\ p}\right)_T = \sum_{j=1}^{NG} n_j \tag{4.109}$$

Thus we have reduced the number of pressure-derivative equations from $(NS + l + 1)$ to $[(NS - NG) + l + 1]$.

The reduced number of effective $[(NS - NG) + l + 1]$ "unknowns" are

$$\left(\frac{\partial n_j}{\partial \ln p}\right)_T \qquad (j = NG + 1, \ldots, NS)$$

$$\left(\frac{\partial \pi_i}{\partial \ln p}\right)_T \qquad (i = 1, \ldots, l)$$

$$\left(\frac{\partial \ln n}{\partial \ln p}\right)_T$$

On solving Eqs. 4.105, 4.108, and 4.109 the derivative $(\partial \ln n/\partial \ln p)_T$ only is required to obtain the second key derivative $(\partial \ln v/\partial \ln p)_T$, Eq. 4.91.

4.5.1.3 Velocity of Sound

In predicting the performance of rocket propulsion systems, the velocity of sound a is one of the important derived thermodynamic properties to be determined. The fundamental definition of velocity of sound is

$$a^2 \equiv \left(\frac{\partial p}{\partial \rho}\right)_s = -v^2\left(\frac{\partial p}{\partial v}\right)_s \tag{4.110}$$

Adopting Bridgman Table

$$a^2 \equiv \left(\frac{\partial p}{\partial \rho}\right)_s = -v^2\frac{(\partial p)_s}{(\partial v)_s} = -v^2\frac{c_{p,e}}{c_{p,e}\left(\frac{\partial v}{\partial p}\right)_T + T\left(\frac{\partial v}{\partial T}\right)_p^2}$$

The above relation can be simplified to

$$a^2 \equiv \left(\frac{\partial p}{\partial \rho}\right)_s$$

$$= -pv\frac{c_{p,e}}{\left(\frac{\partial \ln v}{\partial \ln p}\right)_T\left[c_{p,e} + (pv/T)\left(\left(\frac{\partial \ln v}{\partial \ln T}\right)_p^2 \Big/ \left(\frac{\partial \ln v}{\partial \ln p}\right)_T\right)\right]} \tag{4.111}$$

From Bridgman Table,

$$c_{v,e} \equiv \left(\frac{\partial u}{\partial T}\right)_v = \left[c_{p,e} + (pv/T)\left(\left(\frac{\partial \ln v}{\partial \ln T}\right)_p^2 \Big/ \left(\frac{\partial \ln v}{\partial \ln p}\right)_T\right)\right] \tag{4.112}$$

and noting $c_{p,e}/c_{v,e} \equiv \gamma$,

$$a^2 \equiv \left(\frac{\partial p}{\partial \rho}\right)_s = -pv \frac{\gamma}{\left(\frac{\partial \ln\ v}{\partial \ln\ p}\right)_T} \tag{4.113}$$

Notice $(\partial \ln v/\partial \ln p)_T$ is a negative quantity. Using the symbol,

$$\gamma_s = -\frac{\gamma}{\left(\frac{\partial \ln\ v}{\partial \ln\ p}\right)_T} \tag{4.114}$$

where γ_s is termed as the **isentropic exponent** and γ is the ratio of specific heats.

$$a^2 \equiv \left(\frac{\partial p}{\partial \rho}\right)_s = nR_u T\gamma_s \tag{4.115}$$

Realize that the velocity of sound for a reacting system is given by $\sqrt{nR_u T\gamma_s}$ and not the usual $\sqrt{nR_u T\gamma}$.

Note that for a reacting system the derived thermodynamic properties $c_{p,e}$ (Eqs. 4.95 and 4.101), $c_{v,e}$ (Eq. 4.112), (and hence γ), and γ_s (Eq. 4.114) and finally a (Eq. 4.113) all use the key derivatives of reacting systems which are additionally dependent on partial derivatives of n_j's, π_i's, and n with respect to temperature and pressure. In addition, observe and appreciate that for a non-reacting ideal gas ($pv = nR_uT$), where n_j's and n are constants, $(\partial \ln\ v/\partial \ln\ T)_p = 1$ and $(\partial \ln\ v/\partial \ln\ p)_T = -1$, and hence the above relations Eqs. 4.95, 4.112 and 4.114 get simplified to yield $c_p = \sum_{j=1}^{NS} n_j \bar{c}^0_{p,j}$, $c_v = c_p - nR_u$ and $\gamma_s = c_p/c_v = \gamma$.

In Example 4.7, we learn the calculations of the three key derivatives under frozen and equilibrium conditions for a combustion product mixture containing condensed species. Also we calculate the isentropic exponent γ_s, ratio of specific heats γ, and velocity of sound.

Example 4.7 Gaseous oxygen and aluminum powder at 298.15 K are burned in an adiabatic combustion chamber. For the purpose of simplicity only the species O, O_2, and $Al_2O_3(l)$ are assumed to be significant in the combustion products. Calculated adiabatic flame temperature is 5652.99 K. Mole fractions of the species are $x_O = 0.51785$, $x_{O_2} = 0.18572$, and $x_{Al_2O_3(l)} = 0.29643$. Gaseous and condensed phase molar mass $\overline{m}_c = 44.452$ kg/kgmol. At the adiabatic flame temperature, the calculated molar specific heats at constant pressure and standard state molar enthalpies for the species are the following.

$$\begin{aligned}
\overline{c}^0_{p,O} &= 2.211907685E+01 \text{ kJ}/(\text{kgmolK})\\
\overline{c}^0_{p,O_2} &= 4.365298050E+01 \text{ kJ}/(\text{kgmolK})\\
\overline{c}^0_{p,Al_2O_3(l)} &= 1.629E+02 \text{ kJ}/(\text{kgmolK})
\end{aligned}$$

$$\begin{aligned}
\overline{h}^0_O &= 3.627382172E+05 \text{ kJ}/\text{kgmol}\\
\overline{h}^0_{O_2} &= 2.096830632E+05 \text{ kJ}/\text{kgmol}\\
\overline{h}^0_{Al_2O_3(l)} &= -7.650625507E+05 \text{ kJ}/\text{kgmol}
\end{aligned}$$

Calculate the three key derivatives $(\partial \ln v/\partial \ln T)_p$, $(\partial \ln v/\partial \ln p)_T$, and c_p for frozen and equilibrium conditions. Also calculate for frozen and equilibrium conditions the isentropic exponent γ_s, the ratio of specific heats γ, and the velocity of sound. Discuss.

Solution

For the frozen condition evidently $(\partial \ln v/\partial \ln T)_p = 1$ and $(\partial \ln v/\partial \ln p)_T = -1$ and $c_{p,f} = \sum_{j=1}^{NS} n_j \overline{c}^0_{p,j}$. The a_{ij} matrix for the species is given in Table 4.17. The applicable equations with respect to Eqs. 4.98, 4.102, and 4.103 have to be obtained. With the solved values of the temperature-derivatives $(\partial \pi_1/\partial \ln T)_p$, $(\partial \pi_2/\partial \ln T)_p$, $(\partial n_{Al_2O_3(l)}/\partial \ln T)_p$, and $(\partial \ln n/\partial \ln T)_p$ we can evaluate the first and the third key derivatives, $(\partial \ln v/\partial \ln T)_p$ and $c_{p,e}$ following Eqs. 4.90 and 4.101. Again following the a_{ij} matrix for the considered species Table 4.17, applicable equations with respect to Eqs. 4.105, 4.108, and 4.109 have to be obtained. With the solved value of the pressure-derivative $(\partial \ln n/\partial \ln p)_T$ we evaluate the second key derivative $(\partial \ln v/\partial \ln p)_T$ following Eq. 4.91.

Table 4.17 a_{ij} matrix: Example 4.7

Atom	n_O	n_{O_2}	$n_{Al_2O_3(l)}$
O	$a_{11} = 1$	$a_{12} = 2$	$a_{13} = 3$
Al	$a_{21} = 0$	$a_{22} = 0$	$a_{23} = 2$

Analysis

Temperature Derivatives Following the a_{ij} matrix Table 4.17, we write the applicable equations for Eqs. 4.98, 4.102, and 4.103.

$$(n_O + 4n_{O_2})\left(\frac{\partial \pi_1}{\partial \ln T}\right)_p + 3\left(\frac{\partial n_{Al_2O_3(l)}}{\partial \ln T}\right)_p + (n_O + 2n_{O_2})\left(\frac{\partial \ln n}{\partial \ln T}\right)_p = -\frac{\left(n_O\overline{h}_O^0 + 2n_{O_2}\overline{h}_{O_2}^0\right)}{R_u T}$$

$$2\left(\frac{\partial n_{Al_2O_3(l)}}{\partial \ln T}\right)_p = 0$$

$$3\left(\frac{\partial \pi_1}{\partial \ln T}\right)_p + 2\left(\frac{\partial \pi_2}{\partial \ln T}\right)_p = -\frac{\overline{h}^0_{Al_2O_3(l)}}{R_u T}$$

$$(n_O + 2n_{O_2})\left(\frac{\partial \pi_1}{\partial \ln T}\right)_p = -\frac{\left(n_O\overline{h}_O^0 + n_{O_2}\overline{h}_{O_2}^0\right)}{R_u T}$$

Noting $n_j = x_j/\overline{m}_c$,

$$n_O = 1.164961866E - 02 \text{ kgmol/kg}$$
$$n_{O_2} = 4.177893000E - 03 \text{ kgmol/kg}$$
$$n_{Al_2O_3(l)} = 6.668468221E - 03 \text{ kgmol/kg}$$

On substituting these values and the values of enthalpies and temperature,

$$2.836119066E - 02\left(\frac{\partial \pi_1}{\partial \ln T}\right)_p + 3\left(\frac{\partial n_{Al_2O_3(l)}}{\partial \ln T}\right)_p + 2.000540466E - 02\left(\frac{\partial \ln n}{\partial \ln T}\right)_p = -1.271828777E - 01$$

$$2\left(\frac{\partial n_{Al_2O_3(l)}}{\partial \ln T}\right)_p = 0$$

$$3\left(\frac{\partial \pi_1}{\partial \ln T}\right)_p + 2\left(\frac{\partial \pi_2}{\partial \ln T}\right)_p = 1.627729089E+01$$

$$2.000540466E-02\left(\frac{\partial \pi_1}{\partial \ln T}\right)_p = -1.085445971E-01$$

The solutions are,

$$\left(\frac{\partial \pi_1}{\partial \ln T}\right)_p = -5.425763632E+00$$

$$\left(\frac{\partial \pi_2}{\partial \ln T}\right)_p = 1.627729089E+01$$

$$\left(\frac{\partial n_{Al_2O_3(l)}}{\partial \ln T}\right)_p = 0$$

$$\left(\frac{\partial \ln n}{\partial \ln T}\right)_p = 1.334551319E+00$$

Recall,

$$\left(\frac{\partial \ln v}{\partial \ln T}\right)_p = 1 + \left(\frac{\partial \ln n}{\partial \ln T}\right)_p \tag{4.90}$$

Therefore,

$$\left(\frac{\partial \ln v}{\partial \ln T}\right)_p = 1 + 1.3346 = 2.3346$$

Adopting the a_{ij} matrix from Table 4.17, Eq. 4.101 becomes

$$\frac{c_{p,e}}{R_u} = \left(\frac{n_O\bar{h}^0_O + 2n_{O_2}\bar{h}^0_{O_2}}{R_uT}\right)\left(\frac{\partial \pi_1}{\partial \ln T}\right)_p + \frac{\bar{h}^0_{Al_2O_3(l)}}{R_uT}\left(\frac{\partial n_{Al_2O_3(l)}}{\partial \ln T}\right)_p$$
$$+ \left(\frac{n_O\bar{h}^0_O + n_{O_2}\bar{h}^0_{O_2}}{R_uT}\right)\left(\frac{\partial \ln n}{\partial \ln T}\right)_p$$
$$+ \frac{n_O\bar{c}^0_{p,O} + n_{O_2}\bar{c}^0_{p,O_2} + n_{Al_2O_3(l)}\bar{c}^0_{p,Al_2O_3(l)}}{R_u}$$
$$+ \frac{n_O\bar{h}^{0\,2}_O + n_{O_2}\bar{h}^{0\,2}_{O_2}}{R_u^2T^2}$$

Recall that the second last-term in the equation for $c_{p,e}/R_u$ represents the frozen contribution. Substituting the n_j's, molar specific heats at constant pressure, and standard state molar enthalpies, and the solved temperature-derivatives,

$$c_{p,e} = 3.4536 \text{ kJ}/(\text{kgK})$$
$$c_{p,f} = 1.5263 \text{ kJ}/(\text{kgK})$$

Pressure Derivatives Following the a_{ij} Table 4.17, we write the applicable equations for Eqs. 4.105, 4.108, and 4.109,

$$(n_O + 4n_{O_2})\left(\frac{\partial \pi_1}{\partial \ln p}\right)_T + 3\left(\frac{\partial n_{Al_2O_3(l)}}{\partial \ln p}\right)_T + (n_O + 2n_{O_2})\left(\frac{\partial \ln n}{\partial \ln p}\right)_T$$
$$= n_O + 2n_{O_2}$$

$$2\left(\frac{\partial n_{Al_2O_3(l)}}{\partial \ln p}\right)_T = 0$$

$$3\left(\frac{\partial \pi_1}{\partial \ln p}\right)_T + 2\left(\frac{\partial \pi_2}{\partial \ln p}\right)_T = 0$$

$$(n_O + 2n_{O_2})\left(\frac{\partial \pi_1}{\partial \ln p}\right)_T = n_O + n_{O_2}$$

On substituting the values of n_j's,

$$2.836119066E-02\left(\frac{\partial \pi_1}{\partial \ln\ p}\right)_T+3\left(\frac{\partial n_{Al_2O_3(l)}}{\partial \ln\ p}\right)_T$$

$$+2.000540466E-02\left(\frac{\partial \ln\ n}{\partial \ln\ p}\right)_T=2.000540466E-02$$

$$2\left(\frac{\partial n_{Al_2O_3(l)}}{\partial \ln\ p}\right)_T=0$$

$$3\left(\frac{\partial \pi_1}{\partial \ln\ p}\right)_T+2\left(\frac{\partial \pi_2}{\partial \ln\ p}\right)_T=0$$

$$2.000540466E-02\left(\frac{\partial \pi_1}{\partial \ln\ p}\right)_T=1.582751166E-02$$

The solutions are

$$\left(\frac{\partial \pi_1}{\partial \ln\ p}\right)_T=7.911617850E-01$$

$$\left(\frac{\partial \pi_2}{\partial \ln\ p}\right)_T=-1.186742677E+00$$

$$\left(\frac{\partial n_{Al_2O_3(l)}}{\partial \ln\ p}\right)_T=0$$

$$\left(\frac{\partial \ln\ n}{\partial \ln\ p}\right)_T=-1.216114149E-01$$

The pressure-derivative $(\partial \ln n/\partial \ln p)_T$ only is required to obtain the second key derivative $(\partial \ln v/\partial \ln p)_T$, Eq. 4.91.

$$\left(\frac{\partial \ln\ v}{\partial \ln\ p}\right)_T=-1+\left(\frac{\partial \ln\ n}{\partial \ln\ p}\right)_T \tag{4.91}$$

$$\left(\frac{\partial \ln\ v}{\partial \ln\ p}\right)_T=-1.12161$$

The specific heat at constant volume is given by Eq. 4.112,

$$c_{v,e} \equiv \left(\frac{\partial u}{\partial T}\right)_v = \left[c_{p,e} + nR_u\left(\left(\frac{\partial \ln v}{\partial \ln T}\right)_p^2 \Big/ \left(\frac{\partial \ln v}{\partial \ln p}\right)_T\right)\right] \tag{4.112}$$

Therefore,

$$c_{v,e} = \left[3.4536 + 1.58275E - 02 \times 8.31451\left(2.3346^2/(-1.12161)\right)\right]$$
$$= 2.81411 \text{ kJ/kg}$$

For frozen condition, $(\partial \ln\ v/\partial \ln\ T)_p = 1$ and $(\partial \ln\ v/\partial \ln\ p)_T = -1$.

$$c_{v,f} = c_{p,f} - nR_u$$
$$= [1.5263 - 1.58275E - 02 \times 8.31451] = 1.3947 \text{ kJ/kg}$$

Therefore ratio of specific heats for equilibrium and for frozen conditions are given as

$$\gamma_e = \frac{c_{p,e}}{c_{v,e}} = \frac{3.4536}{2.81411} = 1.2272$$

$$\gamma_f = \frac{c_{p,f}}{c_{v,f}} = \frac{1.5263}{1.3947} = 1.0944$$

Isentropic exponent is given as

$$\gamma_s = -\frac{\gamma}{\left(\frac{\partial \ln\ v}{\partial \ln\ p}\right)_T} \tag{4.114}$$

For equilibrium condition,

$$\gamma_{s,e} = -\frac{\gamma_e}{\left(\frac{\partial \ln\ v}{\partial \ln\ p}\right)_T} = -\frac{1.2272}{-1.12161} = 1.0941$$

And, for frozen condition,

$$\gamma_{s,f} = -\frac{\gamma_f}{\left(\frac{\partial \ln v}{\partial \ln p}\right)_T} = -\frac{1.0944}{-1} = 1.0944$$

Note that the ratio of specific heats for equilibrium condition $\gamma_e = 1.2272$ but its isentropic exponent $\gamma_{s,e} = 1.0941$. Also note that the isentropic exponent for frozen condition is the same as its ratio of specific heats: $\gamma_f = \gamma_{s,f} = 1.0944$.

Velocity of sound can be calculated from Eq. 4.115. Isentropic exponent for the equilibrium system $\gamma_{s,e}$ is different from that for the frozen one $\gamma_{s,f}$. Furthermore, note that the latter $\gamma_{s,f}$ is nothing but the corresponding ratio of specific heats γ_f but the former $\gamma_{s,e}$ is different from the corresponding ratio of specific heats γ_e.

The velocity of sound under equilibrium condition is

$$a_e = \sqrt{nR_uT\gamma_{s,e}}$$

$$= \sqrt{1.5827E-02 \times 8314.51 \times 5652.99 \times 1.0941} = 902.16 \text{ m/s}$$

And, the velocity of sound under frozen condition,

$$a_f = \sqrt{nR_uT\gamma_{s,f}} = \sqrt{nR_uT\gamma_f}$$

$$= \sqrt{1.5827E-02 \times 8314.51 \times 5652.99 \times 1.0944} = 902.29 \text{ m/s}$$

Discussion

Important parameters like specific heat at constant pressure and speed of sound considered for rocket performance calculations assume different values for equilibrium condition and frozen condition. Combustion products accelerate towards the nozzle end from near zero velocity values at the combustion chamber head end. At the nozzle end the Mach numbers are generally in the range of 0.3–0.4. Thereafter, the products accelerate to Mach number unity at the throat. Since the average residence time per unit length in the region up to the throat is relatively long, equilibrium operation is considered applicable up to the throat. Since the supersonic flow downstream of thc throat has relatively short residence time, frozen flow condition is considered applicable. However, despite the supersonic flow having short residence time in the divergent portion of the nozzle, the consideration of equilibrium operation may become necessary there too, if an imminent condensation of certain gaseous species or solidification of a particular liquid one is possible (see Example 4.11 in Sect. 4.7).

The effect of temperature or pressure on volume is more for the reacting mixture than for the inert one. That the values of $c_{p,e}$ and $c_{v,e}$ always correspondingly being greater than those of $c_{p,f}$ and $c_{v,f}$ indicates that the heat capacity of reacting mixture is more than that of inert one.

4.6 Thermodynamic Data

For all equilibrium composition calculations, we must take into account the thermal energy of species. As seen previously, under the common condition of assigned pressure $p = p^*$, we have three other assigned conditions, represented by the trivial equations $T = T^*$, $h = h^*$, and $s = s^*$. And, each of these goes with $p = p^*$ to form three pairs, encompassing the possible ideal operating conditions in open systems (control volumes under the principles of gas dynamics). Rocket propulsion systems, being open ones, are evidently covered under these three conditions. To arrive at the equilibrium composition under the condition of assigned temperature and pressure ($T = T^*$ & $p = p^*$), we need to calculate the thermal energy of *a trial composition* at the assigned T^* & p^* and check whether that composition satisfies the energy minimization and mass balance—Table 4.8: Eqs. 4.67, 4.85, and 4.86 with $\Delta \ln\ T = 0$ (refer Example 4.5). But, to arrive at the equilibrium composition under the condition that the enthalpy of reactants is equal to the enthalpy of products at the assigned pressure ($h = h^*$ & $p = p^*$) we need to calculate the thermal energy of *a trial composition* at *a trial temperature* and the assigned p^* and check whether the trial values satisfy the energy minimization, and mass-, and enthalpy balance—Table 4.8: Eqs. 4.67, 4.85–4.87 for $h = h^*$. Those trial values which satisfy $h = h^*$ & $p = p^*$ are the solutions, and the solution temperature is the **adiabatic flame temperature** at the specified pressure p^* that we search for. Similarly for the assigned entropy and pressure ($s = s^*$ & $p = p^*$), the thermal energy and the entropy of *a trial composition* at *a trial temperature* and the assigned p^* are calculated to check whether that composition satisfies the energy minimization, and mass-, and entropy-balance—Table 4.8: Eqs. 4.67, 4.85, 4.86 and 4.88 for $s = s^*$. In all the three pairs of assigned conditions, the trial composition that satisfies the assigned condition is the equilibrium composition that we seek.

In order to take into account the thermal energy of the reacting species, the **standard state molar enthalpy** $\overline{h}^0(T)$ of each species at the considered temperature T with respect to the standard reference state is to be determined. The standard reference state is at 298.15 K and 1 bar. The standard state molar enthalpy $\overline{h}^0(T)$ is the sum of (1) the enthalpy that takes into account the energy associated with the chemical bonds at the standard reference state, the **enthalpy of formation** or the **heat of formation** $\overline{h}_f^0$ (J/kgmol), and (2) the enthalpy associated with the temperature T, the **sensible enthalpy change** or sensible heat $\Delta\overline{h}_s(T)$ (J/kgmol) from the standard reference state temperature $T_{ref} = 298.15$–T K. Thus we can write the molar standard state enthalpy for species j at T as

$$\overline{h}_j^0(T) = \overline{h}_{f,j}^0 + \Delta\overline{h}_{s,j}(T) \tag{4.116}$$

At the standard reference state, since $\Delta\overline{h}_s = 0$, the molar standard state enthalpy becomes the enthalpy of formation, $\overline{h}_j^0(T_{ref}) = \overline{h}_{f,j}^0$. The enthalpy of formation can be defined as follows. *The enthalpy of formation per mole of a compound is the heat transacted (negative or positive—exothermic or endothermic respectively) when one mole of the compound is formed at the standard reference state of 298.15 K and 1 bar from the respective elements in their naturally occurring states at the standard reference state.*

By convention the enthalpies of formation are zero for elements in their naturally occurring state at the standard reference state. Values of enthalpy of formation of compounds can be measured experimentally or determined by means of statistical thermodynamics. For example, oxygen and hydrogen exist as gaseous diatomic molecules at the standard reference state and hence their enthalpies of formation are zero. Whereas, at the standard reference state, as one kgmole of H_2O vapor is formed from oxygen and hydrogen the exothermic heat transacted, which can be measured, is −241,826 kJ. Therefore the enthalpy of formation for H_2O vapor is −241,826 kJ/kgmol. The heat of condensation that is released at the standard reference state by the H_2O vapor to become liquid is −44,004 kJ/kgmol. Therefore the enthalpy of formation for liquid water at the standard reference state is −285,830 kJ/kgmol. Similarly, given that the enthalpy of formation of oxygen is zero, the molar standard state enthalpy of liquid oxygen at 90.17 K $\overline{h}_{O_2(l)}^0$ (90.17 K) is −12,979 kJ/kgmol, as the sensible enthalpy change required to cool the oxygen gas to its condensation point is −6065 kJ/kgmol and the heat of condensation is −6914 kJ/kgmol. Both the heat exchanges are the heat to be extracted from the gaseous oxygen and hence negative by thermodynamic convention.

The primary inputs required for the equilibrium composition calculations are the energies contained in the reactants and the molar standard state enthalpies. As one of the input data files required for its CEA code, NASA has compiled thermodynamic data of a large number of propellant ingredients and related product species in the file *thermo.inp* which can be downloaded from the URL of [8]. Furthermore, Purdue School of Aeronautics and Astronautics has compiled values of enthalpies of formation of more than 1000 propellant ingredients and additives and these are available at the URL of [11]. Selected thermodynamic data of some propellant ingredients and additives, drawn principally from [7, 8, 11] are given in Tables 4.18 and 4.19.

In order to calculate at the assigned pressure, the thermal energy (the enthalpy) and the entropy of the reacting species, we require the value of the underlying property, namely, the molar heat capacity at constant pressure $\overline{c}_p^0$ (or the molar specific heat at constant pressure). Once we are able to evaluate the value of this basic property we can determine the molar standard state enthalpy $\overline{h}^0$, and the molar standard state entropy $\overline{s}^0$ of the reacting species, because $\overline{h}^0 = \int \overline{c}_p^0 dT$ and $\overline{s}^0 = \int (\overline{c}_p^0/T)dT$. NASA recommends a general form for $\overline{c}_p^0$ in terms of seven least square coefficients, as given below [6].

$$\frac{\overline{c}_p^0}{R_u} = a_1T^{-2} + a_2T^{-1} + a_3 + a_4T + a_5T^2 + a_6T^3 + a_7T^4 \tag{4.117}$$

As $\overline{h}^0 = \int \overline{c}_p^0 dT$ and $\overline{s}^0 = \int (\overline{c}_p^0/T)dT$, substituting for $\overline{c}_p^0$ in these two equations and integrating, we get the equation for molar standard state enthalpy $\overline{h}^0$ in the form,

$$\begin{aligned}\frac{\overline{h}^0}{R_uT} = & -a_1T^{-2} + a_2T^{-1}\ln T + a_3 + a_4\frac{T}{2} + a_5\frac{T^2}{3} \\ & + a_6\frac{T^3}{4} + a_7\frac{T^4}{5} + \frac{a_8}{T}\end{aligned} \tag{4.118}$$

and the equation for molar standard state entropy $\overline{s}^0$ in the form,

$$\begin{aligned}\frac{\overline{s}^0}{R_u} = & -a_1\frac{T^{-2}}{2} - a_2T^{-1} + a_3\ln T + a_4T + a_5\frac{T^2}{2} \\ & + a_6\frac{T^3}{3} + a_7\frac{T^4}{4} + a_9\end{aligned} \tag{4.119}$$

For a large number of reactant and product species, the values of the least square coefficients a_1–a_9 are given in the file *thermo.inp* of [8] up to nine-decimal accuracy generally for two temperature ranges, 200–1000 K and 1000–6000 K. Here in Tables 4.20 and 4.21, with a view to help solving some simple problems, we present these coefficients rounded to fifth decimal place for a few selected species. Hence, the calculated results using these numbers will be relatively less accurate. However, in Example 4.8, the calculations are carried out with the least square coefficients considered up to nine-decimal accuracy. Example 4.9 explains the calculation of the enthalpy per unit mass of a solid propellant.

Table 4.18 Thermodynamic data for propellant ingredients[a]

Propellant ingredient	Phase	Boiling point (K)[b]	Standard state enthalpy (kJ/gmol)	Density (g/cm^3)
Acetonitrile, C_2H_3N	L	–	31.38	0.7793
Acetylene, C_2H_2	g	–	228.2	–
Acetylene, C_2H_2	L	192.35	207.599	0.729
Ammonia, NH_3	g	–	−45.940	
Ammonia, NH_3	L	239.72	−71.555	0.6821
Ammonium perchlorate, NH_4ClO_4	I	–	−295.767	1.95
Ammonium nitrate, NH_4NO_3	IV	–	−365.6	1.725
Aniline, $C_6H_5NH_2$	L	–	31.5	1.02173
BAMO, $C_5H_8N_6O$	*s*	–	*427.31*	*1.3*
Benzene, C_6H_6	L	–	49.08	0.87366
n-Butane, C_4H_{10}	g	–	−125.79	–
n-Butane, C_4H_{10}	L	272.638	−150.664	0.6012
i- Butane, C_4H_{10}	g	–	−134.99	–
i- Butane, C_4H_{10}	L	261.361	−159.664	0.5941
1-Butene, C_4H_8	g	–	−0.54	–
1-Butene, C_4H_8	L	266.92	−25.173	0.6254
CTPB, $C_{72}H_{108}O$	*L*	–	*662.78*	–
Chlorine, Cl_2	L	239.12	−22.55	1.559
Chlorine trifluoride, ClF_3	g	–	−164.6	–
Chlorine trifluoride, ClF_3	L	284.89	−193.386	1.85
Cyanogen, C_2N_2	g	–	309.1	–
Cyanogen, C_2N_2	L	252.05	283.209	0.9577
Diborane, B_2H_6	g	–	36.6	–
Diborane, B_2H_6	L	180.59	16.445	0.4371
Dioctyl adipate, $C_{22}H_{42}O_4$	*L*	–	*−1137*	–
Ethane, C_2H_6	g	–	−83.852	–
Ethane, C_2H_6	L	184.559	−103.819	0.5465
Ethanol, C_2H_5OH	L	–	−277.51	0.78509
Ethylene, C_2H_4	g	–	52.5	–
Ethylene, C_2H_4	L	169.42	33.945	0.569
Ethylene oxide, C_2H_4O	g	–	−52.635	–
Ethylene oxide, C_2H_4O	L	283.65	−78.841	0.8824
Ethylene polymer, $(CH_2)_x$	s	–	−25.6	0.935
Fluorine, F_2	L	85.02	−13.091	1.5021
GAP, $C_3H_5N_3O$	*s*	–	*141*	*1.3*
n-Heptane, C_7H_{16}	L	–	−224.35	0.6795
n-Hexane, C_6H_{14}	L	–	−198.66	0.65484

(continued)

Table 4.18 (continued)

Propellant ingredient	Phase	Boiling point (K)[b]	Standard state enthalpy (kJ/gmol)	Density (g/cm^3)
HMX, $C_4H_8N_8O_8$	*s*	–	*75.63*	*1.899*
Hydrazine, N_2H_4	L	–	50.38	1.004
Hydrogen, H_2	L	20.27	−9.012	0.0709
Hydrogen peroxide, H_2O_2	L	–	−187.78	1.4425
HTPB, $C_{73}H_{110}O_6$	*L*		*−136.08*	–
IRFNA (HNO3:N2O4:H2O:HF = 0.835:0.14:0.02:0.005 by mass), $H_{1.57216}N_{1.62945}O_{4.69505}F_{0.02499}$	L	–	−270.496	1.478
Jet-A, $C_{12}H_{23}$	L	–	−303.403	0.818
JP-4, RP-1, $CH_{1.9423}$	L	–	−22.723	0.773
JP-5, ASTMA1, $CH_{1.9185}$	L	–	−22.183	0.807
Lithium, Li	L	453.69[c]	7.187	0.512
Lithium perchlorate, $LiClO_4$	s	–	−380.7	2.428
Methane, CH_4	g	–	−74.6	–
Methane, CH_4	L	111.643	−89.233	0.4211
Methanol, CH_3OH	L	–	−238.91	0.78664
Monomethyl hydrazine, CH_6N_2	L	–	54.2	0.874
Nitric acid, HNO_3	g	–	−133.913	–
Nitric acid, HNO_3	L	–	−173.013	1.5037
Nitrocellulose (12.6% N), $C_{600}H_{755}N_{245}O_{990}$	*s*	–	*−70353*	*1.55*
Nitrogen, N_2	L	77.352	−12.107	0.8081
Nitrogen tetroxide, N_2O_4	g	–	11.111	–
Nitrogen tetroxide, N_2O_4	L	–	−17.549	1.431
Nitrogen trifluoride, NF_3	g	–	−131.7	–
Nitrogen trifluoride, NF_3	L	144.09	−150.387	1.537
Nitroglycrin, $C_3H_5N_3O_9$	*L*	–	*380*	*1.6*
Nitroguanidine, $CH_4N_4O_2$	*s*	–	*−91.05*	*1.73*
Nitromethane, CH_3NO_2	L	–	−113.1	1.1286
Nitrous oxide, N_2O	g	–	81.6	–
n-Octane, C_8H_{18}	L	–	−250.26	0.69854
i-Octane, C_8H_{18}	L	–	−259.16	0.68784
Oxygen, O_2	L	90.17	−12.979	1.149
Oxygen difluoride, OF_2	g	–	24.5	–
Oxygen difluoride, OF_2	L	128.4	6.672	1.518
Ozone, O_3	g	–	141.8	–
Ozone, O_3	L	161.85	122.527	1.449
Pentaborane, B_5H_9	L	–	42.84	0.6183

(continued)

Table 4.18 (continued)

Propellant ingredient	Phase	Boiling point (K)[b]	Standard state enthalpy (kJ/gmol)	Density (g/cm^3)
n-Pentane, C_5H_{12}	L	–	−173.49	0.6213
Perchloryl fluoride, $ClFO_3$	g	–	−23.8	–
Perchloryl fluoride, $ClFO_3$	L	226.4	−47.436	1.392
PBAN, $C_{653}H_{854}O_{19}N_{72}$	*L*	–	*13165*	–
Polymethyl methacrylate, $(C_5H_8O_2)_x$	*s*	–	*−379.7*	*1.18*
Potassium nitrate, KNO_3	*s*	–	*–494*	*2.11*
Potassium perchlorate, $KClO_4$	*s*	–	*–430*	*2.524*
Propane, C_3H_8	g	–	−104.68	–
Propane, C_3H_8	L	231.076	−128.228	0.581
Propylene, C_3H_6	g	–	20	–
Propylene, C_3H_6	L	225.46	−2.704	0.6093
Propyl nitrate, $C_3H_7NO_3$	L	–	−214.5	1.0538
RDX (Cyclo-trimethylene-trinitramine), $C_3H_6N_6O_6$	*s*	–	*61.37*	*1.816*
Sodium perchlorate, $NaClO_4$	*s*	–	*−384.4*	*2.5*
Toluene, C_7H_8	L	–	12.18	0.8622
Unsymmetrical dimethyl hydrazine (UDMH), $C_2H_8N_2$	L	–	48.9	0.7914
Water, H_2O	L	–	−285.83	1
Xylidine, $C_8H_{11}N$	*L*	–	*–73*	–

[a]Values are from [7]; but the values in *italics* are either from [11] or from different web sites, more particularly from Wikipedia, the free encyclopedia

[b]Boiling points are given only for liquids, which have boiling points below 298.15 K. If the column is blank the temperature of reference is 298.15 K for the molar standard state enthalpy in the next column and hence it is also the enthalpy of formation

[c]Melting point

Example 4.8 The least square coefficients for CO_2 for the temperature range of 200–1000 K are as per below. Calculate its (1) the specific heat at constant pressure at 298.15 K and 1000 K, (2) the molar standard state enthalpy at 298.15 K and check for its value as the enthalpy of formation −393510 kJ/kgmol, (3) the sensible enthalpy change from 298.15 to 1000 K, and (4) the molar standard state entropy at 298.15 and 1000 K.

Table 4.19 Enthalpies of formation for propellant additives[a]

Additive	Molecular formula	Molar mass g/gmol	Standard state enthalpy (kJ/gmol)
Cadelillawax	C_2H_4	28.052	−53.19
Cadelillawax	$C_{69}H_{122}O_3$	999.662	−594.21
Calcium carbonate	$CaCO_3$	100	−1211.85
Copper chromite	$CuCrO_3$	163.55	−112.28
Copper oxide	Cu_2O	143.1	−166.53
Diethyl phthalate	$C_{12}H_{14}O_4$	222.232	−681.88
Diethylene glycol dinitrate (DEGDN)	$C_4H_8N_2O_7$	196.124	−426.91
Dinitrotoluene	$C_7H_6N_2O_4$	182.138	−6251.92
Ferric oxide	Fe_2O_3	159.69	−822.21
Ethyl centralite	$C_{17}H_{20}N_2O$	268.349	−142.66
Lead salicylate	$C_{14}H_{10}O_6Pb$	481.41	−169.28
Lithium Fluoride	LiF	25.939	−610.22
Oxamide	$C_2H_4N_2O_2$	88.072	−507.29
Paraffininc oil	$C_{73}H_{124}$	1001.718	−1538.9
Petrin	$C_5H_9N_3O_{10}$	271.152	−582.28
Potassium sulfate	K_2SO_4	174.26	−1434.1
Tetra-methyl-lead	$C_4H_{12}Pb$	267.326	226.04
Titanium dioxide	TiO_2	79.9	−853.21
Triacetin	$C_9H_{14}O_6$	218.202	−1218.47
Tricalcium phosphate	$Ca_3O_8P_2$	310.18	−4097.79
Triethylene glycol dinitrate (TEGDN)	$C_6H_{12}N_2O_8$	240.176	−657.52
Trimethylolethane trinitrate (TMETN)	$C_5H_9N_3O_9$	255.152	−424.02

[a] Values are either from [11] or from different web sites, more particularly from Wikipedia, the free encyclopedia

$a_1 = 4.943650540\text{E}{+}04$ $\quad a_2 = -6.264116010\text{E}{+}02$

$a_3 = 5.301725240\text{E}{+}00$ $\quad a_4 = 2.503813816\text{E-}03$

$a_5 = -2.127308728\text{E-}07$ $\quad a_6 = -7.689988780\text{E-}10$

$a_7 = 2.849677801\text{E-}13$ $\quad a_8 = -4.528198460\text{E}{+}04$

$a_9 = -7.048279440\text{E}{+}00$

Solution

The functions for $\bar{c}_p^0$, $\bar{h}^0$, and $\bar{s}^0$ are given by Eqs. 4.117, 4.118, and 4.119 respectively. Notice that Eqs. 4.117–4.119 are given in non-dimensional form.

Table 4.20 Thermodynamic data of selected species from [8]

Species	$\overline{h}_f^0$ (kJ/kgmol)	Temperature range (K)	a_1	a_2	a_3	a_4
$Al_2O_3(s)$	−1675700	500–1200	−6.04209E+05	0.00000E+00	1.47548E+01	8.27229E-04
		1200–2327	0.00000E+00	0.00000E+00	1.29377E+01	1.99278E-03
$Al_2O_3(l)$	−1675700	2327–6000	0.00000E+00	0.00000E+00	1.95923E+01	0.00000E+00
$BeO(a)(s)$	−609400	1000–2373	1.02406E+06	1.93747E+03	−2.17021E+00	6.70094E-03
$BeO(b)(s)$	−609400	2373–2851	0.00000E+00	0.00000E+00	6.73521E+00	0.00000E+00
$BeO(l)$	−609400	2851–6000	0.00000E+00	0.00000E+00	1.01028E+01	0.00000E+00
$BeOH$	−99718	200–1000	−3.85302E+04	4.49806E+02	1.53210E+00	1.29991E-02
		1000–6000	8.55033E+05	−2.64654E +03	8.20478E +00	1.11081E-05
CO	−110535	200–1000	1.48905E+04	−2.92229E+02	5.72453E+00	−8.17624E-03
		1000–6000	4.61920E+05	−1.94470E+03	5.91671E+00	−5.66428E-04
CO_2	−393510	200–1000	4.94365E+04	−6.26412E+02	5.30173E+00	2.50381E-03
		1000–6000	1.17696E+05	−1.78879E+03	8.29152E+00	−9.22316E-05
H	217999	200–1000	0.00000E+00	0.00000E+00	2.50000E+00	0.00000E+00
		1000–6000	6.07877E+01	−1.81935E-01	2.50021E+00	−1.22651E-07

Table 4.20 (continued)

Species	Temperature range (K)	a_5	a_6	a_7	a_8	a_9
$Al_2O_3(s)$	500–1200	0.00000E+00	0.00000E+00	0.00000E+00	−2.07924E+05	−8.13603E+01
	1200–2327	0.00000E+00	0.00000E+00	0.00000E+00	−2.06079E+05	−6.96660E+01
$Al_2O_3(l)$	2327–6000	0.00000E+00	0.00000E+00	0.00000E+00	−2.02770E+05	−1.10859E+02
$BeO(a)(s)$	1000–2373	−1.82909E-06	2.75789E-10	−2.71274E-14	−8.27681E+04	1.89661E+01
$BeO(b)(s)$	2373–2851	0.00000E+00	0.00000E+00	0.00000E+00	−7.67447E+04	−3.93753E+01
$BeO(l)$	2851–6000	0.00000E+00	0.00000E+00	0.00000E+00	−7.60024E+04	−6.25381E+01
$BeOH$	200–1000	−1.70898E-05	1.16997E-08	−3.16318E-12	−1.55905E+04	1.54554E+01
	1000–6000	−4.26536E-08	7.92651E-12	−4.64542E-16	3.12122E+03	−2.57164E+01
CO	200–1000	1.45690E-05	−1.08775E-08	3.02794E-12	−1.30313E+04	−7.85924E+00
	1000–6000	1.39881E-07	−1.78768E-11	9.62094E-16	−2.46626E+03	−1.38741E+01
CO_2	200–1000	−2.12731E-07	−7.68999E-10	2.84968E-13	−4.52820E+04	−7.04828E+00
	1000–6000	4.86368E-09	−1.89105E-12	6.33004E-16	−3.90835E+04	−2.65267E+01
H	200–1000	0.00000E+00	0.00000E+00	0.00000E+00	2.54737E+04	−4.46683E-01
	1000–6000	3.73288E-11	−5.68774E-15	3.41021E-19	2.54749E+04	−4.48192E-01

Table 4.21 Thermodynamic data of selected species from [8]

species	$\overline{h}_f^0$ (kJ/kgmol)	Temperature Range (K)	a_1	a_2	a_3	a_4
H_2	0	200–1000	4.07832E+04	−8.00919E+02	8.21470E+00	−1.26971E-02
		1000–6000	5.60813E+05	−8.37150E+02	2.97536E+00	1.25225E-03
H_2O	−241826	200–1000	−3.94796E+04	5.75573E+02	9.31783E-01	7.22271E-03
		1000–6000	1.03497E+06	−2.41270E+03	4.64611E+00	2.29200E-03
$Mg(OH)_2$	−551996	200–1000	5.24589E+04	−1.28906E+03	1.38933E+01	−7.80669E-04
		1000–6000	1.71371E+06	−4.73001E+03	1.44893E+01	1.90782E-04
$MgO(s)$	−601600	500–3100	−1.16952E+05	0.00000E+00	5.51037E+00	1.08593E-03
$MgO(l)$	−601600	3100–6000	0.00000E+00	0.00000E+00	1.01028E+01	0.00000E+00
NH_3	−45940	200–1000	−7.68123E+04	1.27095 E+03	−3.89323E+00	2.14599E-02
		1000–6000	2.45239E+06	−8.04089E+03	1.27135E+01	−3.98019E-04
N_2	0	200–1000	2.21037E+04	−3.81846E+02	6.08274E+00	−8.53091E-03
		1000–6000	5.87712E+05	−2.23925E+03	6.06695E+00	−6.13969E-04
O	249175	200–1000	−7.95361E+03	1.60718E+02	1.96623E+00	1.01367E-03
		1000–6000	2.61902E+05	−7.29872E+02	3.31718E+00	−4.28133E-04
OH	37278	200–1000	−1.99886E+03	9.30014E+01	3.05085E+00	1.52953E-03
		1000–6000	1.01739E+06	−2.50996E+03	5.11655E+00	1.30530E-04
O_2	0	200–1000	−3.42556E+04	4.84700E+02	1.11901E+00	4.29389E-03
		1000–6000	−1.03794E+06	2.34483E+03	1.81973E+00	1.26785E-03

Table 4.21 (continued)

species	Temperature Range (K)	a_5	a_6	a_7	a_8	a_9
H_2	200–1000	1.75361E-05	−1.20286E-08	3.36809E-12	2.68248E+03	−3.04379E+01
	1000–6000	−3.74072E-07	5.93663E-11	−3.60700E-15	5.33982E+03	−2.20277E+00
H_2O	200–1000	−7.34256E-06	4.95504E-09	−1.33693E-12	−3.30397E+04	1.72421E+01
	1000–6000	−6.83683E-07	9.42647E-11	−4.82238E-15	−1.38429E+04	−7.97815E+00
$Mg(OH)_2$	200–1000	−4.15126E-06	6.10947E-09	−2.27414E-12	−6.29509E+04	−5.01535E+01
	1000–6000	−1.22683E-07	2.01534E-11	−1.12899E-15	−3.88772E+04	−5.84050E+01
$MgO(s)$	500–3100	−4.86619E-07	1.25323E-10	0.00000E+00	−7.44348E+04	−2.91156E+01
$MgO(l)$	3100–6000	0.00000E+00	0.00000E+00	0.00000E+00	−7.60937E+04	−6.07689E+01
NH_3	200–1000	−2.18377E-05	1.31739E-08	−3.33232E-12	−1.26489E+04	4.36601E+01
	1000–6000	3.55250E-08	2.53092E-12	−3.32270E-16	4.38619E+04	−6.46233E+01
N_2	200–1000	1.38465E-05	−9.62579E-09	2.51971E-12	7.10846E+02	−1.07600E+01
	1000–6000	1.49181E-07	−1.92311E-11	1.06195E-15	1.28321E+04	−1.58664E+01
O	200–1000	−1.11042E-06	6.51751E-10	−1.58478E-13	2.84036E+04	8.40424E+00
	1000–6000	1.03610E-07	−9.43830E-12	2.72504E-16	3.39243E+04	−6.67959E-01
OH	200–1000	−3.15789E-06	3.31545E-09	−1.13876E-12	2.99121E+03	4.67411E+00
	1000–6000	−8.28432E-08	2.00648E-11	−1.55699E-15	2.01964E+04	−1.10128E+01
O_2	200–1000	−6.83630E-07	−2.02337E-09	1.03904E-12	−3.39145E+03	1.84970E+01
	1000–6000	−2.18807E-07	2.05372E-11	−8.19347E-16	−1.68901E+04	1.73872E+01

By substitution of the value of R_u in J/(kgmolK), kJ/(kgmolK), or J/(gmolK), $\overline{c}_p^0$, $\overline{h}^0$, and $\overline{s}^0$ will have corresponding dimensions and values. Here we will substitute the value of R_u in kJ/(kgmolK): $R_u = 8.31451$ kJ/(kgmolK). The solution lies in the straightforward substitution of given least square coefficients into applicable equations. Solving using excel format will be easy and less cumbersome. We will consider only the answers here.

Analysis

(1) From Eq. 4.117

$$\overline{c}_p^0 = R_u \left(a_1 T^{-2} + a_2 T^{-1} + a_3 + a_4 T + a_5 T^2 + a_6 T^3 + a_7 T^4\right) \quad (4.117a)$$

Substituting the values in the excel spreadsheet and calculating,

$$\overline{c}^0_{p,298.15\,\mathrm{K}} = 3.713538768\mathrm{E}+01\ \mathrm{KJ/(kgmolK)}$$

$$\overline{c}^0_{p,1000\,\mathrm{K}} = 5.430873296\mathrm{E}+01\ \mathrm{KJ/(kgmolK)}$$

These values match with the values given in standard text books.

(2) From Eq. 4.118

$$\overline{h}^0 = R_u T \left(\begin{array}{c} -a_1 T^{-2} + a_2 T^{-1} \ln T + a_3 + a_4 \dfrac{T}{2} + a_5 \dfrac{T^2}{3} \\ +a_6 \dfrac{T^3}{4} + a_7 \dfrac{T^4}{5} + \dfrac{a_8}{T} \end{array} \right) \quad (4.118a)$$

Substituting the values in the excel spreadsheet and solving

$$\overline{h}^0\,(298.15\,\mathrm{K}) = \overline{h}_f^0 = -3.935100000\mathrm{E}+05\ \mathrm{kJ/kgmol}$$

The molar standard state enthalpy at the standard state temperature of 298.15 K is equal to the enthalpy of formation and our calculated value matches with the value given in the example.

(3) The molar standard state enthalpy at 1000 K is calculated as

$$\overline{h}^0\,(1000\,\mathrm{K}) = -3.601101871\mathrm{E}+05\ \mathrm{kJ/kgmol}$$

The sensible enthalpy change can be calculated.

$$\begin{aligned}\Delta \overline{h}_s\,(1000\,\text{K}) &= \overline{h}^0\,(1000\,\text{K}) - \overline{h}_f^0\\ &= -3.601101871E+05 - (-3.935100000\text{E}+05)\\ &= +3.339981297\text{E}+04\ \text{kJ/kgmol}\end{aligned}$$

Notice that the sensible enthalpy change is positive, indicating the heat has to be supplied to CO_2 to reach the molar standard enthalpy of −3.601101871E+05 kJ/kgmol from the enthalpy of formation of −3.9351E+05 kJ/kgmol.

(4) The molar standard state entropy is given by Eq. 4.119,

$$\overline{s}^0 = R_u \begin{pmatrix} -a_1\dfrac{T^{-2}}{2} - a_2T^{-1} + a_3\ln T + a_4T + a_5\dfrac{T^2}{2} \\ +a_6\dfrac{T^3}{3} + a_7\dfrac{T^4}{4} + a_9 \end{pmatrix} \tag{4.119a}$$

Substituting the values in the excel spreadsheet and solving

$$\overline{s}^0\,(298.15\,\text{K}) = 2.137874007\text{E}+02\ \text{KJ/(kgmolK)}$$

$$\overline{s}^0\,(1000\,\text{K}) = 2.692969332\text{E}+02\ \text{KJ/(kgmolK)}$$

These values match with the values given in standard text books.

Discussion

Remember that the molar standard state properties $\overline{c}_p^0$, $\overline{h}^0$, and $\overline{s}^0$ are pure functions of temperature. $\overline{c}_p^0$ is the principal property (the third key derivative: Sect. 4.5), knowing which the other two properties can be calculated. Molar standard state enthalpy at the reference temperature of 298.15 K is the enthalpy of formation. The molar standard state entropy is the one that we determined, but the molar entropy of a gaseous species j in a mixture is given by

$$\overline{s}_j = \overline{s}_j^0\,(T) - R_u \ln\frac{n_j}{n} - R_u \ln\frac{p}{p_{std}} \tag{4.80a}$$

Notice that $p_{std} = 1$ bar and while calculating $\ln(p/p_{std})$, the units of p & p_{std} should be homogeneous, refer Eq. 4.80.

We have considered here a single species CO_2 and therefore avoided the subscript j which is for CO_2. While calculating the thermodynamic properties of the combustion product mixture, the calculation as above has to be done for all the species of the mixture.

Example 4.9 A heterogeneous (composite) propellant contains ammonium perchlorate (AP) NH_4ClO_4 of the enthalpy of formation −295767 kJ/kgmol, hydroxyl terminated polybutadiene polymer (HTPB) of molecular formula $C_{73}H_{110}O_6$ of the enthalpy of formation −136080 kJ/kgmol, and aluminum in the ratio of AP:HTPB:Al = 0.69:0.12:0.19. The propellant at the standard temperature of 298.15 K burns in a rocket chamber at 70 bar. The adiabatic flame temperature, the entropy, and the gaseous molar mass of the combustion product mixture ($\overline{m}$) are calculated respectively as 3626.02 K, 9.3004 kJ/(kgK), and 29.862 kg/kgmol. Calculate for unit mass, the enthalpy, Gibbs energy, density, and number of kgmoles of gaseous species of the combustion products.

Solution
This example is mainly to refresh our understanding of the thermodynamics of equilibrium composition at the assigned state function $h\&p$. The reactants at the specified temperature have the enthalpies of formation and sum of these is equal to h^* and the combustion products at equilibrium in an adiabatic system should have the enthalpy $h = h^*$. That is, the enthalpy of the combustion products is equal to the enthalpy of the reactants, which can be calculated from the enthalpies of formation of the reactants. Gibbs energy $g = h - Ts$. The other required quantities can be calculated from the ideal gas equation of state.

Analysis
Molar mass of AP, NH_4ClO_4

$$14.01 + 4 \times 1.00797 + 35.45 + 4 \times 16 = 117.492 \ \frac{\text{kg}}{\text{kgmol}}$$

Molar mass of HTPB

$$73 \times 12.01 + 110 \times 1.00797 + 6 \times 16 = 1083.607 \ \frac{\text{kg}}{\text{kgmol}}$$

Atomic mass of aluminum 26.98 $\frac{\text{kg}}{\text{kgatom}}$.

Enthalpies of formation for the propellant ingredients are

$$h^0_{f,AP} = \frac{-295767}{117.492} = -2517.337 \ \frac{\text{kJ}}{\text{kg}}$$

$$h^0_{f,HTPB} = \frac{-136080}{1083.607} = -125.581 \ \frac{\text{kJ}}{\text{kg}}$$

$$h^0_{f,Al} = 0$$

Enthalpy of the propellant at the standard reference temperature of 298.15 K

$$0.69 \times (-2517.337) + 0.12 \times (-125.581) + 0.19 \times 0 = -1752.032\ \frac{\text{kJ}}{\text{kg}}$$

At the assigned pair of state functions, $h\&p$: $h = h^*$ and $p = p^*$. The enthalpy of the reactants is equal to the enthalpy of the products. Therefore,

Enthalpy of the combustion products = −1752.032 kJ/kg

Gibbs energy of the combustion products, $g = h - Ts$. Therefore,

$$g = -1752.032 - 3626.02 \times 9.3004 = -35475.469\ \frac{\text{kJ}}{\text{kg}}$$

$$\rho = \frac{\overline{m}p}{R_u T} = \frac{29.862 \times 70 \times 10^5}{8314.51 \times 3626.02} = 6.9335\ \frac{\text{kg}}{\text{m}^3}$$

Number of kgmoles of gaseous species, $n = 1/\overline{m}$. Therefore,

$$n = \frac{1}{\overline{m}} = \frac{1}{29.862} = 0.03349\ \frac{\text{kgmol}}{\text{kg}}$$

Discussion

This simple example is essentially to refresh our understanding of the thermodynamic properties of combustion at the assigned pair of state functions $h\&p$. Note that at this condition the enthalpy of the reactants is equal to the enthalpy of the equilibrium products. At this equilibrium condition recall again the Gibbs energy assumes its minimum value and the entropy reaches its maximum value, Fig. 4.5. It will be interesting to run the CEA program for assigned state function $T\&p$ with a fixed value of p = 70 bar and different values of temperature below the adiabatic flame temperature, say 3000 K&70 bar, 3200 K&70 bar, 3400 K&70 bar, and finally the adiabatic flame temperature condition 3626.02 K&70 bar. At the adiabatic flame temperature condition, the enthalpy will be reaching from a lower value to the enthalpy of the reactants, Gibbs energy will be attaining its lowest value, and the entropy will be reaching its highest value.

It is worthwhile to study the effect of *trial* temperature that is being considered to arrive at the $h = h^*$ condition through Eqs. 4.67, 4.85, and 4.86 through $\mu_j\ (j = 1, \ldots, NG)$, $\mu_j^0\ (j = NG + 1, \ldots, NS)$ and $\overline{h}_j^0\ (j = 1, \ldots, NS)$. But for the assigned state function $T\&p$, with $\Delta \ln\ T = 0$, the *assigned* temperature T influences the values of $\mu_j\ (j = 1, \ldots, NG)$ and $\mu_j^0\ (j = NG + 1, \ldots, NS)$ through Eqs. 4.67, 4.85, and 4.86. Furthermore note $\pi_i = -\lambda_i/(R_u T)$ having temperature T into π_i's and these are existing in all the equations, Eqs. 4.67, 4.85, and 4.86.

4.7 Performance and Internal-Ballistics Design

Total (or stagnation) and static (or stream) properties are required to analyze the dynamics of fluids in open systems. In the absence of work and potential-energy change, the energy balance in an adiabatic open-system can be written as

$$h_0 = h + \frac{u^2}{2} \tag{4.120}$$

where h_0 and h are the total and static enthalpy respectively, and u is the local velocity. In the previous sections, we have taken into account the dependence of specific heats on temperature. That is $\left(\partial h/\partial T\right)_p = c_p\,(T)$ and $\left(\partial u/\partial T\right)_v = c_v\,(T)$ and hence $\left(c_p/c_v\right) = \gamma\,(T)$. Also we have considered reacting and inert mixtures. While the number of kgmoles of species, gaseous and condensed, is a variable in reacting mixtures, but it is a constant in inert ones. For the analysis of utmost simplicity, if we adopt the assumption that the combustion product mixture is inert ($\overline{m}$ is constant) and its specific heats are constant (γ is constant) as in Chap. 3, $c_p = R\gamma/(\gamma - 1)$ and $u^2 = 2c_p\,(T_0 - T)$. Then, we have

$$\frac{T_0}{T} = 1 + \frac{\gamma - 1}{2} M^2 \tag{4.121}$$

where T_0 and T are the total- and static-temperature respectively, and M is the Mach number. Here the velocity of sound $a = \sqrt{\gamma R T}$. The corresponding total and static pressure are related to the temperatures as

$$\frac{p_0}{p} = \left(\frac{T_0}{T}\right)^{\frac{\gamma}{\gamma - 1}} = \left(1 + \frac{\gamma - 1}{2} M^2\right)^{\frac{\gamma}{\gamma - 1}} \tag{4.122}$$

At low Mach numbers, typically $M \approx 0.3$ and below, the differences between the total and static properties are not that significant and the flow can be assumed to be an incompressible one. At higher Mach numbers, on the other hand, the differences between the total and static properties being significant, the flow becomes compressible.

For all practical purposes, we may assume the combustion chamber and the nozzle of a liquid as well as a solid rocket propulsion system to be open adiabatic ones with zero work transfer and no potential-energy change. However, insofar as the gas dynamics in the combustion chamber is concerned, there are differences in flow behaviors between a liquid rocket engine and a solid rocket motor.

In a typical bi-propellant liquid engine, as the oxidizer and fuel are injected into the combustion chamber and ignited, the burning propellant mixture accelerates from the head end, converting chemical energy into thermal energy. The combustion chamber length is designed such that the combustion is essentially complete at the combustion chamber end, which is the nozzle entry, Fig. 4.6. Qualitative variations in the temperature and pressure in the combustion chamber and nozzle of a liquid rocket engine are depicted in Fig. 4.6. As the burning products accelerate from the head end to the nozzle entry, the mean total temperature increases close to the adiabatic flame temperature and thereafter remains more or less constant. Simultaneously, the total pressure falls due to the *mixing and heat addition losses*. Furthermore, due to the acceleration, the static values of temperature and pressure fall from their total values. In typical designs of liquid engines, the nozzle entry Mach number is around 0.3. However, in advanced designs where the **contraction ratio**, the ratio of the combustion chamber exit area to the nozzle throat area, A_c/A_t is significantly less than 2 the nozzle entry Mach number can be higher and even close to unity as in

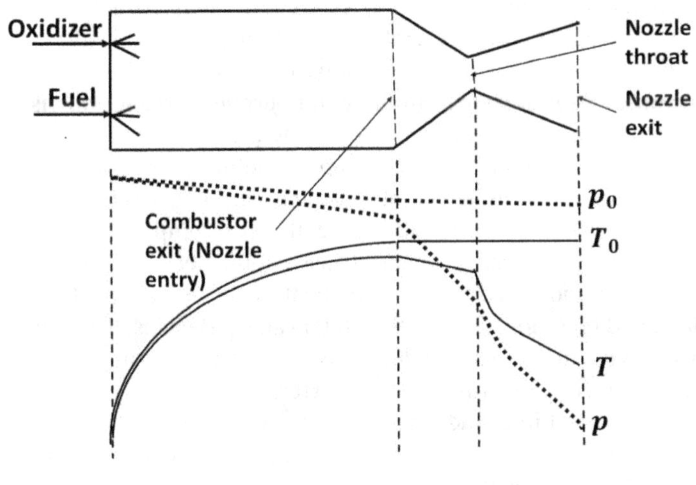

Fig. 4.6 Qualitative variations in temperature and pressure in a liquid rocket engine

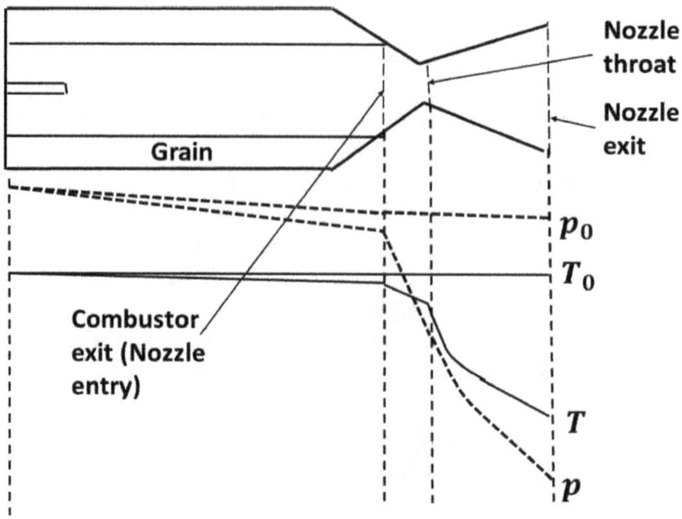

Fig. 4.7 Qualitative variations of temperature and pressure in a solid rocket motor with a tubular grain

tubular combustion chambers, where A_c / A_t tends to the limit of unity. In such cases, the static values there are substantially below the total ones.

The gas dynamics in the combustion chamber of a solid motor is different from that of a liquid engine. In a solid motor, the combustion chamber houses the propellant grain. Let us assume that the grain is of the more frequently adopted tubular configuration, Fig. 4.7. Qualitative variations of the temperature and pressure in the combustion chamber and nozzle of a solid rocket motor with a tubular grain are given in Fig. 4.7. Under typical motor-operating conditions, the combustion, the conversion of chemical energy into thermal energy, is essentially complete well within about one-tenth of a millimeter from the grain burning surface. This is in spite of the fact that some coarse metallic particles in the propellant get expelled from the burning surface and may continue to burn up to one millimeter or at the maximum two millimeters away. Therefore, for all practical purposes, contrary to what exists in the combustion chamber of a liquid engine, the total temperature of the combustion products may be assumed to be constant from the head end to the nozzle exit. Because of the mass addition along the port, the flow accelerates and the static values of temperature and pressure fall from their total values. Besides, the total pressure of the port flow falls due to the *mass addition losses*. In typical designs of solid motors, the nozzle entry Mach number is around 0.3. However, higher Mach numbers at the nozzle entry can prevail in the advanced designs, where the **port–throat ratio**, the ratio of the grain-exit area to nozzle throat area, is appreciably less than 2 or even equal to the limit of unity as in the case of **nozzleless solid motors**.

For both liquid and solid rocket propulsion units, the nozzle flow can be assumed to be adiabatic with constant total temperature but with some loss in total pressure due to friction and species recombination, Figs. 4.6 and 4.7. Because of the high acceleration, the static temperature and pressure fall rapidly. In the case of liquid engines, the combustion products are mostly gaseous but can be laden with carbon particulates if operated with a highly fuel rich propellant combination containing carbon as in the gas generator combustion chambers of the engines operating on gas generator or staged combustion cycle. On the other hand, in solid motors, as the propellants invariably contain metal powders the nozzle flow is heterogeneous with particulates of metal oxides.

4.7.1 Finite Area and Infinite Area Combustion Chambers

We have discussed previously the operational realities in combustion chambers and nozzles of liquid rocket engines and solid rocket motors. In Sect. 3.7 the characteristics of frozen flow and equilibrium flow are considered. We have to now idealize the operations in these systems so that we will be able to predict the propulsion performance without much complications but within the permissible deviation from the reality.

Let us first consider the equilibrium composition for the combustion chamber. In typical designs, as the velocities in the combustion chamber are generally low, the equilibrium composition can be assumed to be the one corresponding to the total conditions. Even under the low velocities, as there is a total pressure loss from head end to the nozzle entry, a suitable mean value of the total pressure may be assumed for the calculations. Such an assumption to calculate the equilibrium composition with respect to the total pressure is known as the assumption of **infinite area combustion chamber (IAC)**, Fig. 4.8. However, as discussed previously, in advanced designs of liquid engines and solid motors, the nozzle entry Mach number could be appreciably higher than 0.3 and even close to unity, and hence static values there are substantially below the total ones. In addition, in such cases, the total pressure at the nozzle entry is significantly lower than that at the head end. Under these circumstances, we have to calculate the equilibrium composition with respect to the static properties at the nozzle entry. Such an assumption to calculate the equilibrium composition with respect to the nozzle entry static properties is known as the assumption of **finite area combustion chamber (FAC)**. Figure 4.8 presents schematic cross sections of IAC and FAC of rocket engines. Our discussion regarding the rocket performance calculation is summarized in Fig. 4.9.

4.7.2 Internal-Ballistics Design

In Chap. 3 under Sect. 3.6 we derived the characteristic parameter equations under the simplifying assumption that c_p and $\overline{m}$ are constant. And under Sect. 3.7, as an improvement, we considered briefly how such equations of utmost simplicity get

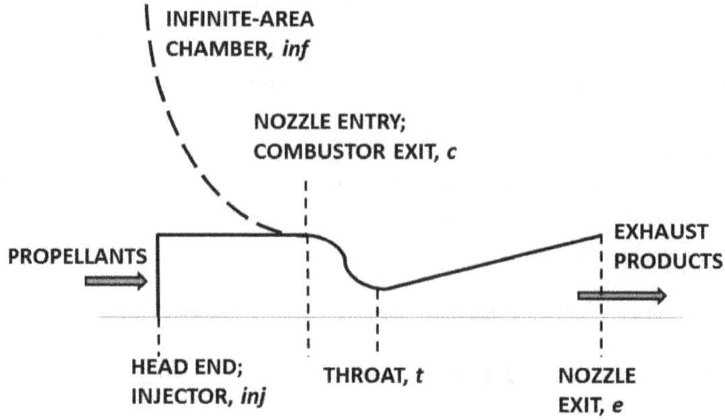

Fig. 4.8 Schematics of rocket combustion chamber cross section. (Adopted from [6])

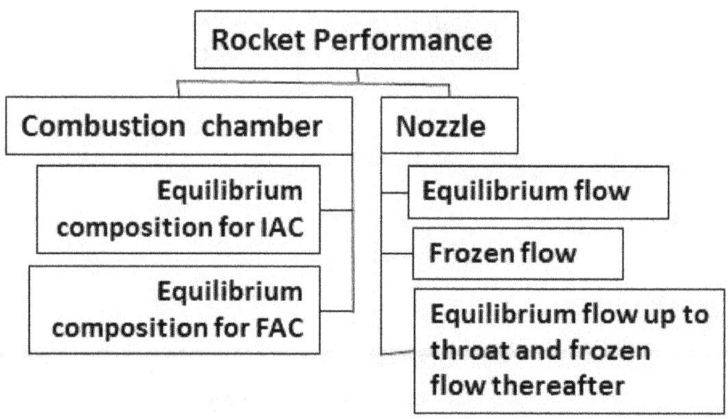

Fig. 4.9 Summary on rocket performance calculations

modified by the rigor of c_p and $\overline{m}$ being variant. In Sect. 4.5 we considered the calculation of the key derivatives $(\partial h/\partial T)_p \equiv c_p$, $(\partial v/\partial T)_p$, and $(\partial v/\partial p)_T$, which are fundamentals to calculate the isentropic exponent γ_s, the ratio of specific heats γ, and the velocity of sound a of reacting compositions. In the present section for the purpose of continuity and ready reference, we list the thrust equations and the characteristic parameter equations with variable c_p and $\overline{m}$. Also briefly comment on how these equations of variable c_p and $\overline{m}$ become the ones of constant c_p and $\overline{m}$. In order to keep the present chapter reasonably independent of the previous one we are renumbering here the previously derived equations of Chap. 3.

From Eq. 4.120, the **nozzle-exit velocity** can be written as

$$u_e = \sqrt{2\,(h_0 - h_e)} \tag{4.123}$$

By Newton's Second law, and considering the components of force and momentum rate only in the flight direction, we get the **thrust equation** for a rocket propulsion system as

$$F = \dot{m}u_e + (p_e - p_a)\,A_e \tag{4.124}$$

where $\dot{m}$ is the choked mass flow rate through the nozzle, p_e is the nozzle-exit plane static pressure, A_e the nozzle-exit plane area, and p_a is the ambient pressure. While $\dot{m}u_e$ is known as **momentum thrust**, $(p_e - p_a)\,A_e$ is known as **pressure thrust**. If the nozzle-exit plane pressure p_e is equal to the ambient pressure p_a, that operating condition is known as **adapted operation** and the thrust equation simplifies to,

$$F^0 = \dot{m}u_e \tag{4.125}$$

Under vacuum operating condition, the thrust in vacuum becomes

$$F_{vac} = \dot{m}u_e + p_e A_e \tag{4.126}$$

We have learned that the performance of rocket engines are characterized by three parameters: characteristic velocity c^*, thrust coefficient C_F, and specific impulse I_{sp}. For assessing the experimental performance of rocket engines, the respective ratios of the experimental characteristic parameters and the theoretical ones are evaluated. These ratios are known as quality factors. The quality factor for c^* namely ξ_c is also known as c^* efficiency, η_{c^*}. The quality factor for C_F is ξ_n. And, the quality factor for specific impulse, namely ξ_s, is sometimes referred to as velocity correction factor.

Characteristic velocity is defined as,

$$c^* \equiv \frac{p_{0n} A_t}{\dot{m}} \tag{4.127}$$

Noting choked-throat mass flux $\dot{m}/A_t = \rho_t a_t = p_t \gamma_{st}/a_t$,

$$c^* = \frac{p_{0n} a_t}{p_t \gamma_{st}} \tag{4.128}$$

where γ_{st} is the isentropic exponent at the throat, Eq. 4.114. We see from this expression that c^* depends on the pressure ratio for the isentropic flow (shifting-equilibrium or frozen flow) to reach sonic speed at the throat and the isentropic exponent there. Thus the calculated value of c^* depends on whether the flow is assumed to be one of shifting-equilibrium or frozen. Furthermore, if we adopt the assumption that the combustion product mixture is inert and its specific heats are constant, c^* can be shown to be.

$$c^* = \frac{\sqrt{RT_0}}{\Gamma} \tag{4.129}$$

where

$$\Gamma = \sqrt{\gamma}\left(\frac{2}{\gamma+1}\right)^{\frac{\gamma+1}{2(\gamma-1)}}$$

Thus, c^* is understood to represent the combustion characteristics of the propellant. With this we define the quality factor ξ_c or the c^* **efficiency** as

$$\xi_c = \eta_{c^*} = \frac{c^*_{expt}}{c^*_{theo}} = \frac{p_{0n} A_t / \dot{m}}{p_{0n} a_t / (p_t \gamma_{st})} \tag{4.130}$$

While the numerator can be experimentally measured, the denominator can be theoretically calculated.

The **thrust coefficient** is defined as,

$$C_F \equiv \frac{F}{p_{0n} A_t} \tag{4.131}$$

Depending on the thrusting mode, the thrust coefficient is termed as thrust coefficient C_F for general mode, **characteristic thrust coefficient** C_F^0 for adapted operation, and **vacuum thrust coefficient** $C_{F,vac}$ for operation in vacuum. From Eq. 4.124 and noting that the throat mass flux $\dot{m}/A_t = \rho_t a_t = p_t \gamma_{st}/a_t$, the thrust coefficient can be written as

$$C_F = \frac{\gamma_{st}\sqrt{2(h_0 - h_e)}}{a_t (p_{0n}/p_t)} + \frac{A_e}{A_t}\left(\frac{p_e}{p_{0n}} - \frac{p_a}{p_{0n}}\right) \tag{4.132}$$

Note that the quantities in the right-hand side of the equation are the calculable ones (γ_{st}, a_t, h_e, and A_e/A_t) for the specified parameters (h_0, p_{0n}, p_{0n}/p_e, and p_a). Furthermore, as the expression for C_F involves quantities that characterize the expansion in the rocket nozzle, the comparison of the experimental value of C_F by Eq. 4.131 with the theoretical one by Eq. 4.132 is enabled and the ratio is known as **nozzle-flow quality factor** or **nozzle-flow efficiency**, ξ_n.

$$\xi_n = \frac{F/(p_{0n} A_t)}{\dfrac{\gamma_{st}\sqrt{2(h_0 - h_e)}}{a_t (p_{0n}/p_t)} + \dfrac{A_e}{A_t}\left(\dfrac{p_e}{p_{0n}} - \dfrac{p_a}{p_{0n}}\right)} \tag{4.133}$$

For the assumption of the simplest situation (that the combustion product mixture is inert and its specific heats are constant), Eq. 4.132 becomes

$$C_F = \Gamma \sqrt{\frac{2\gamma}{\gamma - 1}\left[1 - \left(\frac{p_e}{p_{0n}}\right)^{\frac{\gamma-1}{\gamma}}\right]} + \frac{A_e}{A_t}\left(\frac{p_e}{p_{0n}} - \frac{p_a}{p_{0n}}\right) \tag{4.134}$$

For each of the other two modes of thrusting, Eqs. 4.132 and 4.134 become respectively,

$$C_F^0 = \frac{\gamma_{st}\sqrt{2\left(h_0 - h_e\right)}}{a_t\left(p_{0n}/p_t\right)} \tag{4.135}$$

$$C_F^0 = \Gamma \sqrt{\frac{2\gamma}{\gamma - 1}\left[1 - \left(\frac{p_e}{p_{0n}}\right)^{\frac{\gamma-1}{\gamma}}\right]} \tag{4.136}$$

and,

$$C_{F,vac} = \frac{\gamma_{st}\sqrt{2\left(h_0 - h_e\right)}}{a_t\left(p_{0n}/p_t\right)} + \frac{A_e}{A_t}\frac{p_e}{p_{0n}} \tag{4.137}$$

$$C_{F,vac} = \Gamma \sqrt{\frac{2\gamma}{\gamma - 1}\left[1 - \left(\frac{p_e}{p_{0n}}\right)^{\frac{\gamma-1}{\gamma}}\right]} + \frac{A_e}{A_t}\frac{p_e}{p_{0n}} \tag{4.138}$$

Specific impulse is defined as the thrust per unit mass flow rate of propellant. Therefore, from Eq. 4.124,

$$I_{sp} \equiv \frac{F}{\dot{m}} = u_e + \frac{\left(p_e - p_a\right)A_e}{\dot{m}} = \overline{u}_e \tag{4.139}$$

where $\overline{u}_e$ is known as the **equivalent nozzle-exit velocity** or simply **equivalent exhaust velocity**. Considering again, the other two thrusting modes Eqs. 4.125 and 4.126,

$$I_{sp}^0 = \frac{F^0}{\dot{m}} = u_e \tag{4.140}$$

$$I_{sp,vac} = \frac{F_{vac}}{\dot{m}} = u_e + \frac{p_e A_e}{\dot{m}} = \overline{u}_{e,vac} \tag{4.141}$$

Since $u_e = \sqrt{2\,(h_0 - h_e)}$, these equations for the three modes of specific impulse can be written in terms of the calculable quantities for the specified parameters.

$$I_{sp} = \sqrt{2\,(h_0 - h_e)} + \frac{A_e}{A_t}\frac{p_{0n}}{p_t}\frac{a_t}{\gamma_{st}}\left(\frac{p_e}{p_{0n}} - \frac{p_a}{p_{0n}}\right) \tag{4.142}$$

Equation 4.142 can be written in a more "revealing" form as

$$I_{sp} = \left[\frac{\gamma_{st}\sqrt{2\,(h_0 - h_e)}}{a_t\,(p_{0n}/p_t)} + \frac{A_e}{A_t}\left(\frac{p_e}{p_{0n}} - \frac{p_a}{p_{0n}}\right)\right]\frac{(p_{0n}/p_t)\,a_t}{\gamma_{st}} \tag{4.143}$$

Therefore from Eqs. 4.128 and 4.132,

$$I_{sp} = C_F c^* \tag{4.144}$$

$$I_{sp}^0 = \left[\frac{\gamma_{st}\sqrt{2\,(h_0 - h_e)}}{a_t\,(p_{0n}/p_t)}\right]\frac{(p_{0n}/p_t)\,a_t}{\gamma_{st}} = C_F^0 c^* = u_e \tag{4.145}$$

$$I_{sp,vac} = \left[\frac{\gamma_{st}\sqrt{2\,(h_0 - h_e)}}{a_t\,(p_{0n}/p_t)} + \frac{A_e}{A_t}\frac{p_e}{p_{0n}}\right]\frac{(p_{0n}/p_t)\,a_t}{\gamma_{st}} = C_{F,vac}c^* = \overline{u}_{e,vac} \tag{4.146}$$

Thus, we see the performance parameter I_{sp} represents the performance of the entire propulsion system, that has in it the two major subsystems, combustion chamber and nozzle. The quality factor for specific impulse, namely, ξ_s or the velocity correction factor is the ratio of experimental- to theoretical-value of specific impulse.

$$\xi_s = \frac{I_{sp,\mathrm{exp}t}}{I_{sp,theo}} = \frac{C_{F,\mathrm{exp}t}\,c^*_{\mathrm{exp}t}}{C_{F,theo}\,c^*_{theo}} = \xi_n \xi_c \tag{4.147}$$

As it is possible to separately arrive at the values of ξ_n and ξ_c from the respective experimental results and the theoretical calculations, one can easily identify the subsystem (combustion chamber or nozzle) that needs attention for the performance improvement. The attainment of acceptable combustion efficiency is more difficult in liquid propellant engines as it depends on the atomization, mixing, ignition of propellants, and the completion of combustion within the allowable chamber envelope.

Therefore, during the development of hardware, the role of ξ_c estimation is central to the design and development of liquid engines. Nozzle-exit velocity vectors not being one dimensional but being essentially axi-symmetric (two dimensional), the issues due to two-phase nozzle flows, and in addition the losses caused by surface friction, boundary layer, and heat transfer are reflected in ξ_n.

The following two Examples 4.10 and 4.11 are in a way form the application of what all we have learned in the present chapter and show how the first-cut internal ballistics design is carried out with a reasonable rigor. The examples are lengthy. Nevertheless one should take time to carefully study these and appreciate the design steps involved.

Example 4.10 The Space Shuttle Main Engine (SSME) uses liquid hydrogen and liquid oxygen at an oxidizer–fuel ratio of 6. Combustion chamber pressure is 204 bar. It delivers a thrust of 2090 kN in vacuum with the nozzle area ratio of 77.5 [10].

Let us consider a similar engine in the present example. For the propellant combination of LOX(90.17 K)/LH2(20.27 K) at an oxidizer–fuel ratio of 6 and a chamber pressure of 204 bar, for the assigned state function $h\&p$, combustion chamber conditions are calculated. For the simplicity of calculations, only H, H_2, H_2O, and OH are considered as the significant species. Calculated combustion properties in the chamber are the following. $T_0 = 3619.5$ K. $\overline{m} = 13.645$ kg/kgmol. Enthalpy $h_0 = -986.31$ kJ/kg. Entropy $s = 17.1734$ kJ/(kgK). Isentropic exponent $\gamma_s = 1.1502$. Species mole fractions are $x_H = 0.02650$; $x_{H_2} = 0.24250$; $x_{H_2O} = 0.69140$; and $x_{OH} = 0.03960$.

Assuming a nozzle pressure ratio of 1127, calculate the nozzle throat conditions and the nozzle-exit conditions under shifting-equilibrium flow assumption. Also, evaluate the rocket performance parameters.

For a vacuum thrust of 2090 kN, size the nozzle. Assume a nozzle entry Mach number of 0.1.

Solution

The propellant combination has two elements, hence $l = 2$. There are four significant gaseous species to be considered, hence $NG = 4$. There is no condensed species, hence $NS = NG$.

For the specified oxidizer–fuel ratio of 6, we have to get the molar form of the propellant combination for unit mass of the propellant. Calculate thereafter the values of b_H^o and b_O^o.

At a chosen temperature, the standard state-specific heat at constant pressure, standard state enthalpy, and the standard state entropy for the considered species can be calculated using the polynomials as described previously. The corresponding polynomial coefficients are given in Tables 4.20 and 4.21.

Choked throat pressure is the one for which the velocity of flow u is equal to the velocity of sound a and this has to be obtained by iteration. The initial estimate for the throat pressure and temperature is obtained by assuming the isentropic exponent γ_s to be constant as at the nozzle entry condition.

$$p_t = \frac{p_{0n}}{\left(\dfrac{\gamma_{sn}+1}{2}\right)^{\gamma_{sn}/(\gamma_{sn}-1)}}$$

$$T_t = \frac{T_0}{\left(\dfrac{\gamma_{sn}+1}{2}\right)}$$

For the assigned entropy s and this trial pressure p_t, we have the first trial pair of state functions $s\&p$, for which we have to obtain by iteration the equilibrium composition. For the obtained equilibrium composition and temperature, using the temperature- and pressure-derivative equations, the three key derivatives are evaluated to calculate the flow velocity and the velocity of sound. For the choked condition, these two velocities have to be equal within the allowable error margin. If not, a new trial pressure for the throat has to be obtained from the iteration formula [6],

$$p_{t,k+1} = \left(p\frac{1+\gamma_s M^2}{1+\gamma_s}\right)_{t,k}$$

On convergence (flow velocity = velocity of sound), the properties of the choked flow are calculated. These properties enable us to calculate characteristic velocity c^*, Eq. 4.128.

For the nozzle-exit condition of the pressure ratio of 1127, the exit plane static pressure is 0.181 bar. Assuming that the isentropic exponent evaluated at the throat $\gamma_{s,t}$ to be constant, the trial temperature at the nozzle-exit plane is calculated under isentropic expansion to 0.181 bar. Starting with that trial temperature, this forms the pair of state functions $s\&p$ at the assigned entropy of 17.1734 kJ/(kgK) and the specified exit plane pressure of 0.181 bar. The shifting-equilibrium calculations are to be carried out. Based on the converged equilibrium composition and the temperature at the nozzle exit, using the temperature- and pressure-derivative equations, the three key derivatives are evaluated to calculate rocket performance parameters $C_{F,vac}$ and $I_{sp,vac}$, Eqs. 4.137 and 4.146.

For the specified vacuum thrust of 2090 kN, using the calculated $I_{sp,vac}$ mass flow rate can be calculated, Eq. 4.141. For this mass flow rate, using the evaluated properties at the throat and at the nozzle exit the respective nozzle-areas are calculated. For the specified nozzle entry Mach number, the nozzle-inlet area (combustion chamber exit area) can be calculated.

Analysis

$$H_2 + xO_2 \rightarrow products\ of\ H, H_2, H_2O, \&OH$$

For the oxidizer–fuel ratio of 6,

$$x = \frac{6 \times 2.01588}{31.9988} = 0.37799$$

Therefore,

$$H_2 + 0.37799O_2 \rightarrow products\ of\ H, H_2, H_2O, \&OH$$

This molar form of the propellant combination has a mass of 14.1112 kg. For unit mass of the propellant combination, the molar form of the propellant combination becomes

$$7.08659E-02H_2 + 2.67867E-02O_2$$
$$\rightarrow n_H H + n_{H_2} H_2 + n_{H_2O} H_2O + n_{OH} OH$$

Hence for the unit mass of the propellant combination,

$$b_H^o = 1.41732E-01\ \text{kgatom/kg}$$
$$b_O^o = 5.35734E-02\ \text{kgatom/kg}$$

For the considered species, the a_{ij} matrix is as per Table 4.22,

Nozzle-Throat Calculations To get the initial estimates for the nozzle throat, let us assume that the flow accelerates with the isentropic exponent as at the combustion chamber end, $\gamma_s = 1.1502$. Also assume for the trial that the molar mass and mole fractions of the species as at the combustion chamber end to be same at the throat:

$$\overline{m} = 13.645\ \text{kg/kgmol}$$
$$x_H = 0.02650$$
$$x_{H_2} = 0.24250$$
$$x_{H_2O} = 0.69140$$
$$x_{OH} = 0.03960$$

Noting $n_j = x_j/\overline{m}$,

Table 4.22 a_{ij} matrix: Example 4.10

Atom	n_H	n_{H_2}	n_{H_2O}	n_{OH}
H	$a_{11}=1$	$a_{12}=2$	$a_{13}=2$	$a_{14}=1$
O	$a_{21}=0$	$a_{22}=0$	$a_{23}=1$	$a_{24}=1$

$$\begin{aligned} n_H &= 1.94210E-03 \text{ kgmol/kg} \\ n_{H_2} &= 1.77721E-02 \text{ kgmol/kg} \\ n_{H_2O} &= 5.06706E-02 \text{ kgmol/kg} \\ n_{OH} &= 2.90216E-03 \text{ kgmol/kg} \end{aligned}$$

For these trial n_j's,

$$\begin{aligned} b_H &= 1.94210E-03+2\times 1.77721E-02+2\times 5.06706E-02 \\ &+2.90216E-03 = 0.141730 \text{ kgatom/kg} \end{aligned}$$

$$b_O = 5.06706E-02+2.90216E-03 = 0.0535727 \text{ kgatom/kg}$$

This gives,

$$b_H^o - b_H = 2.22115E-06; \quad b_O^o - b_O = 7.00309E-07$$

For the chosen temperature, the standard state-specific heat at constant pressure, standard state enthalpy, and the standard state entropy for the considered species can be calculated using the polynomials. The corresponding polynomial coefficients are given in Tables 4.20 and 4.21.

Recall,

$$\begin{aligned} \mu_j &= (\overline{g}_j)_{T,p} = \overline{h}_j^0(T) - T\overline{s}_j \\ &= \overline{h}_j^0(T) - T\left[\overline{s}_j^0(T) - R_u\left(\ln\frac{n_j}{n} + \ln p\right)\right] \to (j=1,\ldots,NG)\ p = [\text{bar}] \end{aligned}$$

$$\overline{s}_j = \overline{s}_j^0(T) - R_u\left(\ln\frac{n_j}{n} + \ln\ p\right) \to (j=1,\ldots,NG) \quad p = [\text{bar}]$$

Adopting Eqs. 4.85, 4.86, and 4.88, the four *reduced Gibbs correction equations* are,

$$
\begin{aligned}
&\left(n_H + 4n_{H_2} + 4n_{H_2O} + n_{OH}\right)\pi_1 + \left(2n_{H_2O} + n_{OH}\right)\pi_2 \\
&+ \left(n_H + 2n_{H_2} + 2n_{H_2O} + n_{OH}\right)\Delta \ln\ n \\
&+ \frac{\left(n_H\overline{h}_H^0 + 2n_{H_2}\overline{h}_{H_2}^0 + 2n_{H_2O}\overline{h}_{H_2O}^0 + n_{OH}\overline{h}_{OH}^0\right)}{R_uT}\Delta \ln\ T \\
&= \left(b_H^o - b_H\right) + \frac{\left(n_H\mu_H + 2n_{H_2}\mu_{H_2} + 2n_{H_2O}\mu_{H_2O} + n_{OH}\mu_{OH}\right)}{R_uT}
\end{aligned}
$$

$$
\begin{aligned}
&\left(2n_{H_2O} + n_{OH}\right)\pi_1 + \left(n_{H_2O} + n_{OH}\right)\pi_2 \\
&+ \left(n_{H_2O} + n_{OH}\right)\Delta \ln\ n \\
&+ \frac{\left(n_{H_2O}\overline{h}_{H_2O}^0 + n_{OH}\overline{h}_{OH}^0\right)}{R_uT}\Delta \ln\ T \\
&= \left(b_O^o - b_O\right) + \frac{\left(n_{H_2O}\mu_{H_2O} + n_{OH}\mu_{OH}\right)}{R_uT}
\end{aligned}
$$

$$
\begin{aligned}
&\left(n_H + 2n_{H_2} + 2n_{H_2O} + n_{OH}\right)\pi_1 + \left(n_{H_2O} + n_{OH}\right)\pi_2 \\
&+ \left(n_H + n_{H_2} + n_{H_2O} + n_{OH} - n\right)\Delta \ln\ n \\
&+ \frac{\left(n_H\overline{h}_H^0 + n_{H_2}\overline{h}_{H_2}^0 + n_{H_2O}\overline{h}_{H_2O}^0 + n_{OH}\overline{h}_{OH}^0\right)}{R_uT}\Delta \ln\ T \\
&= n - \left(n_H + n_{H_2} + n_{H_2O} + n_{OH}\right) \\
&+ \frac{\left(n_H\mu_H + n_{H_2}\mu_{H_2} + n_{H_2O}\mu_{H_2O} + n_{OH}\mu_{OH}\right)}{R_uT}
\end{aligned}
$$

$$
\begin{aligned}
&\frac{\left(n_H\overline{s}_H + 2n_{H_2}\overline{s}_{H_2} + 2n_{H_2O}\overline{s}_{H_2O} + n_{OH}\overline{s}_{OH}\right)}{R_u}\pi_1 \\
&+ \frac{\left(n_{H_2O}\overline{s}_{H_2O} + n_{OH}\overline{s}_{OH}\right)}{R_u}\pi_2 \\
&+ \frac{\left(n_H\overline{s}_H + n_{H_2}\overline{s}_{H_2} + n_{H_2O}\overline{s}_{H_2O} + n_{OH}\overline{s}_{OH}\right)}{R_u}\Delta \ln\ n \\
&+ \left[\begin{aligned}&\frac{\left(n_H\overline{c}_{p,H}^0 + n_{H_2}\overline{c}_{p,H_2}^0 + n_{H_2O}\overline{c}_{p,H_2O}^0 + n_{OH}\overline{c}_{p,OH}^0\right)}{R_u} \\ &+ \frac{\left(n_H\overline{h}_H^0\overline{s}_H + n_{H_2}\overline{h}_{H_2}^0\overline{s}_{H_2} + n_{H_2O}\overline{h}_{H_2O}^0\overline{s}_{H_2O} + n_{OH}\overline{h}_{OH}^0\overline{s}_{OH}\right)}{R_u^2T}\end{aligned}\right]\Delta \ln\ T \\
&= \frac{s^* - s}{R_u} + n - \left(n_H + n_{H_2} + n_{H_2O} + n_{OH}\right) \\
&+ \frac{\left(n_H\overline{s}_H\mu_H + n_{H_2}\overline{s}_{H_2}\mu_{H_2} + n_{H_2O}\overline{s}_{H_2O}\mu_{H_2O} + n_{OH}\overline{s}_{OH}\mu_{OH}\right)}{R_u^2T}
\end{aligned}
$$

Table 4.23 Properties of species for the first trial throat conditions of 117.165 bar and 3366 K: Example 4.10

Species	n_j (kgmol/kg)	$\bar{c}^0_{p,j}$ (kJ/(kgmolK))	$\bar{h}^0_j$ (kJ/kgmol)	$\bar{s}_j$ (kJ/(kgmolK))	μ_j (kJ/kgmol)
H	1.94210E-03	20.7863	281768	155.681	−242255
H_2	1.77721E-02	37.8644	102448	179.374	−501325
H_2O	5.06706E-02	57.9134	−93163	257.060	−958428
OH	2.90216E-03	37.6377	140744	248.458	−695564

Table 4.24 Matrix numbers for the four simultaneous correction-equation for the first trial values: Example 4.10

π_1	π_2	$\Delta ln\ n$	$\Delta ln\ T$	RHS
4.02308	1.65331E+00	2.06548E+00	−2.55728E+00	−62.9949
0.278615	1.04243E-01	1.41730E-01	−1.73086E-01	−4.19616
0.141730	5.35727E-02	0.00000E+00	−6.94691E-02	−2.14255
0.104243	5.35727E-02	5.35727E-02	−1.54078E-01	−1.80739

Note $s^* = 17.1734$ kJ/(kgK). Choked throat pressure is the one for which the velocity of flow u is equal to the velocity of sound a and this has to be obtained by iteration. The initial estimate for the throat pressure and throat temperature are obtained assuming that the isentropic exponent γ_s is constant as at the nozzle entry condition.

$$p_t = \frac{204}{(2.1502/2)^{1.1502/0.1502}} = 117.165 \text{ bar}$$

$$T_t = \frac{3619.5}{(2.1502/2)} = 3366 \text{ K}$$

For the first trial pressure of 117.165 bar and the assigned entropy of 17.1734 kJ/(kgK), adopting the polynomials and their coefficients, the calculated values at the estimated throat temperature of 3366 K are given in Table 4.23.
With these values the calculated matrix-numbers corresponding to the four *reduced Gibbs iteration equations* are given in Table 4.24.

Adopting the Gaussian elimination with partial pivoting, the matrix equation after forward elimination is

$$\begin{bmatrix} +4.02308E+00 & +1.65331E+00 & +2.06548E+00 & -2.55728E+00 \\ +0.00000E+00 & +1.07333E-02 & +5.33021E-05 & -8.78158E-02 \\ +0.00000E+00 & +0.00000E+00 & -7.27420E-02 & -1.76028E-02 \\ +0.00000E+00 & +0.00000E+00 & +0.00000E+00 & -7.95838E-02 \end{bmatrix} \times \begin{bmatrix} \pi_1 \\ \pi_2 \\ \Delta \ln n \\ \Delta \ln T \end{bmatrix} = \begin{bmatrix} -6.29949E+01 \\ -1.75105E-01 \\ +4.87106E-04 \\ -8.14209E-04 \end{bmatrix}$$

By back substitution, the solutions are,

$$\begin{aligned} \pi_1 &= -8.97716E+00 \\ \pi_2 &= -1.62305E+01 \\ \Delta \ln n &= -9.17211E-03 \\ \Delta \ln T &= 1.02308E-02 \end{aligned}$$

Using Eq. 4.66,

$$\Delta \ln n_H = \pi_1 + \Delta \ln n + \frac{\overline{h}_H^0}{R_u T} \Delta \ln T - \frac{\mu_H}{R_u T}$$

$$\Delta \ln n_{H_2} = 2\pi_1 + \Delta \ln n + \frac{\overline{h}_{H_2}^0}{R_u T} \Delta \ln T - \frac{\mu_{H_2}}{R_u T}$$

$$\Delta \ln n_{H_2O} = 2\pi_1 + \pi_2 + \Delta \ln n + \frac{\overline{h}_{H_2O}^0}{R_u T} \Delta \ln T - \frac{\mu_{H_2O}}{R_u T}$$

$$\Delta \ln n_{OH} = \pi_1 + \pi_2 + \Delta \ln n + \frac{\overline{h}_{OH}^0}{R_u T} \Delta \ln T - \frac{\mu_{OH}}{R_u T}$$

Substituting the solutions and noting $\Delta n_j = n_j \Delta \ln n_j$,

$$\begin{aligned} n_H &= 1.50080E-03 \text{ kgmol/kg}; & x_H &= 2.06680E-02 \\ n_{H_2} &= 1.75405E-02 \text{ kgmol/kg}; & x_{H_2} &= 2.41556E-01 \\ n_{H_2O} &= 5.15766E-02 \text{ kgmol/kg}; & x_{H_2O} &= 7.10277E-01 \\ n_{OH} &= 1.99685E-03 \text{ kgmol/kg}; & x_{OH} &= 2.74993E-02 \end{aligned}$$

The corresponding molar mass $\overline{m} = 1.37713E + 01$ kg/kgmol. With the solved value of $\Delta \ln\ T = 1.02308E - 02$, the new temperature T is calculated as 3400.44 K with reference to the previous trial temperature of 3366 K.

For the trial throat pressure of 117.165 bar, as the calculated values do not satisfy the convergence criteria of $\left|\delta n_j\right| \Big/ \sum_{j=1}^{NS} n_j \leq 0.5 \times 10^{-5}$, the calculated results of the current iteration become the inputs to the next iteration. After the next three iterations the converged values for the trial throat pressure of 117.165 bar are the following.

$p_t = 117.165$ bar $(assumed\ trial\ pressure)$
$s^* = 17.1734$ kJ/(kgK) $(assigned\ entropy)$
$\overline{m} = 13.7608$ kg/kgmol
$T_t = 3394.71$ K
$n_H = 1.52669E - 03$ kgmol/kg; $n_{H_2} = 1.75700E - 02$ kgmol/kg
$n_{H_2O} = 5.14917E - 02$ kgmol/kg; $n_{OH} = 2.08174E - 03$ kgmol/kg
$n = 7.26701E - 02$ kgmol/kg

$\overline{c}^0_{p,H} = 2.07863E + 01$ kJ/(kgmolK)

$\overline{c}^0_{p,H_2} = 3.79230E + 01$ kJ/(kgmolK)

$\overline{c}^0_{p,H_2O} = 5.79883E + 01$ kJ/(kgmolK)

$\overline{c}^0_{p,OH} = 3.76818E + 01$ kJ/(kgmolK)

$\overline{h}^0_H = 2.82365E + 05$ kJ/kgmol; $\overline{h}^0_{H_2} = 1.03536E + 05$ kJ/kgmol

$\overline{h}^0_{H_2O} = -9.14988E + 04$ kJ/kgmol; $\overline{h}^0_{OH} = 1.41826E + 05$ kJ/kgmol

$h = -2165.99$ kJ/kg.

From this solution set of the first trial value of the throat pressure 117.165 bar we have to check for the equality of local velocity with the local sonic velocity. This requires the evaluation of temperature- and pressure-derivatives. With reference to the temperature-derivative equations Eqs. 4.102 and 4.103,

$$\left(n_H + 4n_{H_2} + 4n_{H_2O} + n_{OH}\right)\left(\frac{\partial \pi_1}{\partial \ln T}\right)_p + \left(2n_{H_2O} + n_{OH}\right)\left(\frac{\partial \pi_2}{\partial \ln T}\right)_p$$
$$+\left(n_H + 2n_{H_2} + 2n_{H_2O} + n_{OH}\right)\left(\frac{\partial \ln n}{\partial \ln T}\right)_p$$
$$= -\left(n_H \overline{h}^0_H + 2n_{H_2}\overline{h}^0_{H_2} + 2n_{H_2O}\overline{h}^0_{H_2O} + n_{OH}\overline{h}^0_{OH}\right)\Big/(R_u T)$$

$$\left(2n_{H_2O} + n_{OH}\right)\left(\frac{\partial \pi_1}{\partial \ln T}\right)_p + \left(n_{H_2O} + n_{OH}\right)\left(\frac{\partial \pi_2}{\partial \ln T}\right)_p$$
$$+\left(n_{H_2O} + n_{OH}\right)\left(\frac{\partial \ln n}{\partial \ln T}\right)_p = -\left(n_{H_2O}\overline{h}^0_{H_2O} + n_{OH}\overline{h}^0_{OH}\right)\Big/(R_u T)$$

$$\left(n_H + 2n_{H_2} + 2n_{H_2O} + n_{OH}\right)\left(\frac{\partial \pi_1}{\partial \ln T}\right)_p + \left(n_{H_2O} + n_{OH}\right)\left(\frac{\partial \pi_2}{\partial \ln T}\right)_p$$
$$= -\left(n_H \overline{h}^0_H + n_{H_2}\overline{h}^0_{H_2} + n_{H_2O}\overline{h}^0_{H_2O} + n_{OH}\overline{h}^0_{OH}\right)\Big/(R_u T)$$

Substituting the values of n_j's and $\overline{h}^0_j$'s of the converged temperature of 3394.71 K, the calculated matrix-numbers corresponding to the three simultaneous temperature-derivative equations are given in Table 4.25.

Adopting Gaussian elimination with partial pivoting and by back substitution, the solutions are

$$\left(\frac{\partial \pi_1}{\partial \ln T}\right)_p = -1.84217E+00; \quad \left(\frac{\partial \pi_2}{\partial \ln T}\right)_p = 6.30598E+00$$
$$\left(\frac{\partial \ln n}{\partial \ln T}\right)_p = 2.27283E-01$$

Substituting the value of $\left(\partial \ln n/\partial \ln T\right)_p$ in Eq. 4.90 we get the value of the first key derivative,

$$\left(\frac{\partial \ln v}{\partial \ln T}\right)_p = 1 + 2.27283E-01 = 1.227283$$

Table 4.25 Matrix numbers of temperature-derivative equations: Example 4.10

$(\partial\pi_1/\partial\ln T)_p$	$(\partial\pi_2/\partial\ln T)_p$	$(\partial\ln n/\partial\ln T)_p$	RHS
2.79855E-01	1.05065E-01	1.41732E-01	1.79211E-01
1.41732E-01	5.35734E-02	0.00000E+00	7.67388E-02
1.05065E-01	5.35734E-02	5.35734E-02	1.56461E-01

Substituting the values of n_j's, $\overline{c}^0_{p,j}$'s, $\overline{h}^0_j$'s, $(\partial\pi_1/\partial\ln T)_p$, $(\partial\pi_2/\partial\ln T)_p$, and $(\partial\ln n/\partial\ln T)_p$ into Eq. 4.101, we get the value of the third key derivative c_p.

$$c_p = 6.33099 \text{ kJ}/(\text{kgK})$$

With reference to the pressure-derivative equations, Eqs. 4.108 and 4.109,

$$\left(n_H + 4n_{H_2} + 4n_{H_2O} + n_{OH}\right)\left(\frac{\partial\pi_1}{\partial\ln p}\right)_T + \left(2n_{H_2O} + n_{OH}\right)\left(\frac{\partial\pi_2}{\partial\ln p}\right)_T$$
$$+\left(n_H + 2n_{H_2} + 2n_{H_2O} + n_{OH}\right)\left(\frac{\partial\ln n}{\partial\ln p}\right)_T = \left(n_H + 2n_{H_2} + 2n_{H_2O} + n_{OH}\right)$$

$$\left(2n_{H_2O} + n_{OH}\right)\left(\frac{\partial\pi_1}{\partial\ln p}\right)_T + \left(n_{H_2O} + n_{OH}\right)\left(\frac{\partial\pi_2}{\partial\ln p}\right)_T$$
$$+\left(n_{H_2O} + n_{OH}\right)\left(\frac{\partial\ln n}{\partial\ln p}\right)_T = \left(n_{H_2O} + n_{OH}\right)$$

$$\left(n_H + 2n_{H_2} + 2n_{H_2O} + n_{OH}\right)\left(\frac{\partial\pi_1}{\partial\ln p}\right)_T + \left(n_{H_2O} + n_{OH}\right)\left(\frac{\partial\pi_2}{\partial\ln p}\right)_T$$
$$= \left(n_H + n_{H_2} + n_{H_2O} + n_{OH}\right)$$

Substituting the values of n_j's, the calculated matrix-numbers corresponding to the three simultaneous pressure-derivative equations are given in Table 4.26.

Table 4.26 Matrix numbers of pressure-derivative equations: Example 4.10

$(\partial\pi_1/\partial\ln\ p)_T$	$(\partial\pi_2/\partial\ln\ p)_T$	$(\partial\ln\ n/\partial\ln\ p)_T$	RHS
2.79855E-01	1.05065E-01	1.41732E-01	1.41732E-01
1.41732E-01	5.35734E-02	0.00000E+00	7.26701E-02
1.05065E-01	5.35734E-02	5.35734E-02	5.35734E-02

Adopting Gaussian elimination with partial pivoting and by back substitution, the solutions are

$$\left(\frac{\partial\pi_1}{\partial\ln\ p}\right)_T = 5.02885E-01;\quad \left(\frac{\partial\pi_2}{\partial\ln\ p}\right)_T = 2.60441E-02$$

$$\left(\frac{\partial\ln\ n}{\partial\ln\ p}\right)_T = -1.22738E-02$$

Substituting the evaluated value of $(\partial\ln\ n/\partial\ln\ p)_T$ into Eq. 4.91, we get the value of the second key derivative.

$$\left(\frac{\partial\ln\ v}{\partial\ln\ p}\right)_T = -1-1.22738E-02 = -1.0122738$$

Substituting the obtained values of $c_p, n, (\partial\ln\ v/\partial\ln\ T)_p$, and $(\partial\ln\ v/\partial\ln\ p)_T$ into Eq. 4.112,

$$c_v = 5.43194\ \text{kJ}/(\text{kgK})$$

Thus we have the ratio of specific heats $\gamma = c_p/c_v = 1.16551$. But the isentropic exponent by Eq. 4.114,

$$\gamma_s = \frac{-\gamma}{(\partial\ln\ v/\partial\ln\ p)_T} = \frac{-1.16551}{-1.01227} = 1.15138$$

Velocity of sound by Eq. 4.115,

$$a = \sqrt{\gamma_s n R_u T} =$$

$$\sqrt{1.15138 \times 7.26701E-02 \times 8314.51 \times 3394.71} = 1536.76\ \text{m/s}$$

And, the flow velocity by Eq. 4.123,

$$u = \sqrt{2 \times (-986.31 + 2165.99) \times 1000} = 1536.02$$
$$M = \frac{1536.02}{1536.76} = 0.999518$$

Although the obtained Mach number is close to unity, for the purpose of completing the procedure, let us get a more improved estimate for the throat pressure [6].

$$p_{t,k+1} = \left(p \frac{1 + \gamma_s M^2}{1 + \gamma_s} \right)_{t,k}$$
$$= \left(117.165 \times \frac{1 + 1.15138 \times 0.999518^2}{2.15138} \right) = 117.1046 \text{ bar}$$

With this improved trial throat pressure of 117.1046 bar, the procedure is repeated. The solution obtained after two iterations is as per the following.

$p_t = 117.1046$ bar (*improved trial pressure*)
$s^* = 17.1734$ kJ/(kgK) (*assigned entropy*)
$\overline{m} = 13.7609$ kg/kgmol
$T_t = 3394.51$ K
$\rho_t = 5.70963$ kg/m^3
$n_H = 1.52632E - 03$ kgmol/kg; $n_{H_2} = 1.75698E - 02$ kgmol/kg
$n_{H_2O} = 5.14924E - 02$ kgmol/kg; $n_{OH} = 2.08104E - 03$ kgmol/kg
$n = 7.26696E - 02$ kgmol/kg

$\overline{c}^0_{p,H} = 2.07863E + 01$ kJ/(kgmolK)
$\overline{c}^0_{p,H_2} = 3.79226E + 01$ kJ/(kgmolK)
$\overline{c}^0_{p,H_2O} = 5.79878E + 01$ kJ/(kgmolK)
$\overline{c}^0_{p,OH} = 3.76815E + 01$ kJ/(kgmolK)
$\overline{h}^0_H = 2.82361E + 05$ kJ/kgmol; $\overline{h}^0_{H_2} = 1.03528E + 05$ kJ/kgmol
$\overline{h}^0_{H_2O} = -9.15104E + 04$ kJ/kgmol; $\overline{h}^0_{OH} = 1.41818E + 05$ kJ/kgmol
$h = -2167.03$ kJ/kg

As detailed previously, substituting the values of n_j's and $\overline{h}^0_j$'s of the converged temperature of 3394.51 K into the temperature-derivative equations and solving them,

$$\left(\frac{\partial\pi_1}{\partial\ln T}\right)_p = -1.84216E+00; \quad \left(\frac{\partial\pi_2}{\partial\ln T}\right)_p = 6.30673E+00$$
$$\left(\frac{\partial\ln n}{\partial\ln T}\right)_p = 2.27232E-01$$

Substituting the value of $(\partial\ln n/\partial\ln T)_p$ in Eq. 4.90, we get the value of the first key derivative.

$$\left(\frac{\partial\ln v}{\partial\ln T}\right)_p = 1 + 2.27232E-01 = 1.227232$$

Substituting the values of n_j's, $\overline{c}^0_{p,j}$'s, $\overline{h}^0_j$'s, $(\partial\pi_1/\partial\ln T)_p$, $(\partial\pi_2/\partial\ln T)_p$, and $(\partial\ln n/\partial\ln T)_p$ into Eq. 4.101, we get the value of the third key derivative c_p.

$$c_p = 6.33050 \text{ kJ}/(\text{kgK})$$

As detailed previously, substituting the values of n_j's of the converged temperature of 3394.51 K into the pressure-derivative equations and solving them,

$$\left(\frac{\partial\pi_1}{\partial\ln p}\right)_T = 5.02885E-01; \quad \left(\frac{\partial\pi_2}{\partial\ln p}\right)_T = 2.60339E-02$$
$$\left(\frac{\partial\ln n}{\partial\ln p}\right)_T = -1.22703E-02$$

Substituting the evaluated value of $(\partial\ln n/\partial\ln p)_T$ into Eq. 4.91, we get the value of the second key derivative.

$$\left(\frac{\partial\ln v}{\partial\ln p}\right)_T = -1 - 1.22703E-02 = -1.0122703$$

Substituting the obtained values of $c_p, n, (\partial\ln v/\partial\ln T)_p$, and $(\partial\ln v/\partial\ln p)_T$ into Eq. 4.112,

$$c_v = 5.43152 \text{ kJ}/(\text{kgK})$$

Thus we have the ratio of specific heats $\gamma = c_p/c_v = 1.16551$. But the isentropic exponent by Eq. 4.114,

$$\gamma_s = \frac{-\gamma}{(\partial\ln v/\partial\ln p)_T} = \frac{-1.16551}{-1.01227} = 1.15138$$

Velocity of sound by Eq. 4.115,

$$a = \sqrt{\gamma_s n R_u T}$$

$$= \sqrt{1.15138 \times 7.26696E - 02 \times 8314.51 \times 3394.51} = 1536.71 \text{ m/s}$$

The flow velocity by Eq. 4.123 and Mach number are

$$u = \sqrt{2 \times (-986.31 + 2167.03) \times 1000} = 1536.70 \text{ m/s}$$

$$M = \frac{1536.70}{1536.71} = 0.999988$$

The reached Mach number is nearly equal to unity and the solution is acceptable. The summary of the solutions at the throat is as per the following.

$$p_0/p_t = 1.7420; \quad p_t = 117.105 \text{ bar}; \quad T_t = 3394.51 \text{ K}$$
$$\overline{m}_t = 13.7609 \text{ kg/kgmol}; \quad \rho_t = 5.7096 \text{ kg/m}^3$$
$$h_t = -2167.03 \text{ kJ/kg}; \quad g_t = -60462.4 \text{ kJ/kg}$$
$$(c_p)_t = 6.33050 \text{ kJ/kgK}; \quad (\gamma_s)_t = 1.15138$$
$$a_t = 1536.71\ m/s; \quad massflux = \rho_t u_t = 8773.96 \text{ kg/m}^2\text{s}$$

With these results, the throat calculations are complete. Next, we have to calculate the nozzle-exit conditions for the exit plane static pressure of 0.181 bar.

Nozzle-Exit-Plane Calculations In the case of nozzle throat calculations we do not know the throat pressure that will yield flow velocity equal to sonic velocity. Therefore, we have to iterate also on the throat pressure. Here for the nozzle-exit plane calculations the exit plane pressure is specified and hence the calculations are not that arduous. Calculations are to be carried out for the assigned pair of state functions $p\&s$ that is 0.181 bar & 17.1734 kJ/(kgK).

For the first iteration, we estimate the exit plane temperature by assuming the isentropic exponent to be constant as at the throat. Therefore,

$$T_e = \frac{T_0}{(p_0/p_e)^{(\gamma_s - 1)/\gamma_s}} = \frac{3619.5}{1127^{(0.15138/1.15138)}} = 1436 \text{ K}$$

To start with, we assume that mole fractions and molar mass to be same as the solved ones at the throat. Therefore the inputs for the first iteration are the following.

$p_e = 0.181$ bar $(specified\ pressure)$
$s^* = 17.1734$ kJ/(kgK) (assigned entropy)
$\overline{m} = 13.7609$ kg/kgmol
$T_e = 1436$ K
$n_H = 1.52632E-03$ kgmol/kg; $n_{H_2} = 1.75698E-02$ kgmol/kg
$n_{H_2O} = 5.14924\ E-02$ kgmol/kg; $n_{OH} = 2.08104E-03$ kgmol/kg
$n = 7.26696E-02$ kgmol/kg

The procedure for the solution of the assigned pair of state functions $p\&s$, the nozzle-exit plane static pressure and the assigned entropy for the isentropic expansion, is the same as detailed previously for the isentropic expansion to the throat. The reduced Gibbs iteration equations remain the same.

By the expansion from the throat temperature of 3395 K to a temperature of about 1000 K (exact value is to be found) considerable amount of recombination is expected by which the dissociated species H and OH may recombine to form the stable species, H_2 and H_2O. Therefore the present trial inputs for the first iteration are not close to the final solution values. These lead to negative mole fraction values for H and OH through the initial iterations. The negative values are replaced with positive trace values, say 1.0E-08. After four iterations, the converged values for the nozzle-exit plane are

$p_e = 0.181$ bar $(specified\ pressure)$
$s^* = 17.1734$ kJ/(kgK) $(assigned\ entropy)$
$\overline{m}_e = 14.1112$ kg/kgmol
$T_e = 1193.47$ K
$\rho_e = 2.57392E-02$ kg/m^3
$n_H = 2.84172E-09$ kgmol/kg; $n_{H_2} = 1.72925E-02$ kgmol/kg
$n_{H_2O} = 5.35734E-02$ kgmol/kg; $n_{OH} = 3.42411E-10$ kgmol/kg
$n = 7.08659E-02$ kgmol/kg

$\overline{c}^0_{p,H} = 2.07863E+01$ kJ/(kgmolK)
$\overline{c}^0_{p,H_2} = 3.09536E+01$ kJ/(kgmolK)
$\overline{c}^0_{p,H_2O} = 4.37583E+01$ kJ/(kgmolK)
$\overline{c}^0_{p,OH} = 3.15648E+01$ kJ/(kgmolK)
$\overline{h}^0_H = 2.36605E+05$ kJ/kgmol; $\overline{h}^0_{H_2} = 2.65852E+04$ kJ/kgmol
$\overline{h}^0_{H_2O} = -2.07604E+05$ kJ/kgmol; $\overline{h}^0_{OH} = 6.42121E+04$ kJ/kgmol
$h_e = -10662.3$ kJ/kg

From these results adopting the temperature- and pressure-derivative equations, we have to evaluate the three key derivatives:

$$(\partial \ln v/\partial \ln T)_p,\ (\partial \ln v/\partial \ln p)_T,\ and\ c_p$$

The temperature- and pressure-derivative equations remain the same as obtained previously. Substituting the converged values of the nozzle-exit plane in the temperature-derivative equations, the solutions are

$$\left(\frac{\partial \pi_1}{\partial \ln T}\right)_p = -1.33956E+00; \quad \left(\frac{\partial \pi_2}{\partial \ln T}\right)_p = 2.36004E+01$$
$$\left(\frac{\partial \ln n}{\partial \ln T}\right)_p = 5.20623E-07$$

Substituting the value of $\left(\partial \ln n / \partial \ln T\right)_p$ in Eq. 4.90 the first key derivative is calculated.

$$\left(\frac{\partial \ln v}{\partial \ln T}\right)_p = 1 + 5.20623E-07 \approx 1$$

Substituting the values of n_j's, $\overline{c}^0_{p,j}$'s, $\overline{h}^0_j$'s, $\left(\partial \pi_1 / \partial \ln T\right)_p$, $\left(\partial \pi_2 / \partial \ln T\right)_p$, and $\left(\partial \ln n / \partial \ln T\right)_p$ into Eq. 4.101 the third key derivative c_p is calculated.

$$c_p = 2.87956 \text{ kJ}/(\text{kgK})$$

As detailed previously, substituting the values of n_j's of the converged temperature 1193.47 K into the pressure-derivative equations and solving them,

$$\left(\frac{\partial \pi_1}{\partial \ln p}\right)_T = 5.00000E-01; \quad \left(\frac{\partial \pi_2}{\partial \ln p}\right)_T = -3.29372E-08$$
$$\left(\frac{\partial \ln n}{\partial \ln p}\right)_T = -1.12329E-08$$

Substituting the evaluated value of $\left(\partial \ln n / \partial \ln p\right)_T$ into Eq. 4.91 the second key derivative is obtained.

$$\left(\frac{\partial \ln v}{\partial \ln p}\right)_T = -1 - 1.12329E-08 \approx -1$$

Substituting the obtained values of $c_p, n, \left(\partial \ln v / \partial \ln T\right)_p$, and $\left(\partial \ln v / \partial \ln p\right)_T$ into Eq. 4.112,

$$c_v = 2.29035 \text{ kJ}/(\text{kgK})$$

Thus we have the ratio of specific heats $\gamma = c_p/c_v = 1.25726$. But the isentropic exponent by Eq. 4.114,

$$\gamma_s = \frac{-\gamma}{(\partial \ln\ v/\partial \ln\ p)_T} = \frac{-1.25726}{-1.} = 1.25726$$

Velocity of sound by Eq. 4.115,

$$\begin{aligned} a &= \sqrt{\gamma_s n R_u T} \\ &= \sqrt{1.25726 \times 7.08659E - 02 \times 8314.51 \times 1193.47} = 940.275 \text{ m/s} \end{aligned}$$

The flow velocity by Eq. 4.123 and Mach number are

$$\begin{aligned} u &= \sqrt{2 \times (-986.31 + 10662.3) \times 1000} = 4399.09 \text{ m/s} \\ M &= \frac{4399.09}{940.275} = 4.6785 \end{aligned}$$

The summary of the solutions at the nozzle exit is as per the following.

$$\begin{aligned} &p_0/p_e = 1127\,(specified)\,; \quad p_e = 0.181 \text{ bar}\,(specified)\,; \quad T_e = 1193.47 \text{ K} \\ &\overline{m}_e = 14.1112 \text{ kg/kgmol}; \quad \rho_e = 2.57392E - 02 \text{ kg/m}^3 \\ &h_e = -10662.3 \text{ kJ/kg}; \quad g_e = -31153.8 \text{ kJ/kg} \\ &\left(c_p\right)_e = 2.87956 \text{ kJ/(kgK)}; \quad (\gamma_s)_e = 1.25726 \\ &a_e = 940.275 \text{ m/s}; \quad massflux = \rho_e u_e = 113.229 \text{ kg/m}^2\text{s} \end{aligned}$$

With these results, the nozzle-exit calculations are complete. Next, we have to calculate the rocket performance parameters.

Rocket Performance Parameters Now we can calculate the rocket performance parameters. Characteristic velocity c^* is given by

$$c^* = \frac{p_{0n} a_t}{p_t (\gamma_s)_t} \tag{4.128}$$

Therefore,

$$c^* = \frac{204 \times 1536.70}{117.105 \times 1.15138} = 2325.04 \text{ m/s}$$

Characteristic thrust coefficient C_F^0 is given by

$$C_F^0 = \frac{(\gamma_s)_t \sqrt{2\,(h_0 - h_e)}}{a_t \left(p_{0n}/p_t\right)} \tag{4.135}$$

And, at the nozzle-exit condition, had there been adapted operation,

$$C_F^0 = \frac{4399.09}{2325.04} = 1.892$$

Vacuum thrust coefficient is given by

$$\begin{aligned}C_{F,vac} &= \frac{(\gamma_s)_t\sqrt{2\,(h_0 - h_e)}}{a_t\left(p_{0n}/p_t\right)} + \frac{A_e}{A_t}\frac{p_e}{p_{0n}} \\ &= \frac{u_e}{c^*} + \frac{\left(A_e/\dot{m}\right)}{\left(A_t/\dot{m}\right)}\frac{p_e}{p_{0n}} = C_F^0 + \frac{\rho_t u_t}{\rho_e u_e}\frac{p_e}{p_{0n}} \\ &= 1.89205 + \frac{8773.96}{113.229} \times \frac{0.181}{204} = 1.96081\end{aligned}$$

By Eq. 4.146,

$$I_{sp,vac} = C_{F,vac}c^* = 1.96081 \times 2325.04 = 4558.95\ \frac{\text{Ns}}{\text{kg}}$$

Nozzle Sizing Nozzle area ratio is given by

$$\frac{A_e}{A_t} = \frac{\rho_t u_t}{\rho_e u_e} = \frac{8773.96}{113.229} = 77.489$$

Mass flow of propellant is

$$\dot{m} = \frac{F_{vac}}{I_{sp,vac}} = \frac{2090 \times 1000}{4558.95} = 458.439\ \frac{\text{kg}}{\text{s}}$$

Therefore throat area and diameter are

$$\begin{aligned}A_t &= \frac{\dot{m}}{\rho_t a_t} = \frac{458.439}{8773.96} = 0.05225\ \text{m}^2 \\ d_t &= 0.2579\ \text{m}\end{aligned}$$

Nozzle-exit area and diameter are

$$\begin{aligned}A_e &= \frac{\dot{m}}{\rho_e u_e} = \frac{458.439}{113.229} = 4.048777\ \text{m}^2 \\ d_e &= 2.2705\ \text{m}\end{aligned}$$

Mach number at the entry to the nozzle is specified as 0.1. Through routine gas-dynamics procedure, using the equation of continuity we get

$$\dot{m} = \frac{p_0 A M \sqrt{\gamma_s}}{\sqrt{(R_u/\overline{m})\, T_0} \left(1 + \frac{\gamma_s - 1}{2} M^2\right)^{\frac{\gamma_s + 1}{2(\gamma_s - 1)}}}$$

With the calculated $\dot{m}$ and the specified values in the combustion chamber p_0, T_0, and $\overline{m}$, the nozzle-inlet area and diameter are calculated for the Mach number of 0.1 as

$$A_n = \frac{458.439 \sqrt{\left(\frac{8314.51}{13.645}\right) \times 3619.5} \left(1 + \frac{0.1502}{2} \times 0.01\right)^{\frac{2.1502}{0.3004}}}{204 \times 10^5 \times 0.1 \times \sqrt{1.1502}}$$

$$A_n = 0.31286 \text{ m}^2$$
$$d_n = 0.631 \text{ m}$$

We can calculate the nozzle entry dimension also by assuming the flow to be incompressible. Since the specified nozzle entry Mach number is very low at 0.1, we can assume the flow to be incompressible. The sonic speed at the combustion chamber conditions a_c is

$$a_c = \sqrt{(\gamma_s)_c T_0 R_u / \overline{m}_c} = \sqrt{1.1502 \times 3619.5 \times 8314.51/13.645} = 1592.73 \text{ m/s}$$

Therefore nozzle entry velocity at M = 0.1 is 159.273 m/s. Gas density at the combustion chamber condition is

$$\rho_c = \frac{p_0 \overline{m}_c}{R_u T_0} = \frac{204 \times 10^5 \times 13.645}{8314.51 \times 3619.5} = 9.2495 \frac{\text{kg}}{\text{m}^3}$$

The mass flux then is $\rho_c u_c = 1473.196 \text{ kg/m}^2\text{s}$. For the engine mass flow rate of 458.439 kg/s, the inlet area and diameter of the nozzle are,

$$A_n = \frac{458.439}{1473.196} = 0.3112 \text{ m}^2$$
$$d_n = 0.6295 \text{ m}$$

The difference in nozzle-inlet dimension between the two calculations is not significant.

Results of all the calculations are given in Table 4.27.

Discussion

This example is to show how the methods that we have learned can be used to

Table 4.27 Summary of calculated values of Example 4.10

	Chamber	Nozzle throat	Nozzle exit
p (bar)	204	117.105	0.181
T (K)	3619.5	3394.51	1193.5
$\rho\ (\text{kg/m}^3)$	9.2495	5.70963	0.025739
$h\ (\text{kJ/kg})$	−986.31	−2167.03	−10662.3
$s\ (\text{kJ/(kgK)})$	17.1734	17.1734	17.1734
$\overline{m}\ (\text{kg/kgmol})$	13.645	13.7609	14.1112
$(\partial \ln v/\partial \ln p)_T$	–	−1.01227	−1
$(\partial \ln v/\partial \ln T)_p$	–	1.22723	1
$c_p\ (\text{kJ/(kgK)})$	–	6.33050	2.87956
γ_s	1.1502	1.15138	1.25726
a (m/s)	1592.7	1536.7	940.3
M	0	1	4.679
x_H	0.02650	0.02100	4.01E-08
x_{H_2}	0.24250	0.24178	0.24402
x_{H_2O}	0.69140	0.70858	0.75598
x_{OH}	0.03960	0.02864	4.83E-09
A_e/A_t	–	1	77.489
c^* (m/s)	–	2325.04	2325.04
C_F^0	–	0.6609	1.892
$C_{F,vac}$	–	–	1.9608
$I_{sp}^0\ (\text{Ns/kg})$	–	–	4399.09
$I_{sp,vac}\ (\text{Ns/kg})$	–	–	4558.95

size a rocket engine. The calculated values more or less match with the data of SSME. The available SSME data are the following. $C_{F,vac} = 1.907$, $I_{sp,vac} = 4442$ Ns/kg, nozzle throat diameter $d_t = 0.26$ m, nozzle-exit diameter $d_e = 2.3$ m, and nozzle area ratio $A_e/A_t = 77.5$. Be alerted that what we have done is only a theoretical exercise to arrive at near practical values. Actual engine development to arrive at the final configuration towards the realization of desired performance parameters involves years of detailed analyses coupled with testing and hardware development.

Observe and appreciate that as the nozzle expansion takes place to a very low pressure and the related low temperature at the nozzle exit the reaction is almost complete and the reaction components in the key derivatives $(\partial \ln v/\partial \ln T)_p$, $(\partial \ln v/\partial \ln p)_T$, and $c_{p,e}$ are becoming insignificant and these derivatives respectively tend to be the frozen ones 1, -1, and $c_{p,f}$. Hence the isentropic exponent tends to be equal to the ratio of specific heats, $\gamma_s = \gamma$, Eqs. 4.90–4.92.

Example 4.11 Oxygen gas and aluminum powder at 298.15 K are burned in an adiabatic combustion chamber at 100 bar. For the purpose of simplicity only O, O_2, and Al_2O_3 (l) are considered to be significant product species. The adopted oxidizer–fuel ratio is 6. The calculated adiabatic flame temperature and other properties of combustion products in the chamber are the following. $T_0 = 3483.09$ K. Entropy $s = 6.9211$ kJ/(kgK). $\overline{m}_c = 38.459$ kg/kgmol. $x_O =$ 4.14112E-02; $x_{O_2} =$ 8.56775E-01; and $x_{Al_2O_3(l)} =$ 1.01814E-01. For frozen condition calculate the following: (1) other properties of combustion products in the chamber—$\overline{m}$, c_p, ratio of specific heats γ, and sonic velocity a; (2) throat conditions; and (3) nozzle-exit conditions for the nozzle-exit static pressure of 1 bar. As the solidification temperature of Al_2O_3 (l) is 2327 K, check the applicability of frozen flow assumption for the nozzle-exit pressure of 1 bar.

Solution

The propellant combination has two elements, hence $l = 2$. There are two significant gaseous species to be considered, hence $NG = 2$. There is one condensed species, hence $NS = 3$.

At a chosen temperature, the standard state-specific heat at constant pressure, standard state enthalpy, and the standard state entropy for the considered species can be calculated using the polynomials as described previously. The corresponding polynomial coefficients are given in Tables 4.20 and 4.21.

Equilibrium composition has been calculated under the specified pair of state functions $h\&p$ in the combustion chamber and the data are given. For this, the energy minimization with elemental-balance constraint and the enthalpy balance have been applied to calculate the equilibrium composition by adopting Eqs. 4.67, and 4.85–4.87.

Now we have to calculate other quantities for the frozen condition, which means that the composition of the product gas is fixed. Therefore, for further calculations, the reaction components vanish in the key derivatives

$$(\partial \ln v/\partial \ln T)_p,\ (\partial \ln v/\partial \ln p)_T,\ and\ c_{p,e}$$

[Eqs. 4.90, 4.91, and 4.101]. Hence $c_{p,e} = c_{p,f}$ and $\gamma_s = \gamma$. Energy minimization is no longer required and the elemental balance has already been obtained for the chamber conditions—hence π_i's are redundant. Consequently, the lone equation to be solved for isentropic frozen nozzle flow is the simplified form Eq. 4.84 where temperature T and pressure p are the only variables—Δn_j, $\Delta \ln n$, and $\left(n - \sum_{j=1}^{NG} n_j\right)$ all vanish. The variable pressure p enters the equation while calculating the entropy $\overline{s}_j$, Eq. 4.80. The simplified equation of Eq. 4.84 is

$$\left(\sum_{j=1}^{NS} \frac{n_j \bar{c}^0_{p,j}}{R_u}\right) \Delta \ln T = \frac{s^* - s}{R_u} \tag{4.84a}$$

Choked throat pressure is the one for which the velocity of flow is equal to the velocity of sound and this has to be obtained by iteration. The initial estimate for the throat pressure and temperature is obtained by assuming the ratio of specific heats to be constant as at the nozzle entry condition.

$$p_t = \frac{p_{0n}}{\left(\frac{\gamma+1}{2}\right)^{\gamma/(\gamma-1)}}$$

$$T_t = \frac{T_0}{\left(\frac{\gamma+1}{2}\right)}$$

For the assigned entropy s^* and this trial pressure p_t, we have the first trial pair of state functions $s^*\&p$, for which we have to solve the temperature to calculate the flow velocity and the velocity of sound. For the choked condition, these two velocities have to be equal within the allowable error margin. If not, a new trial pressure for the throat has to be obtained from the iteration formula [6],

$$p_{t,k+1} = \left(p \frac{1+\gamma M^2}{1+\gamma}\right)_{t,k}$$

On convergence (flow velocity = velocity of sound), the properties of the choked flow are calculated. These properties enable us to calculate characteristic velocity c^*.

The static pressure at the nozzle-exit plane is 1 bar. Assuming that the ratio of specific heats evaluated at the throat to be constant, the trial temperature at the nozzle-exit plane is calculated under isentropic expansion to 1 bar. Starting with that trial temperature, this forms the pair of state functions, the assigned entropy of 6.9211 kJ/(kgK) and the specified exit plane pressure of 1 bar. The converged temperature using Eq. (4.84a) is the nozzle-exit plane temperature. Based on the converged nozzle-exit properties, rocket performance parameters $C_{F,vac}$ and $I_{sp,vac}$ are calculated.

Finally we have to check whether the converged nozzle-exit temperature is not significantly below the solidification temperature of $Al_2O_3(l)$. If no, it is to be alerted that the frozen flow assumption may not be valid for the specified nozzle-exit condition.

Analysis

(1) Combustion Chamber Conditions Noting $n_j = x_j/\overline{m}_c$, n_j's at the nozzle entry are

$$n_O = 1.076750918E - 03 \text{ kgmol/kg}$$
$$n_{O_2} = 2.227737410E - 02 \text{ kgmol/kg}$$
$$n_{Al_2O_3(l)} = 2.647312821E - 03 \text{ kgmol/kg}$$

As described previously, the standard state-specific heat at constant pressure, the standard state enthalpy, and the standard state entropy for the considered species can be calculated using the polynomials for the adiabatic flame temperature of 3483.09 K. The corresponding polynomial coefficients are given in Tables 4.20 and 4.21.

Recall,

$$\mu_j = (\overline{g}_j)_{T,p} = \overline{h}_j^0(T) - T\overline{s}_j$$
$$= \overline{h}_j^0(T) - T\left[\overline{s}_j^0(T) - R_u\left(\ln\frac{n_j}{n} + \ln p\right)\right] \rightarrow (j = 1, \ldots, NG)\ p = [\text{bar}]$$

$$\mu_j(T) = \mu_j^0(T) = \overline{h}_j^0(T) - T\overline{s}_j^0(T) \qquad (j = NG + 1, \ldots, NS)$$

$$\overline{s}_j = \overline{s}_j^0(T) - R_u\left(\ln\frac{n_j}{n} + \ln p\right) \rightarrow (j = 1, \ldots, NG) \quad p = [\text{bar}]$$

$$\overline{s}_j = \overline{s}_j^0(T) \qquad (j = NG + 1, \ldots, NS)$$

Adopting the polynomials and their coefficients, the calculated values at the adiabatic flame temperature of 3483.09 K and chamber pressure of 100 bar are given in Table 4.28.

$$\overline{m} = 1\Big/\sum_{j=1}^{NG} n_j = 42.81899 \text{ kg/kgmol}$$

In the absence of reaction components,

Table 4.28 Properties of species at the adiabatic flame temperature of 3483.09 K and chamber pressure of 100 bar: Example 4.11

Species	n_j (kgmol/kg)	$\overline{c}^0_{p,j}$ (kJ/(kgmolK))	$\overline{h}^0_j$ (kJ/kgmol)	$\overline{s}_j$ (kJ/(kgmolK))	μ_j (kJ/kgmol)
O	1.07675E-03	21.0878	315898	200.135	−381192
O_2	2.22774E-02	40.8750	117652	252.660	−762384
Al_2O_3 (*l*)	2.64731E-03	162.900	−1118540	406.820	−2535530
	Total	1.36454 (kJ/(kgK))	5.26491E-05 (kJ/kg)	6.9211 (kJ/(kgK))	−24106.7 (kJ/kg)

$$c_p = \sum_{j=1}^{NS} n_j \overline{c}^0_{p,j} = 1.36454 \text{ kJ}/(\text{kgK})$$
$$c_v = c_p - R_u/\overline{m} = 1.17036 \text{ kJ}/(\text{kgK})$$
$$\gamma = c_p/c_v = 1.16591$$

$$\rho = \frac{p\overline{m}}{R_u T} = \frac{100 \times 10^5 \times 42.81899}{8314.51 \times 3483.09} = 14.7855 \frac{\text{kg}}{\text{m}^3}$$

$$\rho = \frac{p\overline{m}}{R_u T} = \frac{100 \times 10^5 \times 42.81899}{8314.51 \times 3483.09} = 14.7855 \frac{\text{kg}}{\text{m}^3}$$

Sonic velocity,

$$a = \sqrt{\gamma n R_u T}$$
$$= \sqrt{1.16591 \times 2.33541E - 02 \times 8314.51 \times 3483.09} = 888.01 \text{ m/s}$$

(2) Nozzle-Throat Conditions We have to have trial pressure and temperature to initiate our calculations for the nozzle throat. The condition for the nozzle throat is that the local sonic velocity and the local flow velocity should be equal, that is the local Mach number is to be unity. Adopting the ratio of specific heats calculated for the chamber, the trial throat temperature and pressure are

$$T_t = T_0 \left(\frac{2}{\gamma + 1}\right) = 3483.09 \times \left(\frac{2}{2.16591}\right) = 3216.28 \text{ K}$$

$$p_t = p_0 \left(\frac{2}{\gamma + 1}\right)^{\frac{\gamma}{\gamma - 1}} = 100 \times \left(\frac{2}{2.16591}\right)^{\frac{1.16591}{0.16591}} = 57.119 \text{ bar}$$

Adopting the polynomials and their coefficients [Tables 4.20 and 4.21, $\overline{c}^0_{p,j}$'s and $\overline{h}^0_j$'s (for the trial temperature of 3216.28 K), and $\overline{s}_j$'s (for the trial temperature and pressure of 3216.28 K and 57.119 bar) are calculated. The calculated $\overline{c}^0_{p,j}$'s and s are substituted into Eq. 4.84a to get the first correction of $\Delta T = -1.025$ K, which gives a new trial temperature of 3215.25 K. Against the pressure of 57.119 bar and the new trial temperature of 3215.25 K, iterations are continued [Eq. 4.84a] to arrive at the converged temperature of 3215.26 K. The related values are given in Table 4.29.

From the calculated values,

$$c_p = \sum_{j=1}^{NS} n_j \overline{c}^0_{p,j} = 1.35371 \text{ kJ}/(\text{kgK})$$
$$c_v = c_p - R_u/\overline{m} = 1.15953 \text{ kJ}/(\text{kgK})$$
$$\gamma = c_p/c_v = 1.16746$$

Sonic velocity,

$$a = \sqrt{\gamma n R_u T}$$
$$= \sqrt{1.16746 \times 2.33541E - 02 \times 8314.51 \times 3215.26} = 853.75 \text{ m/}s$$
$$u = \sqrt{2(h_0 - h)} = \sqrt{2 \times (0 + 364035.50)} = 853.27 \text{ m/s}$$
$$M = u/a = 0.999442$$

Although the Mach number is close to unity, let us get an improved trial throat pressure. The improved trial value is given by

$$p_{t,k+1} = \left(p\frac{1+\gamma M^2}{1+\gamma}\right)_{t,k}$$
$$= \left(57.119 \times \frac{1 + 1.16746 \times 0.999442^2}{1 + 1.16746}\right) = 57.0847 \text{ bar}$$

With this improved trial pressure of 57.0847 bar and the previously obtained temperature of 3215.26 K iteration is initiated [Eq. 4.84a] to arrive at the next convergence temperature of 3214.98 K. Thus for the current set of 57.0847 bar and 3214.98 K the calculated values are given in Table 4.30.

Table 4.29 Properties of species at the temperature of 3215.26 K and pressure of 57.119 bar: Example 4.11

Species	n_j (kgmol/kg)	$\bar{c}^0_{p,j}$ (kJ/(kgmolK))	$\bar{h}^0_j$ (kJ/kgmol)	$\bar{s}_j$ (kJ/(kgmolK))	μ_j (kJ/kgmol)
O	1.07675E-03	20.9980	310262	203.108	−342783
O_2	2.22774E-02	40.3931	106768	254.065	−710115
$Al_2O_3\ (l)$	2.64731E-03	162.900	−1162170	393.786	−2428290

Table 4.30 Properties of species at the temperature of 3214.98 K and pressure of 57.0847 bar: Example 4.11

Species	n_j (kgmol/kg)	$\bar{c}^0_{p,j}$ (kJ/(kgmolK))	$\bar{h}^0_j$ (kJ/kgmol)	$\bar{s}_j$ (kJ/(kgmolK))	μ_j (kJ/kgmol)
O	1.07675E-03	20.9979	310256	203.111	−342743
O_2	2.22774E-02	40.3925	106757	254.066	−710061
$Al_2O_3\ (l)$	2.64731E-03	162.900	−1162210	393.772	−2428180
	Total	1.35370 (kJ/(kgK))	−364.411 (kJ/kg)	6.92107 (kJ/(kgK))	−22615.5 (kJ/kg)

From the calculated values,

$$c_p = \sum_{j=1}^{NS} n_j \bar{c}^0_{p,j} = 1.35370\ \text{kJ}/(\text{kgK})$$
$$c_v = c_p - R_u/\overline{m} = 1.15952\ \text{kJ}/(\text{kgK})$$
$$\gamma_t = c_p/c_v = 1.16746$$

Sonic velocity,

$$a = \sqrt{\gamma_t n R_u T}$$
$$= \sqrt{1.16746 \times 2.33541E-02 \times 8314.51 \times 3214.98} = 853.7112\ \text{m/s}$$
$$u = \sqrt{2\,(h_0 - h)} = \sqrt{2 \times (0 + 364410.97)} = 853.7107\text{m/s}$$
$$M = u/a = 0.999999$$

The reached Mach number is close to unity and is accepted as the solution. The characteristic velocity is

$$c^* = \frac{p_{0n} a_t}{p_t \gamma_t} = \frac{100 \times 853.7112}{57.0847 \times 1.16746} = 1281\ \text{m/s}$$
$$\rho_t = \frac{57.0847 \times 10^5 \times 42.819}{8314.51 \times 3214.98} = 9.1441\ \text{kg}/\text{m}^3$$
$$\rho_t u_t = 9.1441 \times 853.7107 = 7806.416\ \text{kg}/\text{m}^2\text{s}$$

Table 4.31 Properties of species at the nozzle temperature of 1770.41 K and pressure of 1 bar: Example 4.11

Species	n_j (kgmol/kg)	$\bar{c}^0_{p,j}$ (kJ/(kgmolK))	$\bar{h}^0_j$ (kJ/kgmol)	$\bar{s}_j$ (kJ/(kgmolK))	μ_j (kJ/kgmol)
O	1.07675E-03	20.8325	280107	224.292	−116984
O_2	2.22774E-02	37.2303	50590.2	264.592	−417847
Al_2O_3 (l)	2.64731E-03	162.900	−1397530	296.585	−1922610
	Total	1.28307 (kJ/(kgK))	−2271.09 (kJ/kg)	6.92107 (kJ/(kgK))	−14524.2 (kJ/kg)

(3) Nozzle-Exit Conditions The nozzle-exit plane pressure is specified as 1 bar. To initiate iterations to find the temperature at the nozzle-exit plane we require a trial temperature. Assuming the ratio of specific heats calculated for the throat to be constant for further expansion, the trial temperature for the nozzle-exit plane under isentropic expansion,

$$T_e = T_0\left(\frac{p_e}{p_0}\right)^{\frac{\gamma-1}{\gamma}} = 3483.09 \times \left(\frac{1}{100}\right)^{\frac{0.16746}{1.16746}} = 1799.23\ \text{K}$$

This trial temperature of 1799.23 K is substantially below the solidification temperature 2327 K of Al_2O_3 (l). Therefore it is to be alerted that the frozen flow assumption may not be valid for the specified nozzle-exit condition. However, despite the fact that the polynomial coefficients for Al_2O_3 (l) are not applicable for temperatures less than the solidification temperature, in order to complete the procedure let us continue with the iterations using Eq. 4.84a.

As per procedure followed previously, the converged temperature for 1 bar is 1770.41 K. The related property values calculated for 1 bar and 1770.41 K are given in Table 4.31.

From the calculated values,

$$c_p = \sum_{j=1}^{NS} n_j \bar{c}^0_{p,j} = 1.28307\ \text{kJ}/(\text{kgK})$$
$$c_v = c_p - R_u/\overline{m} = 1.08889\ \text{kJ}/(\text{kgK})$$
$$\gamma_e = c_p/c_v = 1.17833$$

Sonic velocity,

$$
\begin{aligned}
a_e &= \sqrt{\gamma_e n R_u T_e} \\
&= \sqrt{1.17833 \times 2.33541E-02 \times 8314.51 \times 1770.41} = 636.459 \text{ m/s} \\
u_e &= \sqrt{2(h_0 - h)} = \sqrt{2 \times (0 + 2271090)} = 2131.24 \text{ m/s} \\
M_e &= u_e / a_e = 3.349
\end{aligned}
$$

The other characteristic values are

$$
\begin{aligned}
\rho_e &= \frac{1 \times 10^5 \times 42.819}{8314.51 \times 1770.41} = 0.2909 \text{ kg/m}^3 \\
\rho_e u_e &= 0.2909 \times 2131.24 = 619.95 \text{ kg/m}^2\text{s} \\
\frac{A_e}{A_t} &= \frac{\rho_t u_t}{\rho_e u_e} = \frac{7806.416}{619.95} = 12.592
\end{aligned}
$$

The values of the thrust coefficients and the specific impulses are

$$
\begin{aligned}
C_F^0 &= \frac{\gamma_t \sqrt{2(h_0 - h_e)}}{a_t (p_{0n}/p_t)} = \frac{\sqrt{2(h_0 - h_e)}}{c^*} = \frac{u_e}{c^*} = \frac{2131.24}{1281} = 1.6637 \\
C_{F,vac} &= C_F^0 + \frac{A_e p_e}{A_t p_0} = 1.6637 + \frac{12.592}{100} = 1.7897 \\
I_{sp}^0 &= \sqrt{2(h_0 - h_e)} = C_F^0 c^* = 2131.24 \text{ Ns/kg} \\
I_{sp,vac} &= C_{F,vac} c^* = I_{sp}^0 + p_e/(\rho_e u_e) = 2292.5 \text{ Ns/kg}
\end{aligned}
$$

The summary of our calculations is given in Table 4.32.

Discussion

Under the pair of constant state functions $h\&p$, the calculated values for the combustion chamber are given. For further calculations under the frozen condition the reaction components in the key derivatives vanish. The coefficient of specific heat at constant pressure c_p becomes simply equal to $\sum_{j=1}^{NS} n_j \overline{c}_{p,j}^0$ [Eq. 4.101] and it varies with local temperature. Isentropic exponent γ_s is equal to the ratio of specific heats $\gamma = c_p/c_v$. Consequently, as the combustion products expand through the nozzle with temperature drop, γ varies. Furthermore as n_j's are constant, also $\overline{m}$ and $\overline{m}_c$ remain constant.

The condition for the nozzle throat is that the local sonic velocity and the local flow velocity should be equal, that is the local Mach number is to be unity. As n_j's are fixed, the iteration with trial pressure and temperature is less arduous than in the case of shifting-equilibrium flow. The lone equation to be used for the iteration is the simpler form of Eq. 4.84, namely Eq. 4.84a given previously.

Table 4.32 Summary of calculated values of Example 4.11

	Chamber	Nozzle throat	Nozzle exit	"Nozzle exit"
p (bar)	100	57.0847	1	1
$T\ (K)$	3483.09	3214.98	1770.41	1938.88
$\rho\ (\mathrm{kg/m^3})$	14.7855	9.1441	0.2909	0.2656
$h\ (\mathrm{kJ/kg})$	0	−364.411	−2271.09	−2328.55
$s\ (\mathrm{kJ/(kgK)})$	6.9211	6.9211	6.9211	6.9211
$\overline{m}\ (\mathrm{kg/kgmol})$	42.819	42.819	42.819	42.819
$\overline{m}_c\ (\mathrm{kg/kgmol})$	38.459	38.459	38.459	38.459
$c_p\ (\mathrm{kJ/(kgK)})$	1.3645	1.3537	1.2831	1.2307
γ	1.1659	1.1675	1.1783	1.1873
$a\ (m/s)$	888.01	853.71	636.46	668.59
M	0	1	3.349	3.228
x_O	0.04141	0.04141	0.04141	0.04141
x_{O_2}	0.85678	0.85678	0.85678	0.85678
$x_{Al_2O_3(l)}$	0.10181	0.10181	0.10181	$x_{Al_2O_3(s)}$= 0.10181
A_e/A_t	–	1	12.592	13.619
$c^*\ (m/s)$	–	1281	1281	1281
C_F^0	–	–	1.6637	1.6846
$C_{F,vac}$	–	–	1.7897	1.8208
$I_{sp}^0\ (\mathrm{Ns/kg})$	–	–	2131.24	2158.03
$I_{sp,vac}\ (Ns/\mathrm{kg})$	–	–	2292.5	2332.49

Under frozen flow assumption, we have calculated the nozzle-exit plane properties. The calculated temperature at the nozzle-exit 1771.41 K is substantially below the solidification temperature of $Al_2O_3\ (l)$, 2327 K. Therefore it is to be alerted that the results calculated under the frozen flow assumption may not be valid for the specified nozzle-exit condition. That there are no thermal- and velocity-lag between the gaseous- and condensed-species is the basic assumption in these calculations. And, in the present Example, as the temperature difference between the solidification temperature and the temperature at the nozzle exit is more than 500 °C, even after taking into account the thermal inertia, the liquid particles could solidify somewhere within the divergence, and add the heat of solidification to the nozzle flow, which will enhance the specific impulse value.

As a via media between the shifting-equilibrium flow and frozen flow, we can assume the entire kgmoles of $Al_2O_3\ (l)$ condense into $Al_2O_3\ (s)$ in the nozzle divergence. The nozzle-exit conditions can be calculated by adopting the respective polynomial coefficients for $Al_2O_3\ (s)$, Table 4.20. If done so, the results of calculation are given in the last column of Table 4.32 under the column heading "Nozzle Exit."

Another point that we may note in the multi-phase supersonic nozzle flow is the possibility of condensation shock. In the case of liquid particulates solidifying into solid ones, with their respective densities being more or less equal, impulsive heat release (due to thermal inertia) into the flow through solidification will take place—recall the effect of sudden heat addition in supersonic flow. But in the case of certain gaseous species condensing into liquid particulates, the massive reduction in the volume of the condensing species and their associated sudden heat release are to be considered.

Problems

4.1 Distinguish between shifting-equilibrium flow and frozen-equilibrium flow.

4.2 Generally, the selection of suitable propellant combinations for solid propellants is different from the one for liquid propellants. How?

4.3 Write a brief note on the selection of suitable O/F ratio for the cryogenic propellants LOX/LH2.

4.4 Compare the merits and the demerits between LOX/LH2 cryogenic engines and LOX/Kerosene semi-cryogenic ones on the point of view of their performance.

4.5 What is the meaning of a propellant combination being hypergolic? What is the advantage of a hypergolic propellant on the point of reliability?

4.6 How are solid propellants classified? The composite modified double base propellants are different from modified composite propellants. How?

4.7 Write a brief note on homogeneous (double base) propellants: ingredients and their composition, manufacturing processes, merits & demerits, and applications.

4.8 Write a brief note on heterogeneous (composite) propellants: ingredients and their composition, manufacturing processes, merits & demerits, and applications.

4.9 What is a triangular composition diagram in the selection of composite propellant composition? Briefly explain.

4.10 Prove

$$\overline{m}_c = \overline{m}\left(1 - \sum_{j=NG+1}^{NS} x_j\right)$$

where $\overline{m}_c$ is the molar mass considering gaseous and condensed species, $\overline{m}$ is the molar mass considering only gaseous species, and x_j is the mole fraction.

4.11 What is entropy principle?

4.12 What are the three pairs of state functions applicable for rocket engine performance calculations? Identify the subsystem of the rocket engine or the situation for which each pair is adopted for the performance evaluation.

4.13 From fundamentals, derive the Eq. 4.10. Explain how this is applicable for reversible and irreversible processes as well as for closed and open systems.

$$du + pdv - Tds \leq 0 \tag{4.10}$$

4.14 By entropy principle an arbitrary unit mass system attains equilibrium when,

$$du + pdv - Tds = 0$$

Derive applicable expressions for reactions to attain equilibrium for (1) constant $T\&p$, (2) constant $T\&v$, (3) constant $s\&v$, and (4) constant $s\&p$.

4.15 Consider a non-reacting system of a fixed composition with negligible change in kinetic- and potential-energy and combining the first and second laws of thermodynamics, and in the absence of any other energies, derive Gibbs equations. Explain how these equations are applicable for closed and open systems as well as reversible and irreversible processes.

$$du = Tds - pdv \tag{4.22}$$

$$dg = vdp - sdT \tag{4.23}$$

$$dh = Tds + vdp \tag{4.24}$$

$$df = -sdT - pdv \tag{4.25}$$

4.16 Considering any two energy differentials in a reacting system, say du and dg, derive the derivatives that represent chemical potential μ_j. Discuss how μ_j is an intensive property.

4.17 Chemical potential is also known as partial molar energy. Why?

4.18 What is the necessity to get Gibbs correction equations from the Newton–Raphson equilibrium equations? Discuss the substitution adopted.

4.19 Differentiate between the eight basic thermodynamic properties and derived thermodynamic properties. List the former eight. Give examples of derived thermodynamic properties.

4.20 What are the three key derivatives? How are they used to evaluate the derived properties such as γ and velocity of sound a?

4.21 Distinguish between the isentropic exponent γ_s and the ratio of specific heats γ. They become equal under what condition?

4.22 Define the standard state enthalpy. Distinguish between the enthalpy of formation or heat of formation and sensible enthalpy change or sensible heat.

4.23 Briefly discuss the differences in internal flow behaviors between bi-propellant liquid engines and solid motors.

4.24 Write a brief note on the flow effects in constant area ducts due to mass addition and friction under adiabatic condition, and heat addition on the points of total and static properties, and velocities. Distinguish the effects between subsonic- and supersonic-flows.

4.25 Write a note on the types of injectors adopted in bi-propellant liquid propellant engines. Distinguish their applications for hypergolics and non-hypergolics and also for the Earth storables and cryogenics.

4.26 Write a note on the configurations of combustion chambers adopted in liquid propellant engines.

4.27 What are contraction ratios in liquid engines and port–throat ratios in solid ones? What are their typical values? How do these values affect the mode of performance prediction?

4.28 Write a note on the cycles adopted in turbo-pump fed liquid propellant rocket engines.

4.29 What is the difference between the assumption of an infinite area combustion chamber and the one of a finite area combustion chamber? Discuss their applications.

4.30 Define characteristic velocity c^*. Under the assumption of constant $\overline{m}$ and c_p, derive

$$c^* = \frac{\sqrt{RT_0}}{\Gamma} \tag{4.129}$$

where

$$\Gamma = \sqrt{\gamma}\left(\frac{2}{\gamma+1}\right)^{\frac{\gamma+1}{2(\gamma-1)}}$$

How does the quality factor ξ_c or c^* efficiency η_{c^*} represent the quality of combustion in a rocket chamber?

4.31 Define thrust coefficient C_F. Under the assumption of constant $\overline{m}$ and c_p, derive,

$$C_F = \Gamma\sqrt{\frac{2\gamma}{\gamma-1}\left[1-\left(\frac{p_e}{p_{0n}}\right)^{\frac{\gamma-1}{\gamma}}\right]} + \frac{A_e}{A_t}\left(\frac{p_e}{p_{0n}}-\frac{p_a}{p_{0n}}\right) \tag{4.134}$$

where

$$\Gamma = \sqrt{\gamma}\left(\frac{2}{\gamma+1}\right)^{\frac{\gamma+1}{2(\gamma-1)}}$$

How does the nozzle-flow quality factor or nozzle-flow efficiency ξ_n represent the quality of flow in a rocket nozzle?

4.32 Define specific impulse I_{sp}. What is equivalent nozzle-exit velocity or equivalent exhaust velocity $\overline{u}_e$. How is this equal to the specific impulse? Prove $I_{sp} = C_F c^*$.

4.33 Define specific impulse quality factor or velocity correction ξ_s. Prove $\xi_s = \xi_c \xi_n$.

4.34 During the design and development of combustion chambers for liquid rocket engines, the role of ξ_c estimation is central. Explain.

4.35 Let an iron block of 10 kg mass at a temperature of 1000 K be immersed in an infinite water reservoir of constant temperature 300 K. The iron block system together with the water reservoir surroundings can be considered as an isolated system. The specific heat coefficient for iron is 106 J/(kgK). Calculate the entropy change for the system, surroundings, and the isolated system for the following two temperature changes during the cooling of the iron block: (1) from 1000 to 999 K and (2) from 301 to 300 K. Discuss the verification of Eqs. 4.7–4.9.
Answers:
(1) $\Delta T \rightarrow$ 1000 to 999 K: (i) $\Delta S_u \geq 0 \rightarrow 2.4728$ J/K > 0, (ii) $\Delta Q \leq T\Delta S \rightarrow -1060$ J < -318.159 J, & (iii) $\Delta U + p\Delta V - T\Delta S \leq 0 \rightarrow -741.841$ J < 0.
(2) $\Delta T \rightarrow$ 301–300 K: (i) $\Delta S_u \geq 0 \rightarrow 5.8425 \times 10^{-3}$ J/K > 0, (ii) $\Delta Q \leq T\Delta S \rightarrow -1060$ J < -1058.238 J, & (iii) $\Delta U + p\Delta V - T\Delta S \leq 0 \rightarrow -1.762$ J < 0.

4.36 One kg of air at 350 K and 1 bar in a cylinder–piston arrangement is kept in an atmosphere of 300 K and 1 bar and is allowed to cool. Work out the numerical of the equilibrium process at the cooling of (1) 350–340 K and (2) 310–300 K and demonstrate the entropy principle. Assume for air c_p at 1005 J/(kgK), the ratio of specific heats at 1.4, and molar mass of air at 28.971 kg/kgmol to be constant.
Answers:
(1) $ds_u = 4.36753$ J/(kgK) > 0; $\delta q \leq Tds \rightarrow -10050 < -8739.742$ J/kg;
$du + pdv - Tds = -1308.788$ J/kg < 0
(2) $ds_u = 0.54623$ J/(kgK) > 0; $\delta q \leq Tds \rightarrow -10050 < -9886.132$ J/kg;
$du + pdv - Tds = -162.438$ J/kg < 0.

4.37 Combustion gases of a reciprocating internal-combustion engine are entrapped at a constant volume of 400 cm^3 at a pressure of 6 bar and temperature of 2300 K against the atmospheric condition of 1 bar and 300 K. Normally the entrapped gas will leak through the gap between cylinder wall and piston to atmospheric pressure, and simultaneously get cooled to the atmospheric temperature of 300 K. But, let us now consider the cooling at constant volume without any loss of mass through leaks. By considering the constant volume cooling (1) from 2300 to 2290 K and (2) from 310 to 300 K demonstrate the entropy principle. Assume for the combustion gases c_p at 1860 J/(kgK), the ratio of specific heats γ at 1.2, and the molar mass $\overline{m}$ at 28.9 kg/kgmol as constant.
Answers:
(1) $dS_u = 0.016290\ \text{J/K} > 0$; $\delta Q \leq TdS \rightarrow -5.62182\ \text{J} < -0.73488\ \text{J}$; $dU + pdV - TdS = -4.88694\ \text{J} < 0$
(2) $dS_u = 3.1 \times 10^{-4}\ \text{J/K} > 0$; $\delta Q \leq TdS \rightarrow -5.62182\ \text{J} < -5.53015\ \text{J}$; $dU + pdV - TdS = -0.09167\ \text{J} < 0$.

4.38 From fundamentals prove the identity,

$$\left(\frac{\partial u}{\partial n_j}\right)_{s,v,n_{i\neq j}} = \left(\frac{\partial g}{\partial n_j}\right)_{T,p,n_{i\neq j}} = \left(\frac{\partial h}{\partial n_j}\right)_{s,p,n_{i\neq j}} = \left(\frac{\partial f}{\partial n_j}\right)_{T,v,n_{i\neq j}} \equiv \mu_j$$

to show the chemical potential is an intensive property.

4.39 Gaseous combustion mixture in the chamber of a cryogenic rocket engine burning liquid hydrogen and liquid oxygen contains, among other gaseous species, the gaseous species OH at a mole fraction of 1.6322×10^{-2}. The chamber pressure and temperature are 70 bar and 3200 K. Molar mass of the combustion mixture is 13.197 kg/kgmol. Calculate (1) number of kgmoles of OH per kg of combustion mixture, (2) partial pressure of OH, (3) entropy of OH in kJ/(kgmolK). Also calculate for OH in kJ/kgmol (4) standard Gibbs energy, (5) Gibbs energy, and (6) chemical potential. Standard enthalpy of OH at 3200 K = 136420 kJ/kgmol. Standard entropy of OH at 3200 K = 259.199 kJ/(kgmolK).
Answers:
(1) $n_{OH} = 1.23680 \times 10^{-3}$ kgmol/kg, (2) $p_{OH} = 1.14254$ bar,
(3) $\overline{s}$ (3200 K, 1.14254 bar) = 258.091 kJ/(kgmolK),
(4) $\overline{g}^0$ (3200 K) = −693016.8 kJ/kgmol,
(5) $\overline{g}$ (3200 K, 1.14254 bar) = −689471.4 kJ/kgmol,
(6) $\overline{g}$ (3200 K, 1.14254 bar) = μ (3200 K, 1.14254 bar) = −689471.4 kJ/kgmol.

4.40 A high-temperature solid propellant, containing aluminum powder as one of the propellant's major ingredients, is being analyzed. At the chamber pressure of 100 bar, the adiabatic flame temperature is calculated as 5653 K. The only condensed species found in the combustion products is liquid alumina Al_2O_3 (l) and its mole fraction is 0.2964. The molar mass considering gaseous and condensed species $\overline{m}_c$ is 44.452 kg/kgmol. Standard state enthalpy and entropy of Al_2O_3 (l) can be calculated from the following expressions.

$$\overline{h}^0_{Al_2O_3(l)}(T) = 162.89996T - 1.68593 \times 10^6 \text{ kJ/kgmol}$$
$$\overline{s}^0_{Al_2O_3(l)}(T) = 162.89996 \ln T - 9.21738 \times 10^2 \text{ kJ/(kgmolK)}$$

Calculate (1) $\overline{m}$ the molar mass of the combustion products considering only gaseous species; and (2) the density of the combustion products. In addition calculate for the Al_2O_3 (l) (3) the number of kgmoles per kg of combustion products; (4) the standard state enthalpy; (5) the standard state entropy; (6) the entropy; (7) the standard state Gibbs energy; (8) the Gibbs energy; and (9) the chemical potential.

Answers:
(1) $\overline{m} = 63.178$ kg/kgmol; (2) $\rho = 13.442$ kg/m^3;
(3) $n_{Al_2O_3(l)} = 6.66787 \times 10^{-3}$ kgmol/kg
(4) $\overline{h}^0_{Al_2O_3(l)}$ (5653 K) $= -765057$ kJ/kgmol;
(5) & (6) $\overline{s}^0_{Al_2O_3(l)}$ (5653 K) $= \overline{s}_{Al_2O_3(l)}$ (5653 K) $= 485.708$ kJ/(kgmolK); (7) & (8) $\overline{g}^0_{Al_2O_3(l)} = \overline{g}_{Al_2O_3(l)} = -3510764$ kJ/kgmol;
(9) $\mu_{Al_2O_3(l)} = -3510764$ kJ/kgmol.

4.41 In a cryogenic rocket engine, burning liquid oxygen and liquid hydrogen, the gaseous combustion mixture has the following mole fractions at its nozzle throat. $x_H = 2.10036 \times 10^{-2}$; $x_{H_2} = 2.41777 \times 10^{-1}$; $x_{H_2O} = 7.08583 \times 10^{-1}$, and $x_{OH} = 2.86370 \times 10^{-2}$. The molar mass of the gaseous mixture $\overline{m} = 13.7609$ kg/kgmol. Temperature = 3394.5 K. Pressure = 117.105 bar. The enthalpy and entropy of the species can be calculated through the following polynomials and the values of the respective coefficients are given in Tables 4.20 and 4.21.

$$\frac{\overline{h}^0}{R_u T} = -a_1 T^{-2} + a_2 T^{-1} \ln T + a_3 + a_4 \frac{T}{2} + a_5 \frac{T^2}{3} + a_6 \frac{T^3}{4} + a_7 \frac{T^4}{5} + \frac{a_8}{T}$$

$$\frac{\overline{s}^0}{R_u} = -a_1 \frac{T^{-2}}{2} - a_2 T^{-1} + a_3 \ln T + a_4 T + a_5 \frac{T^2}{2} + a_6 \frac{T^3}{3} + a_7 \frac{T^4}{4} + a_9$$

Calculate for the gaseous mixture (1) the enthalpy, (2) the entropy, and (3) the Gibbs energy.
Answers:
(1) h (3394.5 K) $= -2167.02$ kJ/kg;
(2) s (117.105 bar, 3394.5 K) $= 17.17348$ kJ/(kgK); &
(3) g (117.105 bar, 3394.5 K) $= -60462.4$ kJ/kg.

4.42 Although hypergolic propellant combinations are toxic, they have the advantage of auto ignition. Therefore, they are used in many upper stage propulsion systems for enhanced mission-reliability. One such combination is liquid fuel monomethyl hydrazine (MMH) CH_6N_2 with liquid oxidizer nitrogen tetroxide N_2O_4. Calculate kgatoms of each element per kg of propellant, b_i^o for an oxidizer–fuel ratio of 2.5. Atomic masses of the elements are the following. Carbon: 12.0107 kg/kgatom; hydrogen: 1.00794 kg/kgatom; nitrogen: 14.0067 kg/kgatom; and oxygen: 15.9994 kg/kgatom.

Answers:
$b_C^o = 6.201502 \times 10^{-3}$ kgatom/kg; $b_H^o = 37.209012 \times 10^{-3}$ kgatom/kg
$b_N^o = 27.929104 \times 10^{-3}$ kgatom/kg; $b_O^o = 31.0522 \times 10^{-3}$ kgatom/kg

4.43 A heterogeneous (composite) propellant contains ammonium perchlorate (AP), hydroxyl terminated polybutadiene (HTPB) of molecular form $C_{73}H_{110}O_6$, and aluminum in the mass ratio: AP:HTPB:Al = 0.69:0.12:0.19. Calculate kgatoms of each element per kg of propellant, b_i^o. Atomic masses of the elements are as follows. Al: 26.981538 kg/kgatom; C: 12.0107 kg/kgatom; Cl: 35.453 kg/kgatom; H: 1.00794 kg/kgatom; N: 14.0067 kg/kgatom; & O: 15.9994 kg/kgatom
Answers:
$b_{Al}^o = 7.041852 \times 10^{-3}$ kgatom/kg; $b_C^o = 8.083801 \times 10^{-3}$ kgatom/kg;
$b_{Cl}^o = 5.87289 \times 10^{-3}$ kgatom/kg; $b_H^o = 3.567263 \times 10^{-2}$ kgatom/kg;
$b_N^o = 5.87289 \times 10^{-3}$ kgatom/kg; & $b_O^o = 2.415598 \times 10^{-2}$ kgatom/kg.

4.44 For the specified values of a pair state functions ($T\&p, h\&p, s\&p, u\&v$, etc.), the converged equilibrium composition has to simultaneously satisfy the energy minimization condition given by Eq. 4.45 and the element balance condition given by Eq. 4.54. Evidently, iterations are involved to arrive at the solution. The present example is related to the calculation of equilibrium composition as the combustion products expand in a rocket nozzle for a pair of specified values of entropy and pressure $s\&p$. The propellant combination by mass proportions is $H_2O_2 : H_2O : H_2 = 0.874 : 0.076 : 0.050$. To start the iterations, the first trial set of x_j's are $x_H = 0.001$, $x_{H_2} = 0.001$, $x_{H_2O} = 0.990$, $x_{OH} = 0.001$, and $x_{O_2} = 0.007$. And, the related trial molar mass of the combustion products is 17.5 kg/kgmol. Calculate the element balance errors $\left(b_i^o - b_i\right)$s.
Answers:
$b_H^o - b_H = -3.938 \times 10^{-3}$ kgatom/kg; $b_O^o - b_O = -1.8203 \times 10^{-3}$ kgatom/kg.

4.45 The combustion chamber of an experimental gas-turbine jet engine is operating with hydrogen fuel at an air–fuel mass ratio of 50 at a pressure of 25 bar. For simplicity, assume that the air composition is only of 3.76 moles of nitrogen and one mole of oxygen. The fuel and air are at standard temperature of 298.15 K. Product mixture can be assumed to contain only H_2O, NO, N_2, OH, and O_2. Calculate (1) adiabatic flame temperature, (2) species mole fractions, and (3) molar mass of the product mixture. Refer *themo.inp* for the required thermodynamic data.
Answers:
(1) 1987.4 K; (2) $x_{H_2O} = 2.4977 \times 10^{-1}$; $x_{NO} = 3.5880 \times 10^{-3}$; $x_{N_2} = 0.689$; $x_{OH} = 1.1094 \times 10^{-3}$; and $x_{O_2} = 5.6495 \times 10^{-2}$; (3) 25.736 kg/kgmol.

4.46 The mono-propellant hydrazine N_2H_4 has been finding use in propulsion units for orbit maneuvers and station keeping of satellites. However such chemical propulsion units are increasingly being replaced by electric propulsion units with their energy source from thermo-electric nuclear devices. Furthermore, the electric propulsion units with such energy source, not being energy-limited, are the only power source for interplanetary missions. Be as it may, let us now consider a hydrazine thruster for the present example. The enthalpy of formation of hydrazine $\overline{h}_f^0 = 50380$ kJ/kgmol. The thruster operates at the nozzle entry total pressure of 14 bar with hydrazine being injected at the standard temperature of 298.15 K. At the nozzle entry, the calculated properties of the reaction products are the following. Adiabatic flame temperature = 876.5 K. Molar mass $\overline{m} = 10.75$ kg/kgmol. Product mole fractions are $x_{H_2} = 0.66132$, $x_{NH_3} = 0.00641$, and $x_{N_2} = 0.33226$. Entropy $s = 15.5238$ kJ/(kgK). (1) Check the enthalpy balance at the nozzle entry level. Assuming isentropic shifting-equilibrium flow, (2) calculate for a nozzle pressure ratio of 500 the following for the mixture at the nozzle exit: (1) enthalpy, (2) Gibbs energy, (3) molar mass, (4) mole fractions of the species, and (5) temperature. Refer *themo.inp* for the required thermodynamic data. [Hint: Start with trial mole fractions of $x_{H_2} = 0.46$, $x_{NH_3} = 0.25$, and $x_{N_2} = 0.29$ with ratio of specific heats at 1.14]
Answers:
(1) Reactant enthalpy = nozzle entry stagnation enthalpy $h_{0n} = 1572$ kJ/kg; (2) $h_e = -746.0$ kJ/kg; $g_e = -6135.1$ kJ/kg; (3) $\overline{m}_e = 13.35$ kg/kgmol; (4) $x_{H_2} = 0.4586$, $x_{NH_3} = 0.2497$, and $x_{N2} = 0.2917$; and (5) $T_e = 347.2$ K.

4.47 The Japanese LE-5B upper stage cryogenic engine uses liquid hydrogen and liquid oxygen at an oxidizer–fuel ratio of 5 [10]. Combustion chamber pressure is 36 bar and the nozzle area ratio is 110. It delivers a thrust of 137 kN in vacuum with a specific impulse of 4384 Ns/kg.

Considering only H, H_2, H_2O, and OH as significant species, the theoretical calculation at an oxidizer–fuel ratio of 5 for LH2(20.27 K)/LOX(90.17 K) combination at a chamber pressure 36 bar leads to the following results in the combustion chamber. Adiabatic flame temperature = 3242.5 K. Entropy of combustion products = 19.8571 kJ/(kgK). Molar mass = 11.772 kg/kgmol. The values of mole fractions are $x_H = 0.03133$, $x_{H_2} = 0.35552$, $x_{H_2O} = 0.59102$, & $x_{OH} = 0.02212$. (1) Calculate at the combustion chamber conditions: density, enthalpy, and Gibbs energy. Under the assumption of shifting-equilibrium flow with nozzle pressure ratio of 2050, calculate at the nozzle-exit plane (2) temperature, mole fractions, density, enthalpy, and Gibbs energy. [Hint: To arrive at the first trial value for the nozzle-exit temperature assume the ratio of specific heats γ to be 1.22 and calculate the trial nozzle-exit temperature $T_e = T_c(p_e/p_c)^{(\gamma-1)/\gamma}$. With a high nozzle pressure ratio of 2050, the dissociated species H and OH are expected to get recombined to form more of H_2 and H_2O. Assume therefore the mole fractions x_H and x_{OH} to be in traces at the nozzle exit, say 10^{-6} or less.]
Answers:
(1) density = 1.5719 kg/m^3, enthalpy = −1083.09 kJ/kg, and Gibbs energy = −65470.5 kJ/kg.

(2) temperature = 824 K, molar mass = 12.095 kg/kgmol, mole fractions: x_H = trace, $x_{H_2} = 0.37$, $x_{H_2O} = 0.63$, & x_{OH} = trace, density = 3.1×10^{-3} kg/m^3, enthalpy = −11138 kJ/kg, and Gibbs energy = −27501 kJ/kg.

4.48 Repeat Problem 4.47 under the assumption of frozen-equilibrium flow.
Solution Guidelines:
While the answers under part (1) of Problem 4.47 remain the same, the answers for part (2) have to be calculated. Note under frozen-equilibrium flow, mole fractions, and molar mass are constant ("frozen"). Energy minimization, element balance $\left(b_i^o = b_i\right)$, and enthalpy balance $(h^* = h)$ are applied to calculate combustion chamber conditions. Thereafter, as the mole fractions are frozen, these conservation equations Eqs. 4.66–4.69 become redundant for the frozen-equilibrium flow. Only the entropy balance $(s^* = s)$ Eq. 4.84 in a simplified form has to be satisfied for the frozen-equilibrium flow. The simplified form of Eq. 4.84 is

$$\left(\sum_{j=1}^{NS} \frac{n_j \bar{c}^o_{p,j}}{R_u}\right) \Delta ln\ T = \frac{s^* - s}{R_u} \tag{4.84a}$$

where

$$s = \sum_{j=1}^{NG} n_j \bar{s}_j = \sum_{j=1}^{NG} n_j \left[\bar{s}_j^0 (T) - R_u \ln \tfrac{n_j}{n} - R_u \ln\ p\right]$$
$$+ \sum_{j=NG+1}^{NS} n_j \bar{s}_j^0 (T) \rightarrow p\ in\ [\text{bar}]$$

Thus the simplified equation of Newton–Raphson for isentropic flow with frozen-equilibrium flow becomes a function of only temperature T and pressure p at the specified nozzle section. With the pressure at the nozzle-exit plane being specified, the simplified equation becomes the only function of temperature, which has to be varied to find that temperature that satisfies the entropy balance $(s^* = s)$.
Answers:
(1) density = 1.5719 kg/m^3, enthalpy = −1083.09 kJ/kg, and Gibbs energy = −65470.5 kJ/kg.
(2) temperature = 690 K, molar mass = 11.772 kg/kgmol, mole fractions: x_H = 0.03133, $x_{H_2} = 0.35552$, $x_{H_2O} = 0.59102$, & $x_{OH} = 0.02212$, density = 3.60×10^{-3}kg/m^3, enthalpy = −10407 kJ/kg, and Gibbs energy = −24116 kJ/kg.

4.49 The Space Shuttle Main Engine (SSME) uses liquid hydrogen and liquid oxygen at an oxidizer–fuel ratio of 6. Combustion chamber pressure is 204 bar. It delivers a thrust of 2090 kN in vacuum with the nozzle area ratio of 77.5 [10].

Following the procedure detailed in Sect. 4.4 and Example 4.5, for the propellant combination of LOX(90.17 K)/LH2(20.27 K) at an oxidizer–fuel ratio of 6 and a chamber pressure of 204 bar, combustion chamber conditions are calculated. For the simplicity of calculations, only H, H_2, H_2O, & OH are considered as the significant species. Calculated combustion properties in the chamber are the following. Adiabatic flame temperature $T_0 = 3619.5$ K. $\overline{m} = 13.645$ kg/kgmol.

Enthalpy = −986.31 kJ/kg. Entropy $s = 17.1734$ kJ/(kgK). Species mole fractions are $x_H = 0.02650$, $x_{H_2} = 0.24250$, $x_{H_2O} = 0.69140$, and $x_{OH} = 0.03960$.

Due to heat transfer and other losses, and less efficient combustion, the chamber temperature reached may be lower than the adiabatic flame temperature. Therefore for the propellant combination of LOX/LH2 at the oxidizer–fuel ratio of 6, the chamber pressure of 204 bar, and the product mixture temperature of 3560 K, calculate the following: (1) species mole fractions of the equilibrium composition; (2) molar mass; (3) enthalpy; (4) Gibbs energy; (5) entropy; and (6) density.

Answers:
(1) $x_H = 0.0233$, $x_{H_2} = 0.2425$, $x_{H_2O} = 0.7000$, and $x_{OH} = 0.0342$;
(2) 13.705 kg/kgmol; (3) −1383.3kJ/kg; (4) −62127 kJ/kg;
(5) 17.063 kJ/(kgK); and (6) 9.446 kg/m^3.

4.50 Derive Eqs. 4.97 and 4.98:

$$\left(\frac{\partial \ln n_j}{\partial \ln T}\right)_p = \sum_{i=1}^{l} a_{ij}\left(\frac{\partial \pi_i}{\partial \ln T}\right)_p + \left(\frac{\partial \ln n}{\partial \ln T}\right)_p + \frac{\overline{h}_j^0}{R_u T} \rightarrow j = 1, \ldots, NG \tag{4.97}$$

$$\sum_{i=1}^{l} a_{ij}\left(\frac{\partial \pi_i}{\partial \ln T}\right)_p = -\frac{\overline{h}_j^0}{R_u T} \rightarrow j = NG + 1, \ldots, NS \tag{4.98}$$

4.51 The combustion chamber of a ramjet flying above tropopause at a supersonic condition has the following calculated properties of combustion products. Chamber pressure = 4.26 bar. Adiabatic flame temperature = 2465 K. Molar mass = 28.653 kg/kgmol. $x_{CO} = 1.473 \times 10^{-2}$; $x_{CO_2} = 1.1371 \times 10^{-1}$; $x_{H_2O} = 1.2309 \times 10^{-1}$; $x_{N_2} = 7.3548 \times 10^{-1}$; and $x_{O_2} = 1.298 \times 10^{-2}$. (1) Under equilibrium condition, calculate the values of the three key derivatives $(\partial \ln v / \partial \ln T)_p$, $(\partial \ln v / \partial \ln p)_T$, and $c_{p,e}$. (2) Identify the frozen component $c_{p,f}$ in the value of the third key derivative $c_{p,e}$, the specific heat under equilibrium condition. Also calculate under equilibrium conditions the values of (3) the ratio of specific heats γ_e, (4) the isentropic exponent γ_s, and the speed of sound. In addition, calculate under frozen conditions (5) the ratio of specific heats γ_f and (6) the speed of sound. Refer Tables 4.20 and 4.21 for the thermodynamic data of the species.

Answers:
(1) $(\partial \ln v / \partial \ln T)_p = 1.0702$, $(\partial \ln v / \partial \ln p)_T = -1.0026$, and $c_{p,e} = 2.0014$ kJ/(kgK); (2) $c_{p,f} = 1.4536$ kJ/(kgK);
(3) $\gamma_e = 1.1985$; (4) $\gamma_s = 1.1954$;
(5) speed of sound under equilibrium condition = 924.7 m/s;
and (6) speed of sound under frozen condition = 945.4 m/s.

4.52 The combustion gases of the ramjet of Problem 4.51 expand through the engine's convergent–divergent nozzle under isentropic shifting-equilibrium condition. The nozzle pressure ratio is 22. The calculated equilibrium composition of the gaseous mixture at the nozzle exit has the following properties. Static temperature

= 1368.1 K. Molar mass = 28.866 kg/kgmol. $x_{CO} = 1.0 \times 10^{-6}$; $x_{CO_2} = 1.2939 \times 10^{-1}$; $x_{H_2O} = 1.24 \times 10^{-1}$; $x_{N_2} = 7.4094 \times 10^{-1}$; and $x_{O_2} = 5.66 \times 10^{-3}$. (1) Calculate the values of the three key derivatives under equilibrium condition. (2) Identify the frozen component $c_{p,f}$ in the value of the third key derivative, the specific heat under equilibrium condition $c_{p,e}$. Also calculate under the equilibrium condition the values of (3) the ratio of specific heats γ_e, (4) the isentropic exponent γ_s, and (5) the speed of sound. Refer Tables 4.20 and 4.21 for the thermodynamic data of the species.
Answers:
(1) $\left(\partial \ln v / \partial \ln T\right)_p = 1.0000$, $\left(\partial \ln v / \partial \ln p\right)_T = -1.0000$, and $c_{p,e} = 1.3472$kJ/(kgK)
(2) $c_{p,f} = 1.3454$ kJ/(kgK); (3) $\gamma_e = 1.2720$; (4) $\gamma_s = 1.2720$;
(5) Speed of sound under equilibrium condition = 708.0 m/s; (6) $\gamma_f = 1.2724$; (7) Speed of sound under frozen condition = 708.09 m/s
Observation: When the dissociated species in the reacting mixture are in traces (here the only dissociated species is CO and its mole fraction is in traces), the sum of the reaction components is vanishingly small and hence the key derivatives are more or less equal to those of inert composition and the ratio of specific heats is tending to be equal to the isentropic exponent, $\gamma_e \approx \gamma_s$. Study Example 4.10.

4.53 The combustion gases of the ramjet of Problem 4.51 expand through the engine's convergent–divergent nozzle under isentropic frozen-equilibrium condition. The nozzle pressure ratio is 22. As the gases expand under isentropic frozen-equilibrium condition, the composition remains the same as at the combustion chamber: $x_{CO} = 1.473 \times 10^{-2}$; $x_{CO_2} = 1.1371 \times 10^{-1}$; $x_{H_2O} = 1.2309 \times 10^{-1}$; $x_{N_2} = 7.3548 \times 10^{-1}$; and $x_{O_2} = 1.298 \times 10^{-2}$. Hence molar mass also remains constant at 28.653 kg/kgmol. For the isentropic expansion, the static temperature at the nozzle exit is calculated as 1297.4 K. As the composition is frozen and hence inert, $\left(\partial \ln v / \partial \ln T\right)_p = 1.0000$ and $\left(\partial \ln v / \partial \ln p\right)_T = -1.0000$. Calculate (1) the third key derivative c_p, (2) the ratio of specific heats γ, and (3) the speed of sound.
Answers:
(1) $c_p = 1.3304$ kJ/(kgK); (2) $\gamma = 1.279$; and (3) speed of sound = 693.9 m/s.

4.54 Enthalpy of formation of the polymer hydroxyl terminated polybutadiene (HTPB) $C_{73}H_{110}O_6$ is −136.08 /gmol. Calculate the enthalpy of formation for one kg of HTPB.
Answer:
$h^0_{f,HTPB} = -125.5755$ kJ/kg

4.55 Liquid oxygen at 90.17 K and liquid hydrogen at 20.27 K burn in a rocket chamber with an oxidizer–fuel ratio of 5.6 at 53 bar. The calculated adiabatic flame temperature, entropy, and all-gaseous molar mass are respectively 3392 K, 18.5878 kJ/(kgK), and 12.793 kg/kgmol. Calculate for unit mass of combustion products (1) number of kgmoles of species, (2) enthalpy, and (3) Gibbs energy. Also calculate the density of combustion products.

Answers:
(1) $n = 7.81677 \times 10^{-2}$ kgmol/kg; (2) $h = -1021.5$ kJ/kg; (3) $g = -64071.3$ kJ/kg; and (4) $\rho = 2.4041$ kg/m^3.

4.56 A composite solid propellant containing ammonium perchlorate, hydroxyl terminated polybutadiene, aluminum, and the additive magnesium oxide burns at a pressure of 34 bar. Adiabatic flame temperature is 2723 K. Among many product species, mole fractions of the species H_2 and $Al_2O_3(l)$ are 0.3215 and 0.03672 respectively. Molar mass of the combustion products including condensed species $\overline{m}_c$ is 22.29 kg/kgmol. Total number of moles of gaseous product species is 4.3215×10^{-2} kgmol/kg. For the two product species H_2 and $Al_2O_3(l)$, calculate on molar basis (1) specific heats at constant pressure, (2) enthalpies, (3) entropies, and (4) chemical potentials.
Answers:
(1) $\overline{c}^0_{p,H_2}$ (2723 K) = 36.419 kJ/(kgmolK);
$\overline{c}^0_{p,Al_2O_3(l)}$ (2723 K) = 162.9 kJ/(kgmolK)
(2) $\overline{h}^0_{H_2}$ (2723 K) = 78549.9 kJ/kgmol; $\overline{h}^0_{Al_2O_3(l)}$ (2723 K) = −1242358 kJ/kgmol
(3) $\overline{s}_{H_2}$ (2723 K, 34 bar) = 179.13 kJ/(kgmolK);
$\overline{s}_{Al_2O_3(l)}$ (2723 K) = $\overline{s}^0_{Al_2O_3(l)}$ (2723 K) x = 366.72 kJ/(kgmolK)
(4) μ_{H_2} (2723 K, 34 bar) = −409225 kJ/kgmol;
$\mu_{Al_2O_3(l)}$ (2723 K) = −2240928 kJ/kgmol.

4.57 R-1E Marquardt (USA) N2O4/MMH rocket engine was used as Space Shuttle Orbiter's vernier attitude control and orbit adjust thruster. The engine delivers a nominal thrust of 110 N at the oxidizer–fuel ratio of 1.65 and chamber pressure of 7.4 bar. It operates with the vacuum specific impulse of 2746 Ns/kg with the expansion ratio of 100. Total burn time = 1500 s. The problem that is stated below is for a thruster similar to R-1E.

The calculated equilibrium composition in the combustion chamber of a rocket thruster burning the hypergolic propellant combination nitrogen tetroxide N_2O_4 and monomethyl hydrazine CH_6N_2 at the oxidizer–fuel ratio of 1.65 and the chamber pressure of 7.4 bar has the following properties. Adiabatic flame temperature = 3161 K. All gaseous molar mass = 20.642 kg/kgmol. Specific heat at constant pressure for frozen condition at the chamber c_p is 2.1631 kJ/(kgK). The calculated vacuum specific impulse for a nozzle pressure ratio of 100 under frozen flow assumption is 2966.4 Ns/kg. (1) Calculate the stagnation density of gases in the combustion chamber. (2) Assuming the flow to be isentropic and frozen and further assuming c_p to be constant as at the chamber, evaluate Mach number, static temperature, and pressure at the entry to the nozzle for the nozzle convergence ratio of 1.58. (3) For a specific impulse quality factor ξ_s of 88% and a thrust of 110 N, size the nozzle at its entry, throat, and exit. (4) If the total sum of the burn times with thrust pulses can be 1500 s, estimate the mass of oxidizer and fuel to be stored. Allow 5% for the loss in propellant storage.
PS: Assumption of constant c_p is a drastic simplification of the problem since even under frozen flow assumption c_p has to vary as a function of temperature.

Answers:
(1) $\rho_0 = 0.5812$ kg/m^3;
(2) $M_n = 0.4097$, $T_n = 3101.5$ K, $p_n = 6.68157$ bar, and $\rho_n = 0.5348$ kg/m^3
(3) $d_n = 14.06$ mm, $d_t = 11.2$ mm, $d_e = 37.4$ mm, and $A_e/A_t = 11.15$
(4) $m_o = 39.357$ kg *and* $m_f = 23.852$ kg.

4.58 Repeat Problem 4.57 with the assumption of isentropic frozen flow—the variation of c_p with temperature is to be considered. The calculated equilibrium composition in the combustion chamber of the thruster burning nitrogen tetroxide N_2O_4 and monomethyl hydrazine CH_6N_2 at the oxidizer–fuel ratio of 1.65 and the chamber pressure of 7.4 bar has the following properties. Adiabatic flame temperature = 3161.4 K. All gaseous molar mass $\overline{m} = 20.642$ kg/kgmol. The other calculated properties at the combustion chamber are the following. Specific heat at constant pressure $c_p = 2.1631$ kJ/(kgK). Enthalpy = 325.18 kJ/kg. Entropy = 12.7887 kJ/(kgK). $x_{CO} = 0.13492, x_{CO_2} = 0.03416, x_{H_2} = 0.16666, x_{H_2O} = 0.32559, x_{N_2} = 0.30876$, and $x_{OH} = 0.02992$. (1) Calculate the stagnation density of gases in the combustion chamber. (2) Calculate the throat properties: c_p, γ, static pressure, temperature, and -density; (3) Evaluate c_p, γ, Mach number, and static pressure and temperature at the entry to the nozzle for the nozzle contraction ratio of 1.58. (4) Calculate the vacuum specific impulse for the nozzle pressure ratio of 100. (5) For a specific impulse quality factor ξ_s of 88% and a thrust of 110 N, size the nozzle at its entry, throat, and exit. (6) If the total sum of the burn times with thrust pulses can be 1500 s, estimate the mass of oxidizer and fuel to be stored. Allow 5% for the loss in propellant storage.
Answers:
(1) $\rho_0 = 0.5812$ kg/m^3
(2) $c_{p,t} = 2.1334$ KJ/(kgK), $\gamma_t = 1.2328$, $p_t = 4.1299$ bar, $T_t = 2833.95$ K, and $\rho_t = 0.3618$ kg/m^3
(3) $c_{p,n} = 2.1581$ KJ/(kgK), $\gamma_n = 1.2295$, $M_n = 0.4099$, $p_n = 6.6804$ bar, and $T_n = 3101.67$ K
(4) $I_{sp,vac} = 2966.4$ Ns/kg
(5) $d_n = 14.05$ mm, $d_t = 11.18$ mm, $d_e = 36.29$ mm, and $A_e/A_t = 10.536$
(6) $m_o = 41.323$ kg and $m_f = 25.044$ kg.

4.59 Mitsubishi LE-5B (Japan) LOX/LH2 rocket engine of the H-H rocket vehicle's second stage delivers a vacuum thrust of 137 kN at a vacuum specific impulse of 4384 Ns/kg for 534 s at a chamber pressure of 36 bar, a nozzle area ratio of 110, and an O/F ratio of 5 [10].

Let us consider a similar engine for our internal ballistics analysis. For the propellant combination of LOX(90.17 K)/LH2(20.27 K) at an oxidizer–fuel ratio of 5 and a chamber pressure of 36 bar, combustion chamber conditions are calculated (see Example 4.5 for the procedure). For the simplicity of calculations, only H, H_2, H_2O, and OH are considered as the significant species. Calculated combustion properties in the chamber are the following. $T_0 = 3242.54$ K. All gaseous molar mass $\overline{m} = 11.772$ kg/kgmol. Enthalpy $h_0 = -1083.09$ kJ/kg. Entropy $s = 19.8571$ kJ/(kgK). Isentropic exponent $\gamma_s = 1.1545$. Species mole fractions are $x_H = 0.03133$; $x_{H_2} = 0.35552$; $x_{H_2O} = 0.59102$; and $x_{OH} = 0.02212$. (1) Assuming shifting-equilibrium isentropic flow calculate the nozzle throat conditions and the nozzle-exit conditions for a nozzle area ratio of 110. (2) Also, evaluate the rocket performance parameters: c^*, $C_{F,vac}$, and $I_{sp,vac}$. [Hint: As the nozzle area ratio is very high, the nozzle-exit temperature is expected to be below 1000 K with substantial recombination of dissociated species H and OH into H_2 and H_2O. Therefore start with trace values, say less than 10^{-6}, for mole fractions of H and OH. Adopt thermodata for the species for the temperature range of 200–1000 K.]

The considered engine has to deliver a vacuum thrust of 137 kN for 534 s. Assume that the experimental vacuum specific impulse is 95% of the theoretical one. A design parameter for combustion chamber envelope L^* named as characteristic length is defined as the ratio of chamber volume to throat area, V_c/A_t. The considered engine is to operate on an expanded cycle—one of the propellants (here LH2) first cooling the nozzle and the combustion chamber, subsequently running the turbo-pumps unit, and finally getting injected into the combustion chamber. The recommended value for such an expander-cycle engine with LOX/LH2 combination is $0.56 \leq L^* \leq 0.71$ m [12]. (3) Assuming a nozzle contraction ratio of 2.5 and L^* of 0.6 m, evaluate the engine envelope. (4) Assuming 5% loss in propellant storage, calculate the mass of LOX and LH2 to be stored in the propellant tanks. (5) Assuming 5% of the propellant volume for the ullage, estimate the volumes of the propellant tanks.

Answers:
(1-i) Nozzle throat: $p_t = 20.609$ bar, $T_t = 3031.82$ K,
$x_H = 0.02299$, $x_{H_2} = 0.35876$, $x_{H_2O} = 0.60400$, $x_{OH} = 0.01425$,
$\overline{m}_t = 11.870$ kg/kgmol, $\gamma_t = 1.1698$, $\gamma_{s,t} = 1.1590$, $M_t = 1$, and $c^* = 2364.6$ m/s
(1-ii) Nozzle exit: $p_e = 0.01746$ bar, $T_e = 822.89$ K,
$x_H = trace$, $x_{H_2} = 0.37001$, $x_{H_2O} = 0.62999$, $x_{OH} = trace$,
$\overline{m}_e = 12.095$ kg/kgmol, $\gamma_e = 1.3052$, $\gamma_{s,e} = 1.3052$, and $M_e = 5.220$
(2) $c^* = 2364.6$ m/s, $C_{F,vac} = 1.9501$, and $I_{sp,vac} = 4611.2$ Ns/kg
(3) $d_c = 255.7$ mm, $l_c = 240.0$ mm, $d_t = 161.7$ mm, and $d_e = 1696.2$ mm
(4) $m_p = 17535.4$ kg, $m_{p,LOX} = 14612.9$ kg, and $m_{p,LH2} = 2922.6$ kg
(5) $V_{tank,LOX} = 13.354$ m^3 and $V_{tank,LH2} = 43.282$ m^3.

4.60 Vulcain (France) LOX/LH2 rocket engine that powered the cryogenic core stage of Ariane-5 delivers a vacuum thrust of 1075 kN at a vacuum specific impulse of 4227 Ns/kg for 605 s at a chamber pressure of 102 bar, a nozzle area ratio of 45, and an O/F ratio of 6.2. Unfuelled mass of the engine is 1300 kg [10].

Let us consider a similar engine for our internal ballistics analysis. For the propellant combination of LOX(90.17 K)/LH2(20.27 K) at an oxidizer–fuel ratio of 6.2 and a chamber pressure of 102 bar, combustion chamber conditions are calculated (see Example 4.5 for the procedure). For the simplicity of calculations, only H, H_2, H_2O, and OH are considered as the significant species. Calculated combustion properties in the chamber are the following. $T_0 = 3589.12$ K. All gaseous molar mass $\overline{m} = 13.877$ kg/kgmol. Enthalpy $h_0 = -970.18$ kJ/kg. Entropy $s = 17.3345$ kJ/(kgK). Isentropic exponent $\gamma_s = 1.1435$. Species mole fractions are

$x_H = 0.03342$; $x_{H_2} = 0.21969$; $x_{H_2O} = 0.69250$; and $x_{OH} = 0.05439$. (1) Assuming shifting-equilibrium isentropic flow, calculate the nozzle throat conditions and the nozzle-exit conditions for a nozzle area ratio of 45. (2) Also, evaluate the theoretical rocket performance parameters: c^*, $C_{F,vac}$, and $I_{sp,vac}$.

The considered engine has to deliver a vacuum thrust of 1075 kN for 605 s. Assume the experimental vacuum specific impulse is 95% of the theoretical one. A design parameter for combustion chamber envelope L^* named as characteristic length is defined as the ratio of chamber volume to throat area, V_c/A_t. The recommended value of L^* with LOX/LH2 combination is $0.56 \leq L^* \leq 0.71m$ [12]. (3) Assuming a nozzle contraction ratio of 1.1 and L^* of 0.64 m, evaluate the engine envelope. (4) Assuming 5% loss in propellant storage, calculate the mass of LOX and LH2 to be stored in the propellant tanks. (5) Assuming 5% of the propellant volume for the ullage, estimate the volumes of the propellant tanks.

(6) Assuming the thrust-mass ratio for the engine as 830 N/kg and the propellant tank mass fraction of 0.07 of propellant mass, estimate the cryogenic stage mass.

Answers:
(1-i) Nozzle throat: $p_t = 58.721$ bar, $T_t = 3382.55$ K,
$x_H = 0.02738, x_{H_2} = 0.21827, x_{H_2O} = 0.71302, x_{OH} = 0.04133,$
$\overline{m}_t = 14.0157$ kg/kgmol, $\gamma_t = 1.1627, \gamma_{s,t} = 1.1434, M_t = 1,$ and $c^* = 2301.1$ m/s
(1-ii) Nozzle exit: $p_e = 0.1927$ bar, $T_e = 1452.59$ K,
$x_H = trace, x_{H_2} = 0.21881, x_{H_2O} = 0.78118, x_{OH} = trace,$
$\overline{m}_e = 14.5143$ kg/kgmol, $\gamma_e = 1.2357$, $\gamma_{s,e} = 1.2357$, and $M_e = 4.1946$
(2) $c^* = 2301.1$ m/s, $C_{F,vac} = 1.9336$, and $I_{sp,vac} = 4449.2$ Ns/kg
(3) $d_c = 283.5$ mm, $l_c = 581.8$ mm, $d_t = 270.3$ mm, and $d_e = 1813.1$ mm
(4) $m_p = 161564$ kg, $m_{p,LOX} = 139125$ kg, and $m_{p,LH2} = 22439$ kg
(5) $V_{tank,LOX} = 127.137$ m^3 and $V_{tank,LH2} = 332.319$ m^3.
(6) Estimated stage mass = 174169 kg.

References

Cited

1. Penner SS (1957) Chemistry problems in jet propulsion. Pergamon Press, New York
2. Barrere M, Jaumotte A, De Veubeke BF, Vandenkerckhove J (1960) Rocket propulsion. Elsevier Publishing Company, New York
3. Humble RW, Henry GN, Larson WJ (1995) Space propulsion analysis and design. McGraw-Hill, New York
4. Hill PG, Peterson C (1999) Mechanics and thermodynamics of propulsion, 2nd edn. Addison Wesley, Reading Massachusetts
5. Gordon S, McBride BJ (1976) Computer program for calculating complex chemical equilibrium compositions, rocket performance, incident and reflected shocks and Chapman-Jouguet detonations. NASA SP-273
6. Gordon S, McBride BJ (1994) Computer program for calculation of complex chemical equilibrium compositions and applications: Part I Analysis. NASA RP 1311

7. McBride BJ, Gordon S (1996) Computer program for calculation of complex chemical equilibrium compositions and applications: Part II Users manual and program description. NASA RP 1311
8. Chemical Equilibrium with Applications, Glenn Research Center NASA https://software.nasa.gov/software/LEW-17687-1. Accessed 7 July 2022
9. Krishnan S, Raghavan J (2020) Chemical rockets: performance prediction and internal ballistics design. Springer, Switzerland
10. The Encyclopedia Astronautica http://www.astronautix.com. Accessed 7 July 2022
11. Propulsion Web Page, Purdue School of Aeronautics and Astronautics, Purdue University https://engineering.purdue.edu/~propulsi/propulsion/index.html. Accessed 11 May 2019
12. Huzel DK, Huang DH (1992) Modern engineering for design of liquid-propellant rocket engines, vol 147. Progress in astronautics and aeronautics. American Institute of Aeronautics and Astronautics, Washington DC

Further Reading

13. Heister SD, Anderson WE, Pourpoint TL, Cassady RJ (2019) Rocket propulsion. Cambridge University Press, Cambridge
14. Sutton GP, Biblarz O (2017) Rocket propulsion elements, 9th edn. Wiley, New York
15. Cengal YA, Boles MA (2011) Thermodynamics an engineering approach, 7th edn. McGraw-Hill, Singapore
16. Nag PK (2008) Engineering thermodynamics, 4th edn. Tata McGraw Hill, New Delhi
17. Black WZ, Hartley JG (1985) Thermodynamics. Harper & Row, New York

Chapter 5
Stage Optimization and Trajectories

Synopsis

Adopting multistage rockets is absolutely essential to carry payloads to orbital altitudes or for very long ranges. Of the two methods of multistaging, the parallel staging is more frequently adopted than the serial one. However, the latter one is found to be relevant for lighter payloads. The use of engine multistaging, the third method, appears to be waning. The data of many satellite launch vehicles show that the adoption of LOX/Kerosene and LOX/LH2 is favored against the toxic N2O4/UDMH or HNO3/UDMH systems. The use of all liquid launch vehicles appears to be suitable for large payloads. Nevertheless, for the application to long-range missiles, the ever-ready property of solid systems is seen to be of higher priority than the higher energy content in liquid systems.

The structural coefficient of a stage ε is the ratio of the structural mass m_s to the gross stage mass m. Higher specific impulse and lower ε are always sought after to design a lighter launch vehicle. The trends of ε with the proven stages of different masses and propellant combinations are presented. These indicate that for a chosen propellant combination, realizing a lower ε for a heavier stage mass is possible. But this trend is not applicable for the solid systems as these are of nearly constant high ε's. The properties of multistage vehicles are characterized by payload ratio $\alpha_{(i)}$ and mass ratio $\mu_{(i)}$. $\alpha_{(i)}$ is the ratio of the partial vehicle mass $m_{0(i+1)}$ to the ith gross stage mass $m_{(i)}$. $\mu_{(i)}$ is the ratio of the partial vehicle mass $m_{0(i)}$ to its burnout mass $m_{0b(i)}$. Equations relating $\alpha_{(i)}$ and $\mu_{(i)}$ involving $\varepsilon_{(i)}$ are given.

Given the terminal velocity v_n and the number of stages n with their expected specific impulses $u_{e(i)}$'s and $\varepsilon_{(i)}$'s, the task of optimization is to reach the target altitude with minimum propellants. The preliminary step to this task is loss-free optimization. That is, the flight of the vehicle is considered in a gravity- and drag-free environment. The target altitude is absent in this time-independent procedure.

S. Krishnan and J. Raghavan, *Rocket Propulsion Primer*,
https://doi.org/10.1007/978-981-97-5644-5_5

With the specified v_n as the constraint, we minimize the ratio of the lift-off mass to payload mass $m_{0(1)}/m_l$ (or maximize its inverse, the payload fraction $m_l/m_{0(1)}$). Also equivalently, with $m_{0(1)}/m_l$ as the constraint one can maximize v_n. With the objective function to minimize $m_{0(1)}/m_l$ together with the constraint v_n, the resulting equation with the Lagrange multiplier λ is transcendental one. Therefore, for the given $u_{e(i)}$'s and $\varepsilon_{(i)}$'s, the range and the upper bound of λ are found. Thereafter with v_n as the constraint, the value of λ is numerically obtained to calculate the stage masses $m_{(i)}$'s and their velocity increments $\Delta v_{(i)}$'s. The basic drawback with this loss-free procedure is that it "favors" the most efficient stage (the stage with the highest $u_{e(i)}$ and the lowest $\varepsilon_{(i)}$) by assigning to that stage the heaviest mass and the largest velocity increment. This way, the first stage usually of the lowest u_e gets assigned with the lightest stage mass, contrary to reality. To circumvent this issue, the loss-free procedure can be run with a marked-up terminal velocity (the required terminal velocity plus the estimated total velocity losses). As an improvement over this procedure, the loss-free velocity increments $\Delta v_{(i)i}$'s can be added with individual stage velocity losses $\Delta v_{(i)l}$'s to calculate the scaled-up stage masses. These two procedures give an estimate of lift-off mass of the vehicle $m_{0(1)}$ reasonably close to the final design value. Yet, the values of the stage masses $m_{(i)}$'s and the velocity increments $\Delta v_{(i)}$'s are not satisfactory. Nonetheless, the latter procedure could give good improvements with suitable iterative $m_{(i)}$'s and $\Delta v_{(i)l}$'s drawn from a simple trajectory algorithm. All in all, the loss-free optimization procedure and its two variants are to be taken as preliminary steps to estimate the total vehicle mass and also to choose the number of stages.

In order to calculate the velocity losses due to gravity and drag, we have to first understand the properties of the Earth's atmosphere with altitude. The Earth's atmosphere is of various altitude layers with reducing pressure but with different temperature variations. Approximate limits and ranges for these are troposphere ($0 \le h \le 11$ km & T 300 $\Downarrow$ 200 K), stratosphere ($11 \le h \le 50$ km & T 200 $\Uparrow$ 300 K), mesosphere ($50 \le h \le 85$ km & T 300 $\Downarrow$ 200 K), thermosphere ($85 \le h \le 600$ km & T 200 $\Uparrow$ 900 K), and exosphere ($600 \le h \le 10000$ km & $T \approx 900$ K). Apart from this, there are two other considerations for the classification. The first one is on ionization and the related ionosphere $50 \le h \le 965$ km contains abundant ionized particles. The second consideration is on the air composition. The homosphere $0 \le h \le 120$ km has essentially a constant composition. Neglecting the temperature variations and assuming the gravitational acceleration to be invariant, the density variation in the homosphere can be represented by a semi-empirical equation $\rho = \rho_0 \, exp\,(-h/H)$. Here ρ_0 and H are curve fitted values for a reference location and time. The heterosphere $120 \le h \le 10,000$ km has a widely varying air composition.

The Earth's gravitational acceleration g decreases with altitude. By Newton's law of gravitation $g = g_e\,(r_e/(r_e + h))^2$, where g_e is the standard gravitational acceleration at the Earth's surface, and r_e is the Earth's mean radius. The continuum is characterized by the Knudsen number $Kn \equiv \lambda/L$, where λ is the mean free-path and L is the body's characteristic dimension. Generally, the mesosphere and below ($h \le 85$ km with $Kn \le 0.03$) can be taken as a continuum. Crossing the transition region, the region of $h \geq 150\,\text{km}$ with $Kn > 1$ is no longer a continuum and is known as the one of free molecular flow. The drag force D experienced by the vehicle is $D = (1/2)C_D\rho v^2 A_c$, where C_D is the drag coefficient and A_c the vehicle's characteristic area. In the continuum, C_D varies with Mach number M. Starting from a low value in the subsonic regime, C_D mostly peaks just following Mach number one and thereafter it tapers. Furthermore, given a flight Mach number and the body configuration, C_D is higher for a higher angle of attack, that is, a higher thrust angle γ. Since the atmospheric density ρ reduces exponentially with altitude, for rough calculations however, the drag force can be neglected above the stratosphere, say about $h > 50\,\text{km}$.

Assuming that the Earth's reference system is inertial and the vehicle is symmetrical (hence one of zero lift), the forces acting on the rocket vehicle and the controls needed to optimally steer the vehicle to its destination are considered. The time-dependent two-dimensional equations in polar coordinates are

$$m_0\dot{v} = F\cos\gamma - D - m_0 g\sin\beta,$$

$m_0 v\dot{\beta} = F\sin\gamma + m_0(v^2/(r_e + h))\cos\beta - m_0 g\cos\beta$, $\dot{h} = v\sin\beta$, and $\dot{x} = v\cos\beta$. Here F is the thrust force, γ is the thrust angle, and β is the flight path angle. Adopting these equations, the optimum trajectory to reach the target conditions can be found through a suitable numerical scheme.

The objective of the trajectory optimization is to attain the terminal conditions within the shortest time (to overcome the effect of gravity at the earliest) that too with the lowest velocity losses—the least possible propellant expenditure. Of the velocity losses, the steering loss is the least one and the gravity loss is the most. In between comes the drag loss which is high in the dense atmosphere at low altitudes and becomes negligible above the stratosphere. In view of this, the vehicle has to have a minimum acceleration and hence low velocities until the vehicle clears stratosphere. After the lift-off, the vehicle generally climbs vertically for sometime. On reaching an altitude through the vertical climb, the flight path angle β from its 90° has to be brought down close to 0° at the target altitude. This has to be done through a suitable flight path angle rate $\dot{\beta}$ in association with the thrust angle rate $\dot{\gamma}$. Therefore, with the thrust F in control, the effective variables for the trajectory optimization are β and γ.

The pitch angle $\phi \equiv \beta + \gamma$. In general, a maneuver with $\phi < 90°$ is a pitch maneuver—for the vertical climb $\phi = \beta = 90°$. An optimum trajectory comprises a judicious selection of different elements of the trajectory: vertical climb, constant pitch angle rate $\dot{\phi}$ maneuver, gravity turn, linear tangent steering, and also coasting. The gravity turn happens in the pitch maneuver with $\gamma = 0$. This is due to the gravitational force against the centrifugal force. With the thrust on, the gravity turn at lower altitudes will result in lower climb rate with higher velocities in the denser region causing higher drag loss associated with enhanced dynamic pressure. Therefore, a pitch rate different from gravity-turn rate has to be adopted to lift the vehicle to rarefied regions early. Equations for constant $\dot{\phi}$ maneuver, gravity turn, and bilinear and linear steering laws are derived.

The vertical ascent trajectory of sounding rockets, powered as well as coasting after burnout, is considered. In view of the vertical ascent, the equations of motion in incremental form are $\Delta v = (F/m_0)\Delta t - (D/m_0)\Delta t - g\Delta t$ and $\Delta h = v\Delta t$. Note $(F/m_0)\Delta t = -\overline{u}_e(\Delta m_0/m_0)$, where the negative number Δm_0 represents the reduction in the vehicle mass m_0. Adopting these equations, a suitable numerical scheme can be employed to calculate the ascent trajectory. For the sake of a simple analytical solution, we can assume an averaged constant gravitational acceleration $\overline{g}$ and neglect drag force. Equations in closed form are derived for the burnout-velocity, -altitude, and coasting height.

Worked examples are introduced under the sections. Many concept questions and numerical problems with answers are added at the chapter end.

5.1 Introduction

Depending on their applications, rocket vehicles vary from single-stage systems to complex multistage ones. In warfare applications, rockets can be simple short-range missiles of single stage such as bazookas to complex intermediate-range ballistic missiles (IRBMs) and intercontinental ballistic missiles (ICBMs) of several stages. Similarly, we have simple sounding rockets for near and sub-orbital atmospheric research as well as complex multistage satellite launch vehicles for space exploration. In this chapter, we discuss the details of multistage satellite launch vehicles in operation and their characteristics of structural coefficients and specific impulses. The preliminary step in staging design is the loss-free staging optimization. This is considered in detail along with the corrections for the effects of velocity losses. Details of the atmosphere with its different layers and their characteristics are discussed. Assuming the Earth's reference system to be inertial, the time-dependent two-dimensional equations of motion in polar coordinates are given for the rocket vehicle of zero lift. The different elements of a trajectory, namely, the gravity turn, constant pitch rate maneuver, and bilinear and linear tangent steering are all considered and the respective governing equations are derived and discussed. At the end, the trajectories of sounding rockets are considered with correction for the gravity effect.

5.2 Multistage Rockets

To carry a payload m_l to a large distance or a height with a required terminal velocity v_n, adopting a single huge engine with its enormous quantities of propellants is not practical. During such a single-engine operation as the propellants are being consumed the mass of the vehicle reduces. In order to keep the acceleration stresses within the limits, the huge engine has to be throttled down to inefficient or impossible levels, see Sect. 3.5.8. In addition, for a provided launch mass, as the propellants are being consumed the vehicle has to carry the single large structure all the way thereby reducing the desired distance, or the desired height with the terminal velocity v_n. Therefore, in a multistage rocket vehicle, the individual stages of gross masses $m_{(i)}$'s (with their engines, propellant tanks, propellants, and associated structures)

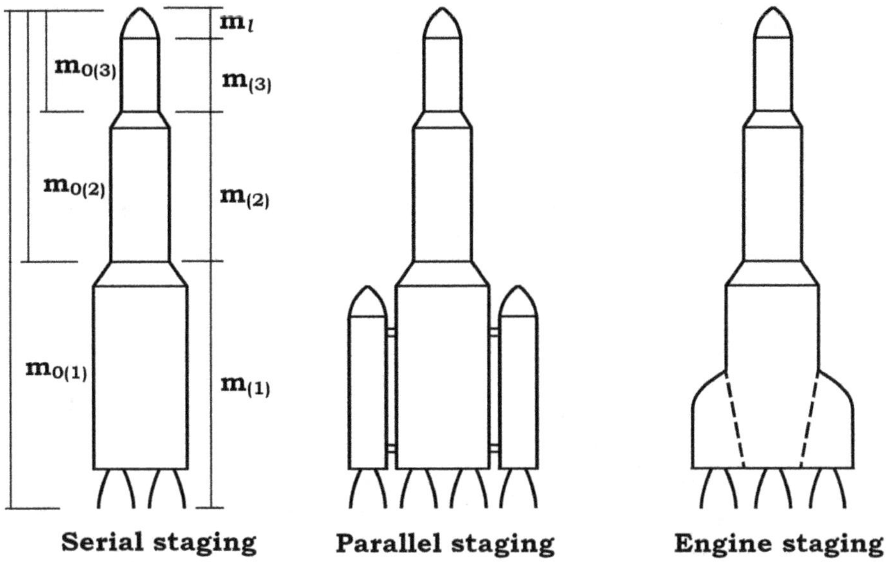

Fig. 5.1 Methods of staging in multistage rocket vehicles

are assembled to carry the payload m_l, Fig. 5.1. And, as the operation of an individual stage is over, its empty structure $m_{s(i)}$ comprising engine, propellant tanks, residual propellants if any, and associated structure is separated.

Principally, there are two methods of staging, one the serial staging and the other the parallel staging, Fig. 5.1. In a serially staged rocket vehicle, the stages are stacked one above the other and the stages come into operation sequentially. During the flight, as and when a particular stage ejects all its propellants combustion products, the empty structure is separated, whereas, in a vehicle with the parallel staging, the bottom stage comprises one core stage and two or more strap-on stages. The assembly of the strap-on units is often referred to as the zeroth stage and the core stage as the first stage. Depending on the limits on the vehicle acceleration and the dynamic pressure to be experienced in the atmosphere ($\rho v^2/2 = \gamma p M^2/2$), thrusting patterns of the strap-ons and the core are designed. Also, there is a third method of staging known as the engine staging. The rocket vehicle Atlas II of Lockheed Martin is a typical example. The first stage of this vehicle has three LOX/RP1 liquid engines, two boosters placed on the sides and one sustainer at the center. They all share the common propellant tanks. At the lift-off, all the three engines are ignited. During the ascent, the shorter thrusting boosters are jettisoned at their burnouts. And, the longer thrusting sustainer is jettisoned after all the propellants are consumed. Then, the second stage takes over. This engine staging is named as "stage-and-a-half" technique by Lockheed Martin. The vehicle Atlas II and its previous variants with engine staging have all been retired after many successful launches, and the engine staging appears to be not in use in the currently active launch vehicles.

In Table 5.1, brief details of many active orbital-launch vehicles are given. The entries against the stage number under the seventh column engine, require some

Table 5.1 Active orbital launch vehicles for low Earth orbit (LEO), Sun-synchronous orbit (SSO), and geosynchronous (or geostationary) transfer orbit (GTO) [1]

No.	Vehicle origin	Orbit	m_l (kg)	$\frac{m_{0(1)}}{m_l}$	Stg.	Engine	Propellant	$m_{(i)}$ (kg)	$\varepsilon(i)$	$I_{sp(i),vac}$ (m/s)	$I_{sp(i),sl}$ (m/s)
1	VLS	LEO	380	131.55	0	4*1*S-43	Solid	8550	0.1553	2550	2206
	Brazil	SSO	80	621.11	1	1*1*S-43TM	Solid	8720	0.1761	2716	1667
					2	1*1*S-40TM	Solid	5664	0.2140	2697	
					3	1*1*S-44	Solid	1025	0.1854	2765	
2	Chang Zheng 2C	LEO	2500	77.80	1	1*4*YF-20A	N2O4/UDMH	153000	0.0654	2854	2559
	China	SSO	2200	88.27	2	1*1*YF-22A/23A	N2O4/UDMH	39000	0.1026	2893	
3	Chang Zheng 2C-III/SD	LEO	2800	75.29	1	1*4*YF-20A	N2O4/UDMH	153000	0.0654	2854	2559
	China	SSO	1400	149.57	2	1*1*YF-22A/23A	N2O4/UDMH	55000	0.0909	2922	
4	Chang Zheng 2D	LEO	3500	67.36	1	1*4*YF-20B	N2O4/UDMH	192700	0.0493	2834	2540
	China				2	1*1*YF-25/23	N2O4/UDMH	39550	0.1011	2893	
5	Chang Zheng 2F	LEO	8400	54.81	0	4*1*YF-20B	N2O4/UDMH	41000	0.0780	2854	2559
	China	GTO	3500	130.14	1	1*4*YF-20B	N2O4/UDMH	196500	0.0483	2834	2559
					2	1*1*YF-25/23	N2O4/UDMH	91500	0.0601	2922	
6	Chang Zheng 3A	LEO	7200	33.44	1	1*4*YF-20B	N2O4/UDMH	179000	0.0503	2834	2540
	China	GTO	2600	90.85	2	1*1*YF-25/23	N2O4/UDMH	33600	0.1190	2913	
					3	1*2*YF-75	LOX/LH2	21000	0.1333	4315	
7	Chang Zheng 3B	LEO	11200	38.41	0	4*1*YF-20B	N2O4/UDMH	41000	0.0780	2854	2559
	China	GTO	5100	83.16	1	1*4*YF-20B	N2O4/UDMH	179000	0.0503	2834	2540
					2	1*1*YF-22A/23A	N2O4/UDMH	55000	0.0909	2922	
					3	1*2*YF-75	LOX/LH2	21000	0.1333	4315	

(continued)

Table 5.1 (continued)

No.	Vehicle origin	Orbit	m_l (kg)	$\frac{m_{0(1)}}{m_l}$	Stg.	Engine	Propellant	$m_{(i)}$ (kg)	$\varepsilon(i)$	$I_{sp(i),vac}$ (m/s)	$I_{sp(i),sl}$ (m/s)
8	Chang Zheng 3C	LEO	–	–	0	2*1*YF-20B	N2O4/UDMH	41000	0.0780	2854	2559
	China	GTO	3800	89.68	1	1*1*YF-20B	N2O4/UDMH	179000	0.0503	2834	2540
					2	1*1*YF-22A/23A	N2O4/UDMH	55000	0.0909	2922	
					3	1*1*YF-75	LOX/LH2	21000	0.1333	4315	
9	Ariane 5ECA	LEO	16000	48.45	0	2*1*P241	Solid	278330	0.1372	2697	2452
	ESA	GTO	10500	73.30	1	1*1*Vulcain 2	LOX/LH2	186000	0.0683	4256	3285
					2	1*1*HM7-B	LOX/LH2	16500	0.1273	4374	
10	Ariane 5ES	LEO	–	–	0	2*1*P241	Solid	278330	0.1372	2697	2452
	ESA	GTO	8000	95.43	1	1*1*Vulcain 2	LOX/LH2	186000	0.0683	4256	3285
					2	1*1*Aestus	N2O4/MMH	12750	0.2157	3177	
11	Ariane 5G	LEO	16000	47.14	0	2*1*P230	Solid	277500	0.1434	2697	2452
	ESA	GTO	6800	109.57	1	1*1* Vulcain	LOX/LH2	170800	0.0744	4217	3334
					2	1*1*Aestus	N2O4/MMH	12500	0.2160	3177	
12	PSLV	LEO	3700	79.62	0	6*1*SLV-1	Solid	10930	0.1839	2569	2246
	India	GTO	800	364.63	1	1*1*PSLV-1	Solid	168200	0.1795	2638	2324
					2	1*1*Viking 4	N2O4/UDMH	45800	0.1157	2873	
					3	1*1*PSLV-3	Solid	8400	0.1310	2883	
					4	1*2*PSLV-4	N2O4/MMH	2920	0.3151	3020	
13	GSLV	LEO	5000	80.44	0	4*1*Viking 2	N2O4/UDMH	45600	0.1228	2756	2354
	India	GTO	2500	159.88	1	1*1*PSLV-1	Solid	157300	0.1799	2609	2324
					2	1*1*Viking 4	N2O4/UDMH	42900	0.1259	2893	
					3	1*1*RD-56M	LOX/LH2	14600	0.1507	4511	

(continued)

Table 5.1 (continued)

No.	Vehicle origin	Orbit	m_l (kg)	$\frac{m_{0(1)}}{m_l}$	Stg.	Engine	Propellant	$m_{(i)}$ (kg)	$\varepsilon(i)$	$I_{sp(i),vac}$ (m/s)	$I_{sp(i),sl}$ (m/s)
14	Shaviyt 1	LEO	225	121.04	1	1*1*LK-1	Solid	13990	0.0886	2628	2452
	Israel				2	1*1*RSA-3-2	Solid	10971	0.1614	2716	
					3	1*1*RSA-3-3	Solid	2048	0.0830	2922	
15	H-IIA	LEO	11730	30.57	0	2*1*SRB-A	Solid	76400	0.1361	2746	2255
	Japan	GTO	5000	70.36	1	2*1*Castor 4AXL	Solid	15200	0.1382	2775	
					2	2*1*Castor 4AXL	Solid	15200	0.1382	2775	
					3	1*1*LE-7A	LOX/LH2	113600	0.1197	4315	
					4	1*1*LE-5B	LOX/LH2	19600	0.1531	4384	
16	M-V	LEO	1800	73.48	1	1*1*M14	Solid	83560	0.1444	2707	2412
	Japan	SSO	1300	101.35	2	1*1*M24	Solid	34470	0.0989	2824	
					3	1*1*M34	Solid	11000	0.0909	2952	
					4	1*1*M-V-4	Solid	1430	0.0825	2922	
17	Soyuz-U	LEO	7200	43.86	0	4*1*RD-107-11D511	LOX/Krsn	44500	0.0850	3040	2589
	Russia	LEO	6220	50.61	1	1*1*RD-108-11D512	LOX/Krsn	105400	0.0652	3050	2403
					2	1*1*RD-0110	LOX/Krsn	25200	0.0935	3236	
18	Soyuz-FG	LEO	7420	42.54	0	4*1*RD-117	LOX/Krsn	44400	0.0858	3040	2589
	Russia				1	1*1*RD-118	LOX/Krsn	105400	0.0652	3050	2403
					2	1*1*RD-0124	LOX/Krsn	25200	0.0935	3521	
19	Soyuz-ST-B	LEO	7800	40.51	0	4*1*RD-117	LOX/Krsn	44400	0.0858	3040	2589
	Russia				1	1*1*RD-118	LOX/Krsn	105400	0.0652	3050	2403
					2	1*1*RD-0124	LOX/Krsn	25200	0.0935	3521	

(continued)

Table 5.1 (continued)

No.	Vehicle origin	Orbit	m_l (kg)	$\frac{m_{0(1)}}{m_l}$	Stg.	Engine	Propellant	$m_{(i)}$ (kg)	$\varepsilon(i)$	$I_{sp(i),vac}$ (m/s)	$I_{sp(i),sl}$ (m/s)
20	Soyuz ST / Fregat ST	LEO	5500	58.22	0	4*1*RD-117	LOX/Krsn	44400	0.0858	3040	2589
	Russia	SSO	4500	70.94	1	1*1*RD-118	LOX/Krsn	105400	0.0652	3050	2403
					2	1*1*RD-0124	LOX/Krsn	25200	0.0935	3521	
					3	1*1*S5.92	N2O4/UDMH	6535	0.1683	3207	
21	Molniya 8K78M	LEO	1800	98.48	0	4*1*RD-107-8D728	LOX/Krsn	43400	0.0869	3079	2520
	Russia	GTO	1600	110.66	1	1*1*RD-108-8D727	LOX/Krsn	100600	0.0676	3089	2432
					2	1*1*RD-0110	LOX/Krsn	24800	0.0797	3236	
					3	1*1*S1.5400A	LOX/Krsn	6660	0.1742	3334	
22	Kosmos 11K65M	LEO	1500	72.56	1	1*1*RD-216	HNO3/UDMH	87200	0.0608	2863	2432
	Russia	SSO	775	139.50	2	1*1*11D49	HNO3/UDMH	20135	0.0713	2971	
23	Rokot	LEO	1800	54.28	1	1*4*RD-0232	N2O4/UDMH	77150	0.0738	3040	2795
	Russia	SSO	1000	96.91	2	1*1*RD-0235	N2O4/UDMH	12195	0.1218	3158	
					3	1*1*S5.98M	N2O4/UDMH	6565	0.2437	3197	
24	Start-1	LEO	632	73.78	1	1*1*MIHT-1	Solid	26000	0.1154	2579	2334
	Russia	LEO	500	93.00	2	1*1*MIHT-2	Solid	13000	0.1154	2746	
					3	1*1*MIHT-3	Solid	6000	0.1667	2746	
					4	1*1*MIHT-4	Solid	1000	0.3000	2893	
25	Proton-M/Briz-M	LEO	21000	33.91	1	1*6*RD-253-14D14	N2O4/UDMH	450400	0.0688	3109	2795
	Russia	GTO	4500	154.59	2	1*4*RD-0210	N2O4/UDMH	167828	0.0698	3207	
					3	1*1*RD-0212	N2O4/UDMH	50747	0.0825	3187	
					4	1*1*S5.98M	N2O4/UDMH	22170	0.1069	3197	

(continued)

Table 5.1 (continued)

No.	Vehicle origin	Orbit	m_l (kg)	$\frac{m_{0(1)}}{m_l}$	Stg.	Engine	Propellant	$m_{(i)}$ (kg)	$\varepsilon(i)$	$I_{sp(i),vac}$ (m/s)	$I_{sp(i),sl}$ (m/s)
26	Angara-1.2PP	LEO	3700	45.65	1	1*1*RD-191	LOX/Krsn	140000	0.0714	3315	3040
	Russia				2	1*1*RD-0124	LOX/Krsn	25200	0.0935	3521	
27	Naro-1	LEO	100	1411	1	1*1*RD-191	LOX/Krsn	140000	0.0714	3315	3040
	South Korea				2	1*1*-	Solid	1000	0.1000	2452	
28	Zenit-2	LEO	13740	33.38	1	1*1*RD-171	LOX/Krsn	354300	0.0807	3305	3050
	Ukraine	SSO	5000	89.98	2	1*1*RD-120	LOX/Krsn	90600	0.0993	3422	
29	Zenit-3SL	LEO	5000	93.44	1	1*1*RD-171	LOX/Krsn	354300	0.0807	3305	3050
	Ukraine	GTO	5250	89.04	2	1*1*RD-120	LOX/Krsn	90600	0.0993	3422	
					3	1*1*RD-58M	LOX/Krsn	17300	0.1572	3452	
30	Delta 4M+(4,2)	LEO	11700	28.19	0	2*1*GEM 60	Solid	33798	0.1139	2697	2383
	USA	GTO	5300	66.47	1	1*1*RS-68	LOX/LH2	226400	0.1182	4119	3579
					2	1*1*RL-10B-2	LOX/LH2	24170	0.1179	4531	
31	Delta 4H	LEO	25800	28.52	0	2*1*RS-68	LOX/LH2	226400	0.1182	4119	3579
	USA	GTO	10843	66.47	1	1*1*RS-68	LOX/LH2	226400	0.1182	4119	3579
					2	1*1*RL-10B-2	LOX/LH2	30710	0.1136	4531	
32	Delta 4M	LEO	8600	30.14	1	1*1*RS-68	LOX/LH2	226400	0.1182	4119	3579
	USA	GTO	3900	65.25	2	1*1*RL-10B-2	LOX/LH2	24170	0.1179	4531	
33	Delta 4M+(4,2)	LEO	11700	28.19	0	2*1*GEM 60	Solid	33798	0.1139	2697	2383
	USA	GTO	5300	61.03	1	1*1*RS-68	LOX/LH2	226400	0.1182	4119	3579
					2	1*1*RL-10B-2	LOX/LH2	24170	0.1179	4531	
34	Delta 4M+(5,2)	LEO	10300	31.89	0	2*1*GEM 60	Solid	33798	0.1139	2697	2383
	USA	GTO	4350	74.14	1	1*1*RS-68	LOX/LH2	226400	0.1182	4119	3579
					2	1*1*RL-10B-2	LOX/LH2	24170	0.1179	4531	

(continued)

Table 5.1 (continued)

No.	Vehicle origin	Orbit	m_l (kg)	$\frac{m_{0(1)}}{m_l}$	Stg.	Engine	Propellant	$m_{(i)}$ (kg)	$\varepsilon(i)$	$I_{sp(i),vac}$ (m/s)	$I_{sp(i),sl}$ (m/s)
35	Delta 4M+(5,4)	LEO	13600	29.36	0	4*1*GEM 60	Solid	33798	0.1139	2697	2383
	USA	GTO	6120	64.03	1	1*1*RS-68	LOX/LH2	226400	0.1182	4119	3579
					2	1*1*RL-10B-2	LOX/LH2	24170	0.1179	4531	
36	Atlas V	LEO	12500	43.71	0	5*1*Aerojet SRB	Solid	40824	0.0980	2697	2403
	USA	GTO	5000	107.77	1	1*1*RD-180	LOX/Krsn	306914	0.0732	3315	3050
					2	1*1*RL-10A-4-2	LOX/LH2	22825	0.0888	4423	
37	Taurus	LEO	1363	52.81	1	1*1*Castor 120	Solid	53020	0.0794	2805	2246
	USA	GTO	431	164.84	2	1*1*Pegasus-1	Solid	13242	0.0822	2795	
					3	1*1*Pegasus-2	Solid	3370	0.1024	2863	
					4	1*1*Pegasus-3	Solid	985	0.2061	2873	
38	Minotaur	LEO	640	56.35	1	1*1*M55/TX-55/Tu-122	Solid	23077	0.0993	2569	2324
	USA	SSO	335	106.75	2	1*1*SR-19	Solid	7032	0.1131	2824	
					3	1*1*Pegasus XL-2	Solid	4331	0.0961	2844	
					4	1*1*Pegasus-3	Solid	985	0.2061	2873	

clarification. The first number refers to the number of stages, the second one to the number of engines, and the third an acronym refers to the engine name. For example, for the orbital launch vehicle VLS of Brazil the zeroth (strap-on) stage has four strap-on stages each containing one solid motor S-43. Similarly, for the Chang Zheng 2C of China, the first stage contains in it one cluster of four YF-20A $N_2O_4/UDMH$ liquid engines. The ratio of the empty stage mass to the gross stage mass is the structural coefficient ε. As the strap-on stages and the first stages operate at low altitudes, also the sea level-specific impulse values $I_{sp,sl}$'s are given.

5.3 Stage Optimization in Loss-Free Environment

Considering the flight of a rocket vehicle in a gravity- and drag-free environment (loss-free environment) is an initial step to the stage optimization. Generally, for an orbital mission type (low Earth orbit: LEO, Sun-synchronous orbit: SSO, geosynchronous transfer orbit: GTO, or translunar orbit), the terminal velocity v_n is known. By experience, the extent of total velocity losses for a mission type is also known. The desired payload mass m_l and the specific impulse values $\overline{u}_{e(i)}$'s and structural coefficients $\varepsilon_{(i)}$'s of the selected types of stages (solid, earth-storable, semi-cryogenic, or cryogenic) are known as well. Thus, for the specified terminal velocity v_n, the loss-free optimization considers the number of stages n, $\overline{u}_{e(i)}$'s, and $\varepsilon_{(i)}$'s to minimize the ratio of lift-off mass to payload mass $m_{0(1)}/m_l$ (or maximize payload fraction $m_l/m_{0(1)}$). Alternatively, for the specified $m_{0(1)}/m_l$, it maximizes the terminal velocity v_n.

The loss-free optimization helps in quickly selecting the number of stages for the given mission or to check whether the stages chosen are capable of giving the needed terminal velocity. In order to account for the velocity losses due to gravity, drag, steering, and fairing separation, the loss-free procedure can be considered for a "marked-up terminal velocity" v_n' ($v_n' = v_n$ + velocity losses) to minimize $m_{0(1)}/m_l$. The procedure with a v_n' could give stage-wise mass distribution $m_{(i)}$'s and velocity increments $\Delta v_{(i)}$'s different from the ones of final design. This is because the loss-free optimization "favors" the more efficient stage (higher $\overline{u}_{e(i)}$ and lower $\varepsilon_{(i)}$) by assigning to it larger stage mass $m_{(i)}$ and higher velocity increment $\Delta v_{(i)}$. Nevertheless, the minimized $m_{0(1)}/m_l$ by the loss-free analysis for v_n' is found to be close to the one by the analysis with velocity losses for the actual v_n. Obviously, after the initial selection through the loss-free step, the numerically involved optimization with velocity losses is to be executed for different trajectories concurrently with the vehicle configurations. However, the stage optimization is final only when all significant variables such as velocity losses due to gravity, drag, steering, and fairing separation, and the structural coefficient variations are included along with due considerations for air loading (dynamic pressure: $\rho v^2/2$), inertial loading (vehicle acceleration), and heating.

5.3.1 Formulation

We restrict our optimization analysis only to the serially staged multistage rocket vehicle. To generalize the analysis, let us assume that there are n stages to the vehicle. We have to first look at the definitions of three fundamental parameters [2]. These are **payload ratio, structural coefficient**, and **mass ratio**. The payload of any particular stage is simply the load above, that is the mass of all stages above including the actual payload m_l. The payload ratio is defined as

$$\alpha_{(i)} \equiv \frac{m_{0(i+1)}}{m_{(i)}} = \frac{m_{0(i+1)}}{m_{0(i)} - m_{0(i+1)}} = \frac{m_{0(i+1)}}{m_{s(i)} + m_{p(i)}} \tag{5.1}$$

Here in $\alpha_{(i)}$, above the mass of the ith stage $m_{(i)}$ is the mass of the vehicle portion (or the partial vehicle) $m_{0(i+1)}$ as the "payload", which has in it the stage masses $m_{(i+1)}$, $m_{(i+2)}$,.., $m_{(n)}$, and the given payload m_l. Finally, this way in α_n, above the mass of the final nth stage $m_{(n)}$ is the mass of the given payload m_l.

$$\alpha_n = \frac{m_l}{m_{(n)}}$$

With reference to the serially staged vehicle shown in Fig. 5.1, $\alpha(1) = m_{0(2)}/m_{(1)}$, $\alpha(2) = m_{0(3)}/m_{(2)}$, and $\alpha(3) = m_l/m_{(3)}$.

Structural coefficient is defined as

$$\varepsilon_{(i)} \equiv \frac{m_{s(i)}}{m_{(i)}} = \frac{m_{s(i)}}{m_{p(i)} + m_{s(i)}} = \frac{m_{s(i)}}{m_{0(i)} - m_{0(i+1)}} \tag{5.2}$$

where the gross stage mass $m_{(i)}$ comprises the structural mass $m_{s(i)}$ and the propellant mass $m_{p(i)}$. For a rocket stage the structural coefficient $\varepsilon_{(i)}$ represents the skill of the designer in designing a light empty stage consisting of engine/motor, propellant-tanks/motor-casing, and other support structures. The structural coefficient $\varepsilon_{(i)}$ depends on the type of propellants stored in the stage and the gross stage mass. The variations of structural coefficients with the gross stage mass for different propellant combinations are shown in Figs. 5.2, 5.3, 5.4 and 5.5. On studying these variations, we find that the structural coefficient $\varepsilon_{(i)}$ reduces with the stage gross mass $m_{(i)}$. Such reductions are more significant in liquid propellant stages than in solid propellant ones. Generally, for stages of equal gross mass, the structural coefficients are higher for solid propellant stages. The points of structural coefficients for the considered propellant combinations are highly dispersed and the R^2 correlations are poor—the poorest one is for solid propellant stages. Among the possible trend lines, the power relationships shown in the figures offer the highest possible correlation. The reasons for the dispersal in the points are place and time—the level of technological advancement in a country and the period. In addition, it depends on the stage purpose. For example, since a final stage of a vehicle has to have some special provisions to inject the payload into the desired orbit, it will have a higher

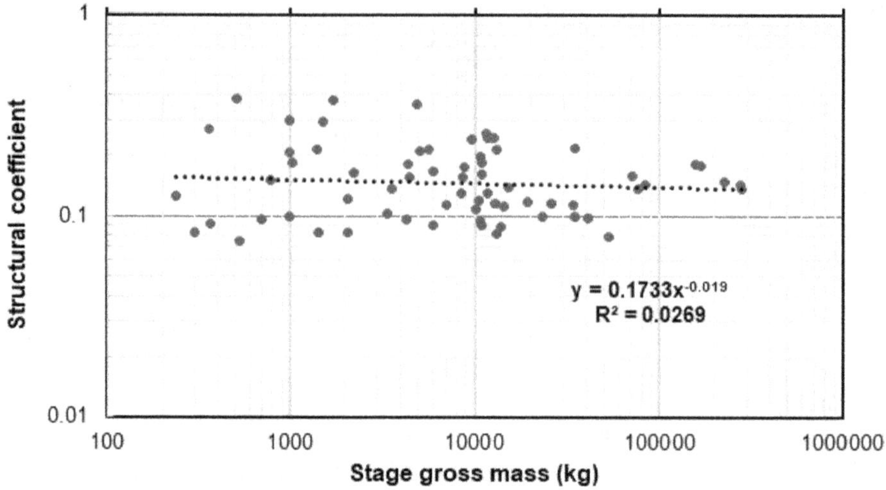

Fig. 5.2 Structural coefficient versus stage gross mass for flight-proven solid propellant stages of different countries and periods

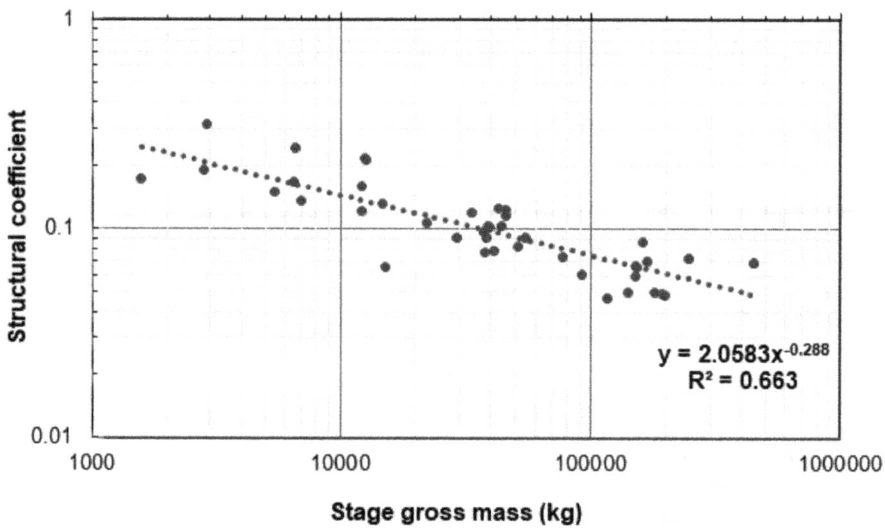

Fig. 5.3 Structural coefficient versus stage gross mass for flight-proven N2O4 stages of different countries and periods

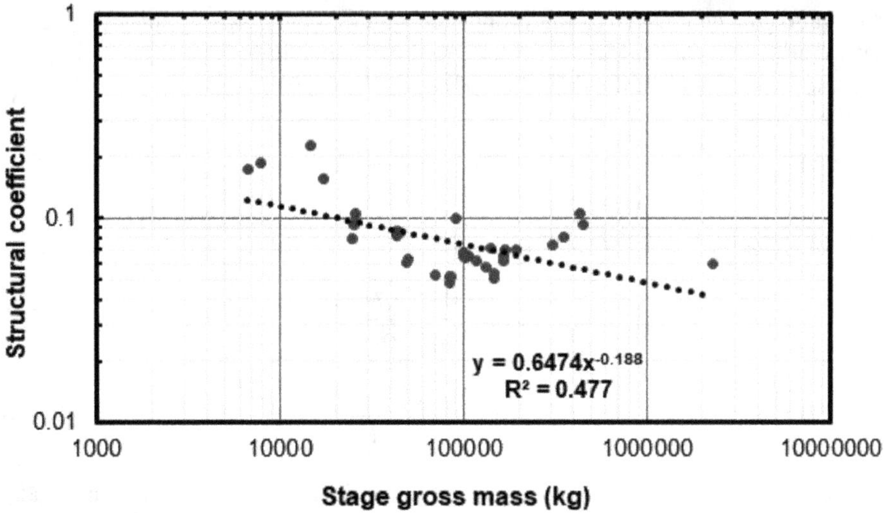

Fig. 5.4 Structural coefficient versus stage gross mass for flight-proven LOX/Kerosene stages of different countries and periods

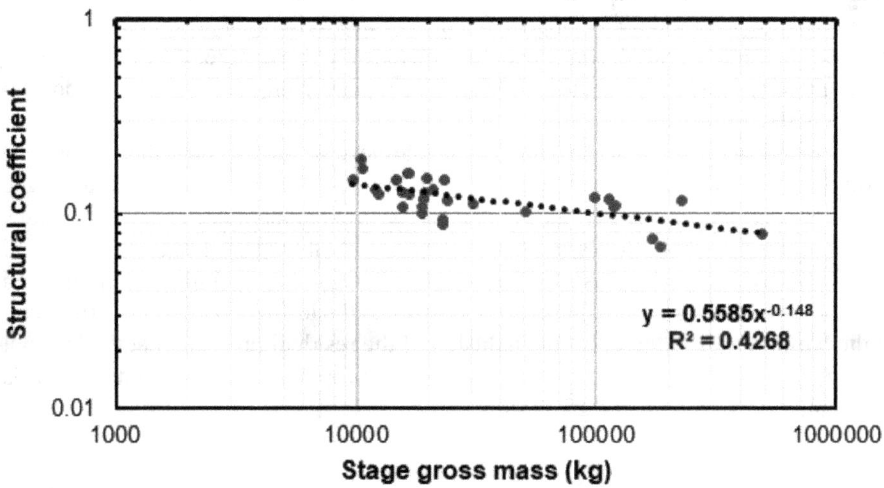

Fig. 5.5 Structural coefficient versus stage gross mass for flight-proven LOX/LH2 stages of different countries and periods

structural coefficient than the one of lower stage of same stage mass and propellant combination. Mass ratio is defined as

$$\mu_{(i)} = \frac{m_{0(i)}}{m_{0b(i)}} = \frac{m_{0(i)}}{m_{0(i+1)} + m_{s(i)}} \tag{5.3}$$

Mass ratio $\mu_{(i)}$ characterizes the propulsion requirements for any mission. Recall that we discussed this in Chap. 1 under Tsiolkovsky equation 1.1.

It can be shown from the definition of payload ratio $\alpha_{(i)}$,

$$\frac{m_{0(i)}}{m_{0(i+1)}} = \frac{1+\alpha_{(i)}}{\alpha_{(i)}} \tag{5.4}$$

Proceeding further with the other two definitions,

$$\mu_{(i)} = \frac{1+\alpha_{(i)}}{\varepsilon_{(i)}+\alpha_{(i)}} \tag{5.5}$$

$$\alpha_{(i)} = \frac{1-\mu_{(i)}\varepsilon_{(i)}}{\mu_{(i)}-1} \tag{5.6}$$

In formulating the optimization procedure, the following assumptions are made. (1) Thrust vector coincides with the flight-velocity vector. (2) Thrust is the only force acting on the vehicle. This indicates that the vehicle is moving in a drag- and gravity-free environment. The assumption of the drag- and gravity-free environment implies that the rocket vehicle is moving in vacuum and the altitude to be reached does not come into consideration. (3) The equivalent exhaust velocity $\overline{u}_e$ (that is the specific impulse I_{sp}) is constant during the entire operation of any stage. As the vehicle by assumption is moving in vacuum, $\overline{u}_e$ (that is the specific impulse I_{sp}) corresponds to the vacuum condition.

As stated previously, let the rocket vehicle have n stages and it has to place a given payload of mass m_l at a terminal velocity v_n at the burnout of its last stage $m_{(n)}$. The total launch mass of the vehicle can be broken down into structural masses $m_{s(i)}$'s and propellant masses $m_{p(i)}$'s of the individual stages of masses $m_{(i)}$'s and finally the given payload mass m_l. The structural mass $m_{s(i)}$ includes the engine and its structure, other associated accessories, and also the leftover or residual propellant at the burnout of the stage. By the inclusion of the residual propellant at the burnout into $m_{s(i)}$, $m_{p(i)}$ will represent only the expelled propellant. The stage mass $m_{(i)}$ is frequently referred to as the gross stage mass.

$$\begin{aligned} m_{0(1)} &= m_{s(1)} + m_{p(1)} + m_{s(2)} + m_{p(2)} \\ &+ .. + m_{s(i)} + m_{p(i)} + .. + m_{s(n)} + m_{p(n)} + m_l \end{aligned} \tag{5.7}$$

The ratio of the total launch mass to the payload mass $m_{0(1)}/m_l$ can be written as

$$\frac{m_{0(1)}}{m_l} = \frac{m_{0(1)}}{m_{0(2)}}\frac{m_{0(2)}}{m_{0(3)}}..\frac{m_{0(i)}}{m_{0(i+1)}}..\frac{m_{0(n-1)}}{m_{0(n)}}\frac{m_{0(n)}}{m_l} \tag{5.8}$$

By Eqs. 5.4 and 5.6,

$$\frac{m_{0(1)}}{m_l} = \prod_{i=1}^{n} \frac{1+\alpha_{(i)}}{\alpha_{(i)}} = \prod_{i=1}^{n} \frac{\mu_{(i)}(1-\varepsilon_{(i)})}{1-\mu_{(i)}\varepsilon_{(i)}} \tag{5.9}$$

Taking natural logarithm of both the sides,

$$ln\,\frac{m_{0(1)}}{m_l} = \sum_{i=1}^{n} ln\,\frac{\mu_{(i)}(1-\varepsilon_{(i)})}{1-\mu_{(i)}\varepsilon_{(i)}} \tag{5.10}$$

Adopting the Tsiolkovsky equation, the prescribed terminal velocity v_n at the burnout of the last stage $m_{(n)}$ is written as

$$v_n = \sum_{i=1}^{n} \Delta\,v_i = \sum_{i=1}^{n} \overline{u}_{e(i)} ln\,\mu_{(i)} \tag{5.11}$$

Now, the optimization problem can be conveniently stated as follows. For the arbitrarily specified n pairs of specific impulse $\overline{u}_{e(i)}$ and structural coefficient $\varepsilon_{(i)}$, by a suitable choice of the mass ratios $\mu_{(i)}$'s, the total launch mass of the rocket vehicle $m_{0(1)}$ of n stages is to be minimized for the given payload m_l, Eq. 5.10. This means that $m_{0(1)}/m_l$ is to be minimized or $m_l/m_{0(1)}$ is to be maximized. This is subject to the constraint that the rocket vehicle attains the prescribed terminal velocity v_n at the burnout of the last stage $m_{(n)}$, Eq. 5.11. Also alternatively, we can say that, for a given constraint of $m_l/m_{0(1)}$ (or $m_{0(1)}/m_l$), the terminal velocity v_n has to be maximized. Starting from the late 1940s to the late 1960s, many studies were done to solve this optimization problem and a selected few of them are [3–11]. Of these, we consider here [4, 5] to present the optimization method.

The optimization can be accomplished by using the method of Lagrange multipliers. For a brief discussion on Lagrange multipliers, see Sect. 4.4.1. Introducing the Lagrange multiplier λ into Eq. 5.11,

$$\lambda\left[\left(\sum_{i=1}^{n} \overline{u}_{e(i)} ln\,\mu_{(i)}\right) - v_n\right] = 0 \tag{5.12}$$

Combining Eq. 5.10 and the constraint Eq. 5.12,

$$ln\,\frac{m_{0(1)}}{m_l} = \sum_{i=1}^{n} ln\,\frac{\mu_{(i)}(1-\varepsilon_{(i)})}{1-\mu_{(i)}\varepsilon_{(i)}} + \lambda\left[\left(\sum_{i=1}^{n} \overline{u}_{e(i)} ln\,\mu_{(i)}\right) - v_n\right] \tag{5.13}$$

That is,

$$ln \frac{m_{0(1)}}{m_l} = \sum_{i=1}^{n} ln\ \mu_{(i)} + \sum_{i=1}^{n} ln\ (1 - \varepsilon_{(i)}) - \sum_{i=1}^{n} ln\ (1 - \mu_{(i)}\varepsilon_{(i)})$$
$$+\lambda\left[\left(\sum_{i=1}^{n} \overline{u}_{e(i)} ln\ \mu_{(i)}\right) - v_n\right] \tag{5.13a}$$

Differentiating Eq. 5.13 with respect to the mass ratio $\mu_{(i)}$, equating to zero, and solving for $\mu_{(i)}$ for the extremum, we get the optimum mass ratio,

$$\mu_{(i)} = \frac{\lambda\overline{u}_{e(i)} + 1}{\lambda\overline{u}_{e(i)}\varepsilon_{(i)}} \qquad i = 1, 2, ..., n \tag{5.14}$$

Equations 5.11 and 5.14 provide a system of $n + 1$ equations for $n + 1$ unknowns, optimum mass ratios $\mu_{(i)}$, $i = 1, .., n$ and λ. To solve these $n + 1$ equations, let us take the natural logarithms of Eq. 5.14 and multiply it by $\overline{u}_{e(i)}$ and sum it over n values of $\mu_{(i)}$'s. We get,

$$\sum_{i=1}^{n} \overline{u}_{e(i)} ln\ \mu_{(i)} = \sum_{i=1}^{n} \overline{u}_{e(i)} ln\ \left(\frac{\lambda\overline{u}_{e(i)} + 1}{\lambda\overline{u}_{e(i)}\varepsilon_{(i)}}\right)$$

By Eq. 5.11,

$$v_n = \sum_{i=1}^{n} \overline{u}_{e(i)} ln\ \left(\frac{\lambda\overline{u}_{e(i)} + 1}{\lambda\overline{u}_{e(i)}\varepsilon_{(i)}}\right) \tag{5.15}$$

For the prescribed terminal velocity v_n and the n pairs of $\overline{u}_{e(i)}$ and $\varepsilon_{(i)}$, evaluating λ from Eq. 5.15, the optimum mass ratios $\mu_{(i)}$'s are found by substituting the evaluated λ into Eq. 5.14. Once the optimum $\mu_{(i)}$ are known, the other stage properties can be evaluated, Eqs. 5.6 and 5.9. But, Eq. 5.15 being a transcendental one with a single unknown λ there must be multiple roots for λ. Therefore, first without considering the constraint v_n, a possible range for λ is to be established for the given n pairs $\overline{u}_{e(i)}$ and $\varepsilon_{(i)}$. This will minimize the iterations necessary to obtain the correct value of λ for the prescribed constraint v_n. Our task thus remains to obtain that root of λ, which will give the absolute minimum mass of the total rocket vehicle $m_{0(1)}$. Before doing this, let us consider two analytically simplified cases of optimization. The **first simple case** is where all stages have equal effective exhaust velocities $\overline{u}_{e(i)}$'s (or, equal specific impulses $I_{sp(i)}$'s) but different structural coefficients $\varepsilon_{(l)}$'s. That is,

$$\overline{u}_{e(i+1)} = \overline{u}_{e(i)} = \overline{u}_e$$

$$\varepsilon_{(i+1)} \neq \varepsilon_{(i)}$$

For this first case of equal effective exhaust velocities and unequal structural coefficients, Eq. 5.14 becomes

$$\mu_{(i)} = \frac{\lambda \overline{u}_e + 1}{\lambda \overline{u}_e \varepsilon_{(i)}} \qquad i = 1, 2, ..., n \tag{5.16}$$

With this simplification, given by Eq. 5.16, Eq. 5.11 becomes

$$v_n = \sum_{i=1}^{n} \overline{u}_e ln \frac{\lambda \overline{u}_e + 1}{\lambda \overline{u}_e \varepsilon_{(i)}} \tag{5.17}$$

Proceeding further,

$$\frac{v_n}{\overline{u}_e} = \sum_{i=1}^{n} ln \frac{\lambda \overline{u}_e + 1}{\lambda \overline{u}_e \varepsilon_{(i)}} = n\ ln \frac{\lambda \overline{u}_e + 1}{\lambda \overline{u}_e} - \sum_{i=1}^{n} ln\ \varepsilon_{(i)}$$

$$\frac{\lambda \overline{u}_e + 1}{\lambda \overline{u}_e} = exp\left(\frac{v_n}{n\overline{u}_e} + \frac{1}{n}\sum_{i=1}^{n} ln\ \varepsilon_{(i)}\right) \tag{5.18}$$

Substituting Eq. 5.18 into Eq. 5.16,

$$\mu_{(i)} = \frac{1}{\varepsilon_{(i)}} exp\left(\frac{v_n}{n\overline{u}_e} + \frac{1}{n}\sum_{i=1}^{n} ln\ \varepsilon_{(i)}\right) \tag{5.19}$$

Using Eq. 5.19, the mass ratios $\mu_{(i)}$'s are all calculated. By Eq. 5.6, the payload ratios $\alpha_{(i)}$'s are known. Starting from the top of the launch vehicle and noting that the payload ratio $\alpha_{(n)} = m_l/m_{(n)}$, the mass of the nth stage $m_{(n)}$ is known for the given payload mass m_l. Thus, the mass of the vehicle portion at the ignition of the nth stage $m_{0(n)} = m_{(n)} + m_l$. Next, noting that the payload ratio $\alpha_{(n-1)} = m_{0(n)}/m_{(n-1)}$, the mass of $(n-1)$th stage $m_{(n-1)}$ is known from the just calculated $m_{0(n)}$. Thus, the mass of the vehicle portion at the ignition of $(n-1)$th stage $m_{0(n-1)} = m_{(n-1)} + m_{0(n)}$. Proceeding downward this way, the minimized total launch mass of the rocket vehicle $m_{0(1)}$ is calculated.

Note from Eq. 5.16,

$$\mu_{(i)}\varepsilon_{(i)} = \left(\frac{m_{0(i)}}{m_{0b(i)}}\right)\varepsilon_{(i)} = \frac{\lambda \overline{u}_e + 1}{\lambda \overline{u}_e} \qquad i = 1, 2, ..., n \tag{5.20}$$

This means that the product $\mu_{(i)}\varepsilon_{(i)}$ has the same value for all the stages. Thus, the optimum mass ratio $\mu_{(i)}$ is larger where the structural coefficient $\varepsilon_{(i)}$ is smaller (better). Furthermore, the stage with a larger mass ratio $\mu_{(i)}$ provides a higher incremental velocity $\Delta\ v_{(i)}$ for the given specific impulse. These observations are demonstrated in Example 5.1.

Next, let us consider the **second simple case**. This is simpler than the first one in that the vehicle has **similar stages** by having equal effective exhaust velocities

$\overline{u}_{e(i)}$'s (or, equal specific impulses $I_{sp(i)}$'s) and also equal structural coefficients $\varepsilon_{(i)}$'s. That is,

$$\overline{u}_{e(i+1)} = \overline{u}_{e(i)} = \overline{u}_e$$
$$\varepsilon_{(i+1)} = \varepsilon_{(i)} = \varepsilon$$

Substituting these into Eq. 5.19,

$$\mu_{(i)} = exp\left(\frac{v_n}{n\overline{u}_e}\right) \tag{5.21}$$

This leads to,

$$ln\ \mu_i = \frac{v_n}{n\overline{u}_e}$$

That is,

$$\Delta\ v_i = \frac{v_n}{n} \tag{5.22}$$

This indicates that all stages will provide equal incremental velocities $\Delta\ v_i$'s if all stages have equal effective exhaust velocities and also equal structural coefficients. This is demonstrated in Example 5.1. In one of the earliest studies on rocket stage optimization, this second simple case was analyzed [3].

Based on what we have learned so far, let us consider in Example 5.1 a small all-solid four-stage launch vehicle, the one more or less similar to one of the earliest launch vehicles SCOUT. However, here in the example we have the structured-parameters $I_{sp(i)}$'s and $\varepsilon_{(i)}$'s for analytical simplicity.

Example 5.1 Consider a small all-solid four-stage launch vehicle. It has to place a payload of mass 100 kg at a terminal velocity of 9500 m/s. Two simple cases are to be considered. In the first case, all stages have equal specific impulse values at 2600 m/s but different structural coefficients, from the first stage to fourth stage at 0.145, 0.155, 0.165, and 0.175 respectively. In the second simple case all stages are similar having equal specific impulse of 2600 m/s and equal structural coefficient of 0.160. Under the optimization procedure, calculate for both the cases the stage masses $m_{(i)}$'s, the stage incremental velocities $\Delta\ v_{(i)}$'s, and the minimized total launch mass of the rocket vehicle $m_{0(1)}$.

Solution

For the first case, all values required for substitution into Eqs. 5.18 and 5.19 are given to calculate $\mu_{(i)}$'s. For the second case $\mu_{(i)}$'s are calculated from Eq. 5.21. By Eq. 5.6, $\alpha_{(i)}$ are calculated. As discussed previously, descending from the top, the optimum stage masses and stage incremental velocities are calculated. And, finally the minimized total launch mass of the rocket vehicle $m_{0(1)}$ is known.

Analysis

First Case

By Eq. 5.18,

$$exp\left(\frac{v_n}{n\bar{u}_e}+\frac{1}{n}\sum_{i=1}^{n} ln\ \varepsilon_{(i)}\right)$$

$$= exp\left[\frac{9500}{4\times 2600}+\frac{1}{4}\times ln\ (0.145\times 0.155\times 0.165\times 0.175)\right]=0.397893$$

By Eq. 5.19

$$\mu_{(1)}=\frac{0.397893}{0.145}=2.74409$$

Similarly, $\mu_{(2)} = 2.56705$, $\mu_{(3)} = 2.41148$, & $\mu_{(4)} = 2.27368$
Note by Eq. 5.20, $\mu_{(1)}\varepsilon_{(1)}$ to $\mu_{(4)}\varepsilon_{(4)}$ are all equal at 0.39789.

By Eq. 5.6,

$$\alpha_{(1)}=\frac{1-\mu_{(1)}\varepsilon_{(1)}}{\mu_{(1)}-1}=\frac{1-0.39789}{2.74409-1}=0.345226$$

Similarly the values of $\alpha_{(2)}$ to $\alpha_{(4)}$ are calculated as given in Table 5.2.

$$\alpha_{(4)}=0.47273=\frac{m_l}{m_{(4)}}=\frac{100}{m_{(4)}}$$

$$m_{(4)}=\frac{100}{0.47273}=211.537\,\text{kg}$$

$$m_{0(4)}=m_{(4)}+m_l=211.537+100=311.537\,\text{kg}$$

$$\alpha_{(3)}=\frac{m_{0(4)}}{m_{(3)}}=\frac{311.537}{m_{(3)}}$$

$$m_{(3)}=\frac{311.537}{0.42658}=730.314\,\text{kg}$$

$$m_{0(3)}=m_{(3)}+m_{0(4)}=730.314+311.537=1041.851\,\text{kg}$$

Proceeding downward this way, the minimized total launch mass of the rocket vehicle $m_{0(1)}$ is calculated as 14625.6 kg. The results of the calculations are summarized in Table 5.2.

Table 5.2 Results of the calculations of Example 5.1: equal specific impulse but unequal structural coefficients, payload mass 100 kg and terminal velocity 9500 m/s

Stg	$I_{sp(i)}$ (m/s)	$\varepsilon_{(i)}$	$\mu_{(i)}$	$\mu_{(i)}\varepsilon_{(i)}$	$\alpha_{(i)}$	$m_{(i)}$ (kg)	$m_{0(i)}$ (kg)	$\Delta v_{(i)}$ (m/s)
1	2600	0.145	2.74409	0.39789	0.34523	10872.3	14625.6	2624.6
2	2600	0.155	2.56705	0.39789	0.38423	2711.5	3753.4	2451.2
3	2600	0.165	2.41148	0.39789	0.42658	730.3	1041.8	2288.6
4	2600	0.175	2.27368	0.39789	0.47273	211.5	311.5	2135.6

Second Case
By Eq. 5.21 $\mu_{(i)}$'s are calculated.

$$\mu_{(1)} = exp\left(\frac{9500}{4 \times 2600}\right) = 2.49294$$

Note $\mu_{(i)}$'s are all equal.

$$\mu_{(1)} \times \varepsilon_{(1)} = 2.49294 \times 0.160 = 0.39887$$

Note $\mu_{(i)}\varepsilon_{(i)}$'s are all equal.
By Eq. 5.6,

$$\alpha_{(1)} = \frac{1 - \mu_{(1)}\varepsilon_{(1)}}{\mu_{(1)} - 1} = \frac{1 - 2.49294 \times 0.160}{2.49294 - 1} = 0.40265$$

Note $\alpha_{(i)}$'s are all equal.

$$\alpha_{(4)} = \frac{m_l}{m_{(4)}} \Rightarrow 0.40265 = \frac{100}{m_{(4)}} \Rightarrow m_{(4)} = 248.4\,\text{kg}$$

$$m_{0(4)} = m_{(4)} + m_l = 248.4 + 100 = 348.4\,\text{kg}$$

Thus, marching downward all the stage masses are all calculated. Also finally, the minimized total launch mass of the four-stage vehicle is calculated as 14726.1 kg.

By Tsiolkovsky equation,

$$\Delta\, v_{(1)} = \overline{u}_{e(1)} ln\ \mu_{(1)} = 2600 \times ln\ 2.49294 = 2375\,\text{m/s}$$

Note the stage velocity increments $\Delta\, v_{(i)}$'s are all equal. The results of the calculations are summarized in Table 5.3.

Table 5.3 Results of the calculations of Example 5.1: equal specific impulse and equal structural coefficients, payload mass 100 kg and terminal velocity 9500 m/s

Stg	$I_{sp(i)}$ (m/s)	$\varepsilon_{(i)}$	$\mu_{(i)}$	$\mu_{(i)}\varepsilon_{(i)}$	$\alpha_{(i)}$	$m_{(i)}$ (kg)	$m_{0(i)}$ (kg)	$\Delta v_{(i)}$ (m/s)
1	2600	0.160	2.49294	0.39887	0.40265	10498.8	14726.1	2375.0
2	2600	0.160	2.49294	0.39887	0.40265	3013.8	4227.3	2375.0
3	2600	0.160	2.49294	0.39887	0.40265	865.2	1213.5	2375.0
4	2600	0.160	2.49294	0.39887	0.40265	248.4	348.4	2375.0

Discussion

In the first simple case, all stages have equal specific impulse values but different structural coefficients—the first stage having the lightest structural coefficient. Under these conditions the factor $\mu_{(i)}\varepsilon_{(i)}$'s are all equal for all stages. This way the stage which has the lightest structural coefficient has the highest mass ratio $\mu_{(i)}$. Thus, that stage gets assigned to deliver the highest incremental velocity $\Delta\ v_{(i)}$ under the uniform specific impulse values. In a similar vein, we will see later that a stage with a higher specific impulse gets assigned with a higher incremental velocity $\Delta\ v_{(i)}$ in the optimization analysis.

The second simple case is analytically simpler one. Here the stages have all equal payload ratios $\alpha_{(i)}$'s, mass ratios $\mu_{(i)}$'s, and stage incremental velocities $\Delta\ v_{(i)}$'s.

5.3.2 *Root of Lagrange Multiplier*

To calculate the optimally minimized total mass of the launch vehicle $m_{0(1)}$ we have derived in the previous Sect. 5.3.1 two important equations, Eqs. 5.14 and 5.15.

$$\mu_{(i)} = \frac{\lambda\overline{u}_{e(i)} + 1}{\lambda\overline{u}_{e(i)}\varepsilon_{(i)}} \qquad i = 1, 2, ..., n \tag{5.14}$$

$$v_n = \sum_{i=1}^{n} \overline{u}_{e(i)} ln \left(\frac{\lambda\overline{u}_{e(i)} + 1}{\lambda\overline{u}_{e(i)}\varepsilon_{(i)}} \right) \tag{5.15}$$

For the prescribed terminal velocity v_n and the n pairs of $\overline{u}_{e(i)}$ and $\varepsilon_{(i)}$, λ is to be evaluated from Eq. 5.15. The optimum mass ratios $\mu_{(i)}$'s are then found by substituting the evaluated λ into Eq. 5.14. Calculating thereafter the payload ratios $\alpha_{(i)}$'s, Eq. 5.6, and proceeding downward with the given payload m_l at the top, the minimized optimum mass of the total launch vehicle $m_{0(1)}$ can be calculated—see Example 5.1. But, Eq. 5.15 being a transcendental one with a single unknown λ there must be multiple roots for λ. Therefore, first without considering the constraint v_n, a possible range for λ is to be established for the given n pairs $\overline{u}_{e(i)}$ and $\varepsilon_{(i)}$. This will minimize the iterations necessary to obtain the correct value of λ for the prescribed terminal velocity v_n. The obtained root of λ for the prescribed v_n will give the absolute minimum mass of the total rocket vehicle $m_{0(1)}$ for the given payload of m_l, or simply said the absolute minimum of $m_{0(1)}/m_l$.

Before considering the root of the Lagrange multiplier λ, let us note the order of values of different parameters. Equivalent exhaust velocity or the specific impulse $\overline{u}_e$ is positive and far greater than unity, $\overline{u}_{e(i)} >> 1$. The mass ratio, the ratio of the vehicle mass at any stage before ignition to its mass after burnout is greater than unity, $\mu_{(i)} = m_{0(i)}/m_{0b(i)} > 1$. The structural coefficient is positive and less than unity, $0 < \varepsilon_{(i)} < 1$. And, $\mu_{(i)}\varepsilon_{(i)} < 1$. By Eq. 5.14, the optimum mass ratio $\mu_{(i)}$ can be written as

$$\mu_{(i)} = \frac{1 + \lambda\overline{u}_{e(i)}}{\lambda\overline{u}_{e(i)}\varepsilon_{(i)}} = \frac{1/\lambda + \overline{u}_{e(i)}}{\overline{u}_{e(i)}\varepsilon_{(i)}} \qquad i = 1, 2, ..., n$$

From this equation, it follows for the ith stage,

$$\lambda = \frac{-1}{\overline{u}_{e(i)}(1 - \mu_{(i)}\varepsilon_{(i)})} \tag{5.23}$$

Since $\mu_{(i)}\varepsilon_i < 1$, it follows that λ should be negative, that is $\lambda < 0$. It should be understood that Eq. 5.23 only indicates $\lambda < 0$ and we cannot evaluate λ from Eq. 5.23 since there are multiple sets of $\overline{u}_{e(i)}$ and $\varepsilon_{(i)}$ with different values for the n stages of the vehicle.

To proceed further with our analysis let us not prescribe a value for the terminal velocity and see what its possible range is. We note that the mass ratio being greater than unity,

$$\mu_{(i)} = \frac{1 + \lambda\overline{u}_{e(i)}}{\lambda\overline{u}_{e(i)}\varepsilon_{(i)}} > 1$$

and remembering $\lambda < 0$, it follows,

$$\lambda < -\frac{1}{\overline{u}_{e(i)}(1 - \varepsilon_{(i)})} \tag{5.24}$$

Equation 5.24 reveals that the Lagrange multiplier λ being negative must be less than the most minimum $-1/[\overline{u}_{e(i)}(1 - \varepsilon_{(i)})]$ of all stages. Therefore, to fix the upper bound for the root of λ we choose the stage that gives its value $-1/[\overline{u}_{e(i)}(1 - \varepsilon_{(i)})]$ as the most minimum. That is,

$$\lambda < -\frac{1}{\min_{i=1}^{n}[\overline{u}_{e(i)}(1 - \varepsilon_{(i)})]} \tag{5.25}$$

Firstly it is seen from Eq. 5.15, that the terminal velocity v_n is continuous for all the values λ given by Eq. 5.25. Secondly the derivative of Eq. 5.15 with respect to λ is always negative for all values of λ, Eq. 5.26.

$$\frac{dv_n}{d\lambda} = \sum_{i=1}^{n} -\frac{1}{\lambda^2 \mu_{(i)} \varepsilon_{(i)}} \tag{5.26}$$

These two observations lead to the following important conclusions. (1) For the given sets of $\overline{u}_{e(i)}$ and $\varepsilon_{(i)}$, the terminal velocity v_n must lie in the range between the limiting terminal velocity given as $\lambda \to -\infty$ and the one given by the upper bound of λ given by Eq. 5.25. (2) For the given sets of $\overline{u}_{e(i)}$ and $\varepsilon_{(i)}$, once we prescribe a value for the terminal velocity v_n there is one and only one value of λ in the range $-\infty < \lambda < -1/\{\min_{i=1}^{n}[\overline{u}_{e(i)}(1-\varepsilon_{(i)})]\}$. Starting with a value of λ at its upper bound as the trial one, the applicable root of λ can be numerically evaluated for the prescribed value of v_n through Eq. 5.15.

As the Lagrange multiplier $\lambda \to -\infty$ Eq. 5.15 becomes,

$$\lim_{\lambda \to -\infty} v_n = \sum_{i=1}^{n} u_{e(i)} ln\, \frac{1+\lambda \overline{u}_{e(i)}}{\lambda \overline{u}_{e(i)} \varepsilon_{(i)}} = \sum_{i=1}^{n} u_{e(i)} ln\, \frac{1/\lambda + \overline{u}_{e(i)}}{\overline{u}_{e(i)} \varepsilon_{(i)}} = \sum_{i=1}^{n} u_{e(i)} ln\, \frac{1}{\varepsilon_{(i)}} \tag{5.27}$$

Similarly at the upper bound of the Lagrange multiplier, by Eq. 5.25

$$\lambda_{ub} = -\frac{1}{\min_{i=1}^{n}[\overline{u}_{e(i)}(1-\varepsilon_{(i)})]} \tag{5.25a}$$

The terminal velocity v_n then becomes,

$$v_n = \sum_{i=1}^{n} \overline{u}_{e(i)} ln\, \frac{1/\lambda + \overline{u}_{e(i)}}{\overline{u}_{e(i)} \varepsilon_i} = \sum_{i=1}^{n} \overline{u}_{e(i)} ln\, \left\{ \frac{1}{\varepsilon_{(i)}} \left[1 - \frac{\min_{i=1}^{n}[\overline{u}_{e(i)}(1-\varepsilon_{(i)})]}{\overline{u}_{e(i)}} \right] \right\} \tag{5.28}$$

Thus, the inequality Eq. 5.29 gives the range of possible terminal velocities by considering the two extremities $\lambda \to -\infty$ and $\lambda_{ub} = -1/\min_{i=1}^{n}[\overline{u}_{e(i)}(1-\varepsilon_{(i)})]$.

$$\sum_{i=1}^{n} u_{e(i)} ln\, \frac{1}{\varepsilon_{(i)}} > v_n > \sum_{i=1}^{n} \overline{u}_{e(i)}\, ln \left\{ \frac{1}{\varepsilon_{(i)}} \left[1 - \frac{\min_{i=1}^{n}[\overline{u}_{e(i)}(1-\varepsilon_{(i)})]}{\overline{u}_{e(i)}} \right] \right\} \tag{5.29}$$

Equation 5.29, therefore, is a necessary and sufficient condition on the prescribed value of v_n for a solution of Eq. 5.15 to exist and to be unique.

It is interesting to examine what would happen to $m_{0(1)}/m_l$ if the Lagrange multiplier tends to the value of $-\infty$. With reference to Eq. 5.14, as $\lambda \to -\infty$, $\mu_{(i)} = 1/\varepsilon_{(i)}$.

$$\lim_{\lambda \to -\infty} \mu_{(i)} = \frac{\overline{u}_{e(i)} + 1/\lambda}{\overline{u}_{e(i)}\varepsilon_{(i)}} = \frac{1}{\varepsilon_{(i)}} \tag{5.30}$$

Using this condition $\mu_{(i)} = 1/\varepsilon_{(i)}$ into Eq. 5.10 we see $m_{0(1)}/m_l \to \infty$. This means that either the payload mass tends to zero or the total launch mass of the rocket vehicle is infinitely large. In practical terms, if we need a higher terminal velocity v_n for the fixed launch vehicle ($m_{0(1)}$ is fixed) the payload mass m_l has to be less or for the given payload the launch vehicle has to be bigger—$m_{0(1)}$ has to be heavier. Similar to the inequality Eq. 5.29 we can obtain an expression for $m_{0(1)}/m_l$ when λ assumes its limiting values $-\infty$ and the upper bound, Eq. 5.25a.

$$\infty > \frac{m_{0(1)}}{m_l} > \prod_{i=1}^{n} \left(1 - \frac{1}{\varepsilon_{(i)}}\right) \left[1 - \frac{\overline{u}_{e(i)}}{\min_{i=1}^{n}[\overline{u}_{e(i)}(1 - \varepsilon_{(i)})]}\right] \tag{5.29a}$$

Equations 5.29 and 5.29a being interrelated, the minimum terminal velocity corresponds to a minimum $m_{0(1)}/m_l$ and the maximum one to $m_{0(1)}/m_l \to \infty$. We know that the required terminal velocity for a LEO (around 7.6 km/s) is lower than that for a GTO (around 10.2 km/s). Thus, we see in Table 5.1, $m_{0(1)}/m_{l\,LEO} < m_{0(1)}/m_{l\,GTO}$ or the payload fraction $m_l/m_{0(1)LEO} > m_l/m_{0(1)GTO}$ for any chosen orbital launch vehicle.

In optimizing a chosen rocket vehicle with a certain set of parameters $\overline{u}_{e(i)}, \varepsilon_{(i)}$, and the number of stages n to meet the prescribed terminal velocity v_n, the evaluation of the inequality Eq. 5.29 is useful. It will reveal whether the prescribed terminal velocity v_n is attainable or not—v_n is falling within the range or not. If yes, whether a reduction in the number of stages or a change in any of the other parameters will also satisfy the desired performance or not can be examined. Although a certain design may fall within the range, Eq. 5.29, the ratio of the total launch mass to the payload mass $m_{0(1)}/m_l$ frequently becomes the criteria. Substituting the optimum mass ratio $\mu_{(i)}$ given in Eq. 5.14 into the equation for $m_{0(1)}/m_l$, Eq. 5.10, and rewriting the summation as a product by eliminating the natural logarithm, we get

$$\frac{m_{0(1)}}{m_l} = \prod_{i=1}^{n} \left[\left(1 - \frac{1}{\varepsilon_{(i)}}\right)(1 + \lambda\overline{u}_{e(i)})\right] \tag{5.31}$$

While considering various options, the expression in Eq. 5.31 gives the facility to readily calculate for the chosen set of parameters and to check whether the ratio of the total launch mass to the payload mass $m_{0(1)}/m_l$ is acceptable or not. Example 5.2 covers the aspects that we have considered on the minimized optimum mass of the total launch vehicle $m_{0(1)}$ by choosing the applicable root of the Lagrange multiplier λ.

Example 5.2 A three-stage launch vehicle with the first two stages of $N_2O_4/UDMH$ propellant combination and the third cryogenic stage of $LOX/LH2$ propellant combination is being considered to have a terminal velocity of 9800 m/s with a payload of 2000 kg. The first stage delivers a specific impulse of 2600 m/s with a structural coefficient of 0.105. And, the second one delivers a specific impulse of 2840 m/s with a structural coefficient of 0.108. The third stage delivers a specific impulse of 4355 m/s with a structural coefficient of 0.163. With no turning in a gravity- and drag-free field, calculate the minimized optimum mass of the total launch vehicle and the other stage properties. Suppose that this mission is considered only with two stages: one with the first and the second stages of $N_2O_4/UDMH$ stages and the other with the first $N_2O_4/UDMH$ stage and the cryogenic one of $LOX/LH2$. Calculate for these two two-stage concepts, the minimized optimum mass of the total launch vehicle and the other stage properties. Compare and discuss all the three choices.

Solution
Using Eq. 5.25, the upper bound of the Lagrange multiplier λ is found. Starting from this upper bound as a trial value, through a numerical procedure the root of λ for the prescribed terminal velocity of 9800 m/s is found, Eq. 5.15. The range of terminal velocities as $\lambda \rightarrow -\infty$ and λ equals the upper bound is calculated to check whether the prescribed terminal velocity of 9800 m/s lies within this range, Eq. 5.29. Adopting the obtained root of λ, the mass ratios $\mu_{(1)}$, $\mu_{(2)}$, and $\mu_{(3)}$ are calculated, Eq. 5.14. Through Eq. 5.9, $m_{0(1)}/m_{0(2)}$, $m_{0(2)}/m_{0(3)}$, and $m_{0(3)}/m_l$ are known. By Eq. 5.6, payload ratios $\alpha_{(1)}$, $\alpha_{(2)}$, and $\alpha_{(3)}$ are calculated. Starting from the top, with the given payload value of 2000 kg, the stage mass $m_{(3)}$ is calculated from $\alpha_{(3)}$ as $m_{(3)} = m_l/\alpha_{(3)}$. $m_{0(3)} = m_l + m_{(3)}$. Thus, marching downward $m_{0(2)}$, and the minimized optimum mass of the total launch vehicle $m_{0(1)}$ are arrived at. Velocity increments $\Delta\, v_{(i)}$'s are calculated using Eq. 5.11.

Similarly, the other two two-stage concepts are evaluated.

Analysis
To fix the upper bound for the root of λ, we calculate the values of $-1/[\overline{u}_{e(i)}(1 - \varepsilon_{(i)})]$.

$$-\frac{-1}{\overline{u}_{e(1)}(1-\varepsilon_{(1)})} = -\frac{-1}{2600 \times (1-0.105)} = -0.000430$$

$$-\frac{-1}{\overline{u}_{e(2)}(1-\varepsilon_{(2)})} = -\frac{-1}{2830 \times (1-0.108)} = -0.000396$$

$$-\frac{-1}{\overline{u}_{e(3)}(1-\varepsilon_{(3)})} = -\frac{-1}{4355 \times (1-0.163)} = -0.000274$$

Therefore, the most negative -0.000430 forms the upper bound of λ. That is,

$$\lambda_{ub} = -\frac{1}{\min_{i=1}^{3}[\overline{u}_{e(i)}(1-\varepsilon_{(i)})]} = -\frac{1}{[\overline{u}_{e(1)}(1-\varepsilon_{(1)})]} = -0.000430$$

The range of terminal velocities as $\lambda \to -\infty$ and $\lambda_{ub} = -0.000430$ is to be calculated to check whether the prescribed terminal velocity of 9800 m/s lies within this range, Eq. 5.29. As $\lambda \to -\infty$,

$$\sum_{i=1}^{n} \overline{u}_{e(i)} ln \, \frac{1}{\varepsilon_{(i)}} = \sum_{i=1}^{3} \overline{u}_{e(i)} ln \, \frac{1}{\varepsilon_{(i)}} =$$

$$2600 \times ln \, \frac{1}{0.105} + 2830 \times ln \, \frac{1}{0.108} + 4355 \times ln \, \frac{1}{0.163} = 20058.375 \text{ m/s}$$

And, at $\lambda_{ub} = -0.000430$,

$$\sum_{i=1}^{n} \overline{u}_{e(i)} ln \left\{ \frac{1}{\varepsilon_{(i)}} \left[1 - \frac{\min_{i=1}^{3}[\overline{u}_{e(i)}(1-\varepsilon_{(i)})]}{\overline{u}_{e(i)}} \right] \right\}$$

$$= 2600 \times ln \left\{ \frac{1}{0.105} \left[1 - \frac{1}{2600 \times 0.000430} \right] \right\}$$

$$+2830 \times ln \left\{ \frac{1}{0.108} \left[1 - \frac{1}{2830 \times 0.000430} \right] \right\}$$

$$+4355 \times ln \left\{ \frac{1}{0.163} \left[1 - \frac{1}{4355 \times 0.000430} \right] \right\} = 5981.432 \text{ m/s}$$

We see that the prescribed terminal velocity of 9800 m/s is within the range of 20058.375 m/s and 5981.432 m/s. Through Eq. 5.29a we obtain the value of $m_{0(1)}/m_l$ when λ assumes its value at the upper bound, Eq. 5.25a.

$$\prod_{i=1}^{n} \left[\left(1 - \frac{1}{\varepsilon_{(i)}} \right) \left(1 - \frac{\overline{u}_{e(i)}}{\min_{i=1}^{n}[\overline{u}_{e(i)}(1-\varepsilon_{(i)})]} \right) \right]$$

$$= \left[\left(1 - \frac{1}{0.105} \right) \times \left(1 - \frac{2600}{2327} \right) \right]$$

$$\times \left[\left(1 - \frac{1}{0.108} \right) \times \left(1 - \frac{2830}{2327} \right) \right]$$

$$\times \left[\left(1 - \frac{1}{0.163} \right) \times \left(1 - \frac{4355}{2327} \right) \right] = 7.9895$$

Next, noting that the upper bound for λ is -0.000430 and starting therefore with values lower than this we have to search for its root through a numerical procedure for $v_n = 9800\,\text{m/s}$, Eq. 5.15. This is easily done using Excel spreadsheet. The root of λ for $v_n = 9800\,\text{m/s}$ is found to be $-4.92517E-04$.

Recall Eq. 5.14.

$$\mu_{(i)} = \frac{\lambda \overline{u}_{e(i)} + 1}{\lambda \overline{u}_{e(i)} \varepsilon_{(i)}} \qquad i = 1, 2, ..., n \tag{5.14}$$

$$\mu_{(1)} = \frac{\lambda \overline{u}_{e(1)} + 1}{\lambda \overline{u}_{e(1)} \varepsilon_{(1)}} = \frac{-4.92517E - 04 \times 2600 + 1}{-4.92517E - 04 \times 2600 \times 0.105} = 2.08650$$

$$\mu_{(2)} = \frac{\lambda \overline{u}_{e(2)} + 1}{\lambda \overline{u}_{e(2)} \varepsilon_{(2)}} = \frac{-4.92517E - 04 \times 2830 + 1}{-4.92517E - 04 \times 2830 \times 0.108} = 2.61620$$

$$\mu_{(3)} = \frac{\lambda \overline{u}_{e(3)} + 1}{\lambda \overline{u}_{e(3)} \varepsilon_{(3)}} = \frac{-4.92517E - 04 \times 4355 + 1}{-4.92517E - 04 \times 4355 \times 0.163} = 3.27473$$

Recall Eq. 5.6.

$$\alpha_{(i)} = \frac{1 - \mu_{(i)} \varepsilon_{(i)}}{\mu_{(i)} - 1} \tag{5.6}$$

$$\alpha_{(1)} = \frac{1 - \mu_{(1)} \varepsilon_{(1)}}{\mu_{(1)} - 1} = \frac{1 - 2.08650 \times 0.105}{2.08650 - 1} = 0.71874$$

$$\alpha_{(2)} = \frac{1 - \mu_{(2)} \varepsilon_{(2)}}{\mu_{(2)} - 1} = \frac{1 - 2.61620 \times 0.108}{2.61620 - 1} = 0.44391$$

$$\alpha_{(3)} = \frac{1 - \mu_{(3)} \varepsilon_{(3)}}{\mu_{(3)} - 1} = \frac{1 - 3.27473 \times 0.163}{3.27473 - 1} = 0.20496$$

Recall the definition of payload ratio, Eq. 5.1.

$$\alpha_{(i)} \equiv \frac{m_{0(i+1)}}{m_{(i)}}$$

Now, we have to start from the top.

$$\alpha_{(3)} = \frac{m_l}{m_{(3)}} \quad \Rightarrow m_{(3)} = \frac{m_l}{\alpha_{(3)}}$$

$$m_{(3)} = \frac{2000}{0.20496} = 9758.2\,\text{kg}$$

$$m_{0(3)} = m_l + m_{(3)} = 2000 + 9758.2 = 11758.2\,\text{kg}$$

$$\alpha_{(2)} = \frac{m_{0(3)}}{m_{(2)}} \quad \Rightarrow m_{(2)} = \frac{m_{0(3)}}{\alpha_{(2)}}$$

$$m_{(2)} = \frac{11758.2}{0.44391} = 26487.7\,\text{kg}$$

$$m_{0(2)} = m_{0(3)} + m_{(2)} = 11758.2 + 26487.7 = 38245.8\,\text{kg}$$

$$\alpha_{(1)} = \frac{m_{0(2)}}{m_{(1)}} \quad \Rightarrow m_{(1)} = \frac{m_{0(2)}}{\alpha_{(1)}}$$

$$m_{(1)} = \frac{38245.8}{0.71874} = 53212.0\,\text{kg}$$

$$m_{0(1)} = m_{0(2)} + m_{(1)} = 38245.8 + 53212.0 = 91457.9\,\text{kg}$$

Thus, the minimized optimum mass of the total launch vehicle $m_{0(1)} =$ 91457.9 kg. And the ratio of the total launch-vehicle mass to the payload mass is 91457.9/2000 = 45.729.

Now the stage incremental velocities can be calculated.

$$\Delta\ v_{(1)} = \overline{u}_{e(1)} ln\ \mu_{(1)} = 2600 \times ln\ 2.08650 = 1912.3\,\text{m/s}$$

$$\Delta\ v_{(2)} = \overline{u}_{e(2)} ln\ \mu_{(2)} = 2830 \times ln\ 2.61620 = 2721.7\,\text{m/s}$$

$$\Delta\ v_{(3)} = \overline{u}_{e(3)} ln\ \mu_{(3)} = 4355 \times ln\ 3.27473 = 5166.0\,\text{m/s}$$

These stage incremental velocities sum-up to the prescribed terminal velocity of 9800 m/s. The summary of these results are given in Table 5.4.

Similarly the stage properties are calculated for the other two two-stage options and the results are summarized in Tables 5.5 and 5.6. In the cases of the two two-stage options the upper bound for λ remains the same at $\lambda = -0.000430$. The vehicle of two $N_2O_4/UDMH$ stages has the applicable root of λ as $-1.04600E-03$. This gives the minimized optimum mass of the total launch vehicle $m_{0(1)}$ as 474602.6 kg for the terminal velocity of 9800 m/s. The range of terminal velocities for this concept is 1410 m/s to 12158 m/s. Next, the vehicle of $N_2O_4/UDMH$ and $LOX/LH2$ stages has the applicable root of λ as $-6.80283E-04$. This gives the minimized optimum mass of the total launch vehicle $m_{0(1)}$ as 132074.4 kg for the terminal velocity of 9800 m/s. The range of terminal velocities for this concept is 4572 m/s to 13760 m/s.

Table 5.4 Example 5.2 results: Three-stage launch vehicle of the first two stages of $N_2O_4/UDMH$ combination and the third stage of $LOX/LH2$ combination with a payload of 2000 kg to reach a terminal velocity of 9800 m/s in a gravity- and drag-free environment with no turning

Stg	$I_{sp(i)}$ (m/s)	$\varepsilon_{(i)}$	$\mu_{(i)}$	$\alpha_{(i)}$	$m_{(i)}$ (kg)	$m_{0(i)}$ (kg)	$\Delta v_{(i)}$ (m/s)
1	2600	0.105	2.0865	0.7187	53212.0	91457.9	1912.3
2	2830	0.108	2.6162	0.4439	26487.7	38245.8	2721.7
3	4355	0.163	3.2747	0.2050	9758.2	11758.2	5166.1

Table 5.5 Example 5.2 results: Two-stage launch vehicle of $N_2O_4/UDMH$ combination with a payload of 2000 kg to reach a terminal velocity of 9800 m/s in a gravity- and drag-free environment with no turning

Stg	$I_{sp(i)}$ (m/s)	$\varepsilon_{(i)}$	$\mu_{(i)}$	$\alpha_{(i)}$	$m_{(i)}$ (kg)	$m_{0(i)}$ (kg)	$\Delta v_{(i)}$ (m/s)
1	2600	0.105	6.0219	0.0732	442223.3	474602.6	4668.0
2	2830	0.108	6.1313	0.0658	30379.3	32379.3	5132.0

Table 5.6 Example 5.2 results: Two-stage launch vehicle of the first stage of $N_2O_4/UDMH$ combination and the second one of $LOX/LH2$ combination with a payload of 2000 kg to reach a terminal velocity of 9800 m/s in a gravity- and drag-free environment with no turning

Stg	$I_{sp(i)}$ (m/s)	$\varepsilon_{(i)}$	$\mu_{(i)}$	$\alpha_{(i)}$	$m_{(i)}$ (kg)	$m_{0(i)}$ (kg)	$\Delta v_{(i)}$ (m/s)
1	2600	0.105	4.1393	0.1801	111918.3	132074.4	3693.4
2	4355	0.163	4.0642	0.1102	18156.1	20156.1	6106.6

Discussion

First and foremost, it should be noted that the stage optimization analyses that we have done so far are for the idealized condition of a rocket-vehicle flight in a gravity- and drag-free environment with no turning. This optimization procedure results are to be taken as the preliminary design values. For the selected trajectories, there are **velocity losses** in flights due to gravity, aerodynamic drag, turning or steering, and fairing separation. Gravity losses are always there until the payload, the satellite, reaches the desired orbit. Considerable aerodynamic drag is there until the vehicle climbs above the dense atmospheric air layer. Mostly the flight in the atmospheric air layer is served by the first stage of the rocket vehicle. Above about 55 km altitude the drag can be considered negligible [6]. Density of air at this altitude is about 2000 times less than the sea-level value. Being a function of flight Mach number, the drag coefficient peaks in the transonic regime and thereafter it tapers. In order to steer the vehicle into the selected trajectory, during the turning of the vehicle the thrust vector no longer coincides with the flight-velocity vector causing turning losses. In order to overcome these velocity losses and reach the prescribed terminal velocity v_n, there should be more energy stored in the stages—more propellant stored in the stages. These result in a rocket vehicle much heavier than the one given by the present idealized analyses. Added to these, the structural coefficient for a given propellant class of the stage can no longer be a constant—see Figs. 5.2, 5.3, 5.4 and 5.5. It reduces with the increase in stage mass, although in a narrow range. Cost of the launch vehicle can also enter into the optimization aspect. Reference [11] considers the aspects of the cost and the variation of structural coefficients with stage-mass and -class. Thus, to arrive at the actual optimized rocket vehicle, there are extensive interactive procedures between the trajectory and the various realities. However, in a subsequent section, we consider qualitatively the velocity losses to arrive at the optimized vehicle somewhat better than the one of the present preliminary design analyses.

Examining the results given in Tables 5.4, 5.5 and 5.6, we find the optimization procedures favoring the stage with a higher specific impulse to assign a higher velocity increment. We earlier noted similar behavior in the assignment of higher velocity increment to the stage with a lighter structural coefficient, Table 5.2. Comparing the results of the two two-stage options, the option with two low energy stages (two $N_2O_4/UDMH$ stages) has much heavier launch vehicle of about 475 tons compared to the vehicle mass of about 132 tons of the option with one low- and one high-energy stages ($N_2O_4/UDMH$ & $LOX/LH2$).

5.3.3 *Alternate Optimization Route*

For the optimization of the rocket vehicle, previously we minimized the total mass of the launch vehicle $m_{0(1)}$ for the given payload mass m_l and the prescribed terminal velocity v_n. Or, we can say in short that we minimized $m_{0(1)}/m_l$ for the prescribed terminal velocity v_n. More or less in a generalized way, we can look at this optimization problem into two alternate routes: either we minimize $m_{0(1)}/m_l$ (or maximize $m_l/m_{0(1)}$) for the prescribed terminal velocity v_n or maximize the terminal velocity v_n for the specified $m_{0(1)}/m_l$. And, we will see that these two routes fall into one and the same optimization exercise [12].

Let us recapitulate what we have learned in Sects. 5.3.1 and 5.3.2 to minimize $m_{0(1)}/m_l$ for the prescribed v_n. Adopting the Tsiolkovsky equation, the prescribed terminal velocity v_n at the burnout of the last stage $m_{(n)}$ is written as

$$v_n = \sum_{i=1}^{n} \Delta v_i = \sum_{i=1}^{n} \overline{u}_{e(i)} ln\ \mu_{(i)} \tag{5.11}$$

For minimizing $m_{0(1)}/m_l$ for the prescribed terminal velocity v_n as the constraint, by introducing the Lagrange multiplier λ we have

$$ln\ \frac{m_{0(1)}}{m_l} = \sum_{i=1}^{n} ln\ \frac{\mu_{(i)}(1-\varepsilon_{(i)})}{1-\mu_{(i)}\varepsilon_{(i)}} + \lambda \left[\left(\sum_{i=1}^{n} \overline{u}_{e(i)} ln\ \mu_{(i)} \right) - v_n \right] \tag{5.13}$$

Differentiating Eq. 5.13 with respect to the mass ratio $\mu_{(i)}$, equating it to zero, and solving for $\mu_{(i)}$ for the extremum, we get the optimum mass ratio,

$$\mu_{(i)} = \frac{\lambda \overline{u}_{e(i)} + 1}{\lambda \overline{u}_{e(i)} \varepsilon_{(i)}} \qquad i = 1, .., n \tag{5.14}$$

Substituting these optimum $\mu_{(i)}$'s in Eq. 5.11,

$$v_n = \sum_{i=1}^{n} \overline{u}_{e(i)} ln \left(\frac{\lambda \overline{u}_{e(i)} + 1}{\lambda \overline{u}_{e(i)} \varepsilon_{(i)}} \right) \tag{5.15}$$

For the prescribed terminal velocity v_n and n pairs of $\overline{u}_{e(i)}$ and $\varepsilon_{(i)}$, evaluating λ from Eq. 5.15, the optimum mass ratios $\mu_{(i)}$'s are found by substituting the evaluated λ into Eq. 5.14. The Lagrange multiplier λ is to be numerically evaluated through iterations for the prescribed v_n, Eq. 5.15. It is seen that λ is negative and has to be less than the upper bound given by Eq. 5.25.

$$\lambda < -\frac{1}{\min_{i=1}^{n}[\overline{u}_{e(i)}(1-\varepsilon_{(i)})]} \tag{5.25}$$

Starting with the trial value of λ at this upper bound, the root of λ for the prescribed terminal velocity v_n is evaluated through iterations, Eq. 5.15. Subsequently, the optimized stage properties are calculated, Eq. 5.14—see Example 5.2.

Now proceeding in a similar way, we consider maximizing v_n for the specified $m_{0(1)}/m_l$. To distinguish the Lagrange multiplier from the previous procedure we assign λ' for it presently. The ratio $m_{0(1)}/m_l$ is written as

$$ln\ \frac{m_{0(1)}}{m_l} = \sum_{i=1}^{n} ln\ \frac{\mu_{(i)}(1-\varepsilon_{(i)})}{1-\mu_{(i)}\varepsilon_{(i)}} \tag{5.10}$$

For maximizing the terminal velocity v_n for the specified $m_{0(1)}/m_l$ as the constraint, we have

$$v_n = \sum_{i=1}^{n} u_{e(i)} ln\ \mu_{(i)} + \lambda'[\sum_{i=1}^{n} ln\ \frac{\mu_{(i)}(1-\varepsilon_{(i)})}{1-\mu_{(i)}\varepsilon_{(i)}} - ln\ \frac{m_{0(1)}}{m_l}] \tag{5.32}$$

Differentiating Eq. 5.32 with respect to the mass ratio $\mu_{(i)}$ and equating it to zero,

$$\sum_{i=1}^{n} \frac{\overline{u}_{e(i)}}{\mu_{(i)}} + \lambda' \sum_{i=1}^{n} \frac{1}{\mu_{(i)}(1-\mu_{(i)}\varepsilon_{(i)})} = 0 \tag{5.33}$$

$$\sum_{i=1}^{n} \frac{1}{\mu_{(i)}(1-\mu_{(i)}\varepsilon_{(i)})} + \lambda \sum_{i=1}^{n} \frac{\overline{u}_{e(i)}}{\mu_{(i)}} = 0 \tag{5.34}$$

For comparison, the differentiated equation of Eq. 5.13 with v_n as the constraint for minimizing $m_{0(1)}/m_l$ is also given, Eq. 5.34. Solving for the optimum $\mu_{(i)}$ from Eq. 5.33 for maximizing v_n with $m_{0(1)}/m_l$ as the constraint,

$$\mu_{(i)} = \frac{\lambda' + \overline{u}_{e(i)}}{\overline{u}_{e(i)}\varepsilon_{(i)}} \quad i = 1, .., n \tag{5.35}$$

$$\mu_{(i)} = \frac{\lambda\overline{u}_{e(i)} + 1}{\lambda\overline{u}_{e(i)}\varepsilon_{(i)}} \quad i = 1, .., n \tag{5.14}$$

For comparison, the equivalent optimum mass ratio $\mu_{(i)}$ relation Eq. 5.14 that we got for minimizing $m_{0(1)}/m_l$ with v_n as the constraint is recalled. Comparing Eqs. 5.14 and 5.35, we deduce that $\lambda' = 1/\lambda$ and understand that either minimizing $m_{0(1)}/m_l$ (or maximizing $m_l/m_{0(1)}$) for the prescribed terminal velocity v_n, Eq. 5.34 or maximizing the terminal velocity v_n for the specified $m_{0(1)}/m_l$, Eq. 5.33 is one and the same optimization exercise.

Equations 5.10 and 5.35 provide a system of $n + 1$ equations for $n + 1$ unknowns, optimum mass ratios $\mu_{(i)}$, $i = 1, .., n$ and λ'. Substituting the optimum mass ratio $\mu_{(i)}$ of Eq. 5.35 into Eqs. 5.10 and 5.11, we get

$$ln\ \frac{m_{0(1)}}{m_l} = \sum_{i=1}^{n} ln \left[\left(1 - \frac{1}{\varepsilon_{(i)}}\right)\left(1 + \frac{\overline{u}_{e(i)}}{\lambda'}\right) \right] \tag{5.36}$$

Or,

$$\frac{m_{0(1)}}{m_l} = \prod_{i=1}^{n} \left[\left(1 - \frac{1}{\varepsilon_{(i)}}\right)\left(1 + \frac{\overline{u}_{e(i)}}{\lambda'}\right) \right] \tag{5.36a}$$

And,

$$v_n = \sum_{i=1}^{n} \overline{u}_{e(i)} ln\ \frac{\lambda' + \overline{u}_{e(i)}}{\overline{u}_{e(i)}\varepsilon_{(i)}} \tag{5.37}$$

For the specified $m_{0(1)}/m_l$ and n pairs of $\overline{u}_{e(i)}$ and $\varepsilon_{(i)}$, evaluating numerically λ' from Eq. 5.36, the optimum mass ratios $\mu_{(i)}$'s are found by substituting the evaluated λ' into Eq. 5.35. To initiate the numerical procedure to evaluate λ', the bounds for it are to be arrived at. Solving for λ' from Eq. 5.35 for the ith stage and noting $\mu_{(i)}\varepsilon_{(i)} < 1$, λ' is found to be negative.

$$\lambda' = -\overline{u}_{e(i)}(1 - \mu_{(i)}\varepsilon_{(i)}) \tag{5.38}$$

Since the mass ratio $\mu_{(i)} > 1$, it follows from Eq. 5.35 that the Lagrange multiplier λ' being negative must be greater than the least minimum $-[\overline{u}_{e(i)}(1 - \varepsilon_{(i)})]$ of all stages. Therefore, to fix the lower bound for the root of λ' we choose the stage that gives its value $[\overline{u}_{e(i)}(1 - \varepsilon_{(i)})]$ as the most minimum. That is,

$$0 > \lambda' > -\min_{i=1}^{n}[\overline{u}_{e(i)}(1 - \varepsilon_{(i)})] \tag{5.39}$$

$$-\infty < \lambda < -\frac{1}{\min_{i=1}^{n}[\overline{u}_{e(i)}(1 - \varepsilon_{(i)})]} \tag{5.25a}$$

For comparison, the equivalent bounds for λ that we got for minimizing $m_{0(1)}/m_l$ with v_n as the constraint are recalled, Eq. 5.25a. Starting with the trial value of λ' at the lower bound, Eq. 5.39, the root of λ' for the specified $m_{0(1)}/m_l$ is evaluated through iterations, Eq. 5.36. Subsequently, the optimized stage properties are calculated, Eq. 5.35. Substituting the bounds of λ', Eq. 5.39, into Eqs. 5.36 and 5.37, the limits of $m_{0(1)}/m_l$ and v_n become the previously obtained limits of Eqs. 5.29 and 5.29a. Furthermore, noting $\lambda' = 1/\lambda$ and replacing λ' with $1/\lambda$ in Eqs. 5.36 and 5.37, these equations become identical to Eqs. 5.31 and 5.15 respectively. Equations 5.15 and 5.31 are repeated below for ready reference.

$$v_n = \sum_{i=1}^{n} \overline{u}_{e(i)} ln \left(\frac{\lambda\overline{u}_{e(i)} + 1}{\lambda\overline{u}_{e(i)}\varepsilon_{(i)}} \right) \tag{5.15}$$

$$\frac{m_{0(1)}}{m_l} = \prod_{i=1}^{n} \left[\left(1 - \frac{1}{\varepsilon_{(i)}} \right) (1 + \lambda\overline{u}_{e(i)}) \right] \tag{5.31}$$

Thus, the unification of the two routes of optimization as a single exercise can be summarized as follows.

1. To minimize $m_{0(1)}/m_l$ (or maximize $m_l/m_{0(1)}$) for a prescribed terminal velocity v_n, the Lagrange multiplier λ is determined from Eq. 5.15 through iterations by starting with a trial value of it at its upper bound Eq. 5.25, and the same is substituted into Eq. 5.31 to give the minimized $m_{0(1)}/m_l$.
2. To maximize the terminal velocity v_n for a specified $m_{0(1)}/m_l$, the Lagrange multiplier λ is determined from Eq. 5.31 through iterations by starting with a trial value of it at its upper bound Eq. 5.25, and the same is substituted into Eq. 5.15 to give the maximized v_n.

Using the data of Saturn V, further analysis of loss-free optimization procedure is given in Example 5.3. More importantly, the procedure to obtain the characteristics of optimum terminal velocities versus optimum $m_{0(1)}/m_l$ is considered in this example.

Example 5.3 The data of a version of Saturn V rocket vehicle are being considered. The vehicle's first stage is a cluster of five LOX/Kerosene F1 engines. The second stage is a cluster of five LOX/LH2 J2 engines and the third one is of one LOX/LH2 J2 engine. The vehicle can deliver a payload of 118000 kg into a LEO, Table 5.7. Also for a translunar orbit, the same vehicle can deliver a payload of 47000 kg, Table 5.8. (1) Considering the corresponding ratios of $m_{0(1)}/m_l$ for a LEO and for a translunar orbit, calculate the optimum terminal velocities and the optimum stage properties under gravity- and drag-free environment without turning. (2) Furthermore, calculate the limiting values of terminal velocities and $m_{0(1)}/m_l$'s. (3) Compare the calculated stage masses with the actual stage masses of the Saturn V. (4) For the given values of specific impulses and structural coefficients, draw the characteristics of optimum terminal velocity versus $m_{0(1)}/m_l$ for ranges of Lambda; consider different options of vehicle configurations : (a) the three-stage, (b) the two-stage option with LOX/Kerosene and LOX/LH2 of the second stage, and (c) the different single stages. Discuss.

Table 5.7 Example 5.3: Selected data of Saturn V rocket vehicle [1], LEO payload = 118000 kg, the last four columns are calculated values, and $\Delta v_{(i)}$'s are for drag- and gravity-free environment with no turning

Stg	$I_{sp(i)}$ (m/s)	$\varepsilon_{(i)}$	$m_{(i)}$ (kg)	$m_{0(i)}$ (kg)	$m_{0b(i)}$ (kg)	$\mu_{(i)}$	$\Delta v_{(i)}$ (m/s)
1	2981	0.05914	2286217	3014895	863896	3.48988	3725.9
2	4129	0.07956	490778	728678	276948	2.63110	3994.4
3	4129	0.11093	119900	237900	131300	1.81188	2454.1
$\sum_{i=1}^{3} m_{(i)}$ (kg) = 2896895				$\sum_{i=1}^{3} \Delta v_{(i)}$ (m/s) = 10174.4			

Table 5.8 Example 5.3: Selected data of Saturn V rocket vehicle [1], translunar payload = 47000 kg, the last four columns are calculated values, and $\Delta v_{(i)}$'s are for drag- and gravity-free environment with no turning

Stg	$I_{sp(i)}$ (m/s)	$\varepsilon_{(i)}$	$m_{(i)}$ (kg)	$m_{0(i)}$ (kg)	$m_{0b(i)}$ (kg)	$\mu_{(i)}$	$\Delta v_{(i)}$(m/s)
1	2981	0.05914	2286217	2943895	792896	3.71284	3910.5
2	4129	0.07956	490778	657678	205948	3.19342	4794.1
3	4129	0.11093	119900	166900	60300	2.76783	4203.6
$\sum_{i=1}^{3} m_{(i)}$ (kg) = 2896895				$\sum_{i=1}^{3} \Delta v_{(i)}$ (m/s) = 12908.2			

Solution

(1) Optimum Terminal Velocities and Stage Properties: The upper bound of Lagrange multiplier λ_{ub} is found through Eq. 5.25a. With the respective payload masses of 118000 kg and 47000 kg calculate the corresponding $m_{0(1)}/m_l$'s. Using Eq. 5.31, through iterations find the root of λ for each of the $m_{0(1)}/m_l$'s. Substitution of the obtained root along with the given stage properties $\overline{u}_{e(i)}$'s and $\varepsilon_{(i)}$'s into Eq. 5.15 we get the optimum terminal velocity for the drag- and gravity-free environment with no turning. Adopting Eqs. 5.14 and 5.6, respectively $\mu_{(i)}$'s and $\alpha_{(i)}$'s can be calculated. As previously explained, descending from the top, optimum stage masses are calculated. Adopting Eq. 5.11, individual-stage incremental velocities $\Delta v_{(i)}$'s can be calculated.
(2) Limiting Values of Terminal Velocities and $m_{0(1)}/m_l$'s: Substituting the obtained roots of λ's and the given $\overline{u}_{e(i)}$'s and $\varepsilon_{(i)}$'s into Eqs. 5.29 and 5.29a, the limiting values of the terminal velocities and the ratios of launch vehicle mass to payload mass can be calculated.
(3) Characteristics of Optimum Terminal Velocity versus $m_{0(1)}/m_l$ for Ranges of Lambda: Adopting Eqs. 5.29 and 5.29a and starting from the λ_{ub}'s, the values of terminal velocities v_n's and the ratios launch vehicle mass to payload mass $m_{0(1)}/m_l$'s can be calculated for different values of λ. A suitable graph for v_n versus $m_{0(1)}/m_l$ can be drawn for the required options of vehicle configurations (a), (b), and (c).

Analysis

(1) Optimum Terminal Velocities and Stage Properties
The upper bound of λ is to be calculated. By Eq. 5.25a

$$-\infty < \lambda < -\frac{1}{\min_{i=1}^{n}[\overline{u}_{e(i)}(1-\varepsilon_{(i)})]} \tag{5.25a}$$

Stage	$-1/[\overline{u}_{e(i)}(1-\varepsilon_{(i)})]$
1	–0.000356546
2	–0.000263124
3	–0.000272406

Therefore, the least value $-3.56546E-4$ of $-1/[\overline{u}_{e(1)}(1-\varepsilon_{(1)})]$ is the upper bound of the Lagrange multiplier, λ_{ub}. With the payload mass of 118000 kg, the launch vehicle mass for the LEO is 3014895 kg, Table 5.7. This gives the value of $m_{0(1)}/m_l$ as 25.54996. And, with the payload mass of 47000 kg the launch vehicle mass for the translunar orbit is 2943895 kg, Table 5.8. This gives the value of $m_{0(1)}/m_l$ as 62.63606.

By Eq. 5.31,

$$\frac{m_{0(1)}}{m_l} = \prod_{i=1}^{n}\left[\left(1-\frac{1}{\varepsilon_{(i)}}\right)\left(1+\lambda\overline{u}_{e(i)}\right)\right] \tag{5.31}$$

Through iterations for the $m_{0(1)}/m_l = 25.54996$ corresponding to the LEO, the root of the Lagrange multiplier $\lambda = -3.60011E - 4$. By Eq. 5.15,

$$v_n = \sum_{i=1}^{n} \overline{u}_{e(i)} ln\left(\frac{\lambda\overline{u}_{e(i)}+1}{\lambda\overline{u}_{e(i)}\varepsilon_{(i)}}\right) \tag{5.15}$$

Substituting the converged value of the $\lambda = -3.60011E - 4$ of the LEO into Eq. 5.15,

$$v_{n,LEO} = 2981 \times \ln\frac{-3.60011E-4\times 2981+1}{-3.60011E-4\times 2981\times 0.05914}$$

$$+4129 \times \ln\frac{-3.60011E-4\times 4129+1}{-3.60011E-4\times 4129\times 0.07956}$$

$$+4129 \times \ln\frac{-3.60011E-4\times 4129+1}{-3.60011E-4\times 4129\times 0.11093} = 10731.3\,\mathrm{m/s}$$

The optimum terminal velocity v_n for the $m_{0(1)}/m_l$ of 25.54996 of LEO is 10731.3 m/s.

Recall Eq. 5.14.

$$\mu_{(i)} = \frac{\lambda\overline{u}_{e(i)}+1}{\lambda\overline{u}_{e(i)}\varepsilon_{(i)}} \qquad i = 1, 2, ..., n \tag{5.14}$$

$$\mu_{(1)} = \frac{\lambda\overline{u}_{e(1)}+1}{\lambda\overline{u}_{e(1)}\varepsilon_{(1)}} = \frac{-3.60011E-4\times 2981+1}{-3.60011E-4\times 2981\times 0.05914} = 1.15310$$

$$\mu_{(2)} = \frac{\lambda\overline{u}_{e(2)}+1}{\lambda\overline{u}_{e(2)}\varepsilon_{(2)}} = \frac{-3.60011E-4\times 4129+1}{-3.60011E-4\times 4129\times 0.07956} = 4.11334$$

$$\mu_{(3)} = \frac{\lambda\overline{u}_{e(3)}+1}{\lambda\overline{u}_{e(3)}\varepsilon_{(3)}} = \frac{-3.60011E-4\times 4129+1}{-3.60011E-4\times 4129\times 0.11093} = 2.95036$$

Recall Eq. 5.6.

$$\alpha_{(i)} = \frac{1 - \mu_{(i)}\varepsilon_{(i)}}{\mu_{(i)} - 1} \tag{5.6}$$

$$\alpha_{(1)} = \frac{1 - \mu_{(1)}\varepsilon_{(1)}}{\mu_{(1)} - 1} = \frac{1 - 1.15310 \times 0.05914}{1.15310 - 1} = 6.08623$$

$$\alpha_{(2)} = \frac{1 - \mu_{(2)}\varepsilon_{(2)}}{\mu_{(2)} - 1} = \frac{1 - 4.11334 \times 0.07956}{4.11334 - 1} = 0.21608$$

$$\alpha_{(3)} = \frac{1 - \mu_{(3)}\varepsilon_{(3)}}{\mu_{(3)} - 1} = \frac{1 - 2.95036 \times 0.11093}{2.95036 - 1} = 0.34492$$

Recall the definition of payload ratio, Eq. 5.1.

$$\alpha_{(i)} \equiv \frac{m_{0(i+1)}}{m_{(i)}}$$

Now, we have to start from the top.

$$\alpha_{(3)} = \frac{m_l}{m_{(3)}} \quad \Rightarrow m_{(3)} = \frac{m_l}{\alpha_{(3)}}$$

$$m_{(3)} = \frac{118000}{0.34492} = 342103.9\,\text{kg}$$

$$m_{0(3)} = m_l + m_{(3)} = 118000 + 342103.9 = 460103.9\,\text{kg}$$

$$\alpha_{(2)} = \frac{m_{0(3)}}{m_{(2)}} \quad \Rightarrow m_{(2)} = \frac{m_{0(3)}}{\alpha_{(2)}}$$

$$m_{(2)} = \frac{460103.9}{0.21608} = 2129327.5\,\text{kg}$$

$$m_{0(2)} = m_{0(3)} + m_{(2)} = 460103.9 + 2129327.5 = 2589431.4\,\text{kg}$$

$$\alpha_{(1)} = \frac{m_{0(2)}}{m_{(1)}} \quad \Rightarrow m_{(1)} = \frac{m_{0(2)}}{\alpha_{(1)}}$$

$$m_{(1)} = \frac{2589431.4}{6.08623} = 425457.7\,\text{kg}$$

$$m_{0(1)} = m_{0(2)} + m_{(1)} = 2589431.4 + 425457.7 = 3014889.2\,\text{kg}$$

Thus, we get the mass of the total launch vehicle $m_{0(1)} = 3014889.2\,\text{kg}$. And the ratio of the total launch-vehicle mass to the payload mass is 3014889.2/ 118000 = 25.55. The small difference between the finally arrived value of $m_{0(1)} = 3014889.2\,\text{kg}$ and the specified one of 3014895 kg is due to the convergence error in determining the root of λ.

Now the stage incremental velocities can be calculated.

$$\Delta\ v_{(1)} = \overline{u}_{e(1)} ln\ \mu_{(1)} = 2981 \times ln\ 1.15310 = 424.7\,\text{m/s}$$

$$\Delta\ v_{(2)} = \overline{u}_{e(2)} ln\ \mu_{(2)} = 4129 \times ln\ 4.11334 = 5839.4\,\text{m/s}$$

$$\Delta\ v_{(3)} = \overline{u}_{e(3)} ln\ \mu_{(3)} = 4129 \times ln\ 2.95036 = 4467.3\,\text{m/s}$$

These stage incremental velocities sum-up to the optimum terminal velocity of 10731.3 m/s that we previously calculated. The summary of these results are given in Table 5.9.

With the payload mass of 47000 kg, the launch vehicle mass for the translunar orbit is 2943895 kg, Table 5.8. This gives the value of $m_{0(1)}/m_l$ as 62.63606. The upper bound of the Lagrange multiplier λ_{ub} remains the same at $-3.56546E-4$. Starting from this upper bound, the root of Lagrange multiplier λ is found through iterations as $-3.79667E-4$ for the $m_{0(1)}/m_l$ value of 62.63606. Following similar procedures the optimum terminal velocity for the translunar orbit and the corresponding stage properties are calculated and given in Table 5.10.

Table 5.9 Example 5.3 results: Optimized stage properties of a three-stage launch vehicle for the payload load of 118000 kg and $m_{0(1)}/m_l$ of 25.54996 to a low Earth orbit in a gravity- and drag-free environment with no turning; values of the specific impulses and structural coefficients are as of Saturn V

Stg	$I_{sp(i)}$ (m/s)	$\varepsilon_{(i)}$	$\mu_{(i)}$	$\alpha_{(i)}$	$m_{(i)}$ (kg)	$m_{0(i)}$ (kg)	$\Delta v_{(i)}$ (m/s)
1	2981	0.05914	1.15310	6.08623	425457.7	3014889.2	424.7
2	4129	0.07956	4.11334	0.21608	2129327.5	2589431.4	5839.4
3	4129	0.11093	2.95036	0.34492	342103.9	460103.9	4467.3
$\sum_{i=1}^{3} m_{(i)}$ (kg) = 2896889.1						$\sum_{i=1}^{3} \Delta v_{(i)}$ (m/s) = 10731.4	

Table 5.10 Example 5.3 results: Optimized stage properties of a three-stage launch vehicle for the payload load of 47000 kg and $m_{0(1)}/m_l$ of 62.63606 to a translunar orbit in a gravity- and drag-free environment with no turning; values of the specific impulses and structural coefficients are as of Saturn V

Stg	$I_{sp(i)}$ (m/s)	$\varepsilon_{(i)}$	$\mu_{(i)}$	$\alpha_{(i)}$	$m_{(i)}$ (kg)	$m_{0(i)}$ (kg)	$\Delta v_{(i)}$ (m/s)
1	2981	0.05914	1.96877	0.91204	1539660.7	2943894.6	2019.4
2	4129	0.07956	4.55110	0.17963	1190397.7	1404234.0	6257.0
3	4129	0.11093	3.26435	0.28171	166836.3	213836.3	4884.9
$\sum_{i=1}^{3} m_{(i)}$ (kg) = 2896894.7						$\sum_{i=1}^{3} \Delta v_{(i)}$ (m/s) = 13161.3	

(2) Limiting Values of Terminal Velocities and $m_{0(1)}/m_l$'s

By Eq. 5.29,

$$\sum_{i=1}^{n} u_{e(i)} ln\, \frac{1}{\varepsilon_{(i)}} > v_n > \sum_{i=1}^{n} \overline{u}_{e(i)} ln\, \left\{ \frac{1}{\varepsilon_{(i)}} \left[1 - \frac{\min_{i=1}^{n}[\overline{u}_{e(i)}(1-\varepsilon_{(i)})]}{\overline{u}_{e(i)}} \right] \right\} \tag{5.29}$$

And by Eq. 5.29a,

$$\infty > \frac{m_{0(1)}}{m_l} > \prod_{i=1}^{n} \left[\left(1 - \frac{1}{\varepsilon_{(i)}} \right) \left(1 - \frac{\overline{u}_{e(i)}}{\min_{i=1}^{n}[\overline{u}_{e(i)}(1-\varepsilon_{(i)})]} \right) \right] \tag{5.29a}$$

Therefore,

$$v_{n,max} = \sum_{i=1}^{3} u_{e(i)} ln\, \frac{1}{\varepsilon_{(i)}} = 2981 \times ln\, \frac{1}{0.05914} + 4129 \times \ln\, \frac{1}{0.07956}$$

$$+4129 \times ln\, \frac{1}{0.11093} = 27960.1\ \mathrm{m/s}$$

$$v_{n,min} = \sum_{i=1}^{3} \overline{u}_{e(i)} \ln\left\{\frac{1}{\varepsilon_{(i)}}\left[1 + \frac{1}{\lambda_{ub}\overline{u}_{e(i)}}\right]\right\}$$

$$= 2981 \times \ln\left\{\frac{1}{0.05914}\left[1 + \frac{1}{(-3.56546E-4) \times 2981}\right]\right\}$$

$$+4129 \times \ln\left\{\frac{1}{0.07956}\left[1 + \frac{1}{(-3.56546E-4) \times 4129}\right]\right\}$$

$$+4129 \times \ln\left\{\frac{1}{0.11093}\left[1 + \frac{1}{(-3.56546E-4) \times 4129}\right]\right\} = 10140.0\,\mathrm{m/s}$$

We already noted that the upper limit of $m_{0(1)}/m_l$ for any given sets of $\varepsilon_{(i)}$ and $\overline{u}_{e(i)}$ is ∞, indicating either $m_l = 0$ or $m_{0(1)} \to \infty$. The lower limit of $m_{0(1)}/m_l$ is now calculated.

$$\left(\frac{m_{0(1)}}{m_l}\right)_{min} = \prod_{i=1}^{3}\left(1 - \frac{1}{\varepsilon_{(i)}}\right)\left[1 + \lambda_{ub}\overline{u}_{e(i)}\right] =$$

$$\left(1 - \frac{1}{0.05914}\right) \times [1 + (-3.56546E-4) \times 2981] \times$$

$$\left(1 - \frac{1}{0.07956}\right) \times [1 + (-3.56546E-4) \times 4129] \times$$

$$\left(1 - \frac{1}{0.11093}\right) \times [1 + (-3.56546E-4) \times 4129] = 20.67265$$

Note that these limiting values on v_n and $m_{0(1)}/m_l$ are dependent only on the vehicle's stage properties $\overline{u}_{e(i)}$'s and $\varepsilon_{(i)}$'s and not on any mission LEO, geosynchronous transfer orbit (GTO), or the Earth escape.

(3) Comparison of Stage Masses and $\Delta v_{(i)}$'s

The comparison of stage masses $m_{(i)}$'s and the stage incremental velocities $\Delta v_{(i)}$'s of Saturn V with those of optimized stages is summarized in Table 5.11. It must be noted that all $\Delta v_{(i)}$'s are calculated for drag- and gravity-free environment with no turning and these do not represent the actual performance values of Saturn V.

Table 5.11 Comparison of stage masses and $\Delta v_{(i)}$'s with optimized ones; all $\Delta v_{(i)}$'s calculated for drag- and gravity-free environment with no turning

Saturn V: zero-drag, -gravity, & -turning $\Delta\overline{v}_{(i)}$'s						Optimum (zero-drag, -gravity, & -turning)			
Stg	$\overline{u}_{e(i)}(m/s)$	$\varepsilon_{(i)}$	$m_{(i)}$ (kg)	$\Delta v_{(i),LEO}$ (m/s)	$\Delta v_{(i),TLO}$ (m/s)	$m_{(i),LEO}$ (kg)	$\Delta v_{(i),LEO}$ (m/s)	$m_{(i),TLO}$ (kg)	$\Delta v_{(i),TLO}$ (m/s)
1	2981	0.05914	2286217	3726	3911	425458	425	1539661	2019
2	4129	0.07956	490778	3994	4794	2129328	5839	1190398	6257
3	4129	0.11093	119900	2454	4204	342104	4467	166836	4885
$\sum_{i=1}^{3}\Delta v_{(i)}$(m/s)				10174	12909	10731		13161	

Table 5.12 Upper bound of Lagrange multiplier λ_{ub} and the respective minimum terminal velocity $v_{n,min}$ and the minimum ratio of launch vehicle mass to payload mass $(m_{0(1)}/m_l)_{min}$

Vehicle	λ_{ub}	$v_{n,min}$ (m/s)	$(m_{0(1)}/m_l)_{min}$
3 stage	–3.56546E-4	10140.1	20.67298
2 stage (LOX/Krsn + LOX/LH2 of 2nd stg)	–3.56546E-4	5756.1	5.46242
Single stg (LOX/Krsn)	–3.56546E-4	0	1
Single stg (LOX/LH2 of 2nd stg)	–2.63124E-4	0	1
Single stg (LOX/LH2 of 3rd stg)	–2.72406E-4	0	1

(4) Characteristics of Optimum Terminal Velocity versus $m_{0(1)}/m_l$ for Ranges of Lambda

The upper bound of λ is given by Eq. 5.25a.

$$\lambda_{ub} = -\frac{1}{\min_{i=1}^{n}[\overline{u}_{e(i)}(1-\varepsilon_{(i)})]} \tag{5.25a}$$

With this upper bound λ_{ub} and its lower values, the terminal velocity v_n and the ratio of launch vehicle mass to payload mass $m_{0(1)}/m_l$ vary as per Eqs. 5.29 and 5.29a. The values of λ_{ub}'s and the respective $v_{n,min}$'s and $(m_{0(1)}/m_l)_{min}$'s are given in Table 5.12. The variations of $m_{0(1)}/m_l$ against v_n are shown in Fig. 5.6.

$$\sum_{i=1}^{n} u_{e(i)} ln\,\frac{1}{\varepsilon_{(i)}} > v_n > \sum_{i=1}^{n}\overline{u}_{e(i)} ln\,\{\frac{1}{\varepsilon_{(i)}}[1-\frac{\min_{i=1}^{n}[\overline{u}_{e(i)}(1-\varepsilon_{(i)})]}{\overline{u}_{e(i)}}]\} \tag{5.29}$$

$$\infty > \frac{m_{0(1)}}{m_l} > \prod_{i=1}^{n}[(1-\frac{1}{\varepsilon_{(i)}})(1-\frac{\overline{u}_{e(i)}}{\min_{i=1}^{n}[\overline{u}_{e(i)}(1-\varepsilon_{(i)})]})] \tag{5.29a}$$

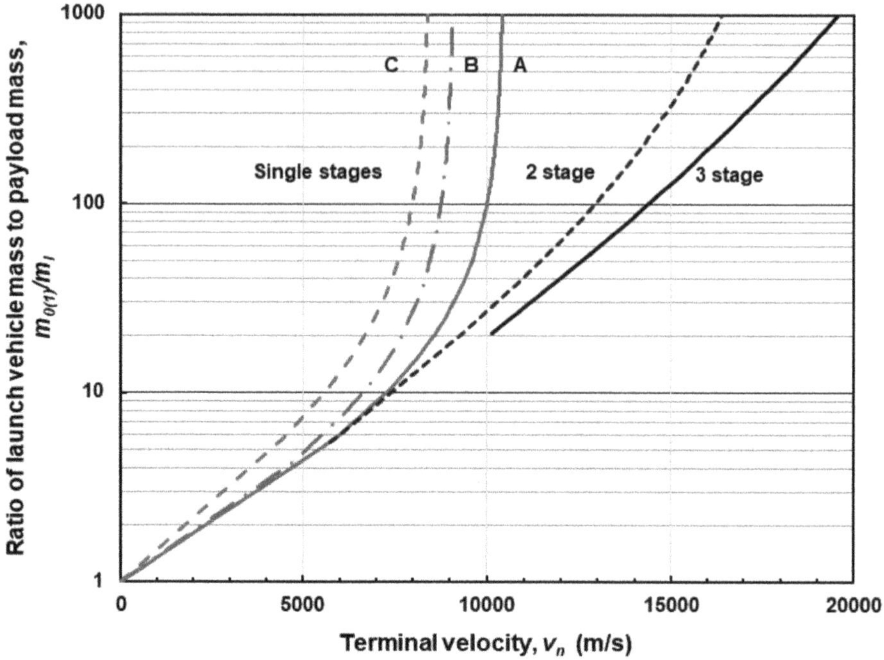

Fig. 5.6 Characteristics of optimization, Example 5.3; 2 stage: LOX/Kerosene + LOX/LH2 of 2nd stage, A: Single stage (LOX/LH2 of 2nd stage), B: Single stage (LOX/LH2 of 3rd stage), and C: Single stage (LOX/Kerosene)

Discussion

Once again we are alerting that all the velocities that we have calculated here are for the drag- and gravity-free environment with no turning. Deducting the velocity losses due to drag, gravity, turning, and fairing separation from these theoretical terminal velocities will give the actual terminal velocities. But the velocity losses are dependent on the chosen trajectories. Generally, for a LEO about 7.6 km/s, for a GTO about 10.3 km/s, and for an escape from the Earth in excess of about 11.2 km/s are the actual terminal-velocity requirements. The greatest velocity losses occur during the first-stage operation as the vehicle has to negotiate the drag in the atmospheric air layer in addition to the other losses. Since the vehicle's upper stages are almost in hard vacuum, the drag losses are almost negligible. Notwithstanding the fact that the actual velocity losses depend on the chosen trajectory of a mission, as a rough estimate the velocity losses for a LEO may amount to about 1.5 km/s and more so for an SSO, a GTO, or an Earth-escape—say less than 2 km/s. It is interesting to note from the calculated results listed in Table 5.11 that the terminal velocities in drag- and gravity-free environment with no turning for the Saturn vehicle (10174 m/s for LEO and 12909 m/s for translunar orbit) as well as the ones calculated for the optimum conditions (10731 m/s for LEO and 13161 m/s for translunar orbit) hover around the actual terminal velocities added with the approximate velocity losses mentioned for the two missions.

As mentioned previously, the optimization exercise for the drag- and gravity-free environment with no turning favors the stage with a lighter structure (low structural coefficient) and a higher energy propellant (higher specific impulse). Thus, we find a higher $\Delta v_{(i)}$ assigned to the second stage with lower structural coefficient than the $\Delta v_{(i)}$ assigned to the third stage (Tables 5.9 and 5.10). This is in spite of the fact that both the stages have the same specific impulse values. Similarly, the optimization procedure, being ignorant of the velocity losses, assigns for the first stage due to its far lower specific impulse the most minimum $m_{(i)}$ and $\Delta v_{(i)}$ for the LEO (Table 5.9) and the most minimum $\Delta v_{(i)}$ for the translunar orbit (Table 5.10). The velocity increments $\Delta v_{(i)}$'s given in Tables 5.7 and 5.8 represent closely the actual velocity increments added with velocity losses. Comparison of these $\Delta v_{(i)}$'s with the corresponding ones of Tables 5.9 and 5.10 reveals the extensive changes that would take place in actual $\Delta v_{(i)}$'s on the consideration of specific trajectory and its associated velocity losses.

Figure 5.6 shows the variation of optimum $m_{0(1)}/m_l$ with respect to v_n for different options of vehicle configurations: (a) the three-stage, (b) the two-stage option with LOX/Kerosene and LOX/LH2 of the second stage, and (c) the different single stages. Such graphs help in quickly arriving at preliminary design configurations. For example a GTO may require an actual injection velocity in excess of 10 km/s. With the velocity losses added to this, a theoretical injection velocity (drag- and gravity-free environment with no turning and fairing-separation losses) of about 12 km/s or more may be the requirement. While such a GTO maneuver appears possible with the three-stage vehicle advantageously with a lower $m_{0(1)}/m_l$ and also with the two-stage vehicle with a higher $m_{0(1)}/m_l$, it is impossible with a single stage system. A LEO may require a theoretical injection velocity (actual injection velocity + velocity losses) close to 10 km/s and this with the two-stage vehicle with the LOX/kerosene of the first stage and the LOX/LH2 of the second stage appears feasible with an acceptable $m_{0(1)}/m_l$. However, the lighter structural coefficient of 0.07956 of the LOX/LH2 of the second stage (see Table 5.11) is a point of concern because it is rather difficult to attain that low a value with the special provisions required for payload injection. On the other hand, the two-stage system with the LOX/Kerosene of the first stage and the LOX/LH2 of the third stage with the heavier structural coefficient of 0.11093 can perform a LEO but with a higher $m_{0(1)}/m_l$. The study of single stage characteristics reveals that it is impossible to attain even the least terminal velocity requirements of LEO launches with single stages. However, the LOX/LH2 single stage (marked A in Fig. 5.6), with its too low a structural coefficient of 0.07956 for the payload injection stage, marginally clears the 10 km/s theoretical injection velocity, although with a very high $m_{0(1)}/m_l$

5.3.4 Choosing Number of Stages

For any desired mission, the payload mass m_l is decided and the required terminal velocity is known. Adding the expected total velocity losses (about 1.5 km/s for a LEO and more for a GTO or an escape from the Earth) to the required terminal velocity, one gets the "marked-up" terminal velocity. Based on the locally available technological expertise and also considering the currently active launch vehicles and their characteristics (see Table 5.1 and Figs. 5.2, 5.3, 5.4 and 5.5) the possible stages of propellant combinations, their structural coefficients $\varepsilon_{(i)}$'s, and specific impulses $\overline{u}_{e(i)}$'s are provisionally selected. For the chosen number of such stages and their values of $\varepsilon_{(i)}$'s, and $\overline{u}_{e(i)}$'s, the minimum and the maximum limits of the theoretical terminal velocities and the lower limit of $m_{0(1)}/m_l$ can be determined through Eqs. 5.29 and 5.29a. If the range between the minimum and the maximum is too wide or too narrow to the marked-up terminal velocity, one can see the possibility of respectively reducing or increasing the number of stages.

$$\sum_{i=1}^{n} u_{e(i)} ln \frac{1}{\varepsilon_{(i)}} > v_n > \sum_{i=1}^{n} \overline{u}_{e(i)} ln \left\{ \frac{1}{\varepsilon_{(i)}} \left[1 - \frac{\min_{i=1}^{n} [\overline{u}_{e(i)}(1-\varepsilon_{(i)})]}{\overline{u}_{e(i)}} \right] \right\} \tag{5.29}$$

$$\infty > \frac{m_{0(1)}}{m_l} > \prod_{i=1}^{n} \left[\left(1 - \frac{1}{\varepsilon_{(i)}}\right) \left(1 - \frac{\overline{u}_{e(i)}}{\min_{i=1}^{n} [\overline{u}_{e(i)}(1-\varepsilon_{(i)})]}\right) \right] \tag{5.29a}$$

Recall Eqs. 5.15 and 5.31—repeated here for ready reference.

$$v_n = \sum_{i=1}^{n} \overline{u}_{e(i)} ln \left(\frac{\lambda \overline{u}_{e(i)} + 1}{\lambda \overline{u}_{e(i)} \varepsilon_{(i)}} \right) \tag{5.15}$$

$$\frac{m_{0(1)}}{m_l} = \prod_{i=1}^{n} \left[\left(1 - \frac{1}{\varepsilon_{(i)}}\right) (1 + \lambda \overline{u}_{e(i)}) \right] \tag{5.31}$$

For the marked-up terminal velocity v_n, the Lagrange multiplier λ is calculated numerically through Eq. 5.15. And, the calculated λ, and the selected $\varepsilon_{(i)}$'s and $\overline{u}_{e(i)}$'s are substituted into Eq. 5.31 to obtain $m_{0(1)}/m_l$. Thus, for the decided payload mass m_l, the optimized minimum total launch vehicle mass $m_{0(1)}$ is estimated. The estimated $m_{0(1)}$ provides a rough budget. Through these iterative steps the stages and their number can be evaluated. For the marked-up v_n, the decided mass m_l, and the selected $\varepsilon_{(i)}$'s and $\overline{u}_{e(i)}$'s, the estimated $m_{0(1)}$ could be close to the one of final design (solve Problems 5.19 & 5.20). But the "optimum" stage masses thus calculated for the marked-up v_n are bound to be different from the final design values. This is because the final stage-masses are governed by the optimum trajectory with the velocity losses with respect to individual stages. Individual stage velocity losses can be calculated only through trajectory analyses, whereas the loss-free calculation is a trajectory-independent procedure.

Generally, an increase in the number of stages for the given $m_{0(1)}/m_l$ results in the increase in the optimum terminal velocity v_n, Fig. 5.6—this can be taken as the marked-up value. Alternatively for *any* given v_n, the increase in the number of stages brings down the optimum $m_{0(1)}/m_l$, Fig. 5.6. This means for the chosen $m_{0(1)}$, the payload m_l is more. Or, for the chosen m_l, $m_{0(1)}$ is less. Although the increase in v_n or in m_l and the decrease $m_{0(1)}$ seem attractive, the price for these comes by way of the enhanced system complexity due to the increase in the number of stages.

The energy content in a propellant combination, represented by the specific impulse $\overline{u}_e$, plays a vital role in deciding the number of stages. Principally there are four types of propellants: solid propellant, nitrogen tetroxide N_2O_4 with a hydrazine based fuel, semi-cryogenic LOX/Kerosene, and cryogenic LOX/LH2. Most solid propellants use ammonium perchlorate as the oxidizer, and aluminum powder and a polybutadiene based polymer (mostly hydroxyl terminated polybutadiene—HTPB) as the fuels. A solid propellant combination, despite its relatively lower specific impulse compared to the liquid one's, its motors are preferred as boosters for the capacity to produce large thrusts without losing the inherent simplicity. Sometimes, because of the inherent simplicity, a small solid motor stage is added at the top of the vehicle to perform the payload injection maneuver. N_2O_4 with the hydrazine based fuels are hypergolic (do not require external igniters) but these are highly toxic and carcinogenic. Therefore, new developments of the engines with these propellant combinations appear to be waning. The other two liquid propellant combinations are of LOX/Kerosene and LOX/LH2. These two types provide higher specific impulse values, Table 5.1. For our present simple analysis with different stage numbers, we consider similar stages ($\overline{u}_{e(i+1)} = \overline{u}_{e(i)}$ & $\varepsilon_{(i+1)} = \varepsilon_{(i)}$) with the three propellant combinations: solid propellant, LOX/Kerosene, and LOX/LH2. $\varepsilon_{(i)}$ values are taken as the midway values of the respective flight-proven stages, Figs. 5.2, 5.4, and 5.5. $\overline{u}_{e(i)}$ values are conservatively taken as the ones corresponding to the active first stages, Table 5.1. Launch vehicles with stages 2, 3, & 4 are considered. The calculated ratios of the total launch vehicle mass to the payload mass $m_{0(1)}/m_l$'s for two marked-up terminal velocities 9500 m/s and 12500 m/s are given in Table 5.13. These marked-up values approximately correspond to the required terminal velocities added with velocity losses respectively for a LEO and a GTO. The calculated results of Table 5.13 are for the simplest optimization assumption, namely similar stages. The results indicate that adopting higher energy propellants renders lower $m_{0(1)}/m_l$ and as already discussed increasing the number of stages brings down $m_{0(1)}/m_l$.

Table 5.13 Effect of stage numbers on $m_{0(1)}/m_l$ for three different propellant combinations: solid propellant, LOX/Kerosene, & LOX/LH2

Stages	Solid propellant		LOX/Kerosene		LOX/LH2	
	$\varepsilon = 0.141$ & $\overline{u}_e = 2700$ m/s		$\varepsilon = 0.074$ & $\overline{u}_e = 3300$ m/s		$\varepsilon = 0.110$ & $\overline{u}_e = 4250$ m/s	
	$v_n = 9.5$ km/s	$v_n = 12.5$ km/s	$v_n = 9.5$ km/s	$v_n = 12.5$ km/s	$v_n = 9.5$ km/s	$v_n = 12.5$ km/s
	$m_{0(1)}/m_l$	$m_{0(1)}/m_l$	$m_{0(1)}/m_l$	$m_{0(1)}/m_l$	$m_{0(1)}/m_l$	$m_{0(1)}/m_l$
2	759	–	32	147	17	55
3	133	1650	27	87	15	38
4	97	604	25	76	14	34

5.3.5 *Scaled-up Stages Considering Velocity Losses*

For the required terminal velocity v_n, payload m_l, and the stage properties (structural coefficients $\varepsilon_{(i)}$'s and specific impulses $\overline{u}_e$'s), the loss-free optimization procedure gives the ideal stages. This result is to be taken as the preliminary step to arrive at the final optimization. The final optimization is carried out through detailed trajectory analyses considering gravity, drag, and turning or steering of the vehicle. We noted that the loss-free optimization procedure "favors" the stage with the higher specific impulse and lower structural coefficient by assigning to it heavier stage mass $m_{(i)}$ and larger velocity increment $\Delta v_{(i)}$—study Examples 5.1–5.3. This way, the first stage, generally of the lowest specific impulse, is assigned with the lowest velocity increment $\Delta v_{(i)}$ and the lightest stage mass $m_{(i)}$. Whereas in reality the first stage, which experiences the most velocity losses, has to be the heaviest one. Furthermore, the loss-free procedure gives a low lift-off mass $m_{0(1)}$. As an improvement, in order to get a lift-off mass $m_{0(1)}$ or $m_{0(1)}/m_l$ closer to the reality, the loss-free procedure is considered with the "marked-up" terminal velocity, which is equal to the required terminal velocity plus the estimated total velocity losses. Although this procedure gives better $m_{0(1)}$ or $m_{0(1)}/m_l$, the stage mass distribution and the stage velocity increments are still far different from the ones of final design. As another step to improve the results, alternatively the individual stage velocity-losses can be added to the ideal stage velocity increments. This way, without losing the simplicity of the loss-free procedure, we get the better lift-off mass $m_{0(1)}$ as well as improved stage mass distribution $m_{(i)}$'s and velocity increments $\Delta v_{(i)}$'s. We consider such a procedure by Koch [13] in the present section.

To adopt the procedure of adding stage velocity losses to the ideal stage velocity-increments, first we have to run the loss-free optimization procedure for the desired payload mass m_l, the required terminal velocity v_n, and the selected number of stages with their structural coefficients $\varepsilon_{(i)}$'s and the specific impulses $\overline{u}_e$'s. Thus, we get the ideal stage velocity-increments $\Delta v_{(i)i}$'s. The individual stage velocity-losses $\Delta v_{(i)l}$'s have to be provided. These can be from previous experience or can come from a linked trajectory software. Thus, each stage has to provide sufficient propellant to cater for the sum $\Delta v_{(i)i} + \Delta v_{(i)l}$. Accordingly as per the energy requirement, the required velocity increment for each stage

$$\Delta v_{(i)r} = \Delta v_{(i)i} + \Delta v_{(i)l}$$

The stage velocity loss includes the gravity loss, drag loss, and the steering loss. The gravity loss, with an essentially constant gravitational acceleration, depends on the angle subtended by the flight velocity vector with the local horizontal, the so-called flight path angle. And, this reduces from its maximum of 90° at the lift-off to a near zero value at the terminal point. Drag loss is mostly experienced by the first stage and it becomes negligible above a certain altitude (say 55 km). The steering loss, the most minimum of the three, is experienced by all the stages.

By the Tsiolkovsky equation

$$\Delta v_{(i)r} = \Delta v_{(i)i} + \Delta v_{(i)l} = \overline{u}_{e(i)} ln\, \frac{m_{0(i)r}}{m_{0b(i)r}} = \overline{u}_{e(i)} ln\, \frac{m_{s(i)r} + m_{p(i)r} + m_{0(i+1)r}}{m_{s(i)r} + m_{0(i+1)r}}$$

Noting $m_{s(i)r} + m_{p(i)r} = m_{s(i)r}/\varepsilon_{(i)}$

$$\Delta v_{(i)i} + \Delta v_{(i)l} = \overline{u}_{e(i)} ln\, \frac{\dfrac{m_{s(i)r}}{\varepsilon_{(i)}} + m_{0(i+1)r}}{m_{s(i)r} + m_{0(i+1)r}} \tag{5.40}$$

Thus, we get

$$m_{s(i)r} = \frac{\left[exp\left(\dfrac{\Delta v_{(i)i} + \Delta v_{(i)l}}{\overline{u}_{e(i)}}\right) - 1\right] m_{0(i+1)r}}{\dfrac{1}{\varepsilon_{(i)}} - exp\left(\dfrac{\Delta v_{(i)i} + \Delta v_{(i)l}}{\overline{u}_{e(i)}}\right)} \tag{5.41}$$

It follows then

$$m_{p(i)r} = \frac{1 - \varepsilon_{(i)}}{\varepsilon_{(i)}} m_{s(i)r}$$

$$m_{(i)r} = m_{s(i)r} + m_{p(i)r}$$

This way the scaled-up stage masses $m_{(1)r}$, .., $m_{(n-1)r}$, & $m_{(n)r}$ are calculated. Starting from the top, noting the payload mass is m_l, the payload ratio

$$\alpha_{(n)r} = \frac{m_l}{m_{(n)r}} \rightarrow \mu_{(n)r} = \frac{1 + \alpha_{(n)r}}{\varepsilon_{(n)} + \alpha_{(n)r}} \rightarrow m_{0(n)r} = m_{(n)r} + m_l$$

$$\alpha_{(n-1)r} = \frac{m_{0(n)r}}{m_{(n-1)r}} \rightarrow \mu_{(n-1)r} = \frac{1 + \alpha_{(n-1)r}}{\varepsilon_{(n-1)} + \alpha_{(n-1)r}}$$

$$\rightarrow m_{0(n-1)r} = m_{(n-1)r} + m_{0(n)r}$$

and so on. Thus, descending from the top, the scaled-up stage masses $m_{(i)r}$'s, the corresponding mass ratios $\mu_{(i)}$'s, and finally the required lift-off mass $m_{0(1)r}$ can be calculated. The solution is reached when

$$\begin{gathered}\sum_{i=1}^{n} \overline{u}_{e(i)} ln\ \mu_{(i)r} = \sum_{i=1}^{n} \left(\Delta v_{(i)i} + \Delta v_{(i)l}\right) \\ \& \\ \left(\Delta v_{(i)i} + \Delta v_{(i)l}\right) - \overline{u}_{e(i)} ln\ \mu_{(i)r} = 0 \quad (i = 1, .., n)\end{gathered} \tag{5.42}$$

If the above conditions are not satisfied the corrections are initiated to the respective stage or stages to reach the solution. As an example let us take a three-stage vehicle.

$$\alpha_{(3)r} = \frac{m_l}{m_{(3)r}} \rightarrow \mu_{(3)r} = \frac{1 + \alpha_{(3)r}}{\varepsilon_{(3)} + \alpha_{(3)r}} \rightarrow m_{0(3)r} = m_{(3)r} + m_l$$

$$\alpha_{(2)r} = \frac{m_{0(3)r}}{m_{(2)r}} \rightarrow \mu_{(2)r} = \frac{1 + \alpha_{(2)r}}{\varepsilon_{(2)} + \alpha_{(2)r}} \rightarrow m_{0(2)r} = m_{(2)r} + m_{0(3)r}$$

$$\alpha_{(1)r} = \frac{m_{0(2)r}}{m_{(1)r}} \rightarrow \mu_{(1)r} = \frac{1 + \alpha_{(1)r}}{\varepsilon_{(1)} + \alpha_{(1)r}} \rightarrow m_{0(1)r} = m_{(1)r} + m_{0(2)r}$$

The solution is reached when

$$\begin{gathered}\sum_{i=1}^{3} \overline{u}_{e(i)} ln\ \mu_{(i)r} = \sum_{i=1}^{3} \left(\Delta v_{(i)i} + \Delta v_{(i)l}\right) \\ \& \\ \left(\Delta v_{(i)i} + \Delta v_{(i)l}\right) - \overline{u}_{e(i)} ln\ \mu_{(i)r} = 0 \quad (i = 1, 2, 3)\end{gathered}$$

This procedure of scaling-up of stages considering velocity losses requires accurate inputs of the losses that can come from the trajectory analysis in an iterative mode. While scaling-up, due care has to be taken on the variation of structural coefficients with respect to the stage masses, Figs. 5.2, 5.3, 5.4 and 5.5. Considering the orbital launch vehicle Ariane 40, the details of the procedure are presented in Example 5.4 [13].

Example 5.4 An orbital launch vehicle with three serial stages similar to the Ariane 40 is considered. Its first two stages are of $N_2O_4/UDMH$ propellants and the third stage is of $LOX/LH2$ propellants. It launches a payload of 1991.8 kg into a GTO with a terminal velocity of 9749.5 m/s. Its first stage has a structural coefficient of 0.1053 and delivers an average specific impulse of 2600 m/s. The second stage has a structural coefficient of 0.1079 and delivers a specific impulse of 2832 m/s. And, the third stage has a structural coefficient of 0.1631 and delivers a specific impulse of 4355 m/s. Stage velocity losses are found to be $\Delta v_{(1)l} = 1310.7$ m/s, $\Delta v_{(2)l} = 353$ m/s, and $\Delta v_{(3)l} = 236$ m/s—totaling to 1899.7 m/s. Following the procedure for calculating the scaled-up stages considering velocity losses, estimate the vehicle's lift-off mass, stage masses, and velocity increments. Compare these properties with those calculated for marked-up terminal velocity of 9749.5 m/s + total velocity losses of 1899.7 m/s under the loss-free optimization procedure.

Solution

The procedure for calculating the scaled-up stages considering the velocity losses has the loss-free optimization procedure as the base. Therefore, we have to first calculate the loss-free optimization results with the given stage properties ($\varepsilon_{(i)}$'s and $\overline{u}_{e(i)}$'s), payload mass of 1991.8 kg, and terminal velocity of 9749.5 m/s. We are asked to compare the scaled-up stage masses considering the velocity losses with the stage properties adopting loss-free optimization procedure for the marked-up terminal velocity of 11649.2 m/s (= 9749.5 m/s + 1899.7 m/s). This forms the second stage of the analysis. Next, we calculate the required properties of the scaled-up stages considering the individual stage velocity losses and compare all the results of these three stages of calculations.

Analysis

Loss-free Optimization for $v_n = 9749.5\,\mathrm{m/s}$

To fix the upper bound for the root of λ, we calculate the values of $-1/[u_{e(i)}(1-\varepsilon_{(i)})]$.

$$-\frac{-1}{u_{e(1)}(1-\varepsilon_{(1)})} = -\frac{-1}{2600\times(1-0.1053)} = -0.0004299$$

$$-\frac{-1}{u_{e(2)}(1-\varepsilon_{(2)})} = -\frac{-1}{2832\times(1-0.1079)} = -0.0003958$$

$$-\frac{-1}{u_{e(3)}(1-\varepsilon_{(3)})} = -\frac{-1}{4355\times(1-0.1631)} = -0.0002744$$

Therefore, the most negative –0.0004299 forms the upper bound of λ. That is,

$$\lambda_{ub} = -\frac{1}{\min_{i=1}^{3}[\overline{u}_{e(i)}(1-\varepsilon_{(i)})]} = -\frac{1}{[\overline{u}_{e(1)}(1-\varepsilon_{(1)})]} = -0.0004299$$

Next, noting that the upper bound for λ is -0.0004299 and starting therefore with values lower than this we have to search for its root through a numerical procedure for $v_n = 9849.5\,\mathrm{m/s}$, Eq. 5.15. This is easily done using Excel spreadsheet. The root of λ for $v_n = 9849.5\,\mathrm{m/s}$ is found to be $-4.91307E-04$.

Recall Eq. 5.14.

$$\mu_{(i)} = \frac{\lambda\overline{u}_{e(i)}+1}{\lambda\overline{u}_{e(i)}\varepsilon_{(i)}} \qquad i = 1, 2, \ldots, n \tag{5.14}$$

$$\mu_{(1)} = \frac{\lambda \overline{u}_{e(1)} + 1}{\lambda \overline{u}_{e(1)} \varepsilon_{(1)}} = \frac{-4.91307E - 04 \times 2600 + 1}{-4.91307E - 04 \times 2600 \times 0.1053} = 2.06229$$

$$\mu_{(2)} = \frac{\lambda \overline{u}_{e(2)} + 1}{\lambda \overline{u}_{e(2)} \varepsilon_{(2)}} = \frac{-4.91307E - 04 \times 2832 + 1}{-4.91307E - 04 \times 2832 \times 0.1079} = 2.60695$$

$$\mu_{(3)} = \frac{\lambda \overline{u}_{e(3)} + 1}{\lambda \overline{u}_{e(3)} \varepsilon_{(3)}} = \frac{-4.91307E - 04 \times 4355 + 1}{-4.91307E - 04 \times 4355 \times 0.1631} = 3.26568$$

Recall Eq. 5.6.

$$\alpha_{(i)} = \frac{1 - \mu_{(i)} \varepsilon_{(i)}}{\mu_{(i)} - 1} \tag{5.6}$$

$$\alpha_{(1)} = \frac{1 - \mu_{(1)} \varepsilon_{(1)}}{\mu_{(1)} - 1} = \frac{1 - 2.06229 \times 0.1053}{2.06229 - 1} = 0.73694$$

$$\alpha_{(2)} = \frac{1 - \mu_{(2)} \varepsilon_{(2)}}{\mu_{(2)} - 1} = \frac{1 - 2.60695 \times 0.1079}{2.60695 - 1} = 0.44725$$

$$\alpha_{(3)} = \frac{1 - \mu_{(3)} \varepsilon_{(3)}}{\mu_{(3)} - 1} = \frac{1 - 3.26568 \times 0.1631}{3.26568 - 1} = 0.20628$$

Recall the definition of payload ratio, Eq. 5.1.

$$\alpha_{(i)} \equiv \frac{m_{0(i+1)}}{m_{(i)}}$$

Now, we have to start from the top.

$$\alpha_{(3)} = \frac{m_l}{m_{(3)}} \quad \Rightarrow m_{(3)} = \frac{m_l}{\alpha_{(3)}}$$

$$m_{(3)} = \frac{1991.8}{0.20628} = 9655.7\,\text{kg}$$

$$m_{0(3)} = m_l + m_{(3)} = 1991.8 + 9655.7 = 11647.5\,\text{kg}$$

$$\alpha_{(2)} = \frac{m_{0(3)}}{m_{(2)}} \quad \Rightarrow m_{(2)} = \frac{m_{0(3)}}{\alpha_{(2)}}$$

$$m_{(2)} = \frac{11647.5}{0.44725} = 26042.6\,\text{kg}$$

$$m_{0(2)} = m_{0(3)} + m_{(2)} = 11647.5 + 26042.6 = 37690.1\,\text{kg}$$

Table 5.14 Example 5.4: Loss-free optimization for $v_n = 9749.5\,\mathrm{m/s}$ and $m_l = 1991.8\,\mathrm{kg}$

Stg	$I_{sp(i)}$ (m/s)	$\varepsilon_{(i)}$	$\mu_{(i)}$	$\alpha_{(i)}$	$m_{(i)}$ (kg)	$m_{0(i)}$ (kg)	$\Delta v_{(i)}$ (m/s)
1	2600	0.1053	2.06229	0.73694	51144.3	88834.4	1881.9
2	2832	0.1079	2.60695	0.44725	26042.6	37690.1	2713.6
3	4355	0.1631	3.26568	0.20628	9655.7	11647.5	5154.0
			$m_l/m_{0(1)} = 0.02242$		$\sum_{i=1}^{3} \Delta v_{(i)}$ (m/s)= 9749.5		

$$\alpha_{(1)} = \frac{m_{0(2)}}{m_{(1)}} \quad \Rightarrow m_{(1)} = \frac{m_{0(2)}}{\alpha_{(1)}}$$

$$m_{(1)} = \frac{37690.1}{0.73694} = 51144.3\,\mathrm{kg}$$

$$m_{0(1)} = m_{0(2)} + m_{(1)} = 37690.1 + 51144.3 = 88834.4\,\mathrm{kg}$$

Thus, the minimized optimum mass of the total launch vehicle $m_{0(1)} = 88834.4\,\mathrm{kg}$. And the ratio of the total launch-vehicle mass to the payload mass is $88834.4/1991.8 = 44.6$.

Now the stage incremental velocities can be calculated.

$$\Delta\ v_{(1)} = \overline{u}_{e(1)} ln\ \mu_{(1)} = 2600 \times ln\ 2.06229 = 1881.9\,\mathrm{m/s}$$

$$\Delta\ v_{(2)} = \overline{u}_{e(2)} ln\ \mu_{(2)} = 2832 \times ln\ 2.60695 = 2713.6\,\mathrm{m/s}$$

$$\Delta\ v_{(3)} = \overline{u}_{e(3)} ln\ \mu_{(3)} = 4355 \times ln\ 3.26568 = 5154.0\,\mathrm{m/s}$$

These stage incremental velocities sum-up to the prescribed terminal velocity of 9749.5 m/s. The summary of these results are given in Table 5.14.

Loss-free Optimization for the marked-up $v_n = 11649.2\,\mathrm{m/s}$

Following a similar procedure, we can calculate the stage properties for the marked-up terminal velocity of 11649.2 m/s. The results of the calculations are given in Table 5.15.

Scaled-up Stages Optimization Considering Stage Velocity Losses

To adopt the optimization procedure of adding stage velocity losses to the ideal stage velocity-increments, we have run the loss-free optimization procedure for the desired payload of 1991.8 kg, the required terminal velocity 9749.5 m/s, and the three serial stages with their $\varepsilon_{(i)}$'s and $\overline{u}_e$'s. The results of our calculations are given in Table 5.14.

Table 5.15 Example 5.4: Loss-free optimization for the marked-up terminal velocity $v_n = 11649.2\,\text{m/s}$ and $m_l = 1991.8\,\text{kg}$

Stg	$I_{sp(i)}$ (m/s)	$\varepsilon_{(i)}$	$\mu_{(i)}$	$\alpha_{(i)}$	$m_{(i)}$ (kg)	$m_{0(i)}$ (kg)	$\Delta v_{(i)}$ (m/s)
1	2600	0.1053	2.84671	0.37918	172325.3	237668.0	2720.0
2	2832	0.1079	3.30976	0.27833	51115.7	65342.7	3389.6
3	4355	0.1631	3.56803	0.16279	12235.3	14227.1	5539.6
			$m_l/m_{0(1)} = 0.00838$		$\sum_{i=1}^{3} \Delta v_{(i)}$ (m/s) = 11649.2		

The required stage velocity increment is the sum of the ideal stage velocity increment calculated through the loss-free optimization step plus the stage velocity loss. That is $\Delta v_{(i)r} = \Delta v_{(i)i} + \Delta v_{(i)l}$.

$$\Delta v_{(1)r} = 1881.9 + 1310.7 = 3192.6\,\text{m/s}$$

$$\Delta v_{(2)r} = 2713.6 + 353 = 3066.6\,\text{m/s}$$

$$\Delta v_{(3)r} = 5154.0 + 236 = 5390.0\,\text{m/s}$$

We have to start from the top. By Eq. 5.41

$$m_{s(3)r} = \frac{\left[exp\left(\dfrac{\Delta v_{(3)r}}{\overline{u}_{e(3)}}\right) - 1\right] m_l}{\dfrac{1}{\varepsilon_{(3)}} - exp\left(\dfrac{\Delta v_{(3)r}}{\overline{u}_{e(3)}}\right)}$$

$$= \frac{\left[exp\left(\dfrac{5390.0}{4355}\right) - 1\right] \times 1991.8}{\dfrac{1}{0.1631} - exp\left(\dfrac{5390.0}{4355}\right)} = 1816.5\,\text{kg}$$

$$m_{(3)r} = \frac{m_{s(3)r}}{\varepsilon_{(3)}} = \frac{1816.5}{0.1631} = 11137.6\,\text{kg}$$

$$m_{0(3)r} = m_{(3)r} + m_l = 11137.6 + 1991.8 = 13129.4\,\text{kg}$$

$$m_{p(3)r} = \left(1 - \varepsilon_{(3)}\right) m_{(3)r} = (1 - 0.1631) \times 11137.6 = 9321.0\,\text{kg}$$

$$m_{0b(3)r} = m_{0(3)r} - m_{p(3)r} = 13129.4 - 9321.0 = 3808.3\,\text{kg}$$

$$\Delta v_{(3)r} = \Delta v_{(3)i} + \Delta v_{(3)l} = \overline{u}_{e(3)} \ln \frac{m_{0(3)r}}{m_{0b(3)r}}$$

$$= 4355 \times ln \frac{13129.4}{3808.3} = 5390.0\,\text{m/s}$$

$$m_{s(2)r} = \frac{\left[exp\left(\frac{\Delta v_{(2)r}}{\overline{u}_{e(2)}}\right) - 1\right] m_{0(3)r}}{\frac{1}{\varepsilon_{(2)}} - exp\left(\frac{\Delta v_{(2)r}}{\overline{u}_{e(2)}}\right)}$$

$$= \frac{\left[exp\left(\frac{3066.6}{2832}\right) - 1\right] \times 13129.4}{\frac{1}{0.1079} - exp\left(\frac{3066.6}{2832}\right)} = 4060.6\,\text{kg}$$

$$m_{(2)r} = \frac{m_{s(2)r}}{\varepsilon_{(2)}} = \frac{4060.6}{0.1079} = 37633.0\,\text{kg}$$

$$m_{0(2)r} = m_{(2)r} + m_{0(3)r} = 37633.0 + 13129.4 = 50762.3\,\text{kg}$$

$$m_{p(2)r} = \left(1 - \varepsilon_{(2)}\right) m_{(2)r} = (1 - 0.1079) \times 37633.0 = 33572.4\,\text{kg}$$

$$m_{0b(2)r} = m_{0(2)r} - m_{p(2)r} = 50762.3 - 33572.4 = 17190.0\,\text{kg}$$

$$\Delta v_{(2)r} = \Delta v_{(2)i} + \Delta v_{(2)l} = \overline{u}_{e(2)} \ln \frac{m_{0(2)r}}{m_{0b(2)r}}$$

$$= 2832 \times ln \frac{50762.3}{17190.0} = 3066.6\,\text{m/s}$$

$$m_{s(1)r} = \frac{\left[exp\left(\dfrac{\Delta v_{(1)r}}{\overline{u}_{e(1)}}\right)-1\right]m_{0(2)r}}{\dfrac{1}{\varepsilon_{(1)}}-exp\left(\dfrac{\Delta v_{(1)r}}{\overline{u}_{e(1)}}\right)}$$

$$= \frac{\left[exp\left(\dfrac{3192.6}{2600}\right)-1\right]\times 50762.3}{\dfrac{1}{0.1053}-exp\left(\dfrac{3192.6}{2600}\right)} = 20147.7\,\text{kg}$$

$$m_{(1)r} = \frac{m_{s(1)r}}{\varepsilon_{(1)}} = \frac{20147.7}{0.1053} = 191336.1\,\text{kg}$$

$$m_{0(1)r} = m_{(1)r} + m_{0(2)r} = 191336.1 + 50762.3 = 242098.4\,\text{kg}$$

$$m_{p(1)r} = \left(1-\varepsilon_{(1)}\right)m_{(1)r} = (1-0.1053)\times 191336.1 = 171188.4\,\text{kg}$$

$$m_{0b(1)r} = m_{0(1)r} - m_{p(1)r} = 242098.4 - 171188.4 = 70910.0\,\text{kg}$$

$$\Delta v_{(1)r} = \Delta v_{(1)i} + \Delta v_{(1)l} = \overline{u}_{e(1)}\ln\frac{m_{0(1)r}}{m_{0b(1)r}}$$

$$= 2600 \times ln\frac{242098.4}{70910.0} = 3192.6\,\text{m/s}$$

The sum of the calculated $\Delta v_{(i)r}$'s through the scaled-up $\mu_{(i)} = (m_{0(i)r}/m_{0b(i)})$'s matches with the $\sum_1^3\left(\Delta v_{(i)i} + \Delta v_{(i)l}\right)$.

$$\sum_1^3 \Delta v_{(i)r} = \sum_1^3 \overline{u}_{e(i)} ln\frac{m_{0(i)r}}{m_{0b(i)r}} = 3192.6 + 3066.6 + 5390.0 = 11649.2\,\text{m/s}$$

The results of the calculations are given Table 5.16.

Table 5.16 Example 5.4: Optimum scaled-up stages considering individual velocity losses $\Delta v_{(1)l} = 1310.7\,\text{m/s}$, $\Delta v_{(2)l} = 353\,\text{m/s}$, and $\Delta v_{(3)l} = 236\,\text{m/s}$ for the required terminal velocity $v_n = 9749.5\,\text{m/s}$ and the payload mass $m_l = 1991.8\,\text{kg}$

Stg	$I_{sp(i)}$ (ms)	$\varepsilon_{(i)}$	$\mu_{(i)}$	$\alpha_{(i)}$	$m_{(i)}$	$m_{0(i)}$ (kg)	$\Delta v_{(i)i}$ (kg)	$\Delta v_{(i)r}$(m/s)
1	2600	0.1053	3.41416	0.26530	191336.1	242098.4	1881.9	3192.6
2	2832	0.1079	2.95302	0.34888	37633.0	50762.3	2713.6	3066.6
3	4355	0.1631	3.44753	0.17884	11137.6	13129.4	5154.0	5390.0
			$m_l/m_{0(1)} = 0.00823$		$\sum_{i=1}^{3} \Delta v_{(i)}$ (m/s) = 9749.5			11649.2

Discussion

As indicated previously the loss-free optimization forms the base for the calculation of the optimum scaled-up stages considering individual-stage velocity losses. The loss-free optimization results are given in Table 5.14. And, we find that these results are far different from the results of the remaining two bases of calculations. The optimized lift-off mass $m_{0(1)}$ is extremely low and hence the ratio of the payload mass to lift-off mass $m_l/m_{0(1)}$ is attractively high. The second base of our calculations is for the marked-up terminal velocity of 11649.2 m/s that is the desired terminal velocity 9749.5 m/s plus the total velocity losses 1899.7 m/s, Table 5.15. We see that the stage masses increase from 27% at the third stage to 240% at the first stage. And, the $m_l/m_{0(1)}$ is significantly reduced from 0.02242 to 0.00838 by about 170% for about 20% increase in the terminal velocity from 9749.5 to 11649.2 m/s. As expected the loss-free ideal optimization favors more efficient stages—higher specific impulse and lower structural coefficient, the effect of the former being more powerful than that of the latter, Tables 5.14 and 5.15. Finally as the third base of the calculations, the scaled-up stage masses are calculated considering the individual-stage velocity losses, Table 5.16. Comparing the calculated results of ideal optimization procedure for the marked-up terminal velocity Table 5.15 and those of the scaled-up stage masses considering individual-stage velocity losses Table 5.16, the values of $m_{0(1)}$ and hence those of $m_l/m_{0(1)}$ are of the same order. But the first stage of higher velocity loss gets larger scaling-up of the mass while the remaining two stages of lower stage velocity losses get reduced stage masses—compare $m_{(i)}$'s of stages in Tables 5.15 and 5.16. The results of Table 5.16 are reported to be very close to the stage masses of Ariane 40 that were obtained through private communications [13].

As the basic data that we assumed for our previous calculation are reported to be through private communications, in order to check these data with those in the open literature, Table 5.17 is presented [1]. Despite the assumed values for the scaled-up stages considering individual-stage velocity losses being

Table 5.17 Example 5.4: Data for Ariane 40 from [1]; GTO payload mass $m_l = 1900$ kg; last four columns are calculated from the data

Stg	$m_{(i)}$ (kg)	$m_{s(i)}$ (kg)	Thrust (N)	Thrstng time (s)	$I_{sp(i)}$ (m/s)	$\varepsilon_{(i)}$	$m_{0(i)}$ (kg)	$\Delta v_{(i)r}$ (m/s)
1	185015	17515	2718000	150	2434	0.0947	236915	2988.0
2	38000	3400	798000	123	2837	0.0895	51900	3116.6
3	12000	1200	63000	720	4200	0.1000	13900	6302.0
			$m_l/m_{0(1)} = 0.00802$			$\sum_{i=1}^{3} \Delta v_{(i)r}$ (m/s) = 12406.6		

different from the open literature data (compare values of m_l's, $\varepsilon_{(i)}$'s, and $I_{sp(i)}$'s), the calculated values reasonably match with the open literature values (compare $\Delta v_{(i)r}$'s, $m_{(i)}$'s, and $m_{0(1)}$'s of Tables 5.16 and 5.17).

The accuracy of the procedure to calculate the scaled-up stages considering the individual-stage velocity losses depends on the inputs: $\varepsilon_{(i)}$'s, $I_{sp(i)}$'s, and most importantly the individual-stage velocity losses. The payload mass m_l, the number of stages with their propellant combinations and more or less constant specific impulses, the required terminal velocity, and the target altitude for injection into the orbit all remain fixed. The stage masses along with their structural coefficients (see Figs. 5.2, 5.3, 5.4 and 5.5) have to be updated during the iterative mode between loss-free optimization results and the scaled-up stage procedure, which requires the individual velocity losses from the trajectory algorithm. To start with, possibly a simple trajectory algorithm can be adopted for the scaled-up stage mass procedure.

5.4 Velocity Losses

Velocity losses occur due to gravity, drag, and turning or steering. The more significant ones are due to gravity and drag. Also due to fairing separation velocity loss occurs, but this is small with respect to the other losses [13]. The greatest losses always occur during the first-stage operation because the vehicle has to negotiate the drag in the dense atmosphere as well as the ever prevailing gravity. In order to calculate the velocity losses due to gravity and drag, we have to first understand the properties of the Earth's atmosphere with altitude and see how these properties affect the gravitational and drag forces.

5.4.1 Atmosphere

The standard molar-composition of dry atmospheric air is nitrogen (78.08%), oxygen (20.95%), argon (0.93%), and carbon dioxide (0.033%). The balance (0.007 %) is of neon, helium, methane, krypton, hydrogen, and xenon in the descending order. Thus, the standard molar mass of air is calculated as 28.9647 kg/kgmol. Noting the universal gas constant R_u is 8314.51 J/(kgmolK), the specific gas constant for the standard air becomes 287.057 J/(kgK). For approximate combustion calculations however, the combustion of a fuel with air is assumed to be the one of the fuel plus the air of composition one mole of oxygen followed by 3.76 moles of nitrogen.

5.4.1.1 Density and Temperature

The Earth's atmosphere can be divided into various altitude-layers with reducing pressure and different temperature variations. Temperature first decreases, then increases, and once again decreases from sea level to about 85 km altitude—all marginally between about 300 K and 200 K. And thereafter above 85 km, it increases exponentially to reach asymptotic values depending on the solar activities and the time during the day. Following all these variations the density of air decreases with altitude. As per NRLMSIS 2.0 model [14], typical variations of temperature and density up to 1000 km altitude at a location and time are given in Fig. 5.7. The first atmospheric layer is **troposphere**. It starts from the Earth's surface and extends 8 to 14.5 km high. This part of the atmosphere is the densest one. Almost all weather is in this region and the temperature decreases with the altitude to a minimum of about 200 K. The **stratosphere** starts just above the troposphere and extends to 50 km high. The ozone layer, which absorbs and scatters the solar ultraviolet radiation, is in this layer. Temperature increases with altitude from its minimum of about 200 K to a maximum close to the standard sea level temperature. The **mesosphere** starts just above the stratosphere and extends to 85 km high. Meteors burn up in this layer. Temperature once again decreases with altitude to a minimum of about 200 K. The **thermosphere** starts just above the mesosphere and extends to about 600 km high. Aurora and satellites occur in this layer. Temperature exponentially increases from its minimum of about 200 K to reach asymptotic values. Depending on the sunspot effects the maximum asymptotic temperature varies from 600–900 K nighttime and 750–1200 K daytime. The **exosphere** is the upper limit of our atmosphere. It extends from the top of the thermosphere up to 10,000 km. Depending on the sunspot effects, the reached asymtotic temperatures during daytime or nighttime are essentially constant.

There are two other considerations for the classification of our atmosphere. The first one is on the ionization of air. The **ionosphere** is an abundant layer of electrons and ionized atoms and molecules that stretches from the **stratopause** of about 50 km above the sea level to the edge of space at about 965 km, overlapping into the mesosphere and thermosphere. This dynamic region grows and shrinks based on solar

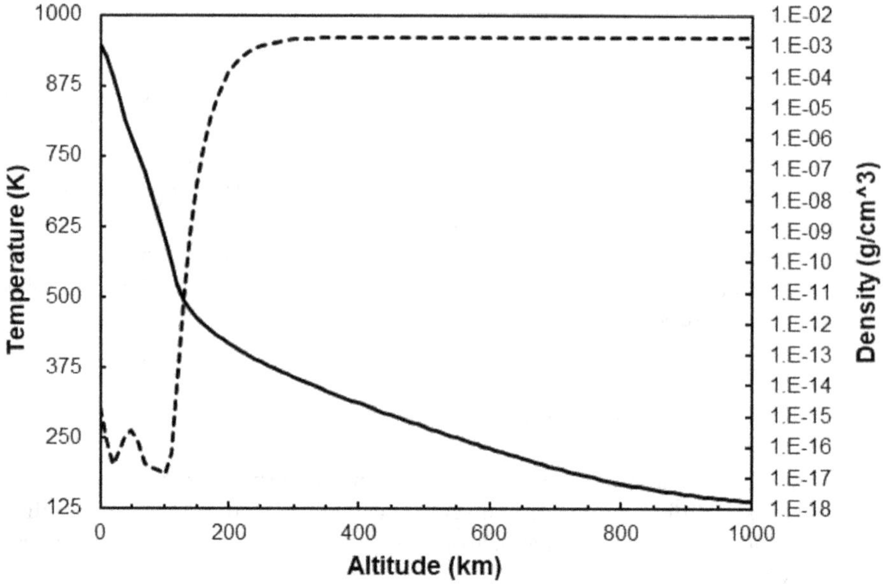

Fig. 5.7 Density and temperature variation up to 1000 km at latitude 14 and longitude 82 on June 10th 2023 6 PM by the NRLMSIS 2.0 model [15]

conditions and divides further into the sub-regions: D, E and F based on what wavelength of solar radiation is absorbed. The ionosphere is a critical link in the chain of Sun-Earth interactions. This region is what makes radio communications possible. The second consideration for the classification is on the air composition. From the surface of the Earth up to about 120 km, the air composition is reasonably constant and hence this region is known as **homosphere**, which includes troposphere, stratosphere, mesosphere, and a small part of thermosphere. The air composition being essentially constant in the homosphere, the specific gas constant for air can be assumed to be constant. As per NRLMSIS 2.0 model [14], typical variations of temperature and density in the homosphere at a location and time are given in Fig. 5.8. Above about 120 km altitude is the **heterosphere** that includes the remaining major part of thermosphere and the exosphere. In the heterosphere the specific gas constant is not constant and it widely varies due to sunspot effects with altitude and time.

5.4.1.2 Gravity and Density

The Earth's gravitational acceleration decreases with altitude. By Newton's law of gravitation, the force between any two bodies of masses m_1 and m_2 is given by

$$F = G\frac{m_1 m_2}{r^2} \tag{5.43}$$

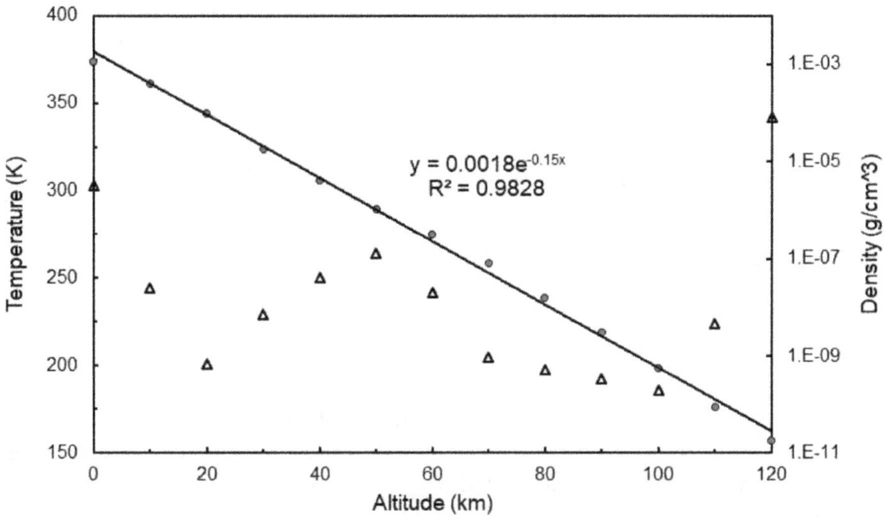

Fig. 5.8 Density and temperature variation in the homosphere at latitude 14 and longitude 82 on June 10th 2023 6 PM by the NRLMSIS 2.0 Model [15]

where G is the gravitational constant $6.67430 \times 10^{-11} m^3/(kgs^2)$ and r is the distance between the two bodies, center to center. Adopting Eq. 5.43 for the force of attraction between the mass of the Earth M and the body of mass m

$$F = G\frac{M}{(r_e + h)^2}m \tag{5.44}$$

where r_e is the Earth's mean radius and h is the altitude from the Earth's surface. Noting $F = mg$

$$g = \frac{GM}{(r_e + h)^2} = \frac{\mu}{(r_e + h)^2} \tag{5.45}$$

where the product GM is the standard gravitational parameter $\mu = 3.986004418 \times 10^{14}\ \mathrm{m^3/s^2}$. With reference to the standard gravitational acceleration at the Earth's surface g_e and adopting Eq. 5.45, the gravitational acceleration g at an altitude h is given by

$$g = g_e\left(\frac{r_e}{r_e + h}\right)^2 \tag{5.46}$$

At about 200 km, typical of a LEO altitude, the acceleration due to gravity g is about 0.94 times the g_e. And at about 1000 km, typical of a Sun synchronous orbit (SSO) altitude, g is around 0.75 times the g_e. At the geosynchronous condition g is far less than g_e.

Atmospheric pressure reduces with altitude. Therefore,

$$\frac{dp}{dh} = -\rho g \tag{5.47}$$

By the equation of state

$$p = \rho RT \tag{5.48}$$

Differentiating Eq. 5.48 with respect to the altitude h and substituting Eq. 5.47, we get

$$\frac{d\rho}{\rho} + \frac{dT}{T} = -\frac{gdh}{RT} \tag{5.49}$$

Thus, knowing the variations of the temperature T and the gravitational acceleration g with altitude h, the density ρ at a given altitude can be calculated. Then with the known temperature T and the calculated ρ, adopting the equation of state Eq. 5.48, pressure p at a given altitude can be evaluated.

In the homosphere Fig. 5.8, from the surface of the Earth up to about 120 km, the temperature varies within narrow limits and the composition is reasonably constant with an essentially invariant specific gas constant R. Combining Eqs. 5.46 and 5.49,

$$\frac{d\rho}{\rho} + \frac{dT}{T} = -\frac{g_e}{RT}\left(\frac{r_e}{r_e + h}\right)^2 dh \tag{5.50}$$

As an approximation, since the temperature varies within narrow limits, dT/T can be neglected and a mean value for the temperature T can be taken to be a constant. Hence,

$$\frac{d\rho}{\rho} = -\left(\frac{r_e}{r_e + h}\right)^2 \frac{dh}{H} \tag{5.51}$$

where H, the so-called **scale height**, is the assumed constant equal to RT/g_e. Integrating Eq. 5.51 with the limits of altitude 0 to h and of densities ρ_0 and ρ,

$$\rho = \rho_0 \, exp\left(-\frac{r_e}{r_e + h}\frac{h}{H}\right) \tag{5.52}$$

Here ρ_0 is the limit for a mean density, yet to be determined. As an additional approximation, since the altitudes under consideration are only up to 120 km, h can be neglected with respect to the Earth's mean radius r_e which is 6371 km. Thus,

$$\rho = \rho_0 \, exp\left(-\frac{h}{H}\right) \tag{5.53}$$

If one fits Eq. 5.53 against the actual density values of a reference atmosphere in the range $0 \leq h \leq 120\,\text{km}$, $\rho_0 = 1.752\,\text{kg/m}^3$ and $H = 6700\ m$ with the error $\Delta\rho/\rho$

being within reasonable limits [12]. Equation 5.53 and the curve fitted values of ρ_0 and H are very convenient to describe the density distribution in the homosphere. Furthermore, for short ranges of altitudes within the homosphere, Eq. 5.53 can be adopted to get suitable curve-fitted values of ρ_0 and H to carry out the flight calculations. As a typical example, the variations of density and temperature with altitude in the homosphere at a location and time are shown in Fig. 5.8. And, the equation of the fitted trend line is $y = 0.0018e^{-0.15x}$ with a high R^2 correlation of 0.9828. Here y is the density ρ in g/cm^3 and x is the altitude in km. If we convert the density and the altitude to kg/m^3 and m respectively, we get $\rho_0 = 1.8\,\text{kg/m}^3$ and $H = 6666.7\,\text{m}$, which are close to the values given by Walter [12] for a reference atmosphere.

5.4.2 Drag Coefficient C_D

Atmosphere can be considered as a continuum up to the altitude below which the **collision frequency** between the air molecules is far greater than the frequency at which the molecules collide with the vehicle surface. This is characterized by the statistical mean average distance between the successive collisions, the **mean free path** λ. The mean free path being proportional to the number of molecules per unit volume $\lambda \propto 1/n \propto T/p$. Typically $\lambda \approx 0.2\ \mu\text{m}$ at sea level, $\approx 80\ \mu\text{m}$ at 50 km altitude, $\approx$ 2 cm at 100 km altitude, and $\approx$ 250 m at 200 km altitude. The continuum is characterized by the non-dimensional **Knudsen number** Kn defined as $Kn \equiv \lambda/L$ where L is the characteristic dimension of the vehicle. Generally, the mesosphere and below can be taken as a continuum and here $Kn \leq 0.03$. The atmosphere from about 85 km (**mesopause**) to 150 km is considered as the one of transition and here $0.03 \leq Kn \leq 1$. Previously we noted that the homosphere extends up to about 120 km altitude wherein the specific gas constant can be taken as constant. Possibly without much additional error the continuum can be extended up to 120 km altitude. At altitudes above 150 km $Kn > 1$ and this region is no longer a continuum and is known as the one of free molecular flow.

The drag force experienced by the vehicle is written as

$$D = C_D \frac{1}{2} \rho v^2 A_c \tag{5.54}$$

where C_D is the **drag coefficient**, ρ the atmospheric density, v the velocity of rocket vehicle, and A_c the characteristic area, which for a rocket vehicle can be taken as the frontal cross-sectional area. In the free molecular flow region the drag coefficient is of the order 2, while in the continuum it is of the order 1. For the free molecular flow range of $150 < h < 600$ km Walter [12] recommends a frequently used rough empirical value of $C_D = 2.2 \pm 0.2$. However for the given body configuration, the exact value of C_D has to be determined through detailed analysis. Despite the fact that in the continuum C_D is of the order 1, it is well known that a skillfully streamlined body will have a much lower C_D than a bluff one. Also in the continuum C_D varies

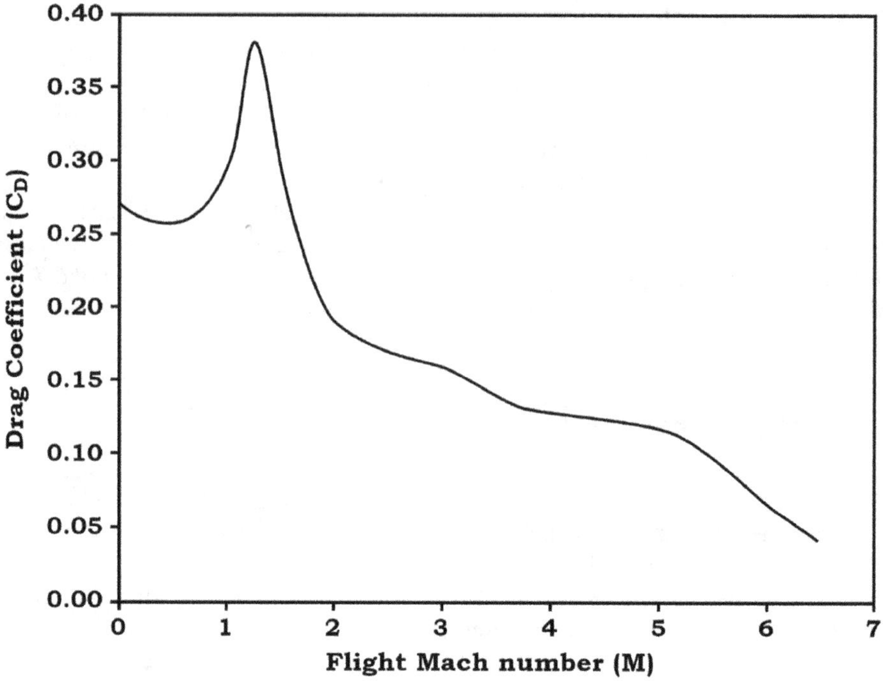

Fig. 5.9 Typical C_D variation with flight Mach number for a well streamlined body

with Mach number M. Starting from a low value in the subsonic regime, C_D mostly peaks just following Mach number one and thereafter it tapers. A typical variation of C_D with Mach number for an extremely streamlined body is shown in Fig. 5.9. However for rocket vehicles the C_D values versus M are much higher. Even for a simple sounding rocket with a conical nose C_D at the peak can be 0.7–0.8. And, for the launch vehicles with serial staging, Fig. 5.1, the peak C_D values can be 1.1–1.2 and for the ones with parallel staging C_D at the peak can be close to 2 and above. Furthermore, given a flight Mach number and body configuration, C_D is higher for a higher angle of attack, that is, a higher thrust angle γ, Fig. 5.10.

With the increasing flight velocity in the supersonic ($M < 5$) and hypersonic ($M > 5$) regimes the drag coefficient tapers. Atmospheric density ρ reduces exponentially with altitude, Eq. 5.53. Although the drag force is proportional to v^2, the vehicle velocity v is increased with altitude only as per the designed launch trajectory. Considering all these, the drag force has to come down for the rocket vehicle for the designed launch trajectory. For example a serially staged rocket vehicle with the characteristic frontal area of 1 meter diameter at an assumed constant $C_D = 0.4$ experiences a drag force of only about 10 N at an altitude of 85 km with $v = 3500\,\text{m/s}$ and only about 0.2 N at an altitude of 120 km with $v = 7000\,\text{m/s}$. Therefore, for a quick first cut trajectory determination drag force can be neglected for second and upper stage flights. For example, MacKay [6], considering different launch trajecto-

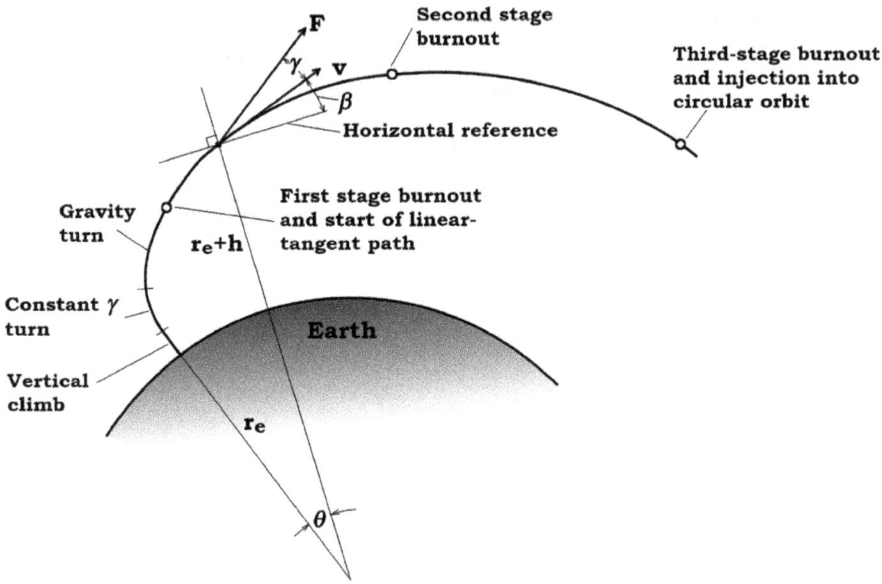

Fig. 5.10 Typical ascent trajectory of a rocket vehicle (adopted from [6])

ries, assumes a constant drag coefficient of 0.4 and neglects drag forces above 55 km altitude for his stage optimization study through performance charts.

5.5 Trajectory

The initial and final conditions of the trajectory to be optimized are known or specified. At the initial time $t = 0$, generally the vehicle velocity $v = 0$, the altitude $h = 0$, and for the vertical lift-off the **flight path angle** $\beta = 90°$ and the **thrust angle** $\gamma = 0°$, Fig. 5.10. At the final time $t = t_f$, the terminal velocity v_n and the corresponding altitude h_f are specified. And, the given payload m_l is to be injected into the orbit with a desired flight path angle β_f, generally equal to 0°. Let us now consider the forces acting on the rocket vehicle and the controls needed to optimally steer the vehicle to its destination.

5.5.1 Equations of Motion

The Earth's rotational speed at the equator is 465 m/s. Therefore, launching the rocket vehicle in the easterly direction at the equator, the vehicle gets this rotational velocity added as its tangential component. The tangential speed of the Earth's rotation at a point on the Earth can be approximated by multiplying the speed at the equator by the cosine of the latitude. Here however, in deriving the equations of motion for a rocket vehicle in a trajectory, we assume that the Earth's reference system is inertial, that is, the Earth's rotation is neglected. Furthermore, we assume that the vehicle is symmetrical with its stages arranged in serial and hence one of zero lift. A typical trajectory and its properties are shown in Fig. 5.10. In this section, we consider the rocket vehicle as one unit without separate consideration to individual stages. The time-dependent two-dimensional equations in polar coordinates are

$$m_0\dot{v} = F\cos\gamma - D - m_0 g\sin\beta \tag{5.55}$$

$$m_0 v\dot{\beta} = F\sin\gamma + m_0\frac{v^2}{r_e+h}\cos\beta - m_0 g\cos\beta \tag{5.56}$$

$$\dot{h} = v\sin\beta \tag{5.57}$$

$$\dot{x} = (r_e+h)\dot{\theta} = v\cos\beta \tag{5.58}$$

Here m_0 is the mass of the rocket vehicle including its stages at the instant and the payload m_l. The thrust force F is given by

$$F = \dot{m}\overline{u}_e = \dot{m}I_{sp} \tag{3.17}$$

where,

$$\overline{u}_e = u_e + \frac{A_e(p_e - p_a)}{\dot{m}} \tag{3.18}$$

Remember that in hard vacuum $p_a = 0$ and the specific impulse corresponds to the vacuum condition, $I_{sp,vac}$. Also note that the propellant ejection rate from the engine $\dot{m}$ is equal to the rate at which mass of the rocket vehicle reduces $-dm_0/dt$. The angle γ is the thrust angle and β is the flight path angle. The local gravitational acceleration g is given by Eq. 5.46. The drag D by Eq. 5.54 is always in the opposite direction to the flight velocity v. The density ρ can be calculated using Eq. 5.53. The term $v^2/(r_e+h)$ in Eq. 5.56 represents the centrifugal acceleration. The angle θ is the one between the instantaneous radius vector $r_e + h$ and the initial radius vector r_e on the Earth's surface. All these supporting equations are given here for ready reference.

$$g = g_e\left(\frac{r_e}{r_e+h}\right)^2 \tag{5.46}$$

$$\rho = \rho_0\, exp\left(-\frac{h}{H}\right) \tag{5.53}$$

$$D = C_D\frac{1}{2}\rho v^2 A_c \tag{5.54}$$

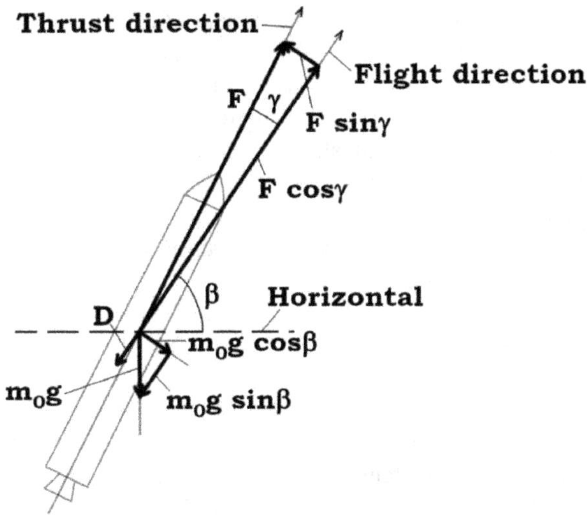

Fig. 5.11 Flight of rocket vehicle

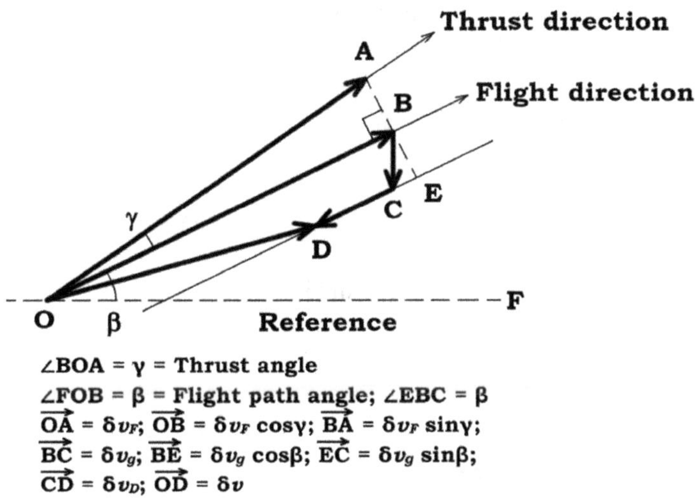

Fig. 5.12 Vector diagram of forces

The flight properties at a location in the atmosphere and the vector diagram of the forces acting on the rocket vehicle are shown in Figs. 5.11 and 5.12. Considering Eq. 5.55 and Figs. 5.11 and 5.12, we see that the rocket vehicle of mass m_0 is accelerating with $\dot{v}$ due to $F\cos\gamma$ against the drag force D (always in the negative direction to the flight velocity) and the $\sin\beta$ component of the gravitational force $m_0 g$. Similarly the rate of change of flight path angle $\dot{\beta}$ is governed positively by the

$\sin\gamma$ component of the thrust and $\cos\beta$ component of centrifugal acceleration, and negatively by the $\cos\beta$ component of the gravitational acceleration.

By a suitable numerical scheme, the trajectory of a vehicle can be calculated adopting Eqs. 5.55–5.58. For example for a small time increment Δt all the elemental components can be calculated as shown in Fig. 5.12. The related equations of motion in incremental form are Eqs. 5.55a–5.58a. For a chosen time instant t and its vehicle's position as typically given in Fig. 5.10, assuming $F(=\dot{m}\overline{u}_e)$, vehicle mass m_0, thrust angle γ, drag D, gravitational acceleration g, and the flight path angle β all to be constant, the elemental increments Δv, $\Delta\beta$, Δh, and Δx can be calculated for a chosen small time increment Δt. For the next time instant $t = t + \Delta t$, the new positional values are $v = v + \Delta v, \beta = \beta + \Delta\beta, h = h + \Delta h$, and $x = x + \Delta x$. Note $(F/m_0)\Delta t = -(\Delta m_0/m_0)\overline{u}_e$ with Δm_0 being a negative number. Therefore, the new reduced vehicle mass $m_0 = m_0 + \Delta m_0$. For the new position thus obtained (x, h, v) calculate the new values of g (Eq. 5.46) and D (Eq. 5.54) to proceed with the next position and the time instant. Evidently, the accuracy of the trajectory calculation depends on the fineness of the chosen Δt. During such trajectory calculations, due care is taken on the maximum permissible dynamic pressure (applicable at low altitudes) and the vehicle acceleration. In arriving at the optimum trajectory with a minimum propellant expenditure within the shortest possible time, the variables for control are the thrust F and the thrust angle γ.

$$\Delta v = \frac{F}{m_0}\cos\,\gamma\,\Delta t - \frac{D}{m_0}\Delta t - g\sin\beta\,\Delta t \tag{5.55a}$$

$$\Delta\beta = \frac{F}{m_0 v}\sin\,\gamma\,\Delta t + \frac{v}{r_e + h}\cos\beta\,\Delta t - \frac{g}{v}\cos\,\beta\,\Delta t \tag{5.56a}$$

$$\Delta h = v\sin\beta\,\Delta t \tag{5.57a}$$

$$\Delta x = v\cos\,\beta\,\Delta t. \tag{5.58a}$$

5.5.2 *Optimum Trajectory*

It can be stated in a nutshell that the *the objective of the trajectory optimization is to attain the specified terminal conditions within the shortest time (to escape from the gravity at the earliest) that too with the lowest possible velocity losses—the least possible propellant expenditure.* In realizing this section, we draw frequent references to [12]. Noting that the thrust $F = \dot{m}\overline{u}_e$, in order to have the shortest trajectory time, the thrust with respect to the time is set at its maximum unless constrained by (1) the maximum permissible inertial load (say maximum **thrust acceleration** F/m_0 of 3–4 g_e) and/or (2) the maximum permissible dynamic pressure $\rho v^2/2$ (say 40 to 60 kPa). Furthermore, the instantaneous vehicle mass $m_0 = m_{0i} - \dot{m}t$, where m_{0i} and $\dot{m}$ are the initial vehicle mass and propellant flow rate respectively. Therefore, the thrust F and the instantaneous vehicle mass m_0 are known functions of time. Considering all these, the effective variables for the trajectory optimization are the flight path angle β and the thrust angle γ, Eqs. 5.55 and 5.56. The altitude h and the downrange distance x are governed by Eqs. 5.57 and 5.58.

To understand the procedure for the trajectory optimization, let us consider first the equation of motion Eq. 5.55 that can be written as

$$dv = \frac{F}{m_0}\cos\gamma\, dt - \frac{D}{m_0}\, dt - g\sin\beta\, dt$$

This in the integral form with the vehicle velocity 0 at lift-off to the final terminal velocity

$$v_n = F\int_0^{t_f}\frac{cos\gamma}{m_0}\, dt - \int_0^{t_f}\frac{D}{m_0}\, dt - \int_0^{t_f} g\sin\beta\, dt$$

which can be written as

$$v_n = F\int_0^{t_f}\frac{dt}{m_0} - F\int_0^{t_f}\frac{1-cos\gamma}{m_0}\, dt - \int_0^{t_f}\frac{D}{m_0}\, dt - \int_0^{t_f} g\sin\beta\, dt$$

Noting $F = \dot{m}\overline{u}_e$ and $\dot{m}dt = -dm_0$,

$$\begin{aligned} v_n &= \overline{u}_e ln\frac{m_{0i}}{m_{0b}} \\ &-2F\int_0^{t_f}\frac{\sin^2(\gamma/2)}{m_0}dt - \frac{1}{H}\int_0^{t_f}\kappa_D v^2 e^{(-h/H)}dt - g_e\int_0^{tf}\frac{r_e^2}{(r_e+h)^2}\sin\beta dt \end{aligned} \tag{5.59}$$

where κ_D is the reduced drag coefficient

$$\kappa_D = \frac{C_D A H}{2m_0}\rho_0 \tag{5.60}$$

While κ_D is appropriately evaluated and used in the trajectory calculations under low speeds, at hypersonic speeds it can be assumed to be a constant, $\kappa_D \approx 25$ [12]. Recall Eq. 5.53 $\rho = \rho_0\, exp(-h/H)$. Equation 5.59 gives an explicit presentation of the velocity losses against the loss-free velocity gain that we get through the Tsiolkovsky equation, the first term of the expression on the right-hand side. Note m_{0i} is the lift-off mass of the vehicle (gross mass of all stages plus the payload mass m_l) and m_{0b} is the mass of the vehicle at the burnout of the last stage, which includes the structural mass of the last stage plus the payload mass m_l. The second term of the expression represents steering loss and the steering loss exists whether the thrust angle γ is positive or negative. The third and fourth terms of the expression respectively represent velocity losses due to drag and gravity.

Of the velocity losses, the least one is due to the steering. Next comes the loss due to the drag which is high in the dense atmosphere at low altitudes and becomes negligible at high altitudes, say above 55–60 km. The gravity loss is the maximum one and it is always there until the vehicle reaches its orbit. In view of these, considering first the drag loss, in order to minimize it the vehicle has to have a minimum acceleration and hence low velocities until the vehicle climbs to about 60 km altitude. After the lift-off, the vehicle generally climbs vertically for some time, $\gamma = 0°$ and $\beta = 90°$. On reaching an altitude through the vertical climb, the flight path angle β from its 90° has to be brought down close to 0° on reaching the specified altitude before injecting the payload into the orbit. This has to be done through a suitable reduction of flight path angle rate $\dot{\beta}$ in association with the thrust angle rate $\dot{\gamma}$ to reach the desired altitude. Therefore, let us next consider the rate of variation of flight path angle $\dot{\beta}$ which has in it the thrust angle γ as a variable, Eq. 5.56.

Equation 5.56 can be written as

$$d\beta = \frac{F}{m_0 v} \sin\gamma \, dt + \frac{v}{(r_e + h)} \cos\beta \, dt - \frac{g_e}{v} \frac{r_e^2}{(r_e + h)^2} \cos\beta \, dt$$

This in the integral form from the flight path angle $\beta = \pi/2$ at the lift-off to the final flight path angle

$$\beta_f = \frac{\pi}{2} + F \int_0^{t_f} \frac{\sin\gamma}{m_0 v} \, dt + \int_0^{t_f} \frac{v}{(r_e + h)} \cos\beta \, dt - \int_0^{t_f} \frac{g_e}{v} \frac{r_e^2}{(r_e + h)^2} \cos\beta \, dt \tag{5.61}$$

The second term in the expression on the right-hand side represents the effect of thrust angle γ variation on the final flight path angle β_f. A negative γ reduces the flight path angle, Eq. 5.56 and Fig. 5.12. The third term is due to the centrifugal force which resists the reduction in the flight path angle. The last term represents the gravity turn, which reduces the flight path angle. Now the procedure for the trajectory optimization is clear. Adopting the two equations Eqs. 5.59 and 5.61 along with Eq. 5.57 ($h = \int_0^{t_f} v \sin\beta \, dt$), through a suitable numerical procedure, judiciously use the effects of steering, centrifugal, and gravity to minimize the velocity losses to attain the trajectory conditions with the least possible propellant expenditure within the shortest possible time.

5.5.2.1 Gravity Turn

On reaching an altitude through the vertical climb a slight side "kick" is momentarily given to the vehicle by a small thrust angle γ by which the vehicle assumes an angle of 2–3° with respect to the instantaneous radius vector $r_e + h$ and the flight direction, Fig. 5.10. After this kick, with the zero thrust angle, due to the gravitational force the angle between the instantaneous radius vector and the flight direction increases. Consequently, the flight path angle β decreases. Such a trajectory turn is known as **gravity turn**. This decrease in the flight path angle is against the opposing centrifugal force, Eq. 5.56. Let us consider the effects of gravity turn. During the gravity turn with the thrust angle γ set at 0, the steering loss is zero, Eq. 5.59. As it is mandatory that the flight path angle β has to decrease close to zero at the trajectory end, the rate at which it decreases during the gravity turn is of interest. Considering Eq. 5.56, during the gravity turn

$$\dot{\beta} = -\left(\frac{g}{v} - \frac{v}{r_e + h}\right)\cos\beta \tag{5.62}$$

In the gravity turn, while the gravitational force through the term g/v is responsible for the reduction of the flight path angle β, the centrifugal force through the term $v/(r_e + h)$ is opposing it, Eq. 5.62. From the lift-off to the payload injection at the end, $\cos\beta$ varies from $0 \to 1$, g/v has the greatest descent from $\infty \to \delta$, and $v/(r_e + h)$ increases marginally from $0 \to \delta$, where $0 < \delta \ll 1$. The net effect is that the reduction rate of flight path angle $\dot{\beta}$ due to the gravity turn is stronger at lower velocities than that at higher ones. At the limit when $\beta = 0$ and $\dot{\beta} = 0$, the opposing centrifugal force is equal to the gravitational force. And, neglecting h increase with time with respect to the Earth's radius r_e, $v = \sqrt{g r_e} \approx 7900\,\mathrm{m/s}$.

5.5.2.2 Pitch Maneuver

The **pitch angle** ϕ is defined as $\phi \equiv \beta + \gamma$. In general, a maneuver with the pitch angle ϕ less than 90° can be termed as a **pitch maneuver**. With the thrust-on if a pitch maneuver is executed through the gravity turn alone ($\gamma = 0$), the flight path angle β will decrease at a faster rate with a lower $\dot{h}$ resulting in higher velocities at lower altitudes of denser atmosphere—higher drag losses possibly added with high dynamic pressures. Therefore, in order to reduce the drag losses it is necessary to reach the rarefied atmosphere within a short time with an acceptable acceleration and low velocities. And so, a pitch maneuver different from the gravity turn is to be selected. That is, the pitch-angle decrease-rate has to be less than that through a gravity turn. Note that the pitch angle decrease is linked to the flight path angle β decrease. And, the flight path angle rate $\dot{\beta}$ in turn has in it the thrust angle γ as a variable, Eq. 5.56. Therefore, the desired pitch angle variation rate, different from that due to gravity turn, has to be obtained by choosing a parametric negative constant for the combination $\dot{\beta} + \dot{\gamma}$. And, this parametric negative constant can be used in numerical trajectory optimization. Thus, the **constant-pitch-rate maneuver** is

$$\dot{\beta} + \dot{\gamma} = \dot{\phi} \tag{5.63}$$

where $\dot{\phi}$ is a negative constant.

Constant Flight Path Angle Rate

Evidently the constant pitch rate maneuver depends on how the thrust angle γ is controlled with time or how the vehicle has to be steered. Before finding this control for the constant pitch rate maneuver let us first see what is the thrust angle control required to achieve a **constant flight path angle rate** or how the flight path angle β must reduce with respect to time at a constant rate. After the lift-off, with a vertical climb for some time, the velocities are still low at the start of the pitch maneuver.

Therefore, the drag and centrifugal forces can be neglected. And, the flight path angle at a time instant t is $\beta = \pi/2 + \dot{\beta}t$, where $\dot{\beta}$ is a negative constant for a constant flight path angle rate. Considering Eq. 5.56 and neglecting centrifugal force

$$v\dot{\beta} = \frac{F}{m_0} \sin\ \gamma - g \cos\ \beta \tag{5.64}$$

Since $F = m_0 a$ where a is the **thrust acceleration**—not the vehicle acceleration

$$v\dot{\beta} = a \sin\ \gamma - g \cos\ \beta$$

We have to find an expression for the thrust angle control with respect to time to achieve the constant flight path angle rate $\dot{\beta}$. From the force equation Eq. 5.55, noting $D \ll F$ (low flight velocities) and the thrust angle γ being very small $\cos\ \gamma = 1$

$$\dot{v} = \frac{F}{m_0} - g \sin\ \beta$$

Noting $m_0 = m_{0i} - \dot{m}t$ where m_{0i} is the initial vehicle mass and t is the time elapsed and the instant under consideration, the above equation can be written as

$$dv = \frac{F}{m_{0i} - \dot{m}t} dt - g \sin\beta dt$$

Noting $d\beta = \dot{\beta}dt$

$$v - v_0 = \int_0^t \frac{F/m_{0i}}{1 - \dot{m}t/m_{0i}} dt - \frac{g}{\dot{\beta}} \int_0^t \sin\beta d\beta$$

$$v\dot{\beta} = \dot{\beta}\left[v_0 - \frac{F}{m_0}\frac{m_0}{\dot{m}} \ln\left(1 - \frac{\dot{m}t}{m_{0i}}\right)\right] + g\cos\beta$$

Substituting this expression for $v\dot{\beta}$ into Eq. 5.64

$$\dot{\beta}\left[v_0 - \frac{F}{m_0}\frac{m_0}{\dot{m}} \ln\left(1 - \frac{\dot{m}t}{m_{0i}}\right)\right] + g\cos\beta = \frac{F}{m_0}\sin\gamma - g\cos\beta$$

$$\sin\gamma = \dot{\beta}\left[v_0\frac{m_0}{F} - \frac{m_0}{\dot{m}} \ln\left(1 - \frac{\dot{m}t}{m_{0i}}\right)\right] + 2\frac{m_0}{F} g\cos\beta$$

Noting $\beta = \frac{\pi}{2} + \dot{\beta}t$, $\cos\beta = \cos(\frac{\pi}{2} + \dot{\beta}t) = -sin\dot{\beta}t$

$$\sin\gamma = \dot{\beta}\left[v_0\frac{m_0}{F} - \frac{m_0}{\dot{m}} \ln\left(1 - \frac{\dot{m}t}{m_{0i}}\right)\right] - 2\frac{m_0}{F} g\sin\dot{\beta}t \tag{5.65}$$

Equation 5.65 gives the expression for the thrust angle control with respect to time to achieve the constant flight path angle rate $\dot{\beta}$ after lift-off. And, this has been derived adopting the force- and angular-equation, Eqs. 5.55 and 5.56 under the simplified conditions of negligible drag- and centrifugal-force (both due to low velocities), and small thrust angle ($\cos\gamma = 1$). After the lift-off as $\dot{\beta}t$ is a small negative number, $\sin\dot{\beta}t = \dot{\beta}t$. Furthermore, $\dot{m}t \ll m_{0i}$ and hence $m_0 \approx m_{0i}$. As already assumed, γ being small, $\sin\gamma = \gamma$. Considering all these, adopting logarithmic series and neglecting higher order terms, Eq. 5.65 can be put in an approximate form as

$$\gamma = \dot{\beta}t\left(1 - \frac{2g}{a_i}\right) + \frac{\dot{\beta}v_0}{a_i} \tag{5.66}$$

where $a_i = F/m_{0i}$, the initial thrust acceleration. Equation 5.66, an approximate form of Eq. 5.65, gives the expression for the thrust angle control with respect to time to achieve the constant flight path angle rate $\dot{\beta}$ after lift-off. Note that for a constant $\dot{\beta}$ the thrust angle γ varies linearly with time. Remember that we want to reduce the flight path angle with time and hence $\dot{\beta}$ is a negative constant. Also, at $t = 0$

$$\gamma(t = 0) = \gamma_0 = \frac{\dot{\beta}v_0}{a_i}$$

Let us next consider the condition for a constant pitch rate trajectory.

Constant Pitch Rate Maneuver

For the constant pitch rate, the pitch angle rate $\dot{\phi}$ is to be a constant. That is $\dot{\beta} + \dot{\gamma} = \dot{\phi}$ a negative constant. The flight path angle rate $\dot{\beta}$ is a function of the thrust angle γ, Eq. 5.56. Therefore,

$$\dot{\beta}(\gamma) + \dot{\gamma} = \dot{\phi}$$

This differential equation can be solved by the simplification of adopting the thrust angle control equation derived for the constant flight path angle rate $\dot{\beta}$, Eq. 5.66. By differentiating Eq. 5.66

$$\dot{\gamma} = \dot{\beta}\left(1 - \frac{2g}{a_i}\right)$$

$$\dot{\phi} = \dot{\beta} + \dot{\gamma} = \dot{\beta}\left(2 - \frac{2g}{a_i}\right)$$

$$\dot{\beta} = \frac{\dot{\phi}}{\left(2 - \frac{2g}{a_{(i)}}\right)}$$

Substituting for $\dot{\beta}$ in Eq. 5.66 and noting $\gamma_0 = \dot{\beta} v_0 / a_i$

$$\frac{\gamma(2 - 2g/a_i)}{(1 - 2g/a_i)} = \dot{\phi} t + \frac{\dot{\beta} v_0 / a_i}{(1 - 2g/a_i)} + \gamma_0$$

Thus, we obtain the thrust angle γ for a constant pitch control with the constant flight path angle rate $\dot{\beta}$ as per the simplification we adopted.

$$\gamma = \frac{(a_i/2 - g)}{(a_i - g)} \left(\dot{\phi} t + \gamma_0\right) + \frac{1}{2} \frac{v_0 \dot{\beta}}{(a_i - g)}. \tag{5.67}$$

5.5.2.3 Linear Tangent Steering Law

The terminal state control considers the initial- and the final-state of the trajectory. The initial state is given by the initial velocity of the vehicle v_0 and the altitude at the lift-off h_0—in most cases both are zeros. The final state to be reached is defined by the terminal velocity v_n, the altitude for payload injection h_f, and the flight path angle β_f. Equations 5.59–5.61 have to be adopted for a numerical optimization procedure to reach the final state with the most minimum propellant expenditure. Even with the Earth's rotation being neglected, the procedure considering the drag losses all through the trajectory becomes quite complex. This is numerically done through a sophisticated proprietary-software involving many trial and error choices. Let us say that such a software including drag losses has found the optimum trajectory conditions for the most minimum propellant expenditure: the flight path angle variation with time $\beta(t)$-function and the time at the trajectory end t_f. On the other hand, the drag losses become negligible at high altitudes. Therefore, neglecting drag losses for altitudes above certain levels, say above 55 km or less [6], Eq. 5.59 becomes

$$v_n = \bar{u}_e ln \frac{m_{0i}}{m_{0b}} - 2F \int_0^{t_f} \frac{\sin^2(\gamma/2)}{m_0} dt - g_e \int_0^{tf} \frac{r_e^2}{(r_e + h)^2} \sin \beta dt \qquad D = 0 \tag{5.68}$$

Equation 5.61 remains unaltered.

$$\beta_f = \frac{\pi}{2} + F \int_0^{t_f} \frac{\sin \gamma}{m_0 v} dt + \int_0^{t_f} \frac{v}{(r_e + h)} \cos \beta \, dt - \int_0^{t_f} \frac{g_e}{v} \frac{r_e^2}{(r_e + h)^2} \cos \beta \, dt \tag{5.61}$$

These two equations neglecting drag losses are amenable to analytical procedures. Therefore, in the absence of drag losses, the variational calculus analytically gives the **bilinear tangent steering law** Eq. 5.69 as the means to provide the time-dependent thrust angle $\gamma(t)$ to realize optimized conditions of $\beta(t)$-function and t_f obtained numerically.

$$\tan(\gamma + \beta) = \frac{c_1 \left(t_f - t\right) + c_2}{c_1' \left(t_f - t\right) + c_2'} \tag{5.69}$$

where c_1, c_2, c_1', and c_2' are the constants attributable to the initial (x_0 & h_0 *and* v_0) and the final (x_f & h_f *and* v_f) states. Generally, the downrange distance $x_f - x_0$ (the downrange distance at which the altitude constraint h_f is attained) is unimportant. Then $c_1' = 0$ and Eq. 5.69 simplifies to the **linear tangent steering law**

$$\tan(\gamma + \beta) = a_1(t_f - t) + b_1 \tag{5.70}$$

where a_1 and b_1 are constants. Considering the x and y components of the vehicle's velocity the linear tangent steering law becomes [12]

$$\tan(\gamma + \beta) = \frac{g\left(t_f - t\right) + v_{y,t_f}}{v_{x,t_f}} \tag{5.71}$$

Note when $t = t_f$

$$\tan\left(\gamma_{t_f} + \beta_{t_f}\right) = v_{y,t_f}/v_{x,t_f} = \tan\beta_{t_f}$$

Thus, the thrust angle or steering angle $\gamma_{t_f} = 0$.

5.5.2.4 Optimum-Trajectory Summary

We summarize here what we have discussed under this section. Under the assumption of non-rotating Earth, we obtained the equations of motion for a rocket vehicle of zero lift. The Earth's rotational speed at the equator is 465 m/s. Its speed at any other point is the equator speed multiplied by the cosine of the latitude at that point. By the assumption of non-rotating Earth, we neglect the tangential component of velocity that will be added to the vehicle velocity for the eastward launch.

The objective of the trajectory optimization is to attain the specified terminal conditions within the shortest time that too with the least possible propellant expenditure. Towards this objective, from the vertical lift-off to the payload injection, the flight path angle has to be brought down from 90° to about 0° through suitable controls of thrust angle in the presence of opposing centrifugal force and supporting gravitational force. Here come the various trajectory elements and controls: the maneuvers of gravity turn, constant flight path angle rate, and constant pitch rate; and the controls of bilinear tangent and linear tangent— all to achieve the optimum trajectory.

The optimum trajectory determination and its execution may be qualitatively characterized as follows. Even with the Earth's rotation being neglected, the optimization procedure considering the drag losses all through the trajectory becomes quite complex. This is numerically done through a sophisticated software involving many trial and error insertions. Considering this numerical procedure is beyond the scope of this book. However, given that such a numerical procedure has provided the optimum time-dependent flight path angle $\beta(t)$-function and the trajectory end-time t_f, the optimum trajectory is generally executed into two parts. The first part is in the dense atmosphere where drag losses are significant along with high dynamic pressures; this, after a brief vertical climb, is through an empirically chosen constant thrust-angle rate $\dot{\gamma}$ maneuver with near-zero thrust angles γ. The second part is in the rarefied atmosphere where the drag losses are insignificant; this one is executed through a bilinear or a linear tangent steering law, analytically obtained through variational calculus.

Generally, the injection velocities and velocity losses for different orbits can be approximately listed as follows. For a LEO or for an SSO, the injection velocity is about 7.6 km/s while it for a GTO is about 10.2 km/s. By and large the total velocity losses for a LEO amount to about 1.5 km/s and for an SSO or a GTO these amount to about 1.9 km/s. However the loss fractions due to gravity, drag, and steering are about 0.82, 0.13, and 0.05 respectively of the total losses for a LEO or an SSO. But, for a GTO these loss fractions are about 0.72, 0.14, and 0.14 respectively.

5.5.3 Sounding Rockets

After considering the various aspects of orbital launch vehicles' ascent trajectory, let us consider the simple ascent trajectory of sounding rockets. The ascent of a sounding rocket, powered as well as coasting after burnout, is nearly vertical. And, the descent is elliptical. Sounding rockets are designed to measure the properties of the upper atmosphere and near space environment (temperature, pressure, turbulence, composition, and radiation) and to perform scientific experiments there. Generally, these rockets are designed to reach around 50 to 150 km above the surface of the Earth—stratosphere edge to thermosphere. However, certain sounding rockets have an apogee between 1000 and 1500 km. The sounding rockets also provide information on the Earth's surface environment, oceans, winds, and weather. The rockets are generally single- or two-stage ones. The second stage can be solid or liquid. In the case of liquid stage, a simple concept of pressure fed regeneratively cooled engine is employed. Nevertheless, sounding rockets reaching the upper atmosphere and space environment may include all subsystems similar to those of orbital launch vehicles: solid motors, liquid engines, stabilizing fins, control and guidance system, payload nose-cone containing instruments, parachute recovery system, range safety destructive system, etc. Measurements are recorded/transmitted during ascent as well as descent.

In view of the near vertical ascent of sounding rockets, the thrust angle $\gamma = 0$, the flight path angle $\beta \approx 90°$, $\dot{\beta} = 0$, and $\dot{x} = 0$. In the form of elemental increments, the equations of motion Eqs. 5.55–5.58 become

$$\Delta v = \frac{F}{m_0}\Delta t - \frac{D}{m_0}\Delta t - g\Delta t \tag{5.72}$$

$$\Delta h = v\Delta t \tag{5.73}$$

Note $(F/m_0)\Delta t = -\overline{u}_e(\Delta m_0/m_0)$, where the negative number Δm_0 represents the reduction in the vehicle mass m_0. Adopting Eqs. 5.72 and 5.73, a suitable numerical scheme can be employed to calculate the ascent trajectory.

For the sake of a simple analytical solution, we can assume an averaged constant gravitational acceleration $\overline{g}$ and neglect drag force. The acceleration due gravity is essentially constant in the normal altitude range of sounding rockets. $g/g_e = 0.955$ at the altitude of 150 km. Although considerable drag exists for the flights at lower altitudes, it is generally considered negligible above 55 km. Furthermore, velocity loss due to the drag is far less than that due to the gravity. For an assumed constant thrust F and an invariant specific impulse $\overline{u}_e$, the propellant flow rate is constant. Hence, the mass of the vehicle at an instant t is $m_0 = m_{0i} - (m_{0i} - m_{0b})(t/t_b)$ where m_{0i} and m_{0b} are the initial and the final vehicle masses, and t_b is the burnout time. With an averaged constant acceleration due to gravity $\overline{g}$, neglecting the drag force per unit mass of the vehicle D/m_0 and integrating the differential form of the Eq. 5.72 from $t = 0$ to an arbitrary instant t

$$v = -\overline{u}_e \, ln \left[1 - \left(1 - \frac{1}{\mu}\right)\frac{t}{t_b}\right] - \overline{g}t \tag{5.74}$$

where the mass ratio $\mu = m_{0i}/m_{0b}$. We get the burnout velocity as

$$v_b = \overline{u}_e \, ln \, \mu - \overline{g}t_b \tag{5.75}$$

Note that in the absence of drag and now neglecting gravity, Eq. 5.75 becomes the Tsiolkovsky equation. Substituting for v from Eq. 5.74 into the differential form of Eq. 5.73 and integrating from the initial time $t = 0$ to the burnout time t_b, we get the burnout altitude

$$h_b = \overline{u}_e t_b \left(1 - \frac{ln \, \mu}{\mu - 1}\right) - \frac{1}{2}\overline{g}t_b^2 \tag{5.76}$$

Equating the kinetic energy of the burnout mass of the vehicle $m_{0b}v_b^2/2$ with the potential energy change from the burnout altitude h_b to the maximum altitude h_{max} we get the coasting height.

$$\frac{m_{0b}\left(\overline{u}_e \, ln \, \mu - \overline{g}t_b\right)^2}{2} = m_{0b}\overline{g}\left(h_{max} - h_b\right)$$

$$\left(h_{max} - h_b\right) = \frac{\left(\overline{u}_e ln \, \mu - \overline{g}t_b\right)^2}{2\overline{g}}$$

Thus, we get

$$h_{max} = h_b + \frac{\left(\overline{u}_e \, ln \, \mu - \overline{g}t_b\right)^2}{2\overline{g}} \tag{5.77}$$

Substituting for h_b from Eq. 5.76 into Eq. 5.77 we get the equation for the maximum altitude h_{max}.

$$h_{max} = \frac{\overline{u}_e^2 (ln\ \mu)^2}{2\overline{g}} - \overline{u}_e t_b \left[\left(\frac{\mu}{\mu - 1} \right) ln\ \mu - 1 \right] \tag{5.78}$$

From Eq. 5.78 we understand that less the thrusting time t_b higher is the maximum altitude h_{max}. But, as we have discussed previously, there are limitations: firstly, the lesser thrusting time resulting in higher acceleration imposes critical inertial-stresses to the vehicle structure and instruments; next, higher acceleration means higher velocity in the denser atmosphere leading to enhanced drag penalty; and finally, to give the higher acceleration the propulsion system has to be bulkier and heavier with enhanced propellant flow rate. So, we recall the optimization slogan, with due care for the inertial stress and aerodynamic load, attain the specified terminal conditions within the shortest time that too with the least possible propellant expenditure.

Example 5.5 A single stage sounding rocket is designed to carry a payload of 38 kg or 68 kg. The vehicle is powered by a low performance solid motor. The motor delivers a constant thrust of 13 kN for 32 s. It is of gross mass 400 kg and empty mass of 111 kg. Calculate the burnout-velocities and -altitudes, and the max altitudes (apogees) of the sounding rocket for the two payloads. Assume an average acceleration due to gravity as 9.6 m/s^2.

Solution

The stage gross mass plus the payload mass is the initial mass of the vehicle. Stage empty mass plus the payload is the burnout mass of the vehicle. The ratio of initial mass to burnout mass of the vehicle is the mass ratio μ. Subtract stage empty mass from the stage gross mass to get the propellant mass. This divided by the thrusting time gives the propellant flow rate. With the given constant thrust and constant propellant flow rate, we get the specific impulse of the motor $\overline{u}_e$. Adopting Eqs. 5.75–5.78, the burnout-velocities and -altitudes, and the maximum altitudes (apogees) are calculated.

Analysis

*Payload 38*kg

Stage gross mass $m = 400$ kg
Stage empty mass $m_s = 111$ kg
Therefore, propellant mass $m_p = 400 - 111 = 289$ kg
Thrusting time = 32 s
Therefore, propellant flow rate $\dot{m} = 289/32 = 9.031$ kg/s
$F = \dot{m}\overline{u}_e \rightarrow \ \overline{u}_e = 13000/9.031 = 1439.45$ m/s

Initial mass of the vehicle = stage gross mass + payload mass = 400 + 38 = 438 kg
Burnout mass of the vehicle = stage empty mass + payload mass = 111 + 38 = 149 kg
Mass ratio $\mu = 438/149 = 2.9396$

Adopting Eq. 5.75, we calculate the vehicle's velocity at the burnout.

$$v_b = \overline{u}_e \, ln \, \mu - \overline{g} t_b = 1439.45 \times ln \, 2.9396 - 9.6 \times 32 = 1244.92 \, \text{m/s}$$

Through Eq. 5.76, we calculate the altitude at the burnout.

$$h_b = \overline{u}_e t_b \left(1 - \frac{ln \, \mu}{\mu - 1}\right) - \frac{1}{2} \overline{g} t_b^2$$

$$= 1439.45 \times 32 \times \left(1 - \frac{ln \, 2.9396}{2.9396 - 1}\right) - \frac{1}{2} \times 9.6 \times 32^2 = 15539.9 \, \text{m}$$

Coasting height is by

$$(h_{max} - h_b) = \frac{(\overline{u}_e ln \, \mu - \overline{g} t_b)^2}{2\overline{g}}$$

$$= \frac{(1439.45 \times ln \, 2.9396 - 9.6 \times 32)^2}{2 \times 9.6} = 80719.5 \, \text{m}$$

Therefore, the max altitude (apogee) reached by the vehicle is 15539.9 + 80719.5 = 96259.4 m.

Payload 68 kg
As detailed above the calculations are repeated for the payload of 68 kg. The calculated results are as per the following.
Mass ratio $\mu = 468/179 = 2.6145$
Vehicle's velocity at the burnout = 1076.2 m/s.
The altitude at the burnout = 13727.5 m.
The coasting height = 60326.2 m.
The apogee = 74053.7 m.

Discussion
The coasting heights are far greater than the burnout altitudes. It is alerted that the calculations neglect the drag losses. Therefore, the burnout velocities, burnout altitudes, coasting heights, and the apogees will all be less in reality.

Note that for the burnout velocities, the calculated respective Mach numbers are highly supersonic at 4.22 and 3.65. These numbers can be calculated for the respective burnout altitudes by reading the ambient temperatures from the standard atmospheric tables.

Problems

5.1 A multistage rocket has to be employed to carry a payload m_l to a large distance or height with a required terminal velocity v_n. Give reasons.

5.2 Distinguish between serial staging, parallel staging, and engine staging in multi-staging of rocket vehicles. Write a brief note.

5.3 Study Table 5.1 and Figs. 5.2, 5.3, 5.4 and 5.5 and write a note of about 200–300 words on multistage rockets on the points of view of (1) m_{01}/m_l and orbits, (2) structural coefficient ε against propellant type and stage mass, and (3) specific impulse values versus propellant types.

5.4 Define payload ratio $\alpha_{(i)}$, structural coefficient $\varepsilon_{(i)}$, and mass ratio $\mu_{(i)}$ with reference to a multistage rocket of n stages.

5.5 Prove the following.

$$\frac{m_{0(i)}}{m_{0(i+1)}} = \frac{1+\alpha_{(i)}}{\alpha_{(i)}}$$

$$\mu_{(i)} = \frac{1+\alpha_{(i)}}{\varepsilon_{(i)}+\alpha_{(i)}}$$

$$\alpha_{(i)} = \frac{1-\mu_{(i)}\varepsilon_{(i)}}{\mu_{(i)}-1}$$

5.6 Derive the following equation.

$$v_n = \overline{u}_e \ln \frac{1}{\varepsilon + (m_l/m_0)(1-\varepsilon)}$$

where m_l/m_0 is the payload fraction of a single-stage rocket, the ratio of the payload mass m_l to the launch vehicle mass m_0 that includes the payload mass.

5.7 Write a brief note of about 200 words on how to choose the number of stages and their types for an orbital launch vehicle.

Table 5.18 Problem 5.17

Launch vehicle	I_{sp} m/s	ε
Solid	2650	0.139
$N_2O_4/UDMH$	2900	0.075
$LOX/RP1$	3200	0.075
$LOX/LH2$	4000	0.102

5.8 Discuss the characteristics of the different layers of the Earth's atmosphere: troposphere, stratosphere, thermosphere, and exosphere.

5.9 What are the features of ionosphere?

5.10 Distinguish between homosphere and heterosphere.

5.11 Define Knudsen number. Explain the continuum concept in our atmosphere.

5.12 What is free molecular flow?

5.13 Define drag coefficient. Discuss its variation in the continuum atmosphere and the atmosphere of free molecular low.

5.14 How does the drag coefficient vary with respect to Mach number and the angle of attack?

5.15 With a neat sketch of a typical ascent trajectory of an orbital launch vehicle, show the thrust angle, flight path angle, and the instantaneous radius vector. Write down the equations of motion. What are the basic assumptions in deriving these equations of motion, Eqs. 5.55–5.58?

5.16 Sketch the vector diagram of the forces acting on the launch vehicle in its ascent trajectory. Indicate the thrust angle and the flight path angle. Identify the drag component and the different components of thrust and gravity.

5.17 For a payload fraction m_l/m_0 of 2%, calculate under loss-free condition the terminal velocity v_n of a solid-propellant single stage vehicle and compare it with the terminal velocities v_n's of liquid-propellant single stage vehicles of $N_2O_4/UDMH$, $LOX/RP1$, and $LOX/LH2$. Based on the data presented in Table 5.1 and Figs. 5.2, 5.3, 5.4 and 5.5 it appears practical to assume the average values of specific impulse and structural coefficient as given in Table 5.18. Noting that for a LEO the terminal velocity including velocity losses being about 9.1 km/s, is it possible to attain such a terminal velocity for these calculated values?
Answers:
$v_{n,solid} = 4920\,\text{m/s}$, $v_{n,N_2O_4/UDMH} = 6872\,\text{m/s}$, $v_{n,LOX/RP1} = 7583\,\text{m/s}$, & $v_{n,LOX/LH2} = 8482\,\text{m/s}$

5.18 Consider two rocket vehicles each to carry a payload of 3 tons in a gravity- and drag-free field. Both the vehicles are of equal lift-off mass of 300 tons. One is of cryogenic propellant stage. And the other is of solid propellant stage. The average specific impulse of the cryogenic one is 50% higher. And, the cryogenic vehicle at its burnout (structural mass + payload) is 10% lighter. The solid-propellant vehicle has an average structural coefficient of 0.125. Find the ratio of the terminal velocities of the two vehicles. Also calculate the average structural coefficient of the cryogenic vehicle.
Answers:
$v_{n,cryo}/v_{n,solid} = 1.579$. $\varepsilon_{cryo} = 0.111$.

5.19 A 3000 kg payload is to be injected into a GTO. Assuming a total velocity loss of 1900 m/s, the marked-up terminal velocity for the injection into a GTO can be taken as 11700 m/s. Let us consider three possible cases of all liquid launch vehicles of the following propellant combinations: N2O4/UDMH, LOX/Kerosene, and LOX/LH2. Considering the all-liquid three-stage vehicles that are currently operational, the following average specific impulses appear reasonable. I_{sp} for N2O4/UDMH = 2880 m/s. I_{sp} for LOX/Kerosene = 3390 m/s. I_{sp} for LOX/LH2 = 4260 m/s. The following are the structural coefficients. For N2O4/UDMH stages $\varepsilon_{(1)} = 0.078$, $\varepsilon_{(2)} = 0.048$, and $\varepsilon_{(3)} = 0.060$. For LOX/Kerosene stages $\varepsilon_{(1)} = 0.081$, $\varepsilon_{(2)} = 0.099$, and $\varepsilon_{(3)} = 0.157$. For LOX/LH2 stages $\varepsilon_{(1)} = 0.118$, $\varepsilon_{(2)} = 0.118$, and $\varepsilon_{(3)} = 0.114$. (1) Under loss-free optimization procedure, calculate the stage properties. (2) As a simpler case of optimization, consider the stages to be similar with $\varepsilon_{N2O4/UDMH} = 0.062$, $\varepsilon_{LOX/Kerosene} = 0.087$, and $\varepsilon_{LOX/LH2} = 0.118$. Calculate the stage properties for the similar stages and compare the results.
Answers:
(1) For N2O4/UDMH: $m_{(1)} = 233500$ kg, $m_{(2)} = 73950$ kg, & $m_{(3)} = 11476$ kg. $\Delta v_{(1)} = 3182$ m/s, $\Delta v_{(2)} = 4580$ m/s, & $\Delta v_{(3)} = 3938$ m/s. For LOX/Kerosene: $m_{(1)} = 191660$ kg, $m_{(2)} = 30971$ kg, & $m_{(3)} = 5342$ kg. $\Delta v_{(1)} = 4875$ m/s, $\Delta v_{(2)} = 4194$ m/s, & $\Delta v_{(3)} = 2631$ m/s. For LOX/LH2: $m_{(1)} = 61115$ kg, $m_{(2)} = 19883$ kg, & $m_{(3)} = 6588$ kg. $\Delta v_{(1)} = 3851$ m/s, $\Delta v_{(2)} = 3851$ m/s, & $\Delta v_{(3)} = 3998$ m/s.
(2) For N2O4/UDMH: $m_{(1)} = 259411$ kg, $m_{(2)} = 54250$ kg, & $m_{(3)} = 11345$ kg. $\Delta v_{(i)}$'s are equal at 3900 m/s. For LOX/Kerosene: $m_{(1)} = 141408$ kg, $m_{(2)} = 35545$ kg, & $m_{(3)} = 8935$ kg. $\Delta v_{(i)}$'s are equal at 3900 m/s. For LOX/LH2: $m_{(1)} = 62195$ kg, $m_{(2)} = 19908$ kg, & $m_{(3)} = 6372$ kg. $\Delta v_{(i)}$'s are equal at 3900 m/s.

5.20 Consider an all-cryogenic two-stage launch vehicle adopting LOX/LH2 propellant combination. It has to place a payload of mass 3900 kg at a marked-up terminal velocity of 12000 m/s into a geosynchronous transfer orbit. Two simple cases are to be considered. In the first case, both the stages have equal specific impulse at 4325 m/s but different structural coefficients, the first stage having 0.1182 and the second one having 0.1179. In the second simple case all stages are similar having equal specific impulse of 4325 m/s and equal structural coefficient of 0.1181. Under the loss-free optimization procedure, calculate for both the cases the stage masses $m_{(i)}$'s, the stage incremental velocities $\Delta\ v_{(i)}$'s, and the minimized total launch mass of the rocket vehicle $m_{0(1)}$.

Answers:
(1) $m_1 = 148734.5\,\text{kg}$, $m_2 = 22254.0\,\text{kg}$, $v_1 = 5994.5\,\text{m/s}$, $v_2 = 6005.5\,\text{m/s}$ & $m_{0(1)} = 174888.5\,\text{kg}$.
(2) $m_1 = 148876.9\,\text{kg}, m_2 = 22224.8\,\text{kg}, v_1 = 6000\,\text{m/s}, \&\ v_2 = 6000\,\text{m/s}\ \&\ m_{0(1)} = 175001.8\,\text{kg}$.

5.21 Consider an all-cryogenic three-stage launch vehicle of LOX/LH2 propellants. For simplicity assume the stages are similar with ε of 0.117 and I_{sp} of 4260 m/s. The vehicle is to be used for an Earth-escape of 11-ton payload. As a first approximation assume that the velocity losses due to gravity, drag, steering, and fairing separation all amount to 2 km/s. (1) Estimate the lift-off and stage masses. (2) At the lift-off, if the acceleration is limited to $0.3g_e$ calculate the burning period of the first stage. For an Earth-escape the terminal velocity has to be slightly in excess of 11185 m/s.
Answers:
(1) $m_{0(1)} = 551925\,\text{kg}$, $m_{(1)} = 402284\,\text{kg}$, $m_{(2)} = 109070\,\text{kg}$, & $m_{(3)} = 29572\,\text{kg}$.
(2) $\Delta t_1 = 215$ s.

5.22 The Russian three-stage orbital launch-vehicle ROKOT of all liquid N2O4/UDMH is capable of placing a 1800 kg payload into a LEO or a 1000 kg payload into an SSO [1]. The vehicle has the following data. STAGE 1. Gross mass: 77150 kg. Empty mass: 5695 kg. Vacuum specific impulse: 3040 m/s. Sea level specific impulse: 2795 m/s. STAGE 2. Gross mass: 12195 kg. Empty mass: 1485 kg. Vacuum specific impulse: 3158 m/s. STAGE 3. Gross mass: 6565 kg. Empty mass: 1600 kg. Vacuum specific impulse: 3197 m/s. Calculate (1) structural coefficients of the stages. For the LEO and the SSO, calculate (2) lift-off mass of the vehicle; neglect vehicle integration mass such as fairings, stage ejection systems, etc.; (3) payload fraction; (4) stage payload ratios; and (5) stage mass ratios of the vehicle. (6) Under the loss-free assumption, calculate for the LEO and the SSO the individual stage velocity increments and terminal velocity. Assume the specific impulse for the first stage to be an average of the vacuum-specific impulse and the sea level specific impulse.
Answers:
(1) $\varepsilon_{(1)} = 0.0738$, $\varepsilon_{(2)} = 0.1218$, and $\varepsilon_{(3)} = 0.2437$.
(2) $m_{0(1),LEO} = 97710\,\text{kg}$ and $m_{0(1),SSO} = 96910\,\text{kg}$
(3) $m_l/m_{(01)\,LEO} = 0.0184$ and $m_l/m_{(01)\,SSO} = 0.0103$
(4) For LEO: $\alpha_{(1)} = 0.2665$, $\alpha_{(2)} = 0.6859$, and $\alpha_{(3)} = 0.2742$. For SSO: $\alpha_{(1)} = 0.2561$, $\alpha_{(2)} = 0.6203$, and $\alpha_{(3)} = 0.1523$.
(5) For LEO: $\mu_{(1)} = 3.7216$, $\mu_{(2)} = 2.0873$, and $\mu_{(3)} = 2.4603$. For SSO: $\mu_{(1)} = 3.8071$, $\mu_{(2)} = 2.1834$, and $\mu_{(3)} = 2.9096$.
(6) For LEO: $\Delta v_{(1)} = 3834\,\text{m/s}$, $\Delta v_{(2)} = 2324\,\text{m/s}$, and $\Delta v_{(3)} = 2878\,\text{m/s}$. $v_n = 9036\,\text{m/s}$. For SSO: $\Delta v_{(1)} = 3900\,\text{m/s}$, $\Delta v_{(2)} = 2466\,\text{m/s}$, and $\Delta v_{(3)} = 3414\,\text{m/s}$. $v_n = 9781\,\text{m/s}$.

5.23 Consider a rocket vehicle of two stages for a LEO transfer with three different options of propellant combinations: (1) all solid, (2) all liquid semi-cryogenic LOX/Kerosene, and (3) all liquid cryogenic LOX/LH2. The marked-up terminal velocity is 9500 m/s and the payload mass is 8600 kg. The all-solid vehicle has $\varepsilon_{(1)} =$

Table 5.19 Answers to Problem 5.23

	Solid	LOX/Krsn	LOX/LH2
λ	−9.81869E-04	−4.88559E-04	−3.79061E-04
$m_{(1)}$ kg	1472784	277526	111924
$\Delta v_{(1)}$ m/s	4229	4716	3768
$m_{(2)}$ kg	130310	43808	37568
$\Delta v_{(2)}$ m/s	5271	4784	5732
$m_{0(1)}$ kg	1611694	329934	158092
$m_l/m_{0(1)}$	0.005	0.026	0.054

0.1154, $\varepsilon_{(2)} = 0.0989, \overline{u}_{e(1)} = 2560\,\text{m/s}$, and $\overline{u}_{e(2)} = 2824\,\text{m/s}$. The semi-cryogenic vehicle has $\varepsilon_{(1)} = 0.0807$, $\varepsilon_{(2)} = 0.0993$, $\overline{u}_{e(1)} = 3178\,\text{m/s}$, and $\overline{u}_{e(2)} = 3422\,\text{m/s}$. And, the cryogenic vehicle has $\varepsilon_{(1)} = 0.1182$, $\varepsilon_{(2)} = 0.1179$, $\overline{u}_{e(1)} = 3849\,\text{m/s}$, and $\overline{u}_{e(2)} = 4531\,\text{m/s}$. Under loss-free assumption, calculate respective Lagrange multipliers. And, compare the stage masses, lift-off mass, and payload fraction $m_l/m_{0(1)}$. Answers in Table 5.19:

5.24 The Ukrainian orbital launch vehicle Zenit-3SL can place a payload of 5250 kg into a GTO. It has the following data [1]. STAGE 1. Gross mass: 354300 kg. Empty mass: 28600 kg. Average specific impulse: 3178 m/s. STAGE 2. Gross mass: 90600 kg. Empty mass: 9000 kg. Vacuum specific impulse: 3422 m/s. STAGE 3. Gross mass: 17300 kg. Empty mass: 2720 kg. Vacuum specific impulse: 3452 m/s. What are the lift-off mass $m_{0(1)}$ and the payload fraction $m_l/m_{0(1)}$? Under the loss-free assumption, calculate the individual stage velocity increments and terminal velocity. Neglect the mass associated with vehicle integration such as fairings, stage separation systems, etc.
Answers:
$m_{0(1)} = 467450\,\text{kg}$ and $m_l/m_{0(1)} = 0.0112$. $\Delta v_{(1)} = 3792\,\text{m/s}$, $\Delta v_{(2)} = 4370\,\text{m/s}$, and $\Delta v_{(3)} = 3590\,\text{m/s}$. $v_n = 11753\,\text{m/s}$.

5.25 For the data of Zenit-3SL given in Problem 5.24 and with the calculated loss-free terminal velocity $v_n = 11753\,\text{m/s}$ as the constraint, calculate under the loss-free assumption the optimum the lift-off mass, and the payload fraction and stage masses. And, compare these optimum values with the actual values of the Zenit-3SL and discuss. Note although the optimum stage masses are different from the actual ones, the optimum lift-off mass is the same order as the actual one.
Answers:
$m_{0(1)} = 454746\,\text{kg}$ and $m_l/m_{0(1)} = 0.0115$. $m_{(1)} = 368411\,\text{kg}$, $m_{(2)} = 69772\,\text{kg}$, and $m_{(3)} = 11313\,\text{kg}$.

5.26 At an altitude of 120 km, an orbital launch vehicle of effective diameter 1 m is accelerating and its velocity at the instant is 4 km/s. The temperature at this altitude is 350 K. The density of the atmosphere at this altitude of the homosphere edge can be estimated through the semi-empirical equation of the type Eq. 5.53.

Table 5.20 Problem 5.28 data and answers

DATA				Answers	
t (s)	h (m)	v (m/s)	F/m_0 in g_e	dyn.pr (kPa)	M
0	0	0	1.53	0	0
21	1150	460	1.88	116.0	1.37
35	3750	560	1.81	132.0	1.72
39	4620	580	1.64	129.2	1.80
48	6920	620	1.64	114.5	1.98
60	10380	690	2.03	94.0	2.32

$$\rho = 1.8 \times exp\left(-\frac{h}{6666.7}\right) \text{ kg/m}^3$$

where h is in meters. Estimate the density and pressure at this altitude. What is the flight Mach number of the vehicle? Although the atmosphere at this altitude is close to vacuum, assuming a drag coefficient of 0.4, estimate the drag force on the vehicle.
Answers:
$\rho = 2.74164E - 08\,\text{kg/m}^3$ and $p = 0.002754\,\text{Pa}$. $M = 10.666$. $D = 0.06891\ N$.

5.27 The lift-off mass of the Space Shuttle was about 2010 ton. At its lift-off the two solid rocket boosters (SRBs) delivered a total thrust of 23040 kN and the three LOX/LH2 cryogenic space shuttle main engines (SSMEs) delivered a total thrust of 6834 kN. What was the acceleration with which the Space Shuttle would have lifted-off?
Answers:
Lift-off acceleration = $5.056\ \text{m/s}^2$.

5.28 Table 5.20 gives the approximate trajectory data of the Space Shuttle until it cleared the altitude of 10 km. At about the 35th second after lift-off the engines were throttled down to keep the dynamic pressure within the limit. Calculate the dynamic pressures at the specified time instants and estimate the maximum dynamic pressure at which the engines were throttled down. Also calculate the flight Mach numbers at the specified time instants. Use temperature and density data from the ICAO standard atmosphere.

5.29 A large satellite launch vehicle of mass 660 tons with a constant thrust of 6300 kN is in its ascent at an altitude of 45 km. Its instantaneous velocity and flight path angle β are respectively 1500 m/s and 70°. The characteristic area can be taken as 60 m^2. The vehicle is initiated with a gravity turn. Estimate its acceleration, climbing rate, and flight path angle rate. Assume $\rho = 1.8exp(-h/6700)$ kg/m^3 and $C_D = 0.4$. Consider the variation of the gravitational acceleration with altitude. If the $I_{sp,vac} =$ 4197 m/s estimate the vehicle mass after 10 seconds of climbing from this instant.

Answers:
$\dot{v} = 0.370\ \mathrm{m/s^2}$. $\dot{h} = 1409.5\ \mathrm{m/s}$. $\dot{\beta} = -0.122°/s$. m_0 after 10 s = 644989 kg.

5.30 A serially arranged three-stage satellite launch vehicle has its first two stages powered by N2O4/UDMH engines, and its last stage is powered by LOX/LH2 engine. After the burnout and separation of its first stage, the vehicle is climbing at an altitude of 50 km with a velocity of 1600 m/s. Its flight path angle β is 71° and thrust angle γ is 5°. The vehicle's effective diameter is 3.35 m. Thrust is 831 kN with a specific impulse of 2913 m/s. The mass of the vehicle at this instant is 61.8 ton. At the burnout and separation of its second stage, the same vehicle is at an altitude of 130 km with a velocity of 4000 m/s. Its flight path angle β is 20° and thrust angle γ is 0°. The vehicle's effective diameter is 3 m. Thrust is 156 kN with a specific impulse of 4315 m/s. The mass of the vehicle at this instant is 28.2 tons. Assume $C_D = 0.4$ for both the instants. Calculate the vehicle acceleration and flight path angle rate at these two instants. Distinguish the effects of thrust, drag, gravity, and centrifugal force on acceleration $\dot{v}$ and flight path angle rate $\dot{\beta}$. Note that drag effects are negligible at high altitudes. Observe centrifugal effect in opposing the flight path angle reduction becomes appreciable at higher velocities, Eq. 5.62. Also, calculate the rate at which the mass of the vehicle is reducing at the two instants. Assume $\rho = 1.8 \times exp(-h/6700)\ \mathrm{kg/m^3}$.
Answers:
(1) At 50 km altitude. $\dot{v} = 4.192\ \mathrm{m/s^2}$. $\dot{\beta} = -0.066°/s$. $\dot{m}_0 = -285.3\ \mathrm{kg/s}$. (2) At 130 km altitude. $\dot{v} = 2.311\ \mathrm{m/s^2}$. $\dot{\beta} = -0.094\ °/s$. $\dot{m}_0 = -36.2\ \mathrm{kg/s}$.

5.31 A sounding rocket of diameter 300 mm is climbing vertically. The burnout occurs at the altitude of 15 km (216.65 K, 0.12112 bar) with a velocity of 1200 m/s and mass of 200 kg. What is the instantaneous deceleration that the vehicle will experience? Assume the gravitational acceleration to be constant and $C_D = 0.1$. Molar mass of air = 28.965 kg/kgmol. Calculate the flight Mach number of the vehicle at its burnout.
Answers:
Deceleration = $-14.762\ \mathrm{m/s^2}$ & M = 4.067.

References

Cited

1. The Encyclopedia Astronautica. http://www.astronautix.com. Cited 13 October 2023
2. Hill PG, Peterson C (1999) Mechanics and thermodynamics of propulsion, 2nd edn. Addison Wesley, Reading Massachusetts
3. Malina FJ, Summerfield M (1947) The problem of escape from the earth by rockets. J Aeronaut Sci 14:471–480
4. Hall HH, Zamballi ED (1958) On the optimization of multistage rockets. Jet Propuls 28:463–465

5. Ragsac RV, Patterson PL (1961) Multistage rocket optimization. ARS J 31:450–452
6. MacKay JS, Weber RJ (1961) Performance charts for multistage rocket boosters. NASA TN D-582
7. Shurmann EEH (1957) Optimum staging technique for multistage rocket vehicles. Jet Propuls 27:863–865
8. Subotowicz M (1958) The optimization of n-step rocket with different construction parameters and propellant specific impulses in each stage. Jet Propuls 28:460–463
9. Weisbord L (1958) A generalized optimization procedure for n-staged missiles. Jet Propuls 28:165–167
10. Builder CH (1959) General solution for optimization of staging of multistage boost vehicles. ARS J 29:497–499
11. Gray JS, Alexander RV (1965) Cost and weight optimization for multistage rockets. J Spacecr 2:80–86
12. Walter U (2019) Astronautics the physics of space flight, 3rd edn. Springer, Switzerland
13. Koch AD (2019) Optimal staging of serially staged rockets with velocity losses and fairing separation. Aerosp Sci Technol 88:65–72
14. Emmert JT et al (2021) NRLMSIS 2.0: a whole-atmosphere empirical model of temperature and neutral species densities. Earth Space Sci 8
15. CCME Instant Run System. https://kauai.ccmc.gsfc.nasa.gov/instantrun/nrlmsis/. Cited 11 August 2023

Further Reading

16. Kluever CA (2018) Space flight dynamics. Wiley, Hoboken NJ
17. Suresh BN, Sivan K (2015) Integrated design for space transportation system. Springer, New Delhi
18. Adimurthy V (2006) Trajectory and mission design of launch vehicles. In: Ramakrishnan CV (ed) Current trends in engineering practice. Lectures under the distinguished visiting professorship scheme of Indian National Academy of Engineering. Narosa Publishing House, New Delhi
19. Curtis HD (2005) Orbital mechanics for engineering students. Elsevier, Amsterdam
20. Wiesel WE (1997) Spaceflight dynamics. McGraw-Hill, New York
21. Adimurthy V (1987) Launch vehicle trajectory optimization. Acta Astronaut 15:845–850
22. Adimurthy V (1976) Launch vehicle trajectory optimization including rotational dynamics. J Spacecr Rocket 13:59–61

Glossary

A	area, m^2
A_c	characteristic area, m^2
a	velocity of sound, m/s; acceleration, m/s^2
$a_1, .., a_9$	least-square polynomial-coefficients for $\overline{c}_p^0$, $\overline{h}^0$, and $\overline{s}^0$
$\vec{B}$	magnetic field intensity
b	kgatom of an element per kg of a mixture, kgatom/kg
C_D	drag coefficient
C_F	thrust coefficient defined by $F/(p_{0n}A_t)$
C_F^0	characteristic thrust coefficient
c^*	characteristic velocity, m/s
c_c	specific heat coefficient of a condensed species, J/(kgK)
c_p	specific heat at constant pressure, J/(kgK)
c_v	specific heat at constant volume, J/(kgK)
D	drag force, N
d	diameter, m
$\vec{\mathbf{E}}$	electric field intensity
F	thrust, N; force, N
$\vec{\mathbf{F}}$	force vector, N
f	fuel-air ratio; specific Helmholtz energy, J/kg
G	gravitational constant, 6.67430×10^{11} $m^3/(kgs^2)$
g	mass-specific Gibbs energy, J/kg; gravitational acceleration, m/s^2
$\overline{g}$	molar-specific Gibbs energy, J/kgmol
H	scale height RT/g_e, m
h	specific enthalpy, J/kg; altitude, m
$\overline{h}_f^0$	molar enthalpy of formation, J/kgmol
I	total impulse, Ns
I_{sp}	specific impulse, Ns/kg or m/s
I_{sp}^0	specific impulse under adapted condition $p_e = p_a$, Ns/kg
Kn	Knudsen number, λ/L
k	non-dimensional index in two-phase flow

S. Krishnan and J. Raghavan, *Rocket Propulsion Primer*,
https://doi.org/10.1007/978-981-97-5644-5

L	characteristic length, m
l	length, m; total number of elements
M	Mach number, Mass of Earth
m	mass, kg; non-dimensional index in two-phase flow
$\dot{m}$	mass flow rate or propellant mass flow rate, kg/s
$\overline{m}$	molecular mass considering only gaseous species, kg/kgmol
$\overline{m}_c$	molecular mass considering both gaseous and condensed species, kg/kgmol
NG	number of gaseous species in a mixture of gaseous and condensed species
NS	total number of species in a mixture of gaseous and condensed species
n	number of kgmoles of gaseous/condensed species per kg mixture of gaseous and condensed species, kgmol/kg; non-dimensional index in two-phase flow
O/F	oxidizer-fuel ratio
P_A	available power, W
P_e	input electric power, W
P_U	useful power, W
p	pressure, Pa
Q	heat, J
q	electric charge
R	specific gas constant, J/(kgK)
R_u	universal gas constant 8314.51 J/(kgmolK)
r	radius, m
S	entropy, J/K
s	mass-specific entropy, J/(kgK)
$\overline{s}$	molar-specific entropy, J/(kgmolK)
T	temperature, K
t	time, s
U	internal energy, J
u	velocity, m/s; specific internal energy, J/kg
$\overline{u}_e$	equivalent exhaust velocity equal to effective specific impulse, m/s or Ns/kg
V	volume, m^3
v	vehicle velocity, m/s; mass-specific volume, m^3/kg
$\overline{v}$	molar-specific volume, m^3/kgmol
$\vec{\mathbf{v}}$	velocity vector, m/s
x_j	mole fraction of the jth species in a mixture of gaseous and condensed species.

Subscripts

0	initial; stagnation or total condition
0(1)	total launch vehicle of stage 1 and above including payload m_l and similarly for 0(2), 0(3), ...

(1)	stage number 1 and similarly for (2), (3), ...
a	ambient condition
a_{ij}	stoichiometric coefficient of element i of species j
b	burnout
c	combustion chamber; condensed species
e	nozzle exit-station; equilibrium condition; dry engine; Earth
$expt$	experimental condition
f	fuel; final; frozen
g	gaseous mixture
i	initial; element i
(i)	stage number, (1), (2), ...
j	species j
k	element k
l	payload
n	nozzle end (or aft end) of combustion chamber, that is, nozzle entry; representing terminal velocity
p	propellant
r	surroundings; reaction; required
s	structure
std	standard
t	nozzle throat station
$theo$	theoretical value
u	isolated system
vac	vacuum condition.

Superscripts

0	standard; adapted
$*$	design/reference condition
o	with respect to the propellant mixture.

Greek Letters

α	nozzle contour angle; payload ratio
α_d	half cone angle of nozzle divergence
β	flight path angles
Γ	a function of ratio of specific heats = $\sqrt{\gamma}\left(2\Big/(\gamma+1)\right)^{\frac{\gamma+1}{2(\gamma-1)}}$
γ	ratio of specific heats; thrust angle
γ_s	isentropic exponent
ε	structural coefficient = $m_s/\left(m_s+m_p\right)$
η	efficiency factor
λ	nozzle divergence factor; Lagrange multiplier; mean free path

μ mass ratio initial mass to burnout mass; standard gravitational parameter, $3.986004418 \times 10^{14}\ \mathrm{m}^3/\mathrm{s}^2$
μ_j chemical potential of species j, J/kgmol
ν velocity ratio u/u_e
ξ_c c^* quality factor, also known as c^* efficiency, η_{c^*}
ξ_n C_F quality factor
ξ_s I_{sp} quality factor or velocity correction factor
π $-\lambda/(R_u T)$
ρ density, $\mathrm{kg/m^3}$
τ parameter for temperature difference between condensed matter and gaseous mixture
ϕ parameter for velocity lag between condensed matter and gaseous mixture; pitch angle, $\phi = \gamma + \beta$
χ mass fraction of the condensed matter in the total flow
ψ nozzle-flow separation parameter, defined by p_a/p_e
ψ_0 nozzle-flow separation constraint.

Index

S. Krishnan and J. Raghavan, *Rocket Propulsion Primer*,
https://doi.org/10.1007/978-981-97-5644-5

Printed in the USA
CPSIA information can be obtained
at www.ICGtesting.com
CBHW050336131024
15650CB00023BA/10

9 789819 756438